VI International Conference on Problems of Mathematical Physics and Mathematical Modelling 2017

Journal of Physics: Conference Series Volume 937

Moscow, Russia
25 - 27 May 2017

Editors:

Valery V. Kozlov
Nikolay A. Kudryashov
Oleg V. Nagornov

ISBN: 978-1-5108-5517-5
ISSN: 1742-6588

Printed from e-media with permission by:

Curran Associates, Inc.
57 Morehouse Lane
Red Hook, NY 12571

Some format issues inherent in the e-media version may also appear in this print version.

Printed by Curran Associates, Inc. (2018)

For permission requests, please contact the Institute of Physics
at the address below.

Institute of Physics
Dirac House, Temple Back
Bristol BS1 6BE UK

Phone: 44 1 17 929 7481
Fax: 44 1 17 920 0979

techtracking@iop.org

Additional copies of this publication are available from:

Curran Associates, Inc.
57 Morehouse Lane
Red Hook, NY 12571 USA
Phone: 845-758-0400
Fax: 845-758-2633
Email: curran@proceedings.com
Web: www.proceedings.com

Table of contents

Volume 937

VI International Conference Problems of Mathematical Physics and Mathematical Modelling

25–27 May 2017, Moscow, Russian Federation

Accepted papers received: 13 December 2017
Published online: 30 December 2017

Preface

Papers

Preface

This annual conference was established at the National Nuclear Research University – Moscow Engineering Institute in the year 2012. The Conference was organized by three leading departments of the MEPhI in the field of mathematical physics.

Each year, more than 100 participants from different countries including Belorussia, Czech Republic, France, Great Britain, Japan, Moldova, Russia, Slovakia, Spain, Sweden and USA take part in the Conference with oral talks.

The aim of this conference is to bring together researchers with interests in: Analytical Theory of Differential Equations, Optimization Problems, Analytical and Solid Mechanics, Fluid Mechanics, Mathematical Modelling.

Valery V. Kozlov
Nikolay A. Kudryashov
Oleg V. Nagornov

IOP Conf. Series: Journal of Physics: Conf. Series **937** (2017) 011002 doi:10.1088/1742-6596/937/1/011002

Peer review statement

All papers published in this volume of *Journal of Physics: Conference Series* have been peer reviewed through processes administered by the proceedings Editors. Reviews were conducted by expert referees to the professional and scientific standards expected of a proceedings journal published by IOP Publishing.

IOP Conf. Series: Journal of Physics: Conf. Series **937** (2017) 012001 doi:10.1088/1742-6596/937/1/012001

Method of construction of the Riemann function for a second-order hyperbolic equation

A V Aksenov

Lomonosov Moscow State University, Moscow, Russia

E-mail: aksenov.av@gmail.com

Abstract. A linear hyperbolic equation of the second order in two independent variables is considered. The Riemann function of the adjoint equation is shown to be invariant with respect to the fundamental solutions transformation group. Symmetries and symmetries of fundamental solutions of the Euler–Poisson–Darboux equation are found. The Riemann function is constructed with the aid of fundamental solutions symmetries. Examples of the application of the algorithm for constructing Riemann function are given.

1. Introduction

In the study [1] B. Riemann suggested the Riemann's principle of integrating that was applied to a hyperbolic second-order partial equation with two independent variables. In order to apply the method it is necessary to construct Riemann function that is a solution of Cauchy special characteristic problem. General method for Riemann function construction does not exist. In [2] an extensive analysis of six certain methods for creation Riemann function of particular types of equations. Ibragimov recommended to find Riemann function with the aid of equation symmetries [3] basing on Ovsyannikov study result [4] in group classification of homogeneous hyperbolic second-order equations. In the present study the method for Riemann function construction that is based on the usage of fundamental solutions symmetries is offered.

2. The Riemann's method

Let us consider the general linear hyperbolic equation of the second order in two independent variables

$$Lu = u_{xy} + a(x,y)u_x + b(x,y)u_y + c(x,y)u = f(x,y)\,. \tag{1}$$

Riemann's method is based on the following identity

$$2(vLu - uL^*v) = (vu_y - uv_y + 2auv)_x + (vu_x - uv_x + 2buv)_y\,,$$

where $L^*v = v_{xy} - (av)_x - (bv)_y + cv$ is the odjoint operator. Riemann's method allows us to convert the integration problem of Eq. (1) to construction of the intermediary Riemann function $v(x, y; x_0, y_0)$, that obeys the following homogeneous adjoint equation of variables x, y

$$L^*R = 0$$

and the following conditions at the characteristics:

$$(R_y - aR)\big|_{x=x_0} = 0\,, \qquad (R_x - bR)\big|_{y=y_0} = 0\,, \quad R(x_0, y_0; x_0, y_0) = 1\,.$$

General solutions of Cauchy problem and Goursat problem are constructed with the aid of Riemann function for Eq. (1) [2].

Riemann function $u = R^*(x, y; x_0, y_0)$ of the adjoint equation satisfies homogeneous equation (4)

$$LR^* = 0$$

and following conditions at characteristics

$$(R^*_y + aR^*)\big|_{x=x_0} = 0\,, \qquad (R^*_x + bR^*)\big|_{y=y_0} = 0\,, \qquad R^*(x_0, y_0; x_0, y_0) = 1\,. \tag{2}$$

Riemann functions R and R^* have properties of reciprocity

$$R^*(x, y; x_0, y_0) = R(x_0, y_0; x, y)\,. \tag{3}$$

3. Symmetries of fundamental solutions

Fundamental solutions of linear partial differential equations are frequently invariant under transformations admitted by the original equation [4]. Below, a fundamental solution is constructed using the algorithm from [5] proposed for finding fundamental solutions of linear partial differential equations. The algorithm makes use of the symmetries admitted by a linear partial differential equation with a delta function on its right-hand side. Let us briefly describe the main result of this work.

Consider the pth-order linear partial differential equation

$$Au \equiv \sum_{\alpha=1}^{p} B_\alpha(x) D^\alpha u = 0\,, \qquad x \in R^m\,. \tag{4}$$

Here, the standard notation is used: $\alpha = (\alpha_1, \ldots, \alpha_m)$ is a multi-index with nonnegative integer components, $\alpha = \alpha_1 + \cdots + \alpha_m$, and

$$D^\alpha \equiv \left(\frac{\partial}{\partial x^1}\right)^{\alpha_1} \cdots \left(\frac{\partial}{\partial x^m}\right)^{\alpha_m}.$$

The fundamental solutions of Eq. (1) are solutions of the equation

$$Au = \delta(x - x_0)\,. \tag{5}$$

It was shown in [6] that Eq. (4) with $p \geqslant 2$ and $m \geqslant 2$ can admit only symmetry operators of the form

$$Y = \sum_{i=1}^{m} \xi^i(x) \frac{\partial}{\partial x^i} + \eta(x, u) \frac{\partial}{\partial u}, \qquad \frac{\partial^2 \eta}{\partial u^2} = 0\,.$$

The basic Lie algebra of symmetry operators of Eq. (4) regarded as a vector space is the direct sum of two subalgebras: one consisting of operators of the form

$$X = \sum_{i=1}^{m} \xi^i(x) \frac{\partial}{\partial x^i} + \zeta(x)\, u \frac{\partial}{\partial u}\,, \tag{6}$$

IOP Conf. Series: Journal of Physics: Conf. Series **937** (2017) 012001 doi:10.1088/1742-6596/937/1/012001

and the infinite-dimensional subalgebra generated by the operators

$$X_\infty = \varphi(x)\frac{\partial}{\partial u}, \tag{7}$$

where $\varphi(x)$ is an arbitrary solution of Eq. (4). Note that operators (7) are symmetry operators of Eq. (5). In what follows, we consider only symmetry operators of form (6). Let denote $\underset{p}{X}$ an extension of order p of symmetry operator (6).

Proposition 1. *The infinitesimal operator given by (6) is a symmetry operator of Eq. (4) if and only if there exists a function $\lambda = \lambda(x)$ satisfying the identity*

$$\underset{p}{X}(Au) \equiv \lambda(x)\, Au \tag{8}$$

for any function $u = u(x)$ from the domain of Eq. (4).

Theorem 1. *The Lie algebra of symmetry operators of Eq. (5) is a subalgebra of the Lie algebra of symmetry operators of Eq. (4) and is defined by the relations*

$$\begin{aligned} &\xi^i(x_0) = 0, \qquad i = 1, \ldots, m\,, \\ &\lambda(x_0) + \sum_{i=1}^{m} \frac{\partial \xi^i(x_0)}{\partial x^i} = 0\,. \end{aligned} \tag{9}$$

Let us describe an algorithm for finding fundamental solutions by applying symmetries [5]:

1. Find a general symmetry operator of Eq. (4) and the corresponding function $\lambda(x)$ satisfying identity (9).
2. Use this operator and relations (9) to obtain the basis for the Lie algebra of symmetry operators of Eq. (5).
3. Construct invariant fundamental solutions with the help of the symmetries of Eq. (5).
4. Obtain new fundamental solutions from the known ones with the help of the symmetries of Eq. (5) (production of solutions).

Remark 1. To find generalized invariant fundamental solutions, we need to search for invariants in the class of generalized functions.

4. The main result

Symmetry operator of homogeneous equation (1) has a form

$$X = \xi^1(x)\,\frac{\partial}{\partial x} + \xi^2(y)\,\frac{\partial}{\partial y} + \zeta(x,y)\,u\,\frac{\partial}{\partial u} \tag{10}$$

and as this takes place the following relations must be hold

$$\begin{aligned} &\frac{\partial \zeta}{\partial x} + \frac{\partial (b\,\xi^1)}{\partial x} + \xi^2\,\frac{\partial b}{\partial y} = 0\,, \\ &\frac{\partial \zeta}{\partial y} + \frac{\partial (a\,\xi^2)}{\partial y} + \xi^1\,\frac{\partial a}{\partial x} = 0\,, \\ &\frac{\partial^2 \zeta}{\partial x \partial y} + a\,\frac{\partial \zeta}{\partial x} + b\,\frac{\partial \zeta}{\partial y} + \frac{\partial (c\,\xi^1)}{\partial x} + \frac{\partial (c\,\xi^2)}{\partial y} = 0\,. \end{aligned} \tag{11}$$

Function $\lambda = \lambda(x, y)$ that satisfies the identity law $\underset{2}{X}(Lu) = \lambda\, Lu$ has the form

$$\lambda = \zeta - \frac{d\,\xi^1}{d\,x} - \frac{d\,\xi^2}{d\,y}\,. \tag{12}$$

IOP Conf. Series: Journal of Physics: Conf. Series **937** (2017) 012001 doi:10.1088/1742-6596/937/1/012001

Let us consider the equation

$$Lu = \delta(x - x')\,\delta(y - y')\,, \tag{13}$$

that describes fundamental solutions of homogeneous Eq. (1). So the symmetry operators of fundamental solutions (or the symmetries of Eq. (13)) satisfy following additional conditions as Theorem 1 takes place

$$\begin{gathered} \xi^1(x') = 0\,, \qquad \xi^2(y') = 0\,, \\ \lambda(x', y') + \frac{d\,\xi^1(x')}{d\,x'} + \frac{d\,\xi^2(y')}{d\,y'} = 0\,. \end{gathered} \tag{14}$$

Show that conditions at characteristics (2) are invariant under symmetry operator (10) at (11), (12) and (14). Note that characteristics $x = x'$, $y = y'$ are invariant under operators of the symmetry of the fundamental solutions. $\zeta(x', y') = 0$ results from relations (12) and the second one of (14). This implies that the latest one of relations (2) is invariant.

Write the invariance condition at the characteristic $x = x'$

$$\underset{1}{X}(u_y + au)\Bigg|_{\substack{x = x' \\ u = R^*}} = 0$$

$$\left\{\left(\zeta - \frac{d\xi^2}{dy}\right)(u_y + au) + u\left[\frac{\partial \zeta}{\partial y} + \frac{\partial(a\,\xi^2)}{\partial y} + \xi^1\,\frac{\partial a}{\partial x}\right]\right\}\Bigg|_{\substack{x = x' \\ u = R^*}} = 0 \tag{15}$$

Invariance condition (15) is realized owing to (11) and (2) at the characteristic $x = x_0$. Invariance of the condition at the characteristic $y = y_0$ is convinced analogous.

Hence the theorem that presents the main result of the section is proved.

Theorem 2. *Symmetries of the fundamental solutions of the second-order linear hyperbolic equation in two independent variables retain Riemann function of adjoint equation invariant.*

Theorem 2 implies that algorithm of the adjoint equation is invariant under the symmetries of fundamental solutions and the solution of the original equation. Hence, Riemann function of the original equation is obtained from the reciprocity relation (3).

Formulate the algorithm of Riemann function construction based on the usage of the fundamental solutions symmetries:

1. Solving for symmetries of linear homogeneous equation (1).
2. Computation of the fundamental solutions symmetries.
3. Construction of invariant solutions with the aid of fundamental solutions symmetries.
4. Extraction of Riemann function from the computed invariant solutions invoking continuity condition of Riemann function and its first derivatives at the point (x_0, y_0) and the condition $R(x_0, y_0; x_0, y_0) = 1$.

Remark 2. The algorithm allows to find Riemann function of hyperbolic equation not going to the characteristic variables. This lays emphasis on the invariant nature of the given method for the construction of Riemann function.

Let us consider the equation

$$Lu = \delta(x - x_0)\,\delta(y - y_0)\,,$$

Remark 3 As Adamar noticed in [7] the fundamental solution of Eq. (4) can be written as

$$u = R^*\theta(x - x_0)\theta(y - y_0)\,.$$

Next, as Theorem 2 takes place, its right part is invariant of fundamental solutions symmetries.
Consider examples of the method application.
Example 1. Consider an equation

$$u_{xy} + u = 0\,. \tag{16}$$

Symmetry operators basis of the finite-dimensional part of Lie algebra of Eq. (16) has the form

$$X_1 = \frac{\partial}{\partial x}\,, \qquad X_1 = \frac{\partial}{\partial y}\,, \qquad X_1 = x\frac{\partial}{\partial x} - y\frac{\partial}{\partial y}\,, \qquad X_4 = u\frac{\partial}{\partial u}\,.$$

Therefore, function $\lambda = a_4$ (see (8)) and equation

$$u_{xy} + u = \delta(x - x_0)\delta(y - y_0)$$

admits the symmetry operator

$$Y = (x - x_0)\frac{\partial}{\partial x} - (y - y_0)\frac{\partial}{\partial y}\,. \tag{17}$$

The solution that is invariant under symmetry operator (17) is found as

$$u = f(z)\,, \qquad z = (x - x_0)(y - y_0)\,.$$

Here function $f = f(z)$ is a solution of the ordinary differential equation

$$zf'' + f' + f = 0\,. \tag{18}$$

The general solution of Eq. (18) has the form

$$f = C_1 J_0(2\sqrt{z}) + C_2 Y_0(2\sqrt{z})\,,$$

where C_1, C_2 are arbitrary constants; $J_0(z)$, $Y_0(z)$ are Bessel functions [8]. The condition $f(0) = 1$ follows

$$R^*(x, y; x_0, y_0) = R(x, y; x_0, y_0) = J_0(2\sqrt{(x - x_0)(y - y_0)})\,.$$

Example 2. Consider an equation

$$u_{xy} + \frac{1}{4(x + y)^2}u = 0\,. \tag{19}$$

Symmetry operators basis of the finite-dimensional part of Lie algebra of (19) has the form

$$X_1 = \frac{\partial}{\partial x} - \frac{\partial}{\partial y}\,, \qquad X_2 = x\frac{\partial}{\partial x} + y\frac{\partial}{\partial y}\,, \qquad X_3 = x^2\frac{\partial}{\partial x} - y^2\frac{\partial}{\partial y}\,, \qquad X_4 = u\frac{\partial}{\partial u}\,.$$

Therefore, function $\lambda = -2a_2 - 2a_3(x - y) + a_4$ and equatiion

$$u_{xy} + \frac{1}{4(x + y)^2}u = \delta(x - x_0)\delta(y - y_0)$$

admits the symmetry operator

$$Y = (x - x_0)(x + y_0)\frac{\partial}{\partial x} - (y - y_0)(y + x_0)\frac{\partial}{\partial y}\,. \tag{20}$$

The solution that is invariant under symmetry operator (20) is found as

$$u = f(z)\,, \qquad z = \frac{(x - x_0)(y - y_0)}{(x_0 + y_0)(x + y)}\,.$$

Here function $f = f(z)$ is a solution of the ordinary differential equation

$$(z^2 + z)f'' + (2z + 1)f' + \frac{1}{4}f = 0\,. \tag{21}$$

The general solution of Eq. (21) has the form

$$f = C_1 EllipticK(i\sqrt{z}) + C_2 EllipticCK(i\sqrt{z})\,,$$

where C_1, C_2 are arbitrary constants; $EllipticK(z)$ is complete elliptic integral of the first kind and $EllipticCK(z)$ is complementary complete elliptic integral of the first kind [8]. From the condition $f(0) = 1$ follows

$$R^*(x, y; x_0, y_0) = R(x, y; x_0, y_0) = \frac{2}{\pi} EllipticK\left(i\sqrt{\frac{(x - x_0)(y - y_0)}{(x_0 + y_0)(x + y)}}\right)\,.$$

5. Riemann's funtion for Euler–Poisson–Darboux equation

Let us consider Euler–Poisson–Darboux equation

$$Lu \equiv \frac{\partial^2 u}{\partial r^2} + \frac{2\alpha}{r}\frac{\partial u}{\partial r} - \frac{\partial^2 u}{\partial z^2} = 0\,. \tag{22}$$

The symmetries of Eq. (22) can be found using the symmetry-finding algorithm from [4].

Proposition 2. *In case of $\alpha \neq 0$ Eq. (22) admits the following basis of the finite part of Lie algebra symmetry operators*

$$Y_1 = \frac{\partial}{\partial z}\,, \qquad Y_2 = r\frac{\partial}{\partial r} + z\frac{\partial}{\partial z}\,, \qquad Y_3 = u\frac{\partial}{\partial u}\,,$$
$$Y_4 = 2rz\frac{\partial}{\partial r} + (r^2 + z^2)\frac{\partial}{\partial z} - 2\alpha z u\frac{\partial}{\partial u}\,.$$

Let us consider the equation

$$Lu = \delta(r - r_0)\delta(z - z_0)\,. \tag{23}$$

We use the results of [5] in order to find the symmetries of Eq. (23).

Proposition 3. *Eq. (23) admits the symmetry operator*

$$Y = 2r(z - z_0)\frac{\partial}{\partial r} + \left[r^2 + (z - z_0)^2 - {r_0}^2\right]\frac{\partial}{\partial z} - 2\alpha(z - z_0)u\frac{\partial}{\partial u}\,. \tag{24}$$

The symmetry operator (24) has two functionally independent invariants

$$\xi = \frac{r^2 - (z - z_0)^2 + {r_0}^2}{2rr_0}\,, \qquad \tau = r^\alpha u\,.$$

Invariant solutions are sought in the form

$$\tau = f(\xi)\,,$$

or

$$u = r^{-\alpha} f(\xi)\,.$$

Proposition 4. *Euler–Poisson–Darboux solutions* (22), *that are invariant under symmetry operator* (24), *have the following form*

$$u = r^{-\alpha}\Big[C_1\,P_{-\alpha}(\xi) + C_2\,Q_{-\alpha}(\xi)\Big]\,,$$

where C_1, C_2 *are arbitrary constants;* $P_{-\alpha}(\xi)$, $Q_{-\alpha}(\xi)$ *are Legendre functions of the first and second kind [8].* The condition $f(0) = 1$ gives us Riemann function R^*.

Proposition 5. *Riemann function of adjoint Eq.* (22) *appears as follows*

$$R^*(r, z; r_0, z_0) = \left(\frac{r}{r_0}\right)^{-\alpha} P_{-\alpha}(\xi)\,. \tag{25}$$

Riemann function (25) describes the solution of characteristic problem of the two centered depression waves interpenetration up to a constant multiplier t_0 [9].

6. Conclusion

The main result of the study is the suggested method for Riemann function construction based on the usage of the symmetries of fundumental solutions. Its effectiveness is performed by examples.

Acknowledgments

This work was supported by RFBR grants 15-01-00361 and 15-01-04677.

References

[1] Riemann B 1948 Proceedings (M.–L.: OGIZ) (in Russian)
[2] Copson E T 1957/58 On the Riemann–Green Function *Archive for Rational Mechanics and Analysis* **1** 324–348
[3] Ibragimov N Kh 1992 Group analysis of ordinary differential equations and the invariance principle in mathematical physics (for the 150th anniversary of Sophus Lie) *Uspekhi Matematicheskikh Nauk* **47** 83–144
[4] Ovsiannikov L V 1982 *Group Analysis of Differential Equations* (New York: Academic Press)
[5] Aksenov A V 1995 Symmetries of linear partial differential equations and fundamental solutions *Doklady Mathematics* **51** 329–331
[6] Bluman G 1990 Simplifying the form of Lie groups admitted by a given differential equation *Journal of Mathematical Analysis and Applications* **145** 52–62
[7] Hadamard J 1923 *Lectures on Cauchy's Problem in Linear Partial Differential Equations* (New Haven: Yale University Press)
[8] Abramowitz M and Stigan I A 1972 *Handbook of Mathematical Functions With Formulas, Graphs, and Mathematical Tables* (Tenth Printing, Washington: National Bureau of Standards)
[9] Courant R and Friedrichs K O 1978 *Supersonic Flow and Shock Waves* (New York: Springer-Velag)

Modelling of the forming devices of high-current pulsed accelerators

G P Averyanov, V V Dmitrieva, N P Kornev

National Research Nuclear University MEPhI (Moscow Engineering Physics Institute), Kashirskoe Highway 31, Moscow, 115409, Russia

E-mails: GPAveryanov@mephi.ru, VVDmitrieva@mephi.ru, NPKornev@mephi.ru

Abstract. The report provides a comparison of the mathematical model of the forming devices of high-current pulsed accelerators based on the solution of unsteady problems for electric circuits with the distributed parameters and the model implementing direct calculation methods obtained on the basis of operator transformations.

1. The first section in your paper

The devices forming the high-voltage pulses of short duration are one of the most important components of high-current pulsed accelerators, with a wide range of applications. Such accelerators consist of three main parts – the primary storage, the sharpening agent (a device forming nanosecond pulses) and the accelerating diode.

As a rule, in such installations, as the primary storage, an impulse voltage generator based on the cascade voltage multiplication circuits (an Arkadiev-Marx generator) is used. A feature of those generators is a parallel connection of capacitors during charging and a serial connection of capacitors during the load operation. Switching is accomplished using spark gaps. At the output of such generators (in an accelerating diode), it is possible to obtain pulsed currents of accelerated electrons of tens of amperes at a pulse duration of several microseconds. Obtaining shorter pulses, and hence higher pulsed power, is difficult due to the influence of parasitic capacitances and inductances in the design of the pulse voltage generator. To form shorter pulses to increase the pulse power with a minimum energy loss, at the output of the pulse voltage generator, a forming device on the basis of a long lines is set up, which ensures operation with a spectrum width up to gigahertz (for a nanosecond pulse duration). The design of these devices (a structure selection and optimization of scheme parameters) is associated with the use of traditional models for the analysis of electrical circuits, including methods of the theory of directed graphs. A difficulty of using traditional methods of analysis of electrical circuits is aggravated by the usage of non-typical circuitry elements (long lines, arresters etc.).

Long lines as forming devices may have different configurations (the double forming line, the alternating wave impedance line, etc.) are the elements with distributed parameters and are described by the evolution equation in partial derivatives. Component equations of these elements require an individual analysis, the complexity of which exceeds the traditional analysis of electrical circuits, being its integral part.

2. Modeling features of forming devices

Mathematical model of formation can be obtained by combining the component and topological equations; and it is necessary to consider the essential feature of high-power pulsed devices, which are the elements of the scheme, which, along with the standard elements with lumped parameters, include also the elements with distributed parameters.

$$\begin{cases} u = u_v + Z(i - i_v) \\ i = i_g + Y(u - u_g) \end{cases} \tag{1}$$

where u_g and i_g are independent initial conditions; a, u_v and i_v are external impacts on the circuit.

As it is known, in general, the topology of the electrical circuit is presented in the form of graphs, the use of which allows to consider the processes in electric circuits on two levels: topological and component. The nature of connections between the circuit elements is determined by the topological properties of the graph elements (branches, nodes, and paths). Topological features determine the order of connections of electric circuit elements and are not associated with their specific parameters. The topology of the circuit is described by topological matrices (an incidence matrix, a matrix of main circuits, a matrix of main cross-sections).

The component equations are mathematical models of the circuit branches and express the voltage and current in each branch through the parameters of the elements of this branch. The most difficult part of the mathematical analysis of electrical circuits is the formation of equations of electrical balance.

Of the greatest interest is the choice of the algorithm allowing to automate this process. During the formation of equations of electrical balance for an extended description of electric circuits is used, and each circuit element is a separate branch of the tree. In highly branched circuits, it is first necessary to find the optimal variant of the construction of the tree of a connected graph, which is the most convenient for analysis of the circuit. For the selected tree, at the subsequent stages of automation of computations, two methods are used – the method of node voltages (2) and the method of loop currents (3).

$$Y_{ij} u_{i0} = j_{i0} \tag{2}$$

where is the matrix of nodal conductances Y_{ij} , the voltage of the i-th node relative to the 0-th node is u_{i0}, and the matrix-column of the nodal currents is j_{i0} .

$$Z_{ij} j_{ii} = e_{ii} \tag{3}$$

here, the matrix of contour resistance is Z_{ij} , the matrix-column of loop currents – j_{ii}, and the matrix-column of contour electromotive force – e_{ii}..

Equations compiled on the basis of the first and second Kirchhoff laws are topological equations for electrical circuits with lumped parameters, but analysis of systems based on lines with distributed parameters requires a different approach, arising from the method of representation of the analyzed system in the form of a directed graph, the branches of which are line segments.

Regions of bifurcation of the branches of the graph are not "nodes" in the standard view of graph theory, so to avoid confusion it is advisable to consider a new mathematical object, marking it as a "connection". Moreover, the change of currents and voltages in the branches of such a graph cannot be described by Kirchhoff laws as for circuits with lumped parameters, because, in a general case, these branches may be transmission lines for which it is required to account for resistance losses and the conductivity of the leakage.

The most effective method of construction of optimal mathematical models of devices of formation of nanosecond high-voltage pulses in high-current accelerators based on the respective topology of electrical circuits is the use of the directed graphs theory, however, the inclusion in the scheme of more complex elements including traditional devices with well-described models and methods, such as diode, transistor, etc., complicates the study of forming device models and requires a preliminary study of the individual nonlinear elements and presenting them in the form of external software modules.

IOP Conf. Series: Journal of Physics: Conf. Series **937** (2017) 012002 doi:10.1088/1742-6596/937/1/012002

The inclusion in the forming device of specific switches and segments of non-uniform transmission lines, which are quadrupoles from the point of view of circuit design, transforms and greatly complicates the structure of the graph; thus there is a need for specialized methods of solution of the unsteady problems describing the transient processes in systems with distributed parameters.

3. Modeling features of subsystems with distributed parameters

As practice has shown, various variants of the method of characteristics (in particular, the Dvorak circuit) for solving telegraph equations are the most effective for solving non-stationary problems related to the description of processes in lines with distributed parameters.

Ultimately, this allows us to consider transmission line segments as quadrupoles, and to incorporate the developed models into specialized CAD systems, while allowing to significantly expand the range of tasks compared to analogue circuit modeling systems (MicroCAP, SPICE, P-CAD, etc.).

The passage of a pulse in a segment of a long line with losses can be described by a system of telegraph equations (4):

$$\begin{cases} -\frac{\partial}{\partial x}u = \left(R + \rho\tau\frac{\partial}{\partial t}\right)i \\ -\rho\frac{\partial}{\partial x}i = \left(\rho G + \tau\frac{\partial}{\partial t}\right)u \end{cases} \tag{4}$$

where $u = u(x,t)$ is a function of voltage and $i = i(x,t)$ is a function of current with the following distributed electrical parameters: L – inductance, C – capacitance, R – resistance, G – conductance, τ is the delay (or electrical length), ρ is the wave resistance.

The solution of the system (4) can be obtained in the form of a recurrent finite-difference schemes for functions of current and voltage (5) depending on time:

$$\begin{cases} u_{k+1}(t+T_k) + \rho_k i_{k+1}(t+T_k) = (1-\rho_k g_k)u_k(t) + (\rho_k - r_k)i_k(t) \\ u_k(t+T_k) - \rho_k i_k(t+T_k) = (1-\rho_k g_k)u_{k+1}(t) + (\rho_k - r_k)i_{k+1}(t) \end{cases} \tag{5}$$

here, the magnitude of current – i_{k+1} and voltage – u_{k+1} in the section of the line with the number *(k+1)* are determined via the amount of current – i_k and voltage u_k, as well as through the electric length T_k, resistance r_k, and conductivity g_k, computed for the previous line segment (with number *k*).

The method of characteristics assumes consideration of a long line as a sequence of sections with smoothly varying wave resistance and the same electrical length.

The analyzed circuit along with the distributed parameters line segments with can contain discrete elements (such as R, G, L, C), as well as keys and programmable sources. Any of the discrete elements can be non-linear and defined as a function of the time and state of the circuit (currents and voltages). In this connection, in the process of computation, there arises the problem of joining distributed parameters lines and blocks of a circuit with lumped parameters.

The Dvorak scheme for solving a system of telegraph equations actually considers distributed parameters line as a sequence of quadrupoles with electrical characteristics calculated at each time step. Thus, the solution of (8) is reduced to solving a system of $2N + 2$ linear algebraic equations by splitting a line with distributed parameters into a sequence of N sections with smoothly varying characteristic impedance. Since the application of the method of nodal voltages or loop currents to calculate the characteristics of discrete elements of the circuit also reduces to the construction of a system of linear algebraic equations, it is possible to construct and solve a generalized system of linear algebraic equations for each time step for the whole scheme. The method of characteristics, which has a high accuracy, is the most suitable for use in specialized CAD systems. To model long lines with small losses in conductors and dielectrics, it is expedient to use direct calculations based on the Laplace transform.

Expressions for the voltage functions $\vec{V}_x(p)$ and $\overleftarrow{V}_x(p)$, represented in an operator form in the system of equations (6) are standard one-dimensional solutions of the Telegraph equation for the line segment:

$$\begin{cases} \vec{V}_x(p) = \dfrac{Z_0}{Z_0 + Z_s} + V_s(p)\Gamma_0 \sum_{k=0}^{\infty} \Gamma_l^k \exp(-\gamma(2kl + x)) \\ \overleftarrow{V}_x(p) = \dfrac{Z_0}{Z_0 + Z_s} + V_s(p)\Gamma_0 \sum_{k=0}^{\infty} \Gamma_l^{k+1} \exp(-\gamma(2(k+1)l - x)) \end{cases} \quad (6)$$

where p is the Laplace operator, $Z_0 = const$ is the characteristic impedance of the line, V_s is the function of the voltage generator, Z_s is the internal resistance of the voltage generator, $\Gamma_0 = const$ is a reflection coefficient at the input of the line, k is the number of the reflected wave, $\Gamma_l = \Gamma_l(p)$ is a coefficient of the load line reflection, and presenting it so it is possible to obtain a universal formula of Laplace transformation:

$$\Gamma_l(p) = \lambda/(p + \beta) + \delta \quad (7)$$

where constants λ, β and δ depend on the parameters of the load and the type of connection (serial or parallel). By substituting (7) into (6), it is simple enough to convert back and get the formula for direct calculation.

The verification of the results of the modeling of transient processes, based on the method of characteristics (in accordance with the modified Dvorak scheme) and the operator method (based on the Laplace transformation), given in Figure 1, shows that the maximum variance in the values of the voltage calculation results does not exceed 5%.

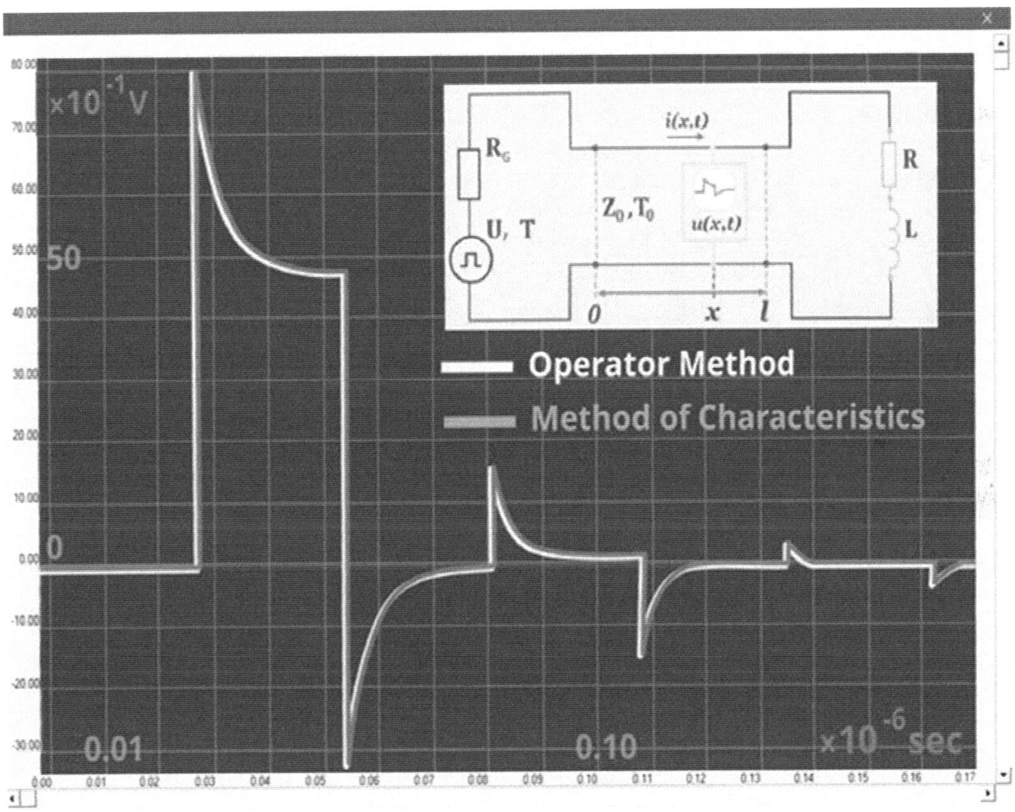

Figure 1. Results of the modelling of transient processes

4. Conclusion

Modeling of transients in shaper circuits can be useful in the development of specialized software for evaluating the factors that cause distortion of pulses in existing installations (including parasitic capacitances and inductances), as well as in the development of virtual powerful pulse technology laboratories of the 5th generation for conducting educational and scientific research in the framework of training, retraining and advanced training of specialists in the field of charged particle beams and accelerator technology.

Presented in this paper, methods of modeling transient processes in distributed-parameter lines, implemented on the basis of two methods for solving telegraph equations, can be used in two successive stages of designing devices for the formation of subnanosecond pulses of high-current high-voltage accelerators.

- At the stage of pre-project studies of transient processes in the lines for preliminary estimation of their main parameters, as well as for estimating the influence of parasitic factors on pulse formation.
- Construction of schemes for the implementation of pulse shapers, as well as optimization and refinement of design parameters of the element base on the basis of preliminary pre-project studies results.

References

[1] Averyanov G. P., Budkin V. A. and Dmitrieva V. V., 2011 (in Russian) CAD in Electrophysics, Part 1. Fundamentals of computer-aided design (Moscow, MEPhI) (Original Russian title: SAPR v Elektrofizike, chast 1. Osnovy avtomatizacii proektirovaniya. M.: MIFI)

[2] Averyanov G. P. et al. 2009 Virtual Electrophysics Laboratories (20th International Workshop on Beam Dynamics and Optimization, BDO, Article number 6890001)

[3] Averyanov G. P. et al. 2014 (in Russian) Virtual Electrophysics Laboratories: New generation (Cybernetics and Physics, Vol. 3, No. 4) pp 153 – 156

[4] Averyanov G. P., et al. 2012 (in Russian), The multimedia laboratory complex "Electrophysics" (Moscow: Proc. Sientific session MEPhI–2012) pp 17 – 18 (Original Russian title: Multimediynjy laboratorniy kompleks «Elektrofizika». M.: Nauchnaya sessiya MIFI-2012. Sbornik nauchnyh trudov)

Modelling dynamic processes in a nuclear reactor by state change modal method

A V Avvakumov[1], V F Strizhov[2], P N Vabishchevich[2,3] and A O Vasilev[3]

[1] National Research Center Kurchatov Institute, Moscow, Russia
[2] Nuclear Safety Institute of RAS, Moscow, Russia
[3] North-Eastern Federal University, Yakutsk, Russia

E-mail: haska87@gmail.com

Abstract. Modelling of dynamic processes in nuclear reactors is carried out, mainly, using the multigroup neutron diffusion approximation. The basic model includes a multidimensional set of coupled parabolic equations and ordinary differential equations. Dynamic processes are modelled by a successive change of the reactor states. It is considered that the transition from one state to another occurs promptly. In the modal method the approximate solution is represented as eigenfunction expansion. The numerical-analytical method is based on the use of dominant time-eigenvalues of a group diffusion model taking into account delayed neutrons.

1. Introduction

To describe neutronic processes occurring in a nuclear reactor [6], a complex integro-differential equation is involved, in which the neutron flux depends on time, energy, spatial and angular variables [11]. As a rule, the group diffusion approximation are used [7] in practical calculations. Currently diffusion models are derived and applied using sophisticated homogenization methodologies (see [9]).

In the theory and practice of neutronics calculations, fast methods to obtain approximate solutions are given great attention. A class of methods for modeling the nonstationary group neutron diffusion is associated with the multiplicative representation of the space-time factorization methods and the quasistatic method [5]. An approximate solution is represented as a product of the time-dependent amplitude the spatial shape function, which is often associated with the fundamental eigenfunction of neutron diffusion equations.

When using the quasistatic method, the problem is significantly simplified, thus, it is doubtful to obtain good accuracy for an approximate solution, in particular, in dynamic regimes with complex flux redistributions. The general approach of the modal method has been developed [10]: the solution is represented as a sum of dominant eigenvalues with time-dependent coefficients.

Nonstationary processes can be described using the approximate solution expansion in time-eigenvalue of α-eigenvalue problem [12]. In a simpler model without delayed neutrons, the modal methods were used in [8]. The principal point is connected with the fact that when using this approach, we deal with an unbound system of the coefficient equations. It should also be noted that the eigenvalues are complex for both λ- and α-eigenvalue problem. To set the initial state, this leads to the need to solve the appropriate adjoint spectral problems.

IOP Conf. Series: Journal of Physics: Conf. Series **937** (2017) 012003 doi:10.1088/1742-6596/937/1/012003

2. Problem statement

The neutron flux is modelled in multigroup diffusion approximation. The neutron dynamics is considered in the bounded convex two-dimensional or three-dimensional area Ω ($\boldsymbol{x} = \{x_1, ..., x_d\} \in \Omega,\ d = 2, 3$) with boundary $\partial\Omega$. The neutron diffusion is described by:

$$V(t)\frac{d\boldsymbol{\phi}}{dt} + (D(t) + S(t))\boldsymbol{\phi} = R(t)\boldsymbol{\phi} + B(t)\boldsymbol{c}, \\ \frac{d\boldsymbol{c}}{dt} + \Lambda(t)\boldsymbol{c} = Q(t)\boldsymbol{\phi}. \tag{1}$$

Where vectors $\boldsymbol{\phi} = \{\phi_1, \phi_2, ..., \phi_G\}$, $\boldsymbol{c} = \{c_1, c_2, ..., c_M\}$ and matrices:

$$V = (v_{gg'}), \quad v_{gg'} = \delta_{gg'} v_g^{-1}, \quad D = (d_{gg'}), \quad d_{gg'} = -\delta_{gg'} \nabla \cdot D_g \nabla,$$
$$S = (s_{gg'}), \quad s_{gg'} = \delta_{gg'}\Sigma_g - \Sigma_{s,g'\to g}, \quad R = (r_{gg'}), \quad r_{gg'} = (1-\beta)\chi_g \nu\Sigma_{fg'},$$
$$B = (b_{gm}), \quad b_{gm} = \widetilde{\chi}_g \lambda_m, \quad \Lambda = (\lambda_{mm'}), \quad \lambda_{mm'} = \lambda_m \delta_{mm'},$$
$$Q = (q_{mg}), \quad q_{mg} = \beta_m \nu\Sigma_{fg}, \quad g, g' = 1, 2, ..., G, \quad m, m' = 1, 2, ..., M,$$

where $\delta_{gg'}$ is the Kronecker symbol. We shall use the set of vectors $\boldsymbol{\phi}$, whose components satisfy the albedo boundary conditions:

$$D_g \frac{\partial \phi_g}{\partial n} + \gamma_g \phi_g = 0, \qquad g = 1, 2, ..., G, \tag{2}$$

where n — outer normal to the boundary $\partial\Omega$.

Here $\phi_g(\boldsymbol{x}, t)$ — neutron flux of g group at point $\boldsymbol{x}$ and time t, G — number of energy groups, M — number of types of delayed neutrons and other terms are standard.

The Cauchy problem is formulated for equations (1) when

$$\boldsymbol{\phi}(0) = \boldsymbol{\phi}^0, \quad \boldsymbol{c}(0) = \boldsymbol{c}^0, \quad \phi^0 = \{\phi_1^0, \phi_2^0, ..., \phi_G^0\}, \quad \boldsymbol{c}^0 = \{c_1^0, c_2^0, ..., c_M^0\}. \tag{3}$$

3. State change modal method

The nuclear reactor is always non-stationary. We will use the following simplified description of the dynamic processes in a nuclear reactor.

In selected time interval, the non-stationary neutron flux is determined by the nuclear reactor state. The reactor state is characterized by the constant coefficients of multigroup diffusion equations (1). Dynamic processes in a nuclear reactor can be considered as a change of states. At a certain time $t = t_s,\ s = 1, 2, ...$ an instantaneous change of state occurs. The state s is defined by the parameters in equations (1):

$$V(t) = V(t_s), \quad D(t) = D(t_s), \quad S(t) = S(t_s), \quad R(t) = R(t_s), \quad B(t) = B(t_s),$$
$$\Lambda(t) = \Lambda(t_s), \quad Q(t) = Q(t_s), \quad t_{s-1} < t \le t_s, \quad s = 1, 2, ...$$

Simulation of the reactor dynamic behavior is considered as a consistent solution of subtasks for the reactor states. The initial condition for the state s (at $t = t_{s-1}$) is the final state of the reactor for the state $s - 1$.

An approximate description of the non-stationary process at a separate stage is based on modal approximation. An approximate solution is considered in the form of decomposition in eigenfunctions of time and α-eigenvalue problem. Finite-element approximation in space is used.

Let's $\boldsymbol{u} = \{\boldsymbol{\phi}, \boldsymbol{c}\}$. Rewrite the system of equations (1) as

$$\boldsymbol{B}\frac{d\boldsymbol{u}}{dt} + \boldsymbol{A}\boldsymbol{u} = 0, \quad t_{s-1} < t \le t_s, \tag{4}$$

IOP Conf. Series: Journal of Physics: Conf. Series **937** (2017) 012003 doi:10.1088/1742-6596/937/1/012003

with constants

$$\boldsymbol{A} = \begin{pmatrix} D(t_s) + S(t_s) - R(t_s) & -B(t_s) \\ -Q(t_s) & \Lambda(t_s) \end{pmatrix}, \quad \boldsymbol{B} = \begin{pmatrix} V(t_s) & 0 \\ 0 & I \end{pmatrix},$$

where I is the identity matrix. Initial conditions:

$$\boldsymbol{u}(t_{s-1}) = \boldsymbol{u}^s. \tag{5}$$

After approximating over the space by the finite element method from (4), (5) we turn to the Cauchy problem for a linear system of ordinary differential equations with constant coefficients:

$$\boldsymbol{B}_h \frac{d\boldsymbol{u}_h}{dt} + \boldsymbol{A}_h \boldsymbol{u}_h = 0, \quad t_{s-1} < t \leq t_s, \tag{6}$$

$$\boldsymbol{u}_h(t_{s-1}) = \boldsymbol{u}_h^s, \tag{7}$$

where h is the discretization parameter. The main feature of the problems we are considering is that the matrices $\boldsymbol{A}_h$ and $\boldsymbol{B}_h$ are real and asymmetric.

The modal approximation corresponds to the representation of the approximate solution ($\boldsymbol{u}_h \approx \boldsymbol{u}_N$) of problem (6), (7) in the following form

$$\boldsymbol{u}_N(\boldsymbol{x}, t) = \sum_{n=1}^{N} a_n(t) \boldsymbol{w}_n(\boldsymbol{x}), \tag{8}$$

where N is the number of dominant eigenvalues, $\boldsymbol{w}_n(\boldsymbol{x})$ are eigenfunctions.

Let us define eigenfunctions and eigenvalues as the solution of the α-eigenvalue problem:

$$\boldsymbol{A}_h \boldsymbol{v} = \lambda \boldsymbol{B}_h \boldsymbol{v}. \tag{9}$$

In general, eigenfunctions and eigenvalues (9) are complex.Then in the representation (8) we obtain

$$a_n(t)\boldsymbol{w}_n(\boldsymbol{x}) = b_n \mathrm{Re}\big(\exp(-\lambda_n(t - t_{s-1}))\boldsymbol{v}_n(\boldsymbol{x})\big),$$
$$a_{n+1}(t)\boldsymbol{w}_{n+1}(\boldsymbol{x}) = b_{n+1} \mathrm{Im}\big(\exp(-\lambda_n(t - t_{s-1}))\boldsymbol{v}_n(\boldsymbol{x})\big).$$

A special attention should be paid to define the coefficients $a_n(t_{s-1}) = b_n, \ n = 1, 2, ..., N$. In the case of real eigenvalues, we have

$$\boldsymbol{u}_h^s(\boldsymbol{x}) = \sum_{n=1}^{N_h} b_n \boldsymbol{v}_n(\boldsymbol{x}).$$

This representation is not very suitable for practical use with modal approximation, when we work only with dominant eigenfunctions.

The initial condition includes two components $\boldsymbol{u}_h^s(\boldsymbol{x}) = (\boldsymbol{\phi}_h^s(\boldsymbol{x}), \boldsymbol{c}_h^s(\boldsymbol{x}))$. Dynamic behaviour of these components is due to different time-scale processes: slow processes for delayed neutrons and fast processes for neutron flux when the reactor state changes. We can model the slow phase of the reactor dynamics with modal approximation: only the function $\boldsymbol{c}_h^s(\boldsymbol{x})$ is approximated. The approximation $\boldsymbol{\phi}_h^s(\boldsymbol{x})$ is not of interest to us; we do not model a fast phase of the state change.

The standard approach for the decomposition of the function $\boldsymbol{u}_h^s(\boldsymbol{x})$ over the system of non-orthogonal functions $\boldsymbol{v}_n(\boldsymbol{x})$, $n = 1, 2, ..., N_h$ consists in using the biorthogonal system of functions [3]. Consider the spectral problem adjoint to (9)

$$\boldsymbol{A}_h^T \widetilde{\boldsymbol{v}} = \lambda \boldsymbol{B}_h^T \widetilde{\boldsymbol{v}}. \tag{10}$$

The eigenfunctions of problems (9) and (10) are orthogonal in the sense of the equality

$$(\boldsymbol{B}_h \boldsymbol{v}_n, \widetilde{\boldsymbol{v}}_m) = 0, \quad m \neq n, \quad m, n = 1, 2, ..., N_h,$$

where $(\cdot, \cdot)$ means corresponding scalar product. In view of this, one can obtain

$$b_n = \frac{1}{(\boldsymbol{B}_h \boldsymbol{v}_n, \widetilde{\boldsymbol{v}}_n)} (\boldsymbol{u}_h^s, \boldsymbol{B}_h \widetilde{\boldsymbol{v}}_n), \quad n = 1, 2, ..., N_h. \tag{11}$$

In the approximate solution of problem (6), (7) only the first N coefficients b_n in (11) are used (see (8)):

$$\boldsymbol{c}_h^s(\boldsymbol{x}) \approx \sum_{n=1}^{N} b_n \boldsymbol{c}_n(\boldsymbol{x}), \tag{12}$$

where $\boldsymbol{v}_n(\boldsymbol{x}) = (\boldsymbol{\phi}_n(\boldsymbol{x}), \boldsymbol{c}_n(\boldsymbol{x}))$. In this case, the spectral problems (9), (10) are solved for N dominant eigenvalues. The solution of the adjoint spectral problem is involved only for calculating the initial condition coefficients.

The state change modal method is based on the following calculating scheme.

Off-line calculation. Calculation of the coefficients of the mathematical model of the multigroup diffusion approximation for the separate reactor states, performed in advance. It can be included calculated dominant eigenvalues and eigenfunctions of the α-eigenvalue problem (9). These data can be supplemented by dominant eigenvalues and eigenvalues of the conjugate eigenvalue problem (10).

On-line calculation. Real-time modeling is performed using the modal solution of the problem (6), (7). The coefficients in the representation (12) are calculated from the initial condition using (11). The solution for other time intervals is determined according to (8).

4. The test: the dynamics of the VVER-1000 reactor during the transition from the supercritical mode to the subcritical mode

A test problem for a VVER-1000 reactor [4] is considered in the two-dimensional approximation using a two-group diffusion approximation with delayed neutrons.

4.1. Supercritical state: α-eigenvalue problem

Below the results of a numerical solution of the α-eigenvalue problem (9) are presented. The dominant eigenvalues $\alpha_n = \lambda_n^{(\alpha)}$, $n = 1, 2, ..., N$ are searched for at

$$\mathrm{Re}\lambda_1^{(\alpha)} \leq \mathrm{Re}\lambda_2^{(\alpha)} \leq ... \leq \mathrm{Re}\lambda_N^{(\alpha)} \leq ... \leq \mathrm{Re}\lambda_{N_h}^{(\alpha)}.$$

Similar calculations of the eigenvalues for the VVER-1000 test problem without delayed neutrons can be found in [1].

The results of solving the spectral problem (9) for the first eigenvalues $\alpha_n = \lambda_n^{(\alpha)}$, $n = 1, 2, ..., N$, $N = 5$ on different computational grids using different finite element approximations are shown in Table 1. The eigenvalues α_2, α_3, α_4, α_5, α_9, α_{10} of the spectral problem (9) are complex with small imaginary parts, the eigenvalues α_1, α_6, α_7, α_8 are real.

In our example, the main eigenvalue is negative and therefore the major harmonic will increase, and all others will fade. This demonstrates the regular mode of the reactor operation. The value $\alpha = \lambda_1^{(\alpha)}$ itself determines the neutron flux amplitude and is directly related to the reactor period in the regular regime.

Table 1. Eigenvalues $\alpha_n = \lambda_n^{(\alpha)}$, $n = 1, 2, ..., 5$

p	κ	α_1	α_2, α_3	α_4, α_5
	6	-0.22557	$0.04241 \mp 3.08808\text{e-}06i$	$0.06588 \mp 4.80449\text{e-}07i$
1	24	-0.82690	$0.03777 \mp 5.37884\text{e-}06i$	$0.06489 \mp 1.37315\text{e-}06i$
	96	-1.74998	$0.03619 \mp 5.69002\text{e-}06i$	$0.06456 \mp 1.40299\text{e-}06i$
	6	-2.10154	$0.03592 \mp 4.96474\text{e-}06i$	$0.06452 \mp 1.21320\text{e-}06i$
2	24	-2.46601	$0.03562 \mp 5.78277\text{e-}06i$	$0.06445 \mp 1.40897\text{e-}06i$
	96	-2.50375	$0.03559 \mp 5.80693\text{e-}06i$	$0.06444 \mp 1.41324\text{e-}06i$

Table 2. Eigenvalues $\alpha_n = \lambda_n^{(\alpha)}$, $n = 1, 2, ..., 10$ for direct and adjoint problems

n	α_n for problem (9)	α_n for problem (10)
1	-2.51280117966	-2.51280117972
2,3	$0.0355815000364 \mp 5.80954455861\text{e-}06$	$0.0355815000365 \mp 5.80954421646\text{e-}06$
4,5	$0.0644427013767 \mp 1.41362187449\text{e-}06$	$0.0644427013767 \mp 1.41362190730\text{e-}06$
6	0.0702618501639	0.0702618501639
7	0.0714652882224	0.0714652882164
8	0.0726456060606	0.0726456060606
9,10	$0.0734708921578 \mp 4.02332269037\text{e-}08$	$0.0734708921578 \mp 4.02332146248\text{e-}08$

4.2. Adjoint spectral problem

Analogous data were obtained for approximate solution of the adjoint spectral problem (10). The eigenvalues of the spectral problems (9) and (10) coincide. Their difference from each other is an indirect measure of the accuracy of the numerical solution. Data on the dominant eigenvalues, which are given in Table 2 ($k = 96$, $p = 3$), show that the eigenvalues of the main and adjoint spectral problems are close to each other with good accuracy.

The spectral problems under consideration are characterized by small imaginary parts of the eigenvalues. Therefore, we can expect that the eigenfunctions of problem (9) are close to orthogonal. For convenience of comparison, the eigenfunctions are normalized in $L_2(\Omega)$. The maximum non-orthogonality (for $(\phi_1^{(1)}, \phi_1^{(7)})$) does not exceed 10 %. The biorthogonality condition of the eigenfunctions of the fundamental functions (see (9)) and the adjoint spectral problems (see (10)) is valid with approximately the same accuracy. This observed error can be related to an approximate calculation of eigenvalues and eigenfunctions.

Within the modal method, we can not rely on high accuracy when considering a relatively small number of dominant eigenvalues. Therefore, in the example under consideration, we can assume that the eigenvalues are real, and the corresponding eigenfunctions are orthogonal. Instead of (11) coefficients are used

$$b_n \approx \frac{1}{(\boldsymbol{c}_n, \boldsymbol{c}_n)} (\boldsymbol{c}_h^s, \boldsymbol{c}_n), \quad n = 1, 2, ..., N, \tag{13}$$

to approximate the initial condition.

4.3. Subcritical state

In the supercritical mode, due to the sufficiently large magnitude of the main eigenvalue, the regular regime of the reactor is rapidly developing, where

$$\boldsymbol{u}(\boldsymbol{x},t) \approx a_1 \exp(-\alpha_1 t)\boldsymbol{v}_1^0(\boldsymbol{x}).$$

Here $\boldsymbol{v}_1^0(\boldsymbol{x})$ is the first mode of the supercritical state. We consider the problem with the transition from this supercritical state at $t_0 = 0$ to the subcritical state.

The subcritical stage is characterized by a 15% increase in the coefficient Σ_2 for material 4 in the VVER-1000 test diffusion constants. The initial state is characterized by specifying the initial conditions at $t_0 = 0$ as

$$\boldsymbol{u}(\boldsymbol{x},0) = \boldsymbol{v}_1^0(\boldsymbol{x}). \quad (14)$$

The calculation results of the dominant eigenvalues for the subcritical state are presented in Tables 3. In this case even the first eigenvalues not significantly differ from each other.

Table 3. Subcritical state: $\alpha_n = \lambda_n^{(\alpha)}$, $n = 1, 2, ..., 5$

p	κ	α_1	α_2, α_3	α_4, α_5
3	96	0.02122	0.05376 ∓ 2.20951e-06i	0.06763 ∓ 1.04756e-06i

For an approximate solution, we use the following formulation:

$$\boldsymbol{u}_N(\boldsymbol{x},t) = \sum_{n=1}^{N} b_n \exp(-\mathrm{Re}\,\alpha_n\, t)\boldsymbol{v}_n(\boldsymbol{x}), \quad (15)$$

where the coefficients b_n, $n = 1, 2, ..., N$ are calculated according to the given initial condition (13). These coefficients for $N = 50$ are shown in Fig.1. As we see, an approximate solution can be described by first mode only.

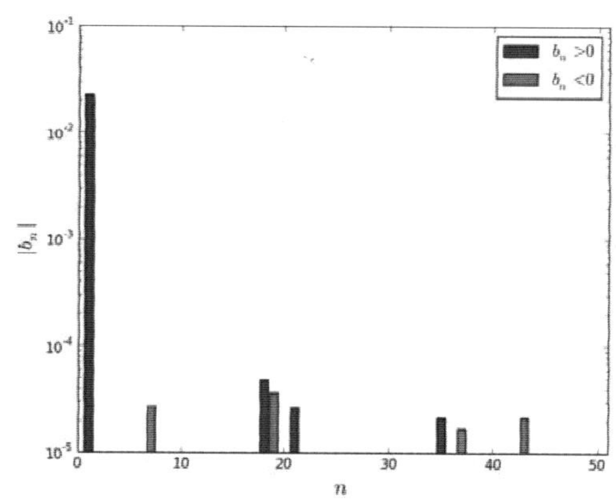

Figure 1. Approximate solution coefficients (15).

IOP Conf. Series: Journal of Physics: Conf. Series **937** (2017) 012003 doi:10.1088/1742-6596/937/1/012003

We distinguish two phases of the dynamic process: fast and slow. In the fast phase, the initial condition (14) is rearranged to the initial condition, which corresponds to (15): from the function $\boldsymbol{u}(\boldsymbol{x}, 0)$ to the function $\boldsymbol{u}_N(\boldsymbol{x}, 0)$. The slow phase is associated with the solution evolution according to (15). Within the state change modal technology, the fast phase is not modeled at all.

The beginning and the end of the fast phase are illustrated through the calculational data shown in Fig. 2. The results were obtained with $N = 10$. We can note very small changes in the topology of the initial and reconstructed initial conditions. Let us pay attention to the substantial restructuring of the solution, which is illustrated by large changes in the neutron flux amplitudes for the first and second groups.

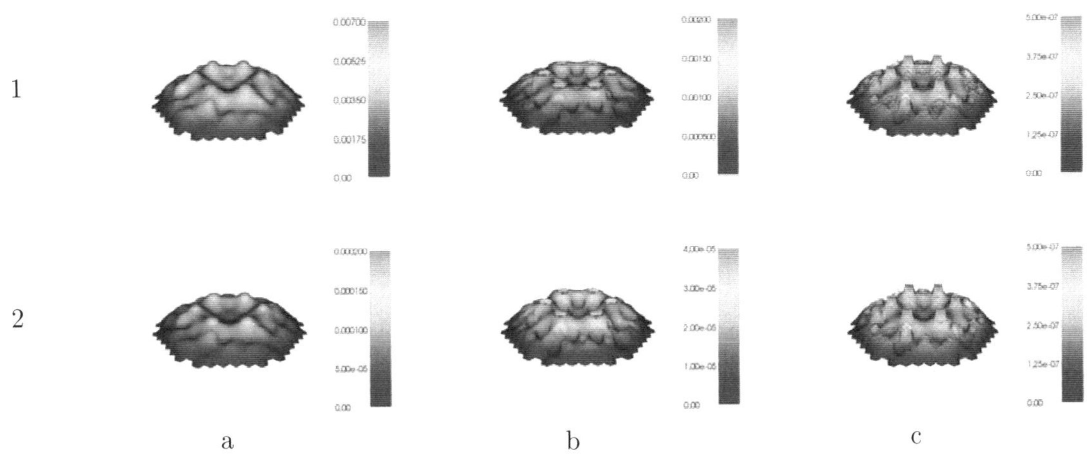

Figure 2. Function $\boldsymbol{u}(\boldsymbol{x}, 0)$ (string 1) and function $\boldsymbol{u}_N(\boldsymbol{x}, 0)$ (ctring 2): a — neutron flux of group 1, b — neutron flux of group 2, c — delayed neutrons source.

4.4. Comparison with the nonstationary problem solution

An approximate solution, obtained using modal approximation, can be compared with the dynamic problem solution (1). Fully implicit scheme on a uniform grid in time with a sufficiently small step $\tau = 0.0025$ is used (see details in [2]). We consider the dynamics of the neutron power of the nuclear reactor P and the delayed neutrons source C at the initial stage during the transition from the critical state to the subcritical one. Here

$$P(t) = \int_\Omega (\nu\Sigma_{f1}\varphi_1 + \nu\Sigma_{f2}\varphi_2) d\boldsymbol{x}, \quad C(t) = \int_\Omega c(\boldsymbol{x}, t) d\boldsymbol{x}.$$

There is a rapid change in neutron power over a short period of time, while the delayed neutrons source changes rather slow. The dynamics of the slow phase is illustrated in Figs. 3. Here the solution of full equations (dynamic in Figs. 3), and the modal approximation solution (modal) are presented. We see that the integral characteristics of the reactor dynamics at slow stage are calculated with good accuracy.

5. Conclusions

The problem of reactor dynamic simulation is considered on the basis of multigroup neutron diffusion equations taken into account delayed neutrons. The modal approximation is used:

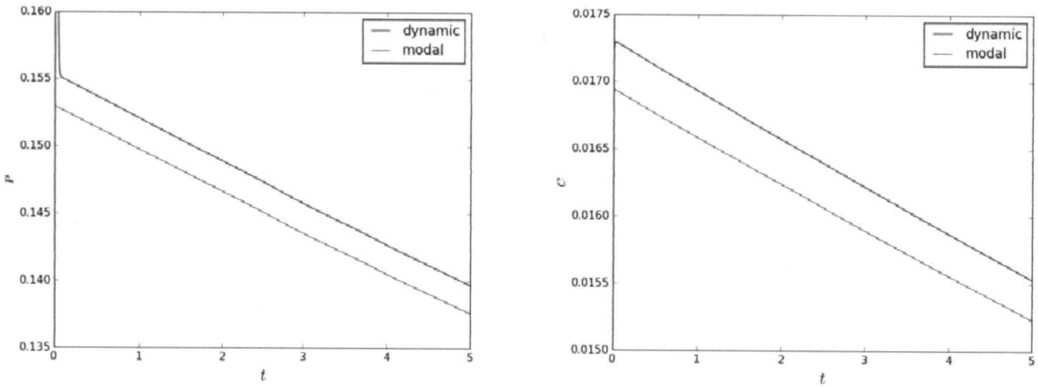

Figure 3. Slow stage of reactor state: neutronic power and delayed neutrons sourse

an approximate solution is represented as an expansion on limited number of dominant eigenfunctions of the α-eigenvalue spectral problem.

Numerical simulation of reactor non-stationary processes is carried out on the basis of a successive change in the states of the reactor. These states are characterized by a set of constant parameters to describe the multigroup neutron flux behavior. The state change modal method was developed. The phase, which described fast transition to the approximate solution, is selected as a set of dominant modes. At a slow phase of the reactor dynamics, the solution is based on the evolution of dominant modes.

The VVER-1000 two-dimensional test problem is used to perform the transition from the supercritical state to the subcritical state. Calculations of dominant modes for a reactor supercritical state are performed. The major mode solution, which determines the reactor regular regime, is used as the initial condition for transition to the subcritical state. Comparison of the calculation results obtained by using the modal approximation and direct dynamics calculation, shows the acceptable accuracy in calculation of neutron power and delayed neutrons source.

Acknowledgements
This work are supported by the Russian Foundation for Basic Research (# 16-08-01215) and by the grant of the Russian Federation Government (# 14.Y26.31.0013).

References
[1] Avvakumov A V, Strizhov V F, Vabishchevich P N and Vasilev A O 2017 *Annals of Nuclear Energy* **99** 68–79
[2] Avvakumov A V, Strizhov V F, Vabishchevich P N and Vasilev A O 2017 *Matem. Mod.* **7** 44–62
[3] Brezinski C 1991 Biorthogonality and Its Applications to Numerical Analysis *CRC Press*
[4] Chao Y A and Shatilla Y A 1995 *Nuclear Science and Engineering* **121** 210–225
[5] Dodds Jr H L 1976 *Nuclear Science and Engineering* **59** 271–276
[6] Duderstadt J J and Hamilton L J 1976 *Nuclear Reactor Analysis* Wiley
[7] Marchuk G I and Lebedev V I 1986 Numerical Methods in the Theory of Neutron Transport *Harwood Academic Pub*
[8] Modak R and Gupta A 2007 *Annals of Nuclear Energy* **34** 213–221
[9] Sanchez R 2009 *Progress in Nuclear Energy* **51** 14–31
[10] Stacey W M 1967 Modal Approximations: Theory and an Application to Reactor Physics *The MIT Press*
[11] Stacey W M 2007 Nuclear Reactor Physics *Wiley*
[12] Verdu G, Ginestar D, Roman J and Vidal V 2010 *Journal of Nuclear Science and Technology* **47** 501–514

About orthogonality of the single systems of special kind

A N Barmenkov, N A Barmenkov

National Research Nuclear University MEPhI (Moscow Engineering Physics Institute), 31 Kashirskoe shosse, 115409 Moscow, Russia

E-mail: anbarmenkov@mail.ru

Abstract. We present an algorithm for construction complete, orthogonal of the single systems of special kind. This sequences may be used in modelling of physical processes.

1. Introduction

The role of the classical trigonometric system and the single trigonometric systems is very important for the resolving of various problems in the theory of different equation. In 1953 K. Shaidukov [1] proved by the method of theory of functions of real variable the completeness in $L_2[0;2\pi]$ of sequence

$$\{\cos(nt+bt);\sin(nt+bt)\}_{n=0}^{\infty}, \tag{1}$$

when $b \le 2/3$ and start of a whole direction of research of such systems. In [2], [3] by the methods of the monograph [4] it is shown that completeness and minimality in $L_p[0;2\pi]$ in a more general system of functions

$$\{\cos(nt+\alpha(t));\sin(nt+\alpha(t))\}_{n=0}^{\infty}. \tag{2}$$

This property depends only on the difference of the $\alpha(t)$ at the endpoints of the segment $[0;2\pi]$, $\alpha(t) \in Lip_\nu[0;2\pi] \cap Var[0;2\pi]$, $0<\nu\le 1$. So that the sequence (2) is complete in.

$L_p[0;2\pi], 1<p<\infty$, when $\dfrac{\alpha(2\pi)-\alpha(0)}{2\pi} \le \dfrac{1}{2}+\dfrac{1}{2p}$. That is system K Shaidukov (1) complete in $L_2[0;2\pi]$ for $\alpha(t)=bt$, $b\le 3/4$. Moreover, if $\alpha(t)\ne const$, $t\in[0;2\pi]$ then system (2) is complete and minimality in $L_p[0;2\pi], 1<p<\infty$, if and only if $\dfrac{1}{2p} < \dfrac{\alpha(2\pi)-\alpha(0)}{2\pi} \le \dfrac{1}{2}+\dfrac{1}{2p}$. In [5] is considered the properties of completeness and orthogonality systems (2). In [6] used a single sequence $\{\sin(nt+bt)\}_{n=0}^{\infty}$ for solving problems of differential equations of mixed type. In this article we consider completeness and orthogonality in $L_2[0;\pi]$ of single systems of that following form

$$\{\sin(nt+\alpha(t))\}_{n=0}^{\infty}, \tag{3}$$

where $\alpha(t)$ is real continuous function. Most of the applications use the complete and orthogonal sequences. Let us consider orthogonality property of systems of the form (3). By method developed in [2], [3] the conditions orthogonality (completeness and minimality) we reduce to the problem o f solvability of some series boundary value problems theory analytic functions. To formulate the main results we recall some definitions.

Define class Hardy H^p: $f(z)\in H^p, 1\le p<\infty, \Leftrightarrow$ if $f(z)$ is analytic function in $|z|<1$ and

$$\sup\int_{-\pi}^{\pi}\left|f(re^{it})\right|^p dt<\infty,\quad (0<r\le 1) \text{ for } p:\ 1\le p<\infty.$$

2. Main results

Theorem 1. Let $\alpha(t)$ a real continuous function on segment $[0;\pi]$. To ensure that the system

$$\{\sin(nt+\alpha(t))\}_{n=0}^{\infty}$$

t o be orthogonal in space $L_2[0;\pi]$ necessary and sufficient the existence analytic function $\phi(z)$ class Hardy H^2 in $|z|<1$ which following condition to be satisfied

$$e^{2i\alpha(t)}=\phi(e^{it}),\ t\in[0;\pi]. \tag{4}$$

Note the well-known complete and orthogonal systems satisfied this condition. For example: 1) For a sequence $\{\sin(nt+\frac{t}{2})\}_{n=0}^{\infty}$, then $\alpha(t)=\frac{t}{2}$ and condition (4) has a form $e^{2i\alpha(t)}\chi_{[0;\pi]}(t)+e^{-2i\alpha(-t)}\chi_{(-\pi;0)}(t)=e^{it}$, where $\chi_{[0;\pi]}(t)$ – characteristic function of segment ($\chi_{[0;\pi]}(t)=1, t\in[0;\pi]; \chi_{[0;\pi]}(t)=0, t\notin[0;\pi]$). In this case the function $\phi(z)$ is equal $\phi(z)=z$. 2) For a sequence $\{\sin(nt+t)\}_{n=0}^{\infty}$, then $\alpha(t)=t$ and condition (4) has $\phi(z)=z^2$.

Theorem 2. Let $\alpha(t)$ a real continuous function on segment $[0;\pi]$. Completeness and minimality of the sequence

$$\{\sin(nt+\alpha(t))\}_{n=1}^{\infty}$$

in space $L_p[0;2\pi]$ are equivalent to the existence of unique solution of a series of task Carleman

$$\phi_k(e^{it})+e^{2i\alpha(t)}\phi(e^{-it})=\frac{-i}{\pi}(e^{2i\alpha(t)}e^{ikt}+e^{-ikt}),\ t\in[0;\pi],$$

$$\phi_k(e^{it})+e^{-2i\alpha(-t)}\phi(e^{-it})=\frac{-i}{\pi}(e^{-2i\alpha(-t)}e^{ikt}+e^{-ikt}),\ t\in(-\pi;0)$$

in Hardy class H^q, $\frac{1}{p}+\frac{1}{q}=1$, $1<p<\infty$, $k=1,2,\ldots$

Consequence. Let $\alpha(t)$ a real continuous function on segment $[0;\pi]$. Completeness of the sequence

$$\{\sin(nt+\alpha(t))\}_{n=1}^{\infty}$$

in space $L_p[0;\pi], 1<p<\infty$ are equivalent to only the existence solution is equal to zero of this homogeneous task Carleman.

Proof of the theorem 1. Orthogonality of sequences of form (3) in $L_2[0;\pi]$ follows

$$0=\int_0^{\pi}\sin(nt+\alpha(t))\sin(kt+\alpha(t))dt=\frac{1}{2}\int_0^{\pi}\cos(n-k)tdt-\frac{1}{2}\int_0^{\pi}\cos((n+k)t+2\alpha(t))dt$$

for arbitrary $n,k=0,1,2,\ldots n\neq k$. Obviously. it is equivalently to equation

IOP Conf. Series: Journal of Physics: Conf. Series **937** (2017) 012004 doi:10.1088/1742-6596/937/1/012004

$$0=\int_0^{\pi}\cos(nt+2\alpha(t))dt=\frac{1}{2}\int_0^{\pi}e^{\text{int}}e^{2i\alpha(t)}dt+\frac{1}{2}\int_0^{\pi}e^{-\text{int}}e^{-2i\alpha(t)}dt \text{ for } n=1,2,3,...$$

Consequently we get

$$0=\int_{-\pi}^{\pi}e^{\text{int}}\{e^{2i\alpha(t)}\chi_{(-\pi;0)}(t)+e^{-2i\alpha(-t)}\chi_{[0;\pi]}(t)\}dt,\ n=1,2,..., \tag{5}$$

where $\chi_{[0;\pi]}(t)$ – characteristic function of segment.

Equation (5) means the existence of a function $\phi(z)$ Hardy class H^1 (see. [5], p.102), for angular boundary values which have the equality

$$e^{2i\alpha(t)}\chi_{[0;\pi]}(t)+e^{-2i\alpha(-t)}\chi_{(-\pi;0)}(t)=\phi(e^{it}),\ t\in(-\pi;\pi].$$

That is another form (4),

$$e^{2i\alpha(t)}=\phi(e^{it}),\ t\in[0;\pi],$$

$$e^{-2i\alpha(-t)}=\phi(e^{it}),\ t\in(-\pi;0),$$

which completes the proof of the theorem 1.

Proof of theorem 2 users task Carleman of special type (see [3], p. 132). Completeness and minimality of the sequence in space $L_p[0;\pi],1<p<\infty$ are equivalent to the existence orthogonal system $\{\varphi_n(t)\}_{n=1}^{\infty}$, $\varphi_n(t)\in L_q[0;\pi]$, $\frac{1}{p}+\frac{1}{q}=1$, such that

$$\delta_n^k=\int_0^{\pi}\sin(nt+\alpha(t))\varphi_k(t)dt=\frac{1}{2i}\int_0^{\pi}e^{\text{int}}e^{i\alpha(t)}\varphi_k(t)dt-\frac{1}{2i}\int_0^{\pi}e^{-\text{int}}e^{-i\alpha(t)}\varphi_k(t)dt,$$

where ($\delta_n^k=1$, if $n=k$; $\delta_n^k=0$, if $n\neq k$), $n,k=1,2,...$ Because $\delta_n^k=\frac{1}{2\pi}\int_{-\pi}^{\pi}e^{\text{int}}e^{-ikt}dt$, analogy to proof of theorem 1 we get

$$\int_{-\pi}^{\pi}e^{\text{int}}(\{e^{i\alpha(t)}\chi_{[0;\pi]}(t)\varphi_k(t)-e^{-i\alpha(-t)}\chi_{(-\pi;0)}\varphi(-t)\}-\frac{2i}{2\pi}e^{-ikt})dt=0,\ n=1,2,...,$$

This equation means the existence of a function $\phi(z)$ Hardy class H^1 (see. [5], p. 102), for angular boundary values which have the equality

$$e^{2i\alpha(t)}\chi_{[0;\pi]}(t)\varphi_k(t)-e^{-2i\alpha(-t)}\chi_{(-\pi;0)}(t)\varphi_k(-t)-\frac{i}{\pi}e^{-ikt}=\phi_k(e^{it})\quad t\in(-\pi;\pi].$$

We get

$$e^{i\alpha(t)}\varphi_k(t)-\frac{i}{\pi}e^{-ikt}=\phi_k(e^{it})\ t\in[0;\pi] \tag{6}$$

$$-e^{-i\alpha(-t)}\varphi_k(-t)-\frac{i}{\pi}e^{-ikt}=\phi_k(e^{it})\ t\in(-\pi;0)$$

$$\varphi_k(-t)=-e^{i\alpha(t)}\{\phi_k(t)+\frac{i}{\pi}e^{-ikt}\}\ k=1,2,...,$$

replace -t to t and substitute into (6). We obtain the ratio

$$\phi_k(e^{it})+e^{2i\alpha(t)}\phi(e^{-it})=\frac{-i}{\pi}(e^{2i\alpha(t)}e^{ikt}+e^{-ikt}),\ t\in[0;\pi],$$

which completes the proof of theorem 2.

IOP Conf. Series: Journal of Physics: Conf. Series **937** (2017) 012004 doi:10.1088/1742-6596/937/1/012004

References

[1] Shaidukov K 1953 *About the completeness of trigonometric systems.* (Moscow: Success of Sciences) chapter 6 p. 143–153.
[2] Barmenkov A N and Kazmin J A 1982 *The completeness of systems of functions of a special form About the completeness of trigonometric system.* (Kiev: Theory of mapping, its generalizations and applications. Collection of scientific articles.Naukova Dumka) p. 23–43.
[3] Barmenkov A N 1983 *About approximation properties of certain systems of functions* Diss..kand.fiz.mat. sciences, Moscow, p. 154.
[4] Danyluk I I 1975 *Irregular boundary value problems on the plane*, M. Science, p. 256.
[5] Barmenkov A N 2017 Journal of Physics: Conference Series, v.788, 012047.
[6] Bitsadze A B 1950 About of system of functions {Moscow: Success. Math. Sciences v. 5. № 4 p. 154–155.

On some control problems of dynamic of reactor

A V Baskakov and N P Volkov

National Research Nuclear University MEPhI (Moscow Engineering Physics Institute), 31, Kashirskoe shosse, 115409 Moscow, Russia

e-mail: npvolkov@mephi.ru

Abstract. The paper analyzes controllability of the transient processes in some problems of nuclear reactor dynamics. In this case, the mathematical model of nuclear reactor dynamics is described by a system of integro-differential equations consisting of the non-stationary anisotropic multi-velocity kinetic equation of neutron transport and the balance equation of delayed neutrons. The paper defines the formulation of the linear problem on control of transient processes in nuclear reactors with application of spatially distributed actions on internal neutron sources, and the formulation of the nonlinear problems on control of transient processes with application of spatially distributed actions on the neutron absorption coefficient and the neutron scattering indicatrix. The required control actions depend on the spatial and velocity coordinates. The theorems on existence and uniqueness of these control actions are proved in the paper. To do this, the control problems mentioned above are reduced to equivalent systems of integral equations. Existence and uniqueness of the solution for this system of integral equations is proved by the method of successive approximations, which makes it possible to construct an iterative scheme for numerical analyses of transient processes in a given nuclear reactor with application of the developed mathematical model. Sufficient conditions for controllability of transient processes are also obtained. In conclusion, a connection is made between the control problems and the observation problems, which, by to the given information, allow us to reconstruct either the function of internal neutron sources, or the neutron absorption coefficient, or the neutron scattering indicatrix....

1. Introduction

The needs for reliable, easily controllable and safe nuclear power facilities permanently increase in the recent time. This growth represents an additional motivation for further developing mathematical theory of nuclear reactors, which is a relatively young and rapidly progressing theory. Some scientific publications [1-12] on mathematical theory of nuclear reactors confirm this fact. When developing the theory, nuclear specialists and technicians are trying to create such models of nuclear reactors whose reliable operation should be maximally independent on human factor effects. The problems related with mathematical modeling of physical processes occurring in nuclear reactors appear on the agenda. These problems should be resolved both for the processes of the reactor designing and for the control processes of the reactor operation.

2. Definition of the linear control problem

In this section we considered the mathematical model that was used to describe reactor dynamics without accounting for temperature feedbacks in Ref. [4]. The mathematical model is based on system of the following integro-differential equations:

1) non-stationary anisotropic multi-velocity equation of neutron transport:

$$\frac{\partial u}{\partial t}(x,v,t)+(v,\nabla_x)u(x,v,t)+\Sigma(x,v,t)u(x,v,t)=\sum_{k=1}^{N} z_k R_k(x,v,t)+$$
$$+\int_V J(x,v,v',t)u(x,v',t)dv'+F(x,v,t); \quad (1)$$

2) balance equation of delayed neutrons:

$$\frac{\partial R_k}{\partial t}(x,v,t)=-z_k R_k(x,v,t)+\int_V J_k(x,v,v',t)u(x,v',t)dv' \quad \forall\ k=\overline{1,N}, \quad (2)$$

where $(x,v,t)\in D=G\times V\times[0,T]$.

In this mathematical model the function $u(x,v,t)$ defines distribution of the neutrons flying through spatial point $x\in G$ with velocity $v\in V$ at time moment $t\in[0,T]$. The functions $\Sigma(x,v,t)$, $J(x,v,v',t)$ and $F(x,v,t)$ characterize properties of the medium where some mass transport processes take place in. The function $\Sigma(x,v,t)$ is a neutron absorption coefficient, the function $J(x,v,v',t)$ is a neutron scattering indicatrix, and the function $F(x,v,t)$ is an intensity of neutron source. Kernels of the scattering integrals $J_k(x,v,v',t)$ characterize distribution of secondary neutrons. The function $R_k(x,v,t)$ characterizes distribution of nuclides – emitters of delayed neutrons in the k-th energy group. The strictly convex region $\mathbf{G}$ is a domain of spatial coordinates. The closed set $\mathbf{V}$ in the spherical layer $\{0<v_0\le|v|\le v_1<\infty\}$ is a set of velocities v with which the emitted particles can move.

It seems natural to assume that the transient process takes place without accounting for external neutron sources, i.e. an external neutron radiation is not able to pass through walls of the reactor vessel. This assumption can be mathematically written as the following boundary condition:

$$u(x,v,t)=0, \quad (x,v,t)\in\gamma_-\times[0,T], \quad (3)$$

where $\gamma_-=\{(x,v)\in\partial\mathbf{G}\times\mathbf{V}:(v,n_x)<0\}$, n_x is an outer normal to the border $\partial\mathbf{G}$ of the domain $\mathbf{G}$ in spatial point x.

The following initial conditions can be also set:

$$u(x,v,0)=\varphi(x,v), \quad (x,v)\in\overline{\mathbf{G}}\times\mathbf{V}, \quad (4)$$

$$R_k(x,v,0)=R_{k0}(x,v), \quad \forall\ k=\overline{1,N}, \quad (x,v)\in\overline{\mathbf{G}}\times V. \quad (5)$$

The linear control problem can be defined by setting the following final condition:

$$u(x,v,t_1)=\psi(x,v), \quad (x,v)\in\overline{\mathbf{G}}\times\mathbf{V}, \quad (6)$$

where time $t_1\in(0,T]$.

The control problem 1. It is required to find sufficient conditions on initial data under which the reactor could be converted from the initial state (4) to the target state (6) for the time interval $t_1\in(0,T]$ by using the time-independent part $f(x,v)$ of the function of neutron sources:

$$F(x,v,t)=f(x,v)g_1(x,v,t), \quad (7)$$

where f is an acceptable control action (distributed stationary control), g_I is an a priori known function (the corrective function).

The control problem 1 is a linear problem from the standpoint of searching for the spatially distributed control action $f(x,v)$. The control problem 1 by its definition is identical to the inverse problem described in Ref. 12].

3. Solution of the linear control problem

To resolve the problem, we used the following functional spaces introduced in Ref. 5]:

1) $H_\infty(\mathbf{D})$ is a space of the functions u, which belong to the class $L_\infty(\mathbf{D})$ together with their generalized derivatives u_t and $(v,\nabla_x)u$ on the set $\mathbf{D}$ and possess the traces $u_{|\Gamma_-}$ on the set $\Gamma_- = \gamma_- \times (0,T)$ from $L_\infty(\Gamma_-)$, i.e.

$$H_\infty(\mathbf{D}) = \left\{ u \in L_\infty(\mathbf{D}) : \frac{\partial u}{\partial t}, (v,\nabla_x)u \in L_\infty(\mathbf{D}), u_{|\Gamma_-} \in L_\infty(\Gamma_-) \right\},$$

which is a Banach space with respect to the following norm:

$$\|u\|_{H_\infty} = \left[\|u\|_{\infty,\mathbf{D}} + \left\| \frac{\partial u}{\partial t} \right\|_{\infty,\mathbf{D}} + \|(v,\nabla_x)u\|_{\infty,\mathbf{D}} + \left\| u_{|\Gamma_-} \right\|_{\infty,\Gamma_-} \right] < \infty .$$

2) $L'_\infty(\mathbf{D})$ is a space of the functions R_k, which belong to the class $L_\infty(\mathbf{D})$ together with their generalized derivatives $\dfrac{\partial R_k}{\partial t}$ on the set $\mathbf{D}$, i.e.

$L'_\infty(\mathbf{D}) = \left\{ R_k \in L_\infty(\mathbf{D}) : \dfrac{\partial R_k}{\partial t} \in L_\infty(\mathbf{D}) \right\}$, which is a Banach space with respect to the following norm:

$$\|R_k\|_{L'_\infty} = \left[\|R_k\|_{\infty,\mathbf{D}} + \left\| \frac{\partial R_k}{\partial t} \right\|_{\infty,\mathbf{D}} \right] < \infty .$$

Where $\|\bullet\|_{\infty,\mathbf{D}}$ is a norm in the space $L_\infty(\mathbf{D})$.

The solution of the linear control problem 1 is presented in the following theorem.

Theorem 1. *Let us assume that* $\Sigma \in L'_\infty(\mathbf{D})$; $J \in L'_\infty(\mathbf{D}\times \mathbf{V})$; φ, $(v,\nabla_x)\varphi \in L_\infty(\mathbf{G}\times\mathbf{V})$, ψ, $(v,\nabla_x)\psi \in L_\infty(\mathbf{G}\times\mathbf{V})$, $\varphi_{|\gamma_-}$, $\psi_{|\gamma_-} \in L_\infty(\gamma_-)$; $g_I \in L'_\infty(\mathbf{D})$,

$|g_I(x,v,t_I)| \geq g_{0I} > 0$; $J_k \in L'_\infty(\mathbf{D}\times\mathbf{V})$; $\mathrm{R}_{k0}(x,v) \in L_\infty(\mathbf{G}\times\mathbf{V})$ $\forall\ k=\overline{1,N}$;

and the following concordance conditions are satisfied.

$$\varphi(x,v)=0 \quad \psi(x,v)=0 \quad (x,v)\in\gamma_- . \qquad (8)$$

Then the reactor can be converted from the initial state (4) to the target state (6) for the time interval t_I *by using the only possible stationary spatially distributed control action* $f(x,v)$ *that belongs to the space* $L_\infty(\mathbf{G}\times\mathbf{V})$. *In this case, the state function* $u(x,v,t)$ *belongs to the class* $H_\infty(\mathbf{D})$.

Proof. To prove the theorem, we used the method proposed in Refs. 7-9] for kinetic equation of neutron transport.

By integrating on the characteristic, the system of equations (1)-(8) can be equivalently reduced to the following system of integral equations:

$$u(y+v\xi,v,t)=\mathcal{A}(y,v,\xi,t)+ \\ +\int_{\eta}^{\xi}[(Pu)(y+v\theta,v,\theta-\xi+t)+f(y+v\theta,v)g_1(y+v\theta,v,\theta-\xi+t)]d\theta \quad , \tag{9}$$

$$f(y+v\xi,v)=[g_1(y+v\xi,v,t_1)]^{-1}\left\{\frac{\partial}{\partial\xi}\psi(y+v\xi,v)-(P\psi)(y+v\xi,v,t_1)+\right. \\ +\int_{\eta}^{\xi}\left[\frac{\partial}{\partial t}(Pu)(y+v\theta,v,\theta-\xi+t_1)+f(y+v\theta,v)\frac{\partial g_1}{\partial t}(y+v\theta,v,\theta-\xi+t_1)\right]d\theta- \\ \left.-\sum_{k=1}^{N}z_k\left[R_{k0}(y+v\xi,v)\exp\{-z_kt_1\}+z_k\int_{\zeta_-}^{\xi}R_{k0}(y+v\theta,v)\exp\{-z_k(\theta-\xi+t_1)\}d\theta\right]\right\}, \tag{10}$$

where $y+v\xi=x$ is a characteristic direct line with the direction vector v, ξ is a parameter from the range $[\zeta_-,\zeta_+]$, while $y+v\zeta_\pm\in\partial\mathbf{G}$, and $(y+v\zeta_-,v)\in\gamma_-$;

$$\mathcal{A}(y,v,\xi,t)=\varphi(y+v(\xi-t),v)+\sum_{k=1}^{N}z_k\int_{\xi-t}^{\xi}R_{k0}(y+v\theta,v)\exp\{-z_k(\theta-\xi+t)\}d\theta \quad ,$$

$\eta=\xi-t$ at $0\le t\le\xi-\zeta_-$, $\xi\in[\zeta_-,\zeta_+]$ and

$$\mathcal{A}(y,v,\xi,t)=\sum_{k=1}^{N}z_k\int_{\zeta_-}^{\xi}R_{k0}(y+v\theta,v)\exp\{-z_k(\theta-\xi+t)\}d\theta \quad , \quad \eta=\zeta_-$$

at $\xi-\zeta_-<t\le T$, $\xi\in[\zeta_-,\zeta_+]$;

$$(Pu)(y+v\xi,v,t)\equiv\sum_{k=1}^{N}z_k\int_0^t\exp\{-z_k(t-\tau)\}\int_{\mathbf{V}}J_k(y+v\xi,v,v',t)u(y+v\xi,v',t)dv'd\tau+ \\ +\int_{\mathbf{V}}J(y+v\xi,v,v',t)u(y+v\xi,v',t)dv'-\Sigma(y+v\xi,v,t)u(y+v\xi,v,t)$$

Hence, the control problem 1 has the sole generalized solution $f\in L_\infty(\mathbf{G}\times\mathbf{V})$ and $u\in H_\infty(\mathbf{D})$, only if the system of integral equations (9), (10) can be unambiguously solved in the mentioned functional class. This is possible under the following condition: $v_0^{-1}\operatorname{diam}\mathbf{G}\le a_1$, where a_1 is a constant depending on the time interval t_1 and on the norm of the operator P.

Farther, we studied a principal capability to resolve the system of integral equations (9), (10). This system can be re-written in the following operator form:

$$\{u, f\} = \mathbf{A}\{u, f\} = \{\mathbf{A}_1\{u, f\}, \mathbf{A}_2\{u, f\}\},$$

where $\mathbf{A}_1$, $\mathbf{A}_2$ are the operators with the domains $H_\infty(\mathbf{D})\times L_\infty(\mathbf{G}\times\mathbf{V})$, which can act on the functions $\{u, f\}$ in the accordance to the rules defined by the right parts of integral equations (9) and (10), respectively. In this case, the operator $\mathbf{A}$ can continuously reflect the Banach space $H_\infty(\mathbf{D})\times L_\infty(\mathbf{G}\times\mathbf{V})$ into itself. Since the image $\mathbf{A}_1\{u, f\}$ has the generalized derivatives belonging to the space $L_\infty(\mathbf{D})$, its trace is an element of the space $L_\infty(\Gamma_-)$. The image $\mathbf{A}_2\{u, f\}$ belongs to the space $L_\infty(\mathbf{G}\times\mathbf{V})$.

Further, we studied compressibility of the operator $\mathbf{A}$ and revealed that its second degree is a contraction operator under the following condition: $v_0^{-1}\mathrm{diam}\,\mathbf{G} \le \boldsymbol{a}_2$, where $\boldsymbol{a}_2$ is a constant depending on the norm of the functions Σ, J, J_k, φ, ψ, g_1 in the respective spaces.

Hence, according to the Banach principle for immovable point, under the condition $v_0^{-1}\mathrm{diam}\,\mathbf{G} \le \boldsymbol{a} = \min\{\boldsymbol{a}_1, \boldsymbol{a}_2\}$ there $\{u, f\}$ is the sole solution for the system of integral equations (8), (9) in the space $H_\infty(\mathbf{D})\times L_\infty(\mathbf{G}\times\mathbf{V})$. Hence, the linear control problem 1 has the sole solution $\{u, f\}$ under the found constraints on initial data, i.e. we obtained sufficient conditions for controllability of transient processes for the given mathematical model of the reactor dynamics.

Thus, the theorem 1 was completely proved.

Remark 1. In the theorem 1 the conditions $\Sigma \in L_\infty^{\prime}(\mathbf{D})$, $J \in L_\infty^{\prime}(\mathbf{D}\times\mathbf{V})$, $g_1 \in L_\infty^{\prime}(\mathbf{D})$, $J_k \in L_\infty^{\prime}(\mathbf{D}\times\mathbf{V})$ $\forall\ k = \overline{1,N}$ can be replaced by the following new conditions: $\Sigma,\ (\boldsymbol{v},\nabla_x)\Sigma \in L_\infty(\mathbf{D})$, $J,\ (\boldsymbol{v},\nabla_x)J \in L_\infty(\mathbf{D}\times\mathbf{V})$, $g_1,\ (\boldsymbol{v},\nabla_x)g_1 \in L_\infty(\mathbf{D})$, $J_k,\ (\boldsymbol{v},\nabla_x)J_k \in L_\infty(\mathbf{D}\times\mathbf{V})$ $\forall\ k = \overline{1,N}$, respectively. Under these new conditions the conclusions derived from the theorem 1 remain to be correct.

4. Non-linear control problems

The control problem 2. Is it possible to convert nuclear reactor from the initial state (4) to the target state (6) for the time interval $t_1 \in (0, \boldsymbol{T}]$ by using the time-independent part $\sigma(\boldsymbol{x},\boldsymbol{v})$ of the neutron absorption coefficient:

$$\Sigma(\boldsymbol{x},\boldsymbol{v},t) = \sigma(\boldsymbol{x},\boldsymbol{v})g_2(\boldsymbol{x},\boldsymbol{v},t), \tag{11}$$

where σ is an acceptable control action (distributed stationary control), and g_2 is an a priori known function (the corrective function)?

The control problem 2 is a non-linear problem from the standpoint of searching for the spatially distributed control action $\sigma(\boldsymbol{x},\boldsymbol{v})$.

Positive response to the question asked above was obtained in the following theorem.

Theorem 2. *Let us assume that* $J \in L_\infty^{\prime}(\mathbf{D}\times\mathbf{V})$; $F \in L_\infty^{\prime}(\mathbf{D})$; $\varphi,\ (\boldsymbol{v},\nabla_x)\varphi \in L_\infty(\mathbf{G}\times\mathbf{V})$, $\psi,\ (\boldsymbol{v},\nabla_x)\psi \in L_\infty(\mathbf{G}\times\mathbf{V})$, $\varphi_{|\gamma_-}$, $\psi_{|\gamma_-} \in L_\infty(\gamma_-)$; $g_2 \in L_\infty^{\prime}(\mathbf{D})$, $|g_2(\boldsymbol{x},\boldsymbol{v},t_1)\psi(\boldsymbol{x},\boldsymbol{v})| \ge g_{02} > 0$; $J_k \in L_\infty^{\prime}(\mathbf{D}\times\mathbf{V})$; $\mathrm{R}_{k0}(\boldsymbol{x},\boldsymbol{v}) \in L_\infty(\mathbf{G}\times\mathbf{V})$ $\forall\ k = \overline{1,N}$; *and the following concordance conditions are satisfied*:

$$\varphi(\boldsymbol{x},\boldsymbol{v}) = 0 \quad \psi(\boldsymbol{x},\boldsymbol{v}) = 0 \quad (\boldsymbol{x},\boldsymbol{v}) \in \gamma_- . \tag{12}$$

Then the reactor can be converted from the initial state (4) to the target state (6) for the time interval t_1 by using the only possible stationary spatially distributed control action $\sigma(x,v)$ that belongs to a compact set Ω from the space $L_\infty(G\times V)$. In this case, the state function $u(x,v,t)$ belongs to the closed limited set U from the class $H_\infty(D)$.

Proof. The two sought functions $\{u,\sigma\}$ can be found in the space $H_\infty(D)\times L_\infty(G\times V)$ on the following set

$$U\times\Omega=\left\{\{u,\sigma\}:\ u\in H_\infty(D),\ \sigma\in L_\infty(G\times V),\ \|u\|_{H_\infty}\le k_1,\ \|\sigma\|_{\infty,D}\le k_2\right\},$$

where constants k_1, k_2 depend on the norms of initial data.

Like the method we used in proving the theorem 1, we reduced the system of equations (1)-(6), (12) to the equivalent system of the second kind integral equations for the two sought functions $\{u,\sigma\}$. This system can be written in the following operator form:

$$\{u,\sigma\}=B\{u,\sigma\}=\{B_1\{u,\sigma\},B_2\{u,\sigma\}\}.$$

In this case, the second degree of the operator B is a contraction operator for all elements of the set $U\times\Omega$, under condition of the following inequality: $v_0^{-1}\mathrm{diam}\,G\le b$, where b is a constant that depends on initial data of the control problem 2.

This means that the derived system of integral equations has the sole solution $\{u,\sigma\}$ in the set $U\times\Omega$. Hence, there is the only one control action $\sigma(x,v)$, which is able to convert nuclear reactor from the initial state (4) to the target state (6) for the time interval t_1. So, the control problem 2 is an unambiguously resolvable problem. Thus, the theorem 2 was completely proved.

Remark 2. The conclusions derived from the theorem 2 remain to the same if the following conditions $J\in L_\infty^t(D\times V)$, $F\in L_\infty^t(D)$, $g_2\in L_\infty^t(D)$, $J_k\in L_\infty^t(D\times V)$ $\forall\ k=\overline{1,N}$ are replaced by the following new conditions $J,\ (v,\nabla_x)J\in L_\infty(D\times V)$, $F,\ (v,\nabla_x)F\in L_\infty(D)$, $g_2,\ (v,\nabla_x)g_2\in L_\infty(D)$, $J_k,\ (v,\nabla_x)J_k\in L_\infty(D\times V)$ $\forall\ k=\overline{1,N}$, respectively.

Now we considered one else type of non-linear control problems. In these control problems the time-independent part $j(x,v)$ of the kernel in the scattering integral

$$J(x,v,v',t)=j(x,v)g_3(x,v,v',t),\tag{13}$$

is used as a spatially distributed control action, where j is an acceptable control action (distributed stationary control), g_3 is an a priori known function (the corrective function).

The control problem 3. Is it possible to convert nuclear reactor from the initial state (4) to the target state (6) for the time interval $t_1\in(0,T]$ by using the time-independent part $j(x,v)$ of the neutron scattering indicatrix $J(x,v,v',t)$?

Positive response to the question asked above was obtained by proving the following theorem.

Theorem 3. *Let us assume that* $\Sigma\in L_\infty^t(D)$; $F\in L_\infty^t(D)$; $\varphi,\ (v,\nabla_x)\varphi\in L_\infty(G\times V)$, $\psi,\ (v,\nabla_x)\psi\in L_\infty(G\times V)$, $\varphi_{|\gamma_-},\ \psi_{|\gamma_-}\in L_\infty(\gamma_-)$; $g_3\in L_\infty^t(D)$,

$|g_3(x,v,t_1)\psi(x,v)| \geq g_{03} > 0$; $J_k \in L^l_\infty(\mathbf{D}\times\mathbf{V})$; $\mathrm{R}_{k0}(x,v) \in L_\infty(\mathbf{G}\times\mathbf{V})$ $\forall\ k=\overline{1,N}$; *and the following concordance conditions are satisfied.*

$$\varphi(x,v)=0 \quad \psi(x,v)=0 \quad (x,v)\in\gamma_- \tag{14}$$

Then the reactor can be converted from the initial state (4) to the target state (6) for the time interval t_1 *by using the only possible stationary spatially distributed control action* $\boldsymbol{j}(x,v)$ *that belongs to a compact set* $\boldsymbol{\Omega}$ *from the space* $L_\infty(\mathbf{G}\times\mathbf{V})$. *In this case, the state function* $u(x,v,t)$ *belongs to the closed limited set* $\mathbf{U}$ *from the class* $H_\infty(\mathbf{D})$.

The theorem 3 can be proved in quite the same manner as we used to prove the theorem 2.

Remark 3. In the theorem 3 the conditions $\Sigma \in L^l_\infty(\mathbf{D})$, $F \in L^l_\infty(\mathbf{D})$, $g_3 \in L^l_\infty(\mathbf{D})$, $J_k \in L^l_\infty(\mathbf{D}\times\mathbf{V})$ $\forall\ k=\overline{1,N}$ can be replaced by the following new conditions: $\Sigma,\ (v,\nabla_x)\Sigma \in L_\infty(\mathbf{D})$, $F,\ (v,\nabla_x)F \in L_\infty(\mathbf{D})$, $g_3,\ (v,\nabla_x)g_3 \in L_\infty(\mathbf{D})$, $J_k,\ (v,\nabla_x)J_k \in L_\infty(\mathbf{D}\times\mathbf{V})$ $\forall\ k=\overline{1,N}$, respectively. Under these new conditions the conclusions derived from the theorem 3 remain to be correct.

5. Conclusion

So, the present paper analyzed one linear and two non-linear problems related with controllability of transient processes in the objects which can be described by equations of nuclear reactor dynamics. The methods proposed in the paper make it possible to construct the spatially distributed control actions, namely the solutions of the defined control problems, as well as to determine the respective phase states of the objects under investigation.

It should be noted here that, as the proofs of the theorems mentioned above are based on the method of successive approximations, this fact allows us to constructively build solutions for three control problems 1, 2, 3. The proposed methods open the possibilities to construct numerical algorithms for determination of the required functions.

In conclusion, we would like to note that the control problems we considered in the paper may be interpreted as the problems of final observation, i.e. as the problems related with restoration of the spatially distributed control actions in the forms (7), (11), (13) and with determination of the respective phase states from the results obtained during observation of the objects under investigation.

The work was performed and the paper was prepared for publishing within the frames of the Program on Competitiveness Enhancement of National Research Nuclear University MEPhI (Moscow Engineering Physics Institute); contract No. 02.a03.21.0005,27.08.2013.

References

[1] Marchuk G I *Methods for calculating nuclear reactors.* Moscow, 1961, pp 667 (In Russian).
[2] Albertoni S and Montagnini B *On the Spectrum of Neutron Transport Equation in Finite Bodies* J. Math. Analys. Appl., 1966, v 13, p 19.
[3] Akcasu Z , Lellouche G S and Shotkin L M *Mathematical Methods in Nuclear Reactor Dynamics* Academic Press, New York – London, 1971, pp 465.
[4] Lewins J *Nuclear Reactor Kinetics and Control* Jefery Lewins. Pergamon Press, 1978. pp 264.
[5] Kryanev A V and Shikhov S B *Problems of nuclear reactor mathematical theory (Non-linear analysis)*. Moscow, Energoatomizdat, 1983, pp 280. (In Russian).
[6] Germogenova T A *Local properties of solutions of the transport equation.* Moscow, Nauka, 1986, pp 272. (In Russian).

[7] Kuznetsov Yu A and Morozov S F *System of integro-differential equations for nuclear reactor kinetics.* Differential equations, 1974, v 10, N. 8, pp 1491–1503. (In Russian).

[8] Prilepko A I and Volkov N P *Inverse problems of finding parameters of a nonstationary kinetic transport equation from supplementary information on traces of the unknown function.*/ Differential equations, 1988. v 24, N 1, pp 136-146. (In Russian).

[9] Volkov N P *On some inverse problems for time-dependent transport equation.* Ill-posed problems in natural sciences. Moscow: TVP-VSP, 1992, pp 431-438.

[10] Volkov N P *Solvability of certain inverse problems for the non-stationary kinetic transport equation.* Computational mathematics and mathematical physics. Pleiades Publishing, 2016, v 56, N 9, pp 1598-1603.

[11] Baskakov A V and Volkov N P *On some conditions for the solvability of control problems for the non-stationary kinetic neutron transport equation.* Electronic journal "MGOU Vestnik" (Series on physics and mathematics), 2013, N 1, pp 1-10. (In Russian).

[12] Volkov N P *Linear Inverse Problem of Dynamics of the Reactor.* Journal of Physics: Conference Series, 2017, v 788, N 1, pp 1-4.

Numerical simulation of equilibrium plasma configurations in toroidal magnetic traps and their cylindrical analogues

K V Brushlinskii[1,2], **I A Kondratyev**[2]

[1] Keldysh Institute for Applied Mathematics RAS, Miusskaya pl., 4, 125047, Moscow, Russia
[2] National Research Nuclear University MEPhI, Kashirskoe highway, 31, 115409, Moscow, Russia

E-mail: mrkondratyev95@gmail.com[2]

Abstract. Some basic characteristics investigations of equilibrium magnetoplasma configurations in toroidal traps for plasma confinement often use mathematical models of their straightened into a cylinder analogues. The aim of this work is a comparative analysis of MHD-models in toroidal and cylindrical configurations. We consider the plasma cylinder with the electric current (Z-pinch) as a simpliest trap type and its toroidal analogue and then extend and define more exactly the cylindrical and toroidal models of the 'Galatea-Belt' trap with two current-carrying conductors, immersed into the plasma. Distinctions between configurations in the traps of different geometry are registered and some stability problems are considered.

1. Introduction

Magnetic traps for plasma confinement are a basic element of various plants being elaborated for nuclear fusion research. The widespread traps are toroidal in shape, for example, tokamaks and stellarators. The galatea-traps, proposed by A.I. Morozov [1, 2], where the conductors with electric current, inducing the magnetic field, are immersed into the plasma, have the same geometry. Their simpliest example is the 'Galatea-Belt' with two ring-shaped conductors inside the torus [3].

Mathematical modeling and computation are of conciderable importance in investigations of equilibrium magnetoplasma configurations in the traps (see [4–7] for example). The models use as a rule magnetohydrodynamic equations. If a trap has any symmetry, its two-dimensional models can be reduced to boundary-value problems with the scalar Grad-Shafranov equation [8–10] for the magnetic flux function. Some conceptual questions, concerning toroidal trap configuration may be explained, solving these problems in the straight cylinder, i.e. in the torus of infinite radius. In this case quantitive distinctions between the toroidal and cylindrical configurations become to be of interest and must be investigated. Some first results concerning numerical simulation of configurations in the 'Belt' and in the cylinder with two straight conductors and their comparative analysis are presented in [11, 12].

In this paper distinctions between toroidal and cylindrical configurations are considered in the more general form. We investigate some configurations in the plasma cylinder with the electric current (Z-pinch) and its toroidal analogue, that may be regarded as a basis of tokamaks.

IOP Conf. Series: Journal of Physics: Conf. Series **937** (2017) 012006 doi:10.1088/1742-6596/937/1/012006

Toroidal configurations are removed from the symmetry axis and distorted in comparison with the cylindrical ones. This effect is the stronger, the smaller is the torus radius. When the radius increases, the configurations become nearer the cylindrical ones. Computations are performed for different electric current distributions in the pinch. The configurations considered above are stable with respect to two-dimensional perturbations.

The computer investigations of 'Galatea-Belt' configurations, mentioned above, are continued. The mathematical model became more convenient for result interpretation after the more appropriate choice of the magnetic flux function unit of measurement. In the series of computations we have determined the maximum plasma pressure values, that admit stable plasma configurations with given electric current in the conductors.

2. Mathematical models of plasma equilibrium in cylinders

General questions of comparison plasma configurations in torus and their cylindrical analogues are convenient to be considered by means of the example of plasma cylinder and torus, which is corresponding to it. The simpliest and most famous example is a round plasma cylinder with electric current parallel to its axis (Z-pinch). We consider its MHD-model as a special case of the model of arbitrary cross-section cylinder, homogeneous in the axial direction. Equilibrium plasma configurations correspond to MHD-equations of plasmastatics [4, 5, 9]

$$\nabla p = \frac{1}{c}\mathbf{j} \times \mathbf{H}; \quad \frac{c}{4\pi}\mathbf{j} = \mathrm{rot}\mathbf{H}; \quad \mathrm{div}\mathbf{H} = 0 \tag{1}$$

for the three variables distributed in space: pressure p, magnetic field strength $\mathbf{H}$ and electric current density $\mathbf{j}$. In two-dimensional plane symmetric problems without longitudinal field ($H_z \equiv 0,\ \partial/\partial z \equiv 0$) they are reduced to one scalar second order equation for the magnetic flux function Ψ - the plane version of Grad-Shafranov equation

$$\triangle\Psi + 4\pi\frac{dp}{d\Psi} = 0, \tag{2}$$

where $H_x = \frac{\partial\Psi}{\partial y}$, $H_y = -\frac{\partial\Psi}{\partial x}$.

Boundary problem with equation (2) should be solved in the region of (x, y) plane, which corresponds the cross-section of the cylinder with the z = const plane with given boundary conditions.

The pressure $p = p(\Psi)$ is the given function of variable Ψ, which describes its distribution between isobaric magnetic surfaces $\Psi(x, y)$ = const.

Two-dimensional boundary problems are considered in square space regions corresponding to square cross-sections of toroidal traps.

It's natural to consider problems in round cylinder that are homogeneous in azimuth and therefore one-dimensional in cylindrical coordinates (r, φ, z) with the one spatial coordinate r and unique non-trivial component of magnetic field is H_φ. The equation (1) became

$$\frac{dp}{dr} = -\frac{1}{c}jH; \quad j = \frac{4\pi}{c}\frac{dHr}{rdr}, \tag{3}$$

where $H = H_\varphi$, $j = j_z$ in the circle with radius R. Knowing one of the unknown functions p, H, j, we find two other functions by integration of equations (3). Thus model doesn't use the equation with Grad-Shafranov type (2), although it is formally valid in the form

$$\frac{1}{r}\frac{d}{dr}\left(r\frac{d\Psi}{dr}\right) + 4\pi\frac{dp}{d\Psi} = 0,$$

where the magnetic flux function Ψ and the dependence $p(\Psi)$ is concerned with the solutions of equations (3) in the following manner:

$$H = -\frac{d\Psi}{dr}; \quad \frac{dp}{d\Psi} = \frac{dp}{dr}\Big/\frac{d\Psi}{dr} = -\frac{dp}{dr}\Big/H. \tag{4}$$

The function $p(\Psi)$, which was gained here for the round Z-pinch, is used below in the models of two-dimensional configurations in square cross-section of the cylinder. These models are convenient to be compared with models of toroidal configurations with square cross-section too.

Boundary problems with the eq. (2) are considered in square $|x| < R$, $|y| < R$. The bound of the square is considered to be opaque for the magnetic field, thus there is $H_n = 0$ on it, where $\mathbf{n}$ is a surface normal to the bound Γ. It follows, that boundary condition is $\Psi = \Psi_\Gamma$. The value of constant Ψ_Γ may be equal to zero, because it doesn't matter in the final solution of the problem.

Problems in torus may be homogeneous only in azimuth direction φ, as at least two-dimensional. Square cross-sections were chosen because our problems are considered to be solved in cylindrical coordinates, and any other geometry would make the coordinate system and mathematical apparatus much more complicated [9]. The arcticle [11] shows, that configurations in round and square cylinder are insignificantly different only on the periphery.

The 'Galatea-Belt' magnetic trap is a torus with two ring-shaped conductors with current immersed into plasma. Its straightened analogue is a cylinder with two conductors, parallel to the axis. Plasma configurations were observed numerically in its round cross-section in [13], and for the reasons above, in its square cross-section in [11,12]. As a continuation and development of recent work we consider square cross-section. Magnetic field, which retains plasma, is formed by conductors with the current. The mathematical model takes this into account in terms of boundary problem with Grad-Shafranov type equation

$$\triangle\Psi + 4\pi\frac{dp}{d\Psi} + \frac{4\pi}{c}j^{ex} = 0, \tag{5}$$

where $j^{ex}(x, y)$ is a given function, which describes the distribution of 'external' current. It is located in the places of crossing of conductors with the (x, y)-plane:

$$j^{ex}(x, y) = \sum_{k=1}^{2} j_0 \exp\left(-\frac{(x - x_k)^2 + y^2}{r_c^2}\right) \tag{6}$$

where $x_1 = x_0, x_2 = -x_0$ - coordinates of the conductor centres, r_c is conditional radius of conductors. The coefficient j_0 is determined from the condition

$$\int\int j^{ex}dxdy = J_c,$$

where the integral is evaluated in the vicinity of each conductor, and J_c is a current value in the conductor [4,11–13].

Isobaric function $p(\Psi)$ is selected on the basis of requirements to concentrate plasma near the central part of space region restricting it touch the conductors. It must be nonmonotonic with a maximum in the centre of the square and the separatrix $\Psi = \Psi_0$ passing through it. For example, this conditions may be satisfied by the function

$$p = p_0 \exp\left(-\frac{(\Psi - \Psi_0)^2}{q^2}\right) \tag{7}$$

where $\Psi_0 = \Psi(0, 0)$.

Boundary problems with the eq. (5) are considered in square $|x| < X, |y| < X$ with condition, as above, $\Psi = 0$ on the bound.

IOP Conf. Series: Journal of Physics: Conf. Series **937** (2017) 012006 doi:10.1088/1742-6596/937/1/012006

3. Plasma equilibrium in toroidal traps

Plasma and magnetic field configurations with toroidal electric current, are homogeneous in azimuth coordinate, are considered, as said above, in torus of square cross-section. The mathematical apparatus of the models is different from the problem above in the coordinate system: cylindrical coordinates (r, z) are used instead of Cartesian ones (x, y).

The Grad-Shafranov equation, proposed in the original papers [8–10] for the axisymmetric problems, is in application for the problem in plasma torus without azimuth field ($H_\varphi \equiv 0$) of the following form

$$\triangle^*\Psi + 4\pi r^2 \frac{dp}{d\Psi} = 0. \tag{8}$$

where $\triangle^*\Psi \equiv r\frac{\partial}{\partial r}\left(\frac{1}{r}\frac{\partial\Psi}{\partial r}\right) + \frac{\partial^2\Psi}{\partial z^2}$, $rH_r = -\frac{\partial\Psi}{\partial z}$, $rH_z = \frac{\partial\Psi}{\partial r}$.

Boundary problems are considered in the square $|z| < R, |r - r_0| < R$, where r_0 is a big torus radius. The dependence $p(\Psi)$ is the same as above, obtained in the one-dimensional problem in the round Z-pinch (4). In the toroidal model of 'Galatea-Belt' the term $4\pi r j^{ex}$ was added in the left part of equation (8), where

$$j^{ex}(z, r) = \sum_{k=1}^{2} j_0 \exp\left(- \frac{(r - r_0)^2 + (z - z_k)^2}{r_c^2} \right) \tag{9}$$

is an 'external' current in ring-shaped conductors with axis (r_0, z_k), $z_1 = z_0$, $z_2 = -z_0$. The pressure $p(\Psi)$ is given by the same formula (7), where parameter Ψ_0 is a value of the desired function in the centre of the configuration and on separatrix, including the singular magnetic field point:

$$\Psi_0 = \max_r \Psi(r, 0) \tag{10}$$

The problems are considered in the square $|r - r_0| < 2z_0$, $z < 2z_0$, on the bound of it we assume $\Psi = 0$, like in cylindrical geometry.

4. About methods of solving the problems

Problems, which are considered above, are solved in dimensionless variables: the units of measurement of all values are composed from the given dimensional constants. In both series of problems such constants are characteristic size r_u - radius R of pinch or the half of distance between the conductors in 'Belt' z_0 (x_0 in plane problems), and characteristic electric current value J in pinch or in each conductor in 'Belt'. The units which are composed from them are

$$r_u; \quad H_u = \frac{2J}{cr_u}; \quad j_u = \frac{cH_u}{4\pi r_u}; \quad p_u = \frac{H_u^2}{4\pi}. \tag{11}$$

Flux function Ψ has different dimension in plane (2) and toroidal (8) problems, and therefore we assume $\Psi_u = H_u r_u$ in plane, but in torus we have to add the factor with dimension of length to its unit. In [11, 12] it was assumed, that $\Psi_u = H_u r_u^2$, and it turned out to be not convenient in the interpretation of the result physical meaning, because dimensionless value of Ψ is proportional to the big torus radius in this case, but parameter q in formula (7) measured in the same units remains the same. In this paper

$$\Psi_u = H_u r_u r_0 \tag{12}$$

has been chosen as a unit of Ψ, r_0 is a torus radius. As a result, the order of dimensionless value of Ψ remains limited together with the parameter q, when the radius r_0 increases. The comparison of toroidal and cylindrical configurations becomes less complicated.

IOP Conf. Series: Journal of Physics: Conf. Series **937** (2017) 012006 doi:10.1088/1742-6596/937/1/012006

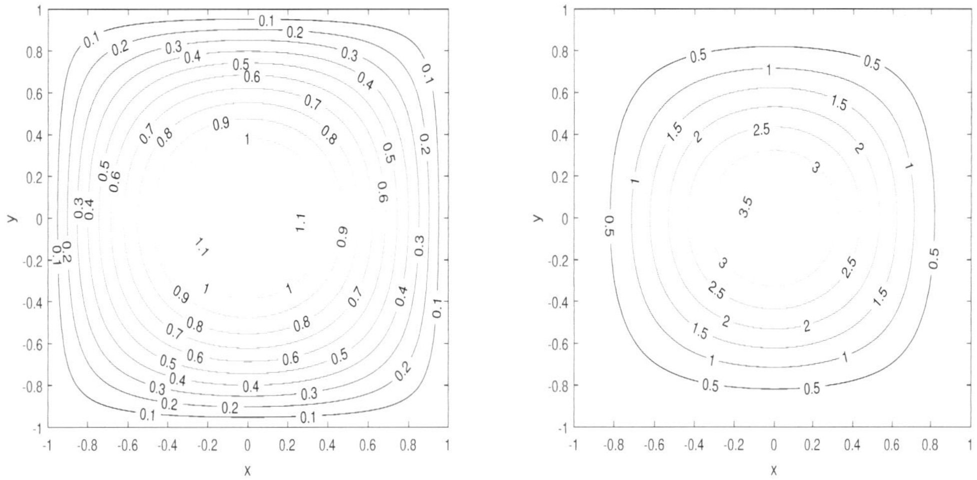

a) *b*)

Figure 1. Equilibrium configurations in the plasma cylinder with 'parabolic' current *a*) magnetic field ($\Psi(r, z) = \text{const}$), *b*) plasma pressure ($p = \text{const}$)

It this case dimesionless Grad-Shafranov equation has the form

$$\triangle^* \Psi + \left(\frac{r}{r_0}\right)^2 \frac{dp}{d\Psi} + \frac{r}{r_0} j^{ex}(r, z) = 0. \tag{13}$$

The difference analogue of Grad-Shafranov type equations is solved numerically by means of iterative relaxation method, which is described and discussed in [4, 5, 11–13].

5. Computation results

To compare equilibrium plasma and field configurations in a straight plasma cylinder and its toroidal extension, obtained in computer models above, we begin through the following simple example. The given electric current ('parabolic', dimensionless)

$$j(r) = 3.64(1 - 0.9r^2)$$

is distributed in the circular cylinder $r < 1$. It follows from the equations (3)

$$H(r) = 1.82r(1 - 0.45r^2); \quad \Psi(r) = 0.91(1 - r^2)(0.775 - 0.225r^2);$$
$$p(r) = 3.3(1 - r^2)(0.28 - 0.72r^2 + 0.135r^4)$$

and according to (4)

$$\frac{dp}{d\Psi} = j = 4(\sqrt{3.27\Psi + 1} - 0.91) \tag{14}$$

The configuration in the plasma cylinder with square cross-section is derived by solving the two-dimensional problem with the equation (2), where the isobaric function $p(\Psi)$ is taken from the one-dimensional problem (14). The same problem is solved in the toroidal geometry with several major axis dimensionless values r_0. Computation results are presented by means of magnetic force lines $\Psi = \text{const}$ and isobares $p = \text{const}$ in the straigt cylinder and torus at

IOP Conf. Series: Journal of Physics: Conf. Series **937** (2017) 012006 doi:10.1088/1742-6596/937/1/012006

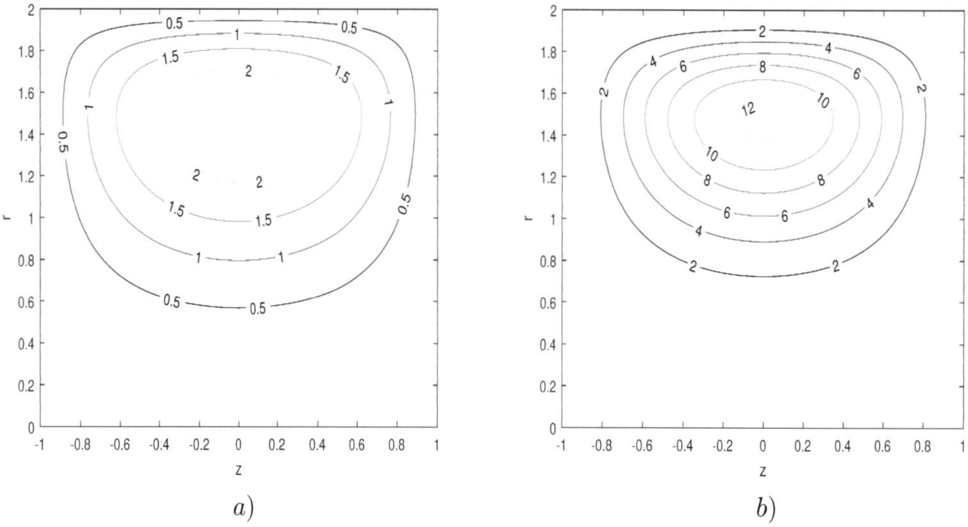

Figure 2. Equilibrium configurations in the plasma torus with 'parabolic' current a) magnetic field ($\Psi(r,z) = \text{const}$), b) plasma pressure ($p = \text{const}$)

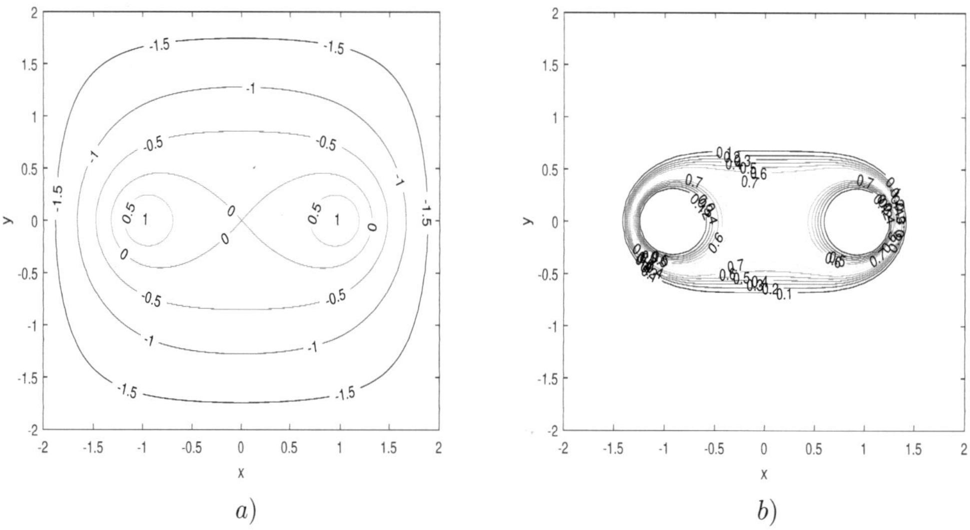

Figure 3. Equilibrium configurations in the straightened analogue of 'Galatea-belt' a) magnetic field ($\Psi(r,z) = \text{const}$), b) plasma pressure ($p = \text{const}$)

the fig. 1, 2. The torus radius value $r_0 = 1$ is chosen as the least admissible one in order that distinctions between the cylinder and the torus would be the most visible. The toroidal cofiguration is deformed and its magnetic axis is removed at the distance $\delta r = 0.48$. The plasma volume is less than in the cylinder and therefore the maximum values of pressure p and poloidal magnetic flux Ψ are greater. In the computation series with several current distributions $j(r)$ a detailed qualitative and quantitative information about distinctions between magnetoplasma

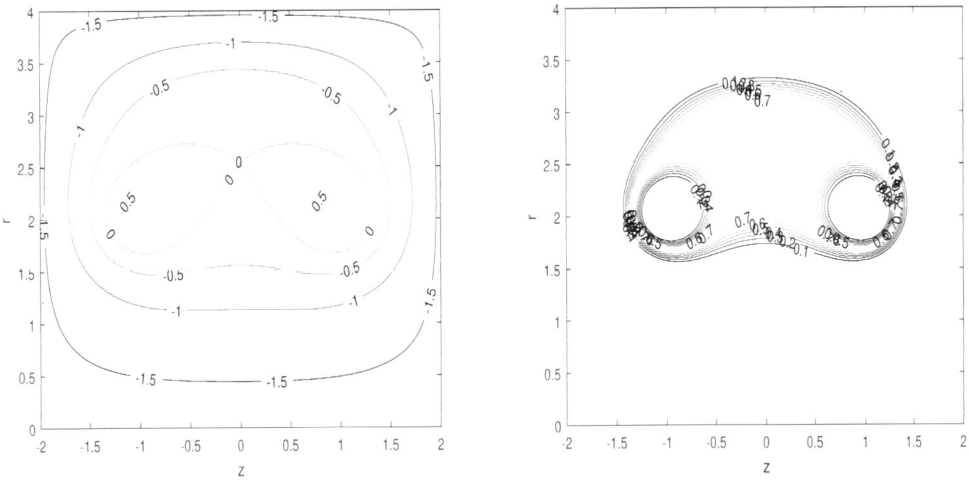

a) *b)*

Figure 4. Equilibrium configurations in the toroidal 'Galatea-Belt' *a)* magnetic field ($\Psi(r, z)$ = const), *b)* plasma pressure (p = const)

configurations under consideration is obtained. It allows to evaluate the errors arising when we use straight models instead of toroidal ones.

The mathematical model of equilibrium configurations in the 'Galatea-Belt' is released in solving of boundary problems with Grad-Shafranov equation (5) and (13) in the square regions. The results are presented at the fig. 3, 4. The value $r_0 = 2$ in toroidal variant is again the least admissible. The plasma and field configurations in the cylinder are symmetric with respect to the coordinate axis. The plasma is concentrated in the central region having a curvilinear quadrangular form with sides convex inwards and thin strips along the field separatrix around the conductors. Because of the more convenient choice of the unit Ψ_u (12) the values of Ψ in the cylinder and in the torus have the same order here, that admit easily explain the results. The configurations in torus are deformed as above: the magnetic axis is removed at $\delta r = 0.52$. The plasma volume in the torus is rather greater than in the cylinder, then the magnetic flux is less.

It is known, that the model concerned admits the relaxation to the equilibrium only under the restriction $p_0 < p_0^{cr}$ in (7) for the pressure maximum (dimesionless, i.e. refered to the magnetic pressure). In the computations under the norming (13) we obtained, that the critical value p_0^{cr} decreases together with the radius r_0, i.e. when the torus curvature increases. The restriction $p_0 < p_0^{cr}$ corresponds to the 'diffusion stability' of configuration, that is one of the necessary MHD-stability conditions (see [4, 5]).

Acknowledgments
We appreciate dr. A.S. Goldich, who took part in computer investigations [11, 12] and in some discussions at the initial stage of the work presented here.

The research was supported by the Russian Scientific Foundation (project No16-11-10278).

References

[1] Morozov A I 1992 *Sov. J. Plasma Phys.* **18** 159
[2] Morozov A I and Savel'ev V V 1998 *Phys. Uspekhi* **41** 11 1049

[3] Morozov A I and Frank A G 1994 *Plasma Phys. Reports* **20** 879
[4] Brushlinskii K V 2009 *Mathematical and Computational Problems of Magnetogasodynamics* (Moscow: BINOM)
[5] Brushlinskii K V 2017 *Mathematical Foundation of Computational Fluid and Plasma Mechanics* (Dolgoprudny: Intellect)
[6] Ren J, Liu Y Q, Liu Y., Medvedev S Yu, Wang Z and Xia G 2016 *Plasma Phys. and Controlled Fusion* **58** 11 1
[7] Medvedev S Yu, Martynov A A, Drozdov V V, Ivanov A A and Poshekhonov Yu Yu 2017 *ibid.* **59** 1
[8] Shafranov V D 1958 *Sov. Phys. JETP* **6** 545
[9] Shafranov V D 1966 *Rev. Plasma Phys.* **2** 103
[10] Grad H and Rubin H 1959 *Proc.* 2nd *United nations Int. Conf. on the Peaceful Uses of Atomic Energy. Geneva* vol 31 (New York: Columbia Univ. Press) p 190
[11] Brushlinskii K V and Goldich A S 2016 *Differential Equations* **52** 7 845
[12] Brushlinskii K V and Goldich A S 2017 *J. Phys.: Conf. Series* **788** 1-6
[13] Brushlinskii K V and Ignatov P A 2010 *Comp. Math. and Math. Phys.* **50** 12 2071

Numerical simulation of plasma flows in curved coaxial ducts with longitudinal magnetic field

K V Brushlinskii[1,2,4] **and E V Styopin**[1,2,3,5]

[1] Keldysh Institute of Applied Mathematics, Russian Academy of Sciences, Miusskaya sq. 4, 125047 Moscow, Russia
[2] National Research Nuclear University MEPhI, Kashirskoe sh. 31, 115409 Moscow, Russia
[3] Author to whom any correspondence should be addressed.

[4] brush@keldysh.ru
[5] styopin.evgeniy@gmail.com

Abstract. The results of numerical researches of plasma flows in coaxial plasma accelerators ducts of a various geometry with an external longitudinal magnetic field are presented. The non-steady-state two-dimensional MHD-model of relaxation of transonic super-Alfven flows with acceleration is considered. The paper focuses on the cumulative influence of a duct geometry and longitudinal magnetic field on flow characteristics.

1. Introduction

1.1. Plasma accelerators. History and applications

Physical processes in plasma are diverse and have a wide range of effective technological applications. The design and research of plasma accelerators, in which electromagnetic energy transforms into kinetic energy of plasma fluxes emerging from them, is the major direction in the area of plasma processes using. The idea of plasma acceleration in crossed electric and magnetic fields belongs to Alexey Morozov [1]. The successful development of these devices was launched in the 1950s in USSR under Lev Artsimovich [2] and in Gersh Budker's laboratory. The development, research and design of different varieties of plasma accelerators over the following decades were mainly conducted by the same Alexey Morozov [3].

On the one hand, these include powerful high-current quasi-stationary plasma accelerators (QSPA) (figure 1), which demonstrated record velocities and energies of plasma fluxes emerging from them [4]. Its thrust is commensurate with rocket engine thrust, and, in terms of its technological applications, these devices can be used as electric jets for air and space flights [5]. They have a number of advantages over liquid jets working on chemical fuel, but require a powerful and compact source of electric power. On another the hand – low-power small stationary plasma thrusters (SPT) (figure 2) [6] with a long lifetime, which actively apply for orbital corrections, stabilization and maneuvers of artificial satellites.

Plasma accelerators scope of application is not limited to cosmic and transport directions. They also can be used as a magnetoplasma compressor, which is used to generate high-temperature plasma and

IOP Conf. Series: Journal of Physics: Conf. Series **937** (2017) 012007 doi:10.1088/1742-6596/937/1/012007

for high-energy plasma fluxes injection into magnetic traps developed at the issue of thermonuclear fusion. Finally, these devices can be used for effective processing and surface properties modification of various materials.

1.2. Longitudinal magnetic field

A flow acceleration mechanism in plasma accelerator ducts is based on an interaction between an electric discharge current passing through plasma with own transverse magnetic field generated by an electric current flowing into the plant. The further technological development of the plasma acceleration scheme is the imposition of a longitudinal magnetic field induced by conductors external to the accelerator.

Figure 1. Quasi-stationary plasma accelerator

Figure 2. Stationary plasma thruster

The existence of a longitudinal magnetic field complicated the picture of plasma motion in the accelerator duct. Depending on the value of an external magnetic field, in terms of a classification, there are two regimes of plasma flows significantly varying on their characteristics [7]: super-Alfven and sub-Alfven (in relation to Alfven velocity corresponding to a longitudinal magnetic field) in a sufficiently weak and strong longitudinal magnetic field, respectively. In an accelerator duct at the same time there may be areas of super-Alfven and sub-Alfven flows.

1.3. Mathematical modelling and its role in plasma processes research

Mathematical modelling of plasma processes and numerical solution of corresponding mathematical problems play an important role in the successful developments and researches of plasma accelerators. This approach allows to determine basic qualitative regularities in properties of these processes, and quantitative results provide an opportunity to reduce costs of expensive experiments.

Plasma flows in plasma accelerator ducts in own transverse magnetic field have been well understood in the previous studies [8]. In [9] it has been demonstrated that, in terms of acceleration effectiveness, ducts with a curved central electrode are preferable to accelerators with a curved external electrode. Plasma flows with an external longitudinal magnetic field are less studied. The first steps in this field lied in the two-dimensional calculations of flows in the duct with the weakly curved central electrode [10] (see also [11]). The paper [12] is focused on the researches of ionization and radiation transport processes in plasma accelerator ducts. The two-dimensional MHD-model of flows in plasma accelerators as in injectors for magnetic traps is considered in [13].

1.4. Objective of the work

The objective of the present work is mathematical modelling and numerical simulation of accelerating plasma flows in QSPA-type plasma accelerator ducts of a various geometry formed by markedly curved electrodes with a longitudinal magnetic field.

IOP Conf. Series: Journal of Physics: Conf. Series **937** (2017) 012007 doi:10.1088/1742-6596/937/1/012007

2. Mathematical model

2.1. Object of simulation

The accelerator nozzle-type duct is formed by two coaxial electrodes connected to a capacitor bank (figure 3).

Plasma flowing in the duct completes the electrical circuit, and the electric current with a density **j** arise. It interacts with own transverse azimuthal magnetic field $\mathbf{H}_{\perp}$ and creates Ampere's force $\mathbf{F}_{\parallel} = (1/c)\,\mathbf{j}\times\mathbf{H}_{\perp}$, which causes plasma acceleration along the duct axis (figure 4).

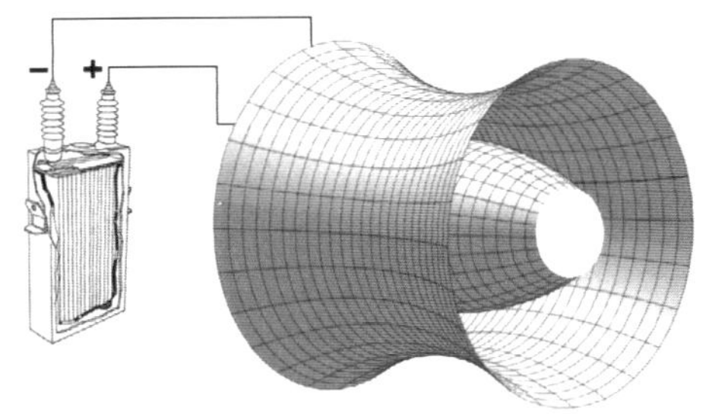

Figure 3. Schematic view of coaxial plasma accelerator

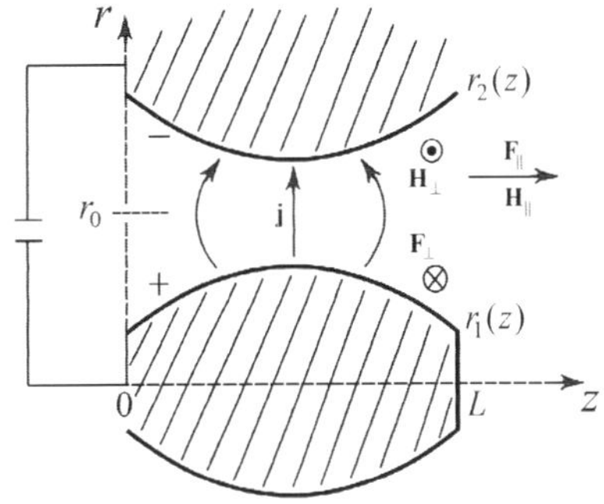

Figure 4. Scheme of coaxial plasma accelerator. Section by plane $\varphi = \mathrm{const}$

It is assumed that a longitudinal magnetic field $\mathbf{H}_{\parallel}$ induced by external conductors can exist in the system. In the interaction with the discharge current **j** the longitudinal magnetic field causes plasma rotation in the transverse direction of the duct with the force $\mathbf{F}_{\perp} = (1/c)\,\mathbf{j}\times\mathbf{H}_{\parallel}$.

So the object of the simulation in the present paper is plasma flow in plasma accelerator ducts in the external longitudinal magnetic field.

2.2. Formulation of the problem

In the present paper, plasma is considered as a continuous medium consisting of ions and electrons with common macro-parameters. Plasma flux injected into the accelerator is axisymmetric $(\partial/\partial\varphi \equiv 0)$. Two-dimensional problems therefore are formulated in the duct section by the plane $\varphi = \mathrm{const}$ in a cylindrical coordinate system (r,φ,z).

Plasma with a specified density ρ_0, temperature T_0 and pressure p_0 enters the duct through its input section $(z=0)$ without rotation $\upsilon_{\varphi 0} = 0$. The discharge current in the input section of the duct has strictly a radial direction, therefore azimuthal magnetic field at the duct input is set as $H_{\varphi 0} = r_0 H_0 / r$, where $H_0 = 2J/cr_0$ – a characteristic value of the magnetic field intensity in the input section, J – a value of the full electric current flowing through the central electrode, r_0 – a characteristic radius of the duct. The longitudinal magnetic field can be determined as a constant value H_{z0} in the duct input. At the exit of the duct $(z=1)$ the flow is supersonic (in relation to fast magnetosonic velocity), so there are no boundary conditions.

The initial conditions of the problem can be quite arbitrary as long as they provide acceleration of plasma, because steady-state flows are of particular interest.

The electrodes $r_1(z)$ and $r_2(z)$ forming the accelerator duct are assumed to be solid $\upsilon_n = 0$ and equipotentional $H_n = 0$, where n – a normal line to the electrodes surface.

2.3. Method of calculation

For the numerical solution of the problem, it is convenient to use a dimensionless form of variables. The units of measurement are the dimensional quantities participating in the formulation of the problem. In order to get rid of curved boundaries of the calculation area, new dimensional coordinates (z, y), which transform the duct area to a square (figure 5), have been introduced:

$$z = z,\ r = (1-y)r_1(z) + yr_2(z) \tag{1}$$

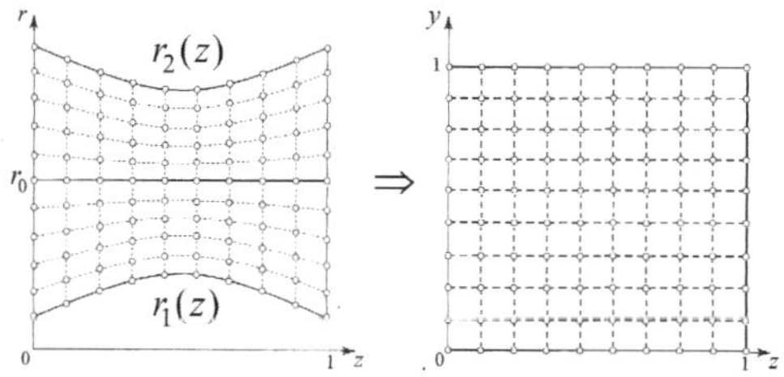

Figure 5. Transformation of curved calculation area of a duct to a square

The apparatus of modelling is based on the numerical solution of magnetohydrodynamics (MHD) equations. Axisymmetric flows of a continuous electrically conductive medium (of neglecting its viscosity and thermal conductivity and on the assumption of its infinite electrical conductivity) are described by the following system of equations [10] (see also [11]) in coordinates (1):

$$\begin{cases}
\frac{\partial}{\partial t}\rho rR + \frac{\partial}{\partial z}\rho\upsilon_z rR + \frac{\partial}{\partial y}\rho\upsilon_y r = 0 \\
\frac{\partial}{\partial t}\rho\upsilon_z rR + \frac{\partial}{\partial z}\left(\rho\upsilon_z^2 + P - H_z^2\right)rR + \frac{\partial}{\partial y}\left(\rho\upsilon_z\upsilon_y - r_z P - H_z H_y\right)r = 0 \\
\frac{\partial}{\partial t}\rho\upsilon_r rR + \frac{\partial}{\partial z}\left(\rho\upsilon_z\upsilon_r - H_z H_r\right)rR + \frac{\partial}{\partial y}\left(\rho\upsilon_r\upsilon_y + P - H_r H_y\right)r = \left(\rho\upsilon_\varphi^2 + P - H_\varphi^2\right)R \\
\frac{\partial}{\partial t}\rho\upsilon_\varphi rR + \frac{\partial}{\partial z}\left(\rho\upsilon_z\upsilon_\varphi - H_\varphi H_z\right)rR + \frac{\partial}{\partial y}\left(\rho\upsilon_\varphi\upsilon_y - H_\varphi H_y\right)r = \left(-\rho\upsilon_r\upsilon_\varphi + H_\varphi H_r\right)R \\
\frac{\partial}{\partial t}\rho\varepsilon rR + \frac{\partial}{\partial z}\rho\varepsilon\upsilon_z rR + \frac{\partial}{\partial y}\rho\varepsilon\upsilon_y r + p\left(\frac{\partial\upsilon_z rR}{\partial z} + \frac{\partial\upsilon_y r}{\partial y}\right) = 0 \\
\frac{\partial}{\partial t}H_\varphi R + \frac{\partial}{\partial z}\left(H_\varphi\upsilon_z - H_z\upsilon_\varphi\right)R + \frac{\partial}{\partial y}\left(H_\varphi\upsilon_y - H_y\upsilon_\varphi\right) = 0 \\
\frac{\partial}{\partial t}H_z rR + \frac{\partial}{\partial y}Er = 0 \\
\frac{\partial}{\partial t}H_y r - \frac{\partial}{\partial z}Er = 0,
\end{cases} \tag{2}$$

$$p = (\gamma - 1)\rho\varepsilon = \frac{\beta}{2}\rho T, \quad P = p + \frac{H_z^2 + H_r^2 + H_\varphi^2}{2},$$

где $\upsilon_y \equiv \upsilon_r - \upsilon_z r_z,\ H_y \equiv H_r - H_z r_z, E \equiv E_\varphi \equiv H_z\upsilon_y - H_y\upsilon_z,$
$R(z) = r_2(z) - r_1(z) = \partial r/\partial y,\ r_z = \partial r/\partial z = (1-y)r_1'(z) + yr_2'(z).$

The dimensionless parameter $\beta = 8\pi p_0/H_0^2$ traditionally participates in modelling of plasma processes. It determines, how many times a characteristic magnetic pressure of a problem differs from a gas-dynamic pressure. Let us introduce also the concept of a velocity along a trajectory $\upsilon_l = \sqrt{\upsilon_z^2 + \upsilon_r^2}$.

The problems are solved numerically until steady-state flow regimes would be achieved. As a numerical method of a solution of the non-steady-state two-dimensional MHD-problems, FCT algorithm [14] with a fully multidimensional flux limiter [15] was selected.

IOP Conf. Series: Journal of Physics: Conf. Series **937** (2017) 012007 doi:10.1088/1742-6596/937/1/012007

3. Numerical results

In the present paper the results of numerical researches of plasma acceleration in coaxial ducts with a longitudinal magnetic field in the two-dimensional formulation of the problem are presented.

The main focus is on the cumulative influence of a duct geometry and longitudinal magnetic field on parameters and properties of transonic super-Alfven flows with acceleration in a sufficiently weak longitudinal magnetic field. The studies are a continuation and development of the previous works [10] with calculations in a longitudinal magnetic field in accelerator ducts of a relatively smooth geometry realized in the experiments on QSPA in 1990s [4].

Numerical results, as in [9], are set out in two basic series differing by electrodes geometry. In them $\beta = 0.05$, which corresponds to a sufficiently strong azimuthal magnetic field and electric current, and $H_{z0} = 0.3$ – the value close to the upper bound of the longitudinal field range acceptable to transonic super-Alfven flows.

A steady-state flow in a duct is characterized by distributions of plasma parameters presented on figures 6 and 7 by solid lines of levels. The influence of a longitudinal magnetic field can be seen clearly in the comparison with the results of similar calculations without it, which are presented by dashed lines.

3.1. Plasma accelerators with curved central electrode

We consider the plasma accelerator with the following configuration:

$$0 < z < 1,\ r_1(z) = 0.55 - (z - 0.5)^2,\ r_2(z) = 0.8 \tag{3}$$

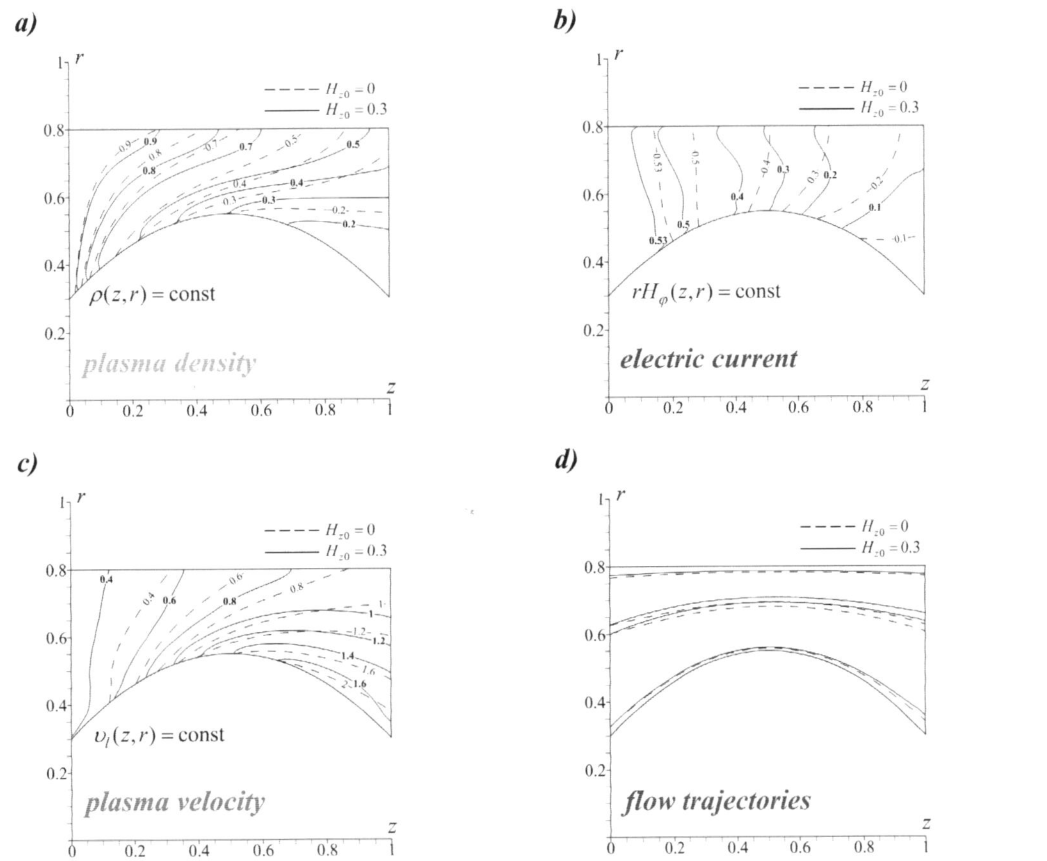

Figure 6. Plasma parameters distributions ***(a, b, c)*** and tubes of flow trajectories ***(d)*** in the duct with the curved central electrode

The longitudinal magnetic field pushes plasma up to the external electrode, resulting in increasing of plasma flow density (figure 6a), and hence, of its temperature in all area of the duct. The longitudinal field deflects electric current lines towards the duct input section and turns them counterclockwise from the radial direction (figure 6b).

Plasma flux acceleration is more intensively near the curved central electrode (figure 6c) by increasing an input value of an azimuthal magnetic field [9]. The longitudinal magnetic field slightly reduces acceleration properties of the duct: the input plasma velocity increases, but the output velocity decreases, because a part of accumulated electromagnetic energy transforms into energy of plasma rotation in the transverse direction of the duct.

Flow tubes are limited by trajectories, which are also level lines of a longitudinal magnetic field. The medium tube even without a longitudinal field is asymmetrical relative to the duct minimal section (figure 6d). This result is followed from the curviness of the central electrode and plasma motion from left to right. The longitudinal field strengthens the trajectories deviation to the external electrode.

3.2. Plasma accelerators with curved external electrode

We consider the plasma accelerator with the following configuration:

$$0 < z < 1,\ r_1(z) = 0.3,\ r_2(z) = 0.55 + (z - 0.5)^2 \tag{4}$$

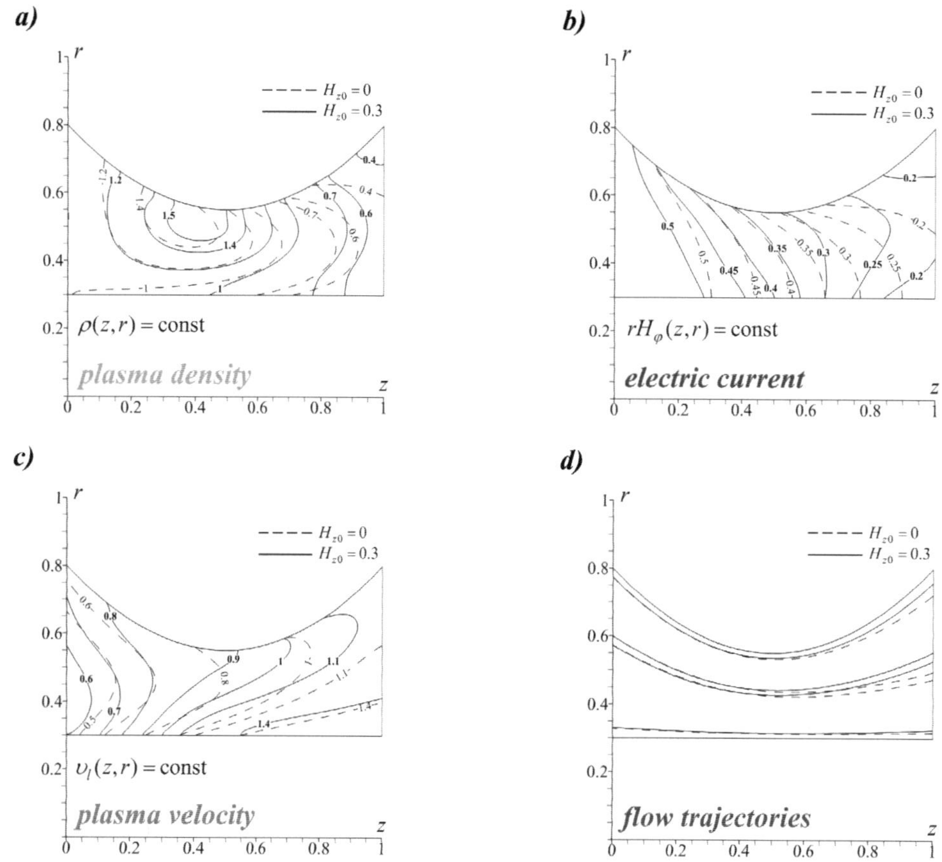

Figure 7. Plasma parameters distributions (***a, b, c***) and tubes of flow trajectories (***d***) in the duct with the curved external electrode

In this geometry, plasma is pushed up to the external duct electrode to such an extent that a zone of flux compression appears and then grows in the input and central area of the duct: plasma density rises to the maximum value which is higher than the input value (figure 7a). This result has been already noted in [9]. This is mostly due to the electrode curviness, and the longitudinal magnetic field additionally enhances this effect. The electric current distribution in this duct geometry is characterized by the marked deviation which is counterclockwise from the radial direction (figure 7b). This deviation intensifies in the longitudinal field.

Plasma acceleration near the external electrode is less intensively due to decreasing of an input value of an azimuthal magnetic field (figure 7c). In so doing, the longitudinal magnetic field, as in the previous case 3.1, in general, reduces acceleration properties of the duct.

Trajectories and thin flow tubes formed by them have a tendency to «uncurling» in the output area of the duct (figure 7d). They deviate down from those symmetrical relative to the minimum section. The longitudinal field seeks to restore the mentioned symmetry.

4. Conclusions

In the present work the non-steady-state two-dimensional MHD-model of relaxation of transonic super-Alfen flows with acceleration in plasma accelerator ducts of a various geometry with an external longitudinal magnetic field has been considered and implemented in calculations.

It has been determined that a longitudinal magnetic field in both of duct configurations leads to the following effects:

- It pushes plasma up to an external electrode. In terms of technical applications, this result can be used to reduce an amount of plasma flux artificially injected through an external electrode to solve «anode crisis» problem in QSPA;
- It deflects an electric current towards a duct input and turns its lines counterclockwise from the radial axis neutralizing at a specific polarity an influence of Hall effect negative for acceleration;
- It slightly reduces acceleration properties of a duct and causes a flux rotation in its transverse direction.

Acknowledgments
The study was performed by the grant from the Russian Science Foundation (project № 16-11-10278).

References

[1] Morozov A I 1957 *JETP* **5** (2) 215–20
[2] Artsimovich L A, Luk'ianov S Iu, Podgornyi I M, Chuvatin S A 1958 *JETP* **6** (1) 1–5
[3] Morozov A I 2012 *Introduction to Plasma Dynamics* (Boca Raton: CRC Press)
[4] Volkov Ya F, Kulik N Y, Marinin V V *et al* 1992 *Sov. J. Plasma Phys.* **18** (11) 718–23
[5] Morozov A I 1978 *Physical Principles of Space Electric Propulsion Thrusters* (Moscow: Atomzidat)
[6] Morozov A I and Savelyev V V 2000 *Reviews of Plasma Physics* vol 21 ed by B B Kadomtsev and V D Shafranov (NY: Kluwer Academic, Plenum Publishers) pp 203–391
[7] Morozov A I and Solov'ev L S 1980 vol 8 ed by B B Kadomtsev *Reviews of Plasma Physics* (NY, London: Consultants Bureau) pp 1–103
[8] Brushlinskii K V and Morozov A I 1980 vol 8 ed by B B Kadomtsev *Reviews of Plasma Physics* (NY, London: Consultants Bureau) pp 105–97
[9] Brushlinskii K V and Styopin E V 2017 *J. Phys.: Conf. Ser.* **788** (1) 012009
[10] Brushlinskii K V and Zhdanova N S 2008 *Plasma Physics Reports* **34** (12) 1037–45
[11] Brushlinskii K V 2009 *Mathematical and Computational Problems in Magnetohydrodynamics* (Moscow: Binom)
[12] Kozlov A N and Konovalov V S 2017 *Communications in Nonlinear Science and Numerical Simulation* **51** 169–79
[13] Kozlov A N 2017 *Plasma Phys. Control. Fusion* **59** (11) 115004
[14] Oran E and Boris J 1987 *Numerical Simulation of Reactive Flow* (NY, Amsterdam, London: Elsevier)
[15] Zalesak S T 1979 *J. Comput. Phys.* **31** 335–62

IOP Conf. Series: Journal of Physics: Conf. Series **937** (2017) 012008 doi:10.1088/1742-6596/937/1/012008

Mathematical model comparing of the multi-level economics systems

S M Brykalov[1] **and A V Kryanev**[2,3]

[1] JSC "Afrikantov OKBM", 603074, Nizhny Novgorod, Russia

[2] National Research Nuclear University "MEPhI", 115409, Moscow, Russia

[3] avkryanev@mephi.ru

Abstract. The mathematical model (scheme) of a multi-level comparison of the economic system, characterized by the system of indices, is worked out. In the mathematical model of the multi-level comparison of the economic systems, the indicators of peer review and forecasting of the economic system under consideration can be used. The model can take into account the uncertainty in the estimated values of the parameters or expert estimations. The model uses the multi-criteria approach based on the Pareto solutions.

1. Introduction

The presented report is a logical continuation of the previous papers of the authors and is devoted to analysis of multi-level economic system [1, 2]. In the analysis of the CES it is necessary to have a mathematical scheme to compare several CES or more options for the development of one of CES and to answer the question - which one is more efficient and reliable. These mathematical models are considered also for complex technical systems (CTS) [3] to determine their technical proficiency and to compare them with other similar CTS [3-5]. As in the previous papers [1, 2], in this report we use MLS, which is characterized by a big number of criteria [5-11].

2. The mathematical scheme clotting particular indices in integrated indicators for the multi-level economic systems

We consider $i=1,...,n$ the CES, each of which is characterized on the lower level by $j=1,..,m_0$ the particular indices (the more value of indices, the more effective CES), and on every levels by the integral indicators $R_{1ij}, j=1,..,m_1, \ldots, R_{iMj}, j=1,..,m_M$, respectively.

Thus, each integral index of the level consists of the index of the previous level.

Let $R_{1ijk}, j=1,..,m_1, k=1,...,n_{1j}, \ldots, R_{Mijk}, j=1,..,m_M, k=1,...,n_{Mj}$ are the particular initial indicators which are included in the j- th integral index of the first level and in the j- th integral index of the $M-th$ level, respectively.

To solve the problems of the forecasting of the particular indicators, the methods of the metric analysis [12-15] can be used.

There are the data values on the lower level:

$$R_{1ijk}, k = 1,...,n_{1j}, i = 1,...,n,\ j = 1,...,m_1, \quad (1)$$

R_{1ijk} is the value of the j - *th* particular index of the *i- th* CES.

Similar to paper [2], experts consider the priority of indicators of each level, with the help of which the partial indicators are reduced to integral indicators of the first level. Then, similarly to the paper [2], the integral indicators of the previous level are sequentially rolled up to the integral indicators of the subsequent level up to the last *M-th* level.

For example, for a three - level economic system $R_{1ijk}, j = 1,..,m_1, k = 1,...,n_{1j}$ and $R_{2ijk}, j = 1,..,m_2, k = 1,...,n_{2j}$ are the particular first-level indicators that are included in the *j- th* integral index of the second level, and in the *j- th* integral index of the third level, respectively (see [1, 2]).

To solve the problems of the forecasting of the particular first-level indicators, the methods of the metric analysis [12-15] can be used.

There are the data baseline values the first level (1),
R_{1ijk} is the value of the j - *th* private index of the *i- th* CES.

Let:

$$1^o.\ V_{1jkl},\ 0 < V_{1jkl} < 1,\ \sum_{k=1}^{n_{1j}} V_{1jkl} = 1,\ j = 1,...,m_2 \quad (2)$$

are the priority factors of particular indices of the l - *th* expert to j *-th* of the integral index of the second level, $l = \overline{1,L}$.

$$2^o.\ V_{2jkl},\ 0 < V_{2jkl} < 1,\ \sum_{k=1}^{n_{2j}} V_{1jkl} = 1,\ j = 1,...,m_3 \quad (3)$$

are the priority factors of integral indicators of the second level of the l - *th* expert to j - *th* of the integral index of the third level, $l = \overline{1,L}$.

$$3^o.\ V_{3jl},\ 0 < V_{3jl} < 1,\ \sum_{j=1}^{m_3} V_{3jl} = 1 \quad (4)$$

are the priority factors of integral indicators in the third level of the l – *th* expert.

$$4^o.\ \omega_l,\ l = 1,...,L,\ 0 < \omega_l < 1,\ \sum_{l=1}^{L} \omega_l = 1 \quad (5)$$

are the boost factors of experts.

Then we perform the following operations.

1) We produce the normalization of the values of particular indices

$$R^H_{1ijk} = \frac{R_{1ijk}}{R_{1jkMAX}},\ i=1,...,n;\ j = 1,...,m_1;\ k = 1,...,n_{1j}, \quad (6)$$

where $R_{1jkMAX} = \max\{R_{1ijk}, i = 1,...,n\}$.

2) We calculate the coefficients of the priority of private indicators for each group of the integral index of the second level:

$$V_{1jk} = \sum_{l=1}^{L} \omega_l \cdot V_{1jkl},\ j = 1,..,m_2, k = 1,...,n_{1j} \quad (7)$$

3) We calculate the normalized value of the evaluation of integrated indicators of the second

level: $$R^H_{2ijk} = \sum_{k=1}^{n_{1j}} V_{1jk} \cdot R^H_{1ijk},\ j = 1,...,m_2,\ k = 1,..., n_{2j}. \tag{8}$$

4) We calculate the coefficients of the priority of integrated indicators of the second level for each group of the integral index of the third level:

$$V_{2jk} = \sum_{l=1}^{L} \omega_l \cdot V_{2jkl},\ j = 1,.., m_3, k = 1,..., n_{2j} \tag{9}$$

5) We calculate the normalized values of the evaluations of integrated indicators in the third level: $$R^H_{i3j} = \sum_{k=1}^{n_{2j}} V_{2jk} \cdot R^H_{2ijk},\ j = 1,...,m_2 \tag{10}$$

6) We calculate the coefficient priority for the integral indicators in the third level:

$$V_{3j} = \sum_{l=1}^{L} \omega_l \cdot V_{3jl},\ j = 1,.., m_3. \tag{11}$$

7) We calculate the normalized value of the complex criteria for each CES:

$$R^H_i = \sum_{j=1}^{m_3} V_{3j} \cdot R^H_{3ij}, i = 1,...,n. \tag{12}$$

8) We produce a ranking of SES on the values of complex criteria:

$$R^H_{(1)} \geq R^H_{(2)} \geq ... \geq R^H_{(n)}. \tag{13}$$

CES with number $i = (1)$ is the best CES by the integrated efficacy.

We use fuzzy numbers [16, 17] to account for uncertainties in expert estimates of the indicators. In this case, each SES is characterized by two complex indicators: a complex indicator of efficiency and a complex index of uncertainty.

3. The mathematical definition of the uncertainty of the multi-level economic system integrated indicators

In the case of the performance of expert evaluations and values to account for uncertainties in the values of particular indicators we use the triangular fuzzy numbers [16, 17]. It is necessary to produce the normalization (6) for each of three numbers $(R_{lijk\,\min}, R^*_{lijk}, R_{lijk\,\max})$, i.e. go to the normalized values

$$(R^H_{lijk\,\min}, R^{H*}_{lijk}, R^H_{lijk\,\max}), \tag{14}$$

where

$$R^H_{lijk\,\min} = \frac{R_{lijk\,\min}}{R_{jkMAX}},\ R^{H*}_{lijk} = \frac{R^*_{lijk}}{R_{jkMAX}}, R^H_{ljk\,\max} = \frac{R_{lijk\,\max}}{R_{jkMAX}}, R_{jkMAX} = \max\{R^*_{lijk}, l = 1,...,L, i = 1,...,n\}. \tag{15}$$

Then the normalized score of each of the integral indicator for the i - th CES will be presented by the triangular fuzzy number

$$(R^H_{ij\,\min}, R^{H*}_{ij}, R^H_{ij\,\max}), \tag{16}$$

where (see the formula (5), (7))

$$R^H_{ij\,\min} = \sum_{k=1}^{n_j} V_{jk} \cdot R^H_{ijk\,\min},\ R^{H*}_{ij} = \sum_{k=1}^{n_j} V_{jk} \cdot R^{H*}_{ijk},\ R^H_{ij\,\max} = \sum_{k=1}^{n_j} V_{jk} \cdot R^H_{ijk\,\max},\quad i = 1,...,n \tag{17}$$

$$R^H_{ijk\,\min} = \sum_{l=1}^{L} \omega_l \cdot R^H_{lijk\,\min},\ R^{H*}_{ijk} = \sum_{l=1}^{L} \omega_l \cdot R^{H*}_{lijk}, R^H_{ijk\,\max} = \sum_{l=1}^{L} \omega_l \cdot R^H_{lijk\,\max},\ j = 1,...,m,\ k = 1,..., n_j, \tag{18}$$

We calculate the normalized value of the integrated indicator of efficiency for each CES presented in the form of a triangular fuzzy number

$$(R^H_{i\min}, R^{H*}_i, R^H_{i\max}),\ i = 1,...,n \tag{19}$$

where

$$R^H_{i\min} = \sum_{j=1}^{m_M} W_j \cdot R^H_{ij\min},\ R^{H*}_i = \sum_{j=1}^{m_M} W_j \cdot R^{H*}_{ij},\ R^H_{i\max} = \sum_{j=1}^{m_M} W_j \cdot R^H_{ij\max}. \tag{20}$$

4. Choosing the best CES as the solutions Pareto of two-criteria problem: an integrated efficiency criterion and an integrated criterion of uncertainty

In the case of the triangular fuzzy numbers to describe uncertainties in the values of indicators of two complex criteria for each SES the complex criterion of efficiency is defined by equality

$$K_{ieff} = R^{H*}_i, \qquad i = 1,...,n, \tag{21}$$

the complex criterion equality uncertainty is defined by equality

$$K_{iunc} = R^H_{i\max} - R^H_{i\min}, \qquad i = 1,...,n. \tag{22}$$

The final selection of the best CES is carried out among the Pareto solutions with respect to two criteria

$$K_{ieff} \text{ - max}, \tag{23}$$

$$K_{iunc} \text{ - min}. \tag{24}$$

On the plane (K_{ieff}, K_{iunc}) the Pareto points form a "south-eastern" part of the boundary of the set of points of criterions K_{ieff} and K_{iunc}.

Conclusion

The paper presents the mathematical model for the comparing of the multi-level economic CES or the comparison of several variants of the same CES which is characterized by the criteria system. This mathematical comparison circuit may use the design or expert assessment indicators and predict some or all of the considered indicators of CES. The model provides the account of uncertainties in the estimated values and expert assessment indicators or forecast values. The model uses a multi-criteria approach to the possibility of taking into account both quantitative and uncertainty indicators. The final decision is taken among the Pareto solutions of the multi-criteria problem with two integral criteria: the complex criterion of efficiency and complex criterion of risk for integral indicators of the final level. The mathematical model presented in this report can be used by offices and enterprises of various branches [18, 19].

References

[1] Brykalov S M and Kryanev A V Mathematical scheme of the three-level evaluation of the economic system 2017 *Applied Mathematical Sciences* **11** 693-701.

[2] Brykalov S M and Kryanev A V Mathematical scheme comparing of the three-level economical systems 2017 *Applied Mathematical Sciences* **11** 703-709.

[3] Semenov S S, Voronov E M, Poltavsky A V, Kryanev A V *Methods of decision-making in tasks assessing the quality and technical level of complex technical systems* 2015 (Moscow: URSS).

[4] Mesarevich M, Mako D, Takahara I *The theory of hierarchical multilevel systems* 1973 (Moscow: Mir Publ.).
[5] Urlov F F, Shapkin E I *Selecting effective strategic decisions on the basis of multi-level and multi-criteria approach* 2007 (N. Novgorod: Tutorial).
[6] Nogin V D *Decision-making in many criteria* 2007 (Sant Peterburg: Publisher UTAS).
[7] IAEA-TECDOC-1478. *Selection of decommissioning strategies: Issues and factors* 2005 IAEA.
[8] Rosental R E *Principles of multi objective optimization* 1984 (California: Monterey).
[9] Postnikov V M, Spiridonov S B Methods of selection criteria for local weighting coefficients 2015 *Science and education, Electronic journal of Bauman MSTU* **6** 267–287.
[10] Saati T L *Decisions Making With Dependendence And Feedback: The Analytic Network Process* 1996 (Pittsburgh: University of Pittsburgh ed.).
[11] Kryanev A V, Bochkarev V V, Hanbekova D T, Ustinova U G *Ranking nuclear and radiation hazardous objects on the basis of expert evaluations* 2015 (Moscow: Scientific session MEPhI-2015 **3** 170).
[12] Kryanev A V, Udumyan D K Metric Analysis, Properties and Applications as a Tool for Interpolation 2014 *International Journal of Mathematical Analysis* **8** 2221 – 2228.
[13] Kryanev A V, Udumyan D K Metric Analysis, Properties and Applications as a Tool for Smoothing 2014 *International Journal of Mathematical Analysis* **8** 2337 – 2346.
[14] Kryanev A V, Udumyan D K Metric Analysis, Properties and Applications as a Tool for Forecasting 2014 *International Journal of Mathematical Analysis* **8** 2971 – 2978.
[15] Kryanev A V, Udumyan D K, Lukin G V, Ivanov V V Metric Analysis Approach for Interpolation and Forecasting of Time Processes 2014 *Applied Mathematical Sciences* **8** 1053 – 1060.
[16] Orlovskii S A *Decision-making problems with fuzzy initial information* 1981 (Moscow: Science ed.).
[17] Pegat A *Fuzzy modeling and control* 2009(Moscow: BINOM ed.).
[18] Brikalov S M The implementation of the principle of decomposition of the formation of a multi-level system of key performance indicators 2013 *Integral* **5, 6**.
[19] Brikalov S M, Urlov F F Balanced scorecard and key performance indicators: a terminological analysis, problems and directions of development Economy and business 2015 **5** 570-575.

Point vortices in the plane: positive-dimensional configurations.

M V Demina, N A Kudryashov, J E Semenova

Department of Applied Mathematics, National Research Nuclear University MEPHI, Kashirskoe highway, 31, 115409, Moscow, Russia

E-mail: mvdemina@mephi.ru, nakudr@gmail.com, uesemenova@gmail.com

Abstract. The problem of constructing and classifying stationary equilibria of point vortices in the plane is studied. An ordinary differential equation that enables one to find positions of point vortices with circulations Γ_1, Γ_2, and Γ_3 in stationary equilibrium is obtained. A necessary condition of an equilibrium existing is derived. The case of point vortex systems consisting of $n+2$ point vortices with n vortices of circulation Γ_1 and two vortices of circulations $\Gamma_2 = a\Gamma_1$ and $\Gamma_3 = b\Gamma_1$, where a and b are integers, is considered in detail. The properties of polynomial solutions of the corresponding ordinary differential equation are investigated. A set of positive–dimensional equilibrium configurations is found. A continuous free parameter is presented in the coefficients of corresponding polynomial solutions. These free parameters affect the positions of the roots and hence the vortex positions. Stationary equilibrium that could be derived from each other by rotation, extension, parallel translation is considered as equivalent. All found configurations seem to be new.

1. Introduction

The motion of M point vortices with circulations (or strengths) Γ_1, ... , Γ_M at positions z_1, ... , z_M is governed by the Helmholtz's equations

$$\frac{dz_k^*}{dt} = \frac{1}{2\pi i}\sum_{j=1}^{M}{}' \frac{\Gamma_j}{z_k - z_j}, \quad k = 1, \ldots, M, \tag{1}$$

where the prime means that the case $j = k$ is excluded and the symbol * stands for complex conjugation. This system induces complicated dynamics and is not integrable in the case $M > 3$. Nevertheless some particular "motions" including stationary, translating, and rotating equilibria are successfully studied. A convenient approach applicable to such types of motion is the so–called "polynomial method". According to this method, polynomials with roots at vortex positions are introduced. In 1964 Tkachenko obtained a differential equation satisfied by generating polynomials of stationary vortex arrangements with equal in absolute value circulations [1]. Now this equation is known as the Tkachenko equation. A simpler derivation of the Tkachenko equation was proposed by Aref [2]. Further the polynomial method was generalized to the case of vortices with generic choice of circulations [3, 4, 5]. In addition it was shown that the polynomial method can be successfully used to construct rotating equilibria and configurations of collapse and scattering [6].

IOP Conf. Series: Journal of Physics: Conf. Series **937** (2017) 012009 doi:10.1088/1742-6596/937/1/012009

2. Equation of stationary equilibria of point vortices

Consider a system of M point vortices with the circulations $\Gamma_\gamma,\ \gamma = 1, ..., N$ from equations (1). We set $\frac{dz_k^*}{dt} = 0$ in the equations of motion (1) and subdivide the vortices into groups according to the values of circulations. Suppose vortices at positions $a_1^{(\gamma)}, ..., a_l^{(\gamma)}$ have circulations $\Gamma_\gamma,\ \gamma = 1, ..., N$.

$$\sum_{\gamma=1}^{N} \sum_{i=1}^{l_\gamma}{}' \frac{\Gamma_\gamma}{a_{i_0}^{\gamma_0} - a_i^{\gamma}} = 0,\ i_0 = 1, ..., l_{\gamma_0},\ \gamma_0 = 1, ..., N \tag{2}$$

A convenient tool for analyzing such a situation is to introduce polynomials with roots at the positions of vortices (2). In our case, we have

$$P_\gamma(z) = \prod_{i=1}^{l_\gamma} (z - a_i^{\gamma}),\ \gamma = 1, ..., N \tag{3}$$

Thus, we see that roots of the polynomial Pj(z) give positions of vortices with circulation Γ_γ.

Furthermore, the following approach is proposed which consists in introducing new functions $\widetilde{P}(z)$ and $\widetilde{Q}(z)$, according to the rules:

$$\widetilde{P}(z) = \prod_{\gamma=1}^{N} P_\gamma^{\frac{\Gamma_\gamma(\Gamma_\gamma+1)}{2}}(z),\ \widetilde{Q}(z) = \prod_{\gamma=1}^{N} P_\gamma^{\frac{\Gamma_\gamma(\Gamma_\gamma-1)}{2}}(z) \tag{4}$$

Using the equalities

$$\frac{d^2}{dz^2} \ln\left\{\widetilde{P}(z)\widetilde{Q}(z)\right\} + \left(\frac{d}{dz} \ln\left\{\frac{\widetilde{P}(z)}{\widetilde{Q}(z)}\right\}\right)^2 = 0 \tag{5}$$

present one can (5) in the more convenient form

$$\widetilde{P}_{zz}\widetilde{Q} - 2\widetilde{P}_z\widetilde{Q}_z + \widetilde{P}\widetilde{Q}_{zz} = 0 \tag{6}$$

Equation (6) is known as the Tkachenko equation. There exists necessary condition of an equilibrium in the arrangements of vortices:

$$\sum_{\gamma=1}^{N} l_\gamma \Gamma_\gamma^2 - \left[\sum_{\gamma=1}^{N} l_\gamma \Gamma_\gamma\right]^2 = 0 \tag{7}$$

Using the polynomial method we derive the ordinary differential equation

$$(z^2 - 1)\frac{d^2}{dz^2}w(z) + 2\{(a+b)z + (b-a)\}\frac{d}{dz}w(z) + 2abw(z) = 0 \tag{8}$$

describing stationary equilibria of point vortex systems consisting of $n+2$ point vortices with n vortices of circulation Γ_1 and two vortices of circulations $\Gamma_2 = a\Gamma_1$ and $\Gamma_3 = b\Gamma_1$, where a and b are integers. In (8) $w(z)$ is a polynomial of degree n.

$$w_n(z) = C_n z^n + C_{n-1} z^{n-1} + C_{x-2} z^{n-2} + .. + C_1 z + C_0, \tag{9}$$

We set $C_n = 1$. Thus, $w(z)$ is unitary polynomial.

3. Theorem on the existence of a polynomial solution of equation (8)

Suppose that solution of equation (8) has the form:

$$w(z) = ... + C_n z^n + C_{n-1} z^{n+-1} + ... + C_{k+1} z^{k+1} + C_k z^k + C_{k-1} z^{k-1} + \\ + ... + C_1 z + C_0 + C_{-1} z^{-1} + C_{-2} z^{-2} + ... \quad (10)$$

Substituting the series (10) in equation (8) and taking into account the summands for different degrees z, we get:

$$... + B_n z^n + B_{n-1} z^{n+-1} + ... + B_{k+1} z^{k+1} + B_k z^k + B_{k-1} z^{k-1} + ... + B_1 z + B_0 + B_{-1} z^{-1} + ... = 0 \quad (11)$$

Coefficient at z^k takes the form:

$$B_k = C_k(2ab + 2k(a+b) + k(k-1)) + C_{k+1} 2(k+1)(b-a) - C_{k+2}(k+1)(k+2) = 0 \quad (12)$$

Consider a natural number n. We assume that the solution of the equation is a polynomial of degree n. Let $C_n = 1$. Taking into account that $w(z)$ is a polynomial of degree n and coefficients $C_{n+1} = 0$ and $C_{n+2} = 0$, we obtain:

$$B_n = 2ab + 2n(a+b) + n(n-1) = 0 \quad (13)$$

Let us to prove the following lemma:

Lemma 1. *The Diophantine equation* (13) *is a necessary condition for the existence of a polynomial solution of equation* (8) *of degree* n.

We set a is free parameter in (13) and find the dependence of b on a. Further consider all the coefficients B_k from $k = n$ to $k = 0$ consistently, and express at each step C_k through the two preceding factors C_{k+1} and C_{k+2}. If $k < 0$ then C_{k+1}, C_{k+2} are equal 0. Eventually, expression (10) is a polynomial of degree n and equation (8) has one polynomial solution.

Lemma 2. *Solution of the equation* (8) *has not roots at the points* $z = \pm 1$.

Proof. The proof is by reductio ad absurdum. Consider $z = 1$. Assume that the point $z = 1$ is root of a polynomial of multiplicity $r \in \mathbb{N}$.

$$w(z) = (z-1)^r \psi(z), \quad \psi(1) \neq 0 \quad (14)$$

Twice differentiating $w(z)$ with respect to z we substitute the result into equation (8). Presenting similar terms with respect to the degrees of expression $(z-1)$. Finding the Taylor series in a neighborhood of the point $z = 1$, we obtain:

$$2r(r-1)\psi(1) + 4br\psi(1) = 2r\psi(1)((r-1) + 2b) = 0 \quad (15)$$

If $r = 0$ then the condition of equality to zero is always satisfied. If $r \neq 0$, we have:

$$r = 1 - 2b \quad (16)$$

r is positive integer (r is degree of multiplicity of the root of a polynomial). If b has a half-integer negative value then condition (16) is true. We do not consider such cases because we have an

IOP Conf. Series: Journal of Physics: Conf. Series **937** (2017) 012009 doi:10.1088/1742-6596/937/1/012009

interest in cases when b is positive integer. Thus, only one case suits us: $r = 0$. Contradiction is obtained as it follows that $z = 1$ isn't polynomial root $w(z)$.

Similar arguments are valid when $z = -1$. The lemma is proved.

Lemma 3. *If the polynomial is a solution of the equation* (8)*, then all its roots are simple.*

Proof. Suppose $w_n(z)$ is the solution of equation (8). Let $w_n(z)$ be given by the form (9).

Expanding the polynomial $w_n(z)$ by the Taylor formula at the point z_0. z_0 is assumed equal to one of the roots of the given polynomial. We construct a general formula for the derived degree q of equation (8). Everyone can verify the validity of the formula for $q = k - 2$ by induction:

$$(z^2-1)w^{(k)}(z) + 2((k-2+a+b)z + 2(b-a))w^{(k-1)}(z) + ((k-2)((k-3)+2(a+b))+ \\ +2ab)w^{(k-2)}(z) = 0 \quad k \geqslant 2 \tag{17}$$

The proof is by reductio ad absurdum. Assume that z_0 is not simple root.

$$w(z_0) = 0 \quad w'(z_0) = 0. \tag{18}$$

Each successive derivative of the polynomial (9) can be expressed through the two preceding ones using the formulas (17). It follows from (18) and Lemma 2 that all subsequent derivatives of w_k will be zero. We see that all coëfficients in the Taylor expression (3.9) are equal to zero. As a result, function w_n is equal to zero identically. This contradiction proves the lemma. Hence, all the roots of the polynomial are simple.

It is a well-known fact [10] that two linearly independent solutions of an ordinary differential equation of the second order

$$y'' + p(x)y' + q(x)y = 0 \tag{19}$$

are related by:

$$y_2(x) = y_1(x)\int e^{-\int p(x)dx}\frac{dx}{y_1^2(x)} \tag{20}$$

The general solution of the equation (19) has the form:

$$y = C_1y_1 + C_2y_2, \quad \text{where } C_1, C_2 \text{ are constants} \tag{21}$$

It is a special case when the equation (8) has two polynomial solutions. In this case, we get an infinite number of stationary equilibrium of point vortices. It follows from (21). We prove the following theorem.

Theorem. *If the Diophantine equation*

$$n(n-1) + 2(a+b)n + 2ab = 0 \tag{22}$$

with fixed integer values of the parameters a and b has two integer solutions, then the general solution of ordinary differential equation (8) *is polynomial.*

Proof. Representing the equation (8) in the form (19)

$$w''(z) + \frac{(2(a+b)z + 2(b-a))}{(z^2-1)}w'(z) + \frac{2ab}{z^2-1}w(z) = 0 \tag{23}$$

IOP Conf. Series: Journal of Physics: Conf. Series **937** (2017) 012009 doi:10.1088/1742-6596/937/1/012009

and substituting (20) in this equation, we get:

$$w_2(z) = w_1(z) \int (z+1)^{2a}(z-1)^{2b} \frac{dz}{w_1^2(z)} \tag{24}$$

We want to show that there exists the polynomial solutions of equation (8). This solution is a polynomial of a lesser degree (of two polynomial solutions).

We must to prove that the solution $w_2(z)$ has no singularity. Consider expression (24). All the roots of the polynomial $w_1(z)$ are simple.(It follows from Lemma (3)) Consequently, the generalized Laurent series in a neighborhood of z_0 has the form

$$w_1^2 = K^2(z-z_0)^2, \quad K \neq 0 \tag{25}$$

Substituting the series (25) into (24):

$$w_2 = K(z-z_0) \int \left(\frac{\kappa_{-2}}{(z-z_0)^2} + \frac{\kappa_{-1}}{(z-z_0)} + \kappa_0 + ... \right) dz \tag{26}$$

Represent in the form:

$$w_2 = K(z-z_0) \left(-\frac{\widetilde{\kappa}_{-2}}{z-z_0} + \widetilde{\kappa}_{-1} \ln(z-z_0) + ... \right) \tag{27}$$

The form of the expression shows that the solution $w_2(z)$ can have poles. We apply the analysis by the theory of residues. This shows that we obtain a rational function after integrating. This function becomes a polynomial when multiplied by $w_1(z)$ (This follows from the formula (24)). Consequently, the equation (8) has a polynomial solution. The theorem is proved.

4. Classification of configurations

In this paper, we study the problem of finding, reserching and classifying stationary equilibrium of point vortices in the plane. We concidered the case when stationary equilibrium of point vortices consists of n point vortices with circulations $\Gamma_2 = a\Gamma_1$ and another one with circulations $\Gamma_3 = b\Gamma_1$, where a and b are integers. The values of circulations are found for which the number of nonequivalent stationary equilibrium of point vortices is infinite. Some examples are given in Tab.1

As a consequence of this theorem we see that a continuous free parameter is present in the coefficients of corresponding polynomial solutions whenever the condition of the theorem is valid. These free parameters affect the positions of the roots and hence the vortex positions. Using this theorem we find new positive–dimensional equilibrium configurations. Until recently only few sets of such configurations have been known.

Some examples are given in Fig.1 .

$$w_2^{(-5,-3)}(z) = 7z^2 + 4z + 1; \tag{28}$$

$$w_{15}^{(-5,-3)}(z) = 91 - 182z + 455z^2 - 455z^3 + 91z^4 + 273z^5 - 221z^6 - 65z^7 + (\frac{455}{3})z^8 - \\ -(\frac{91}{3})z^9 - (\frac{1547}{33})z^{10} + (\frac{273}{11})z^{11} + (\frac{35}{11})z^{12} - 5z^{13} + z^{14};$$

Thus, the general solution of the equation (8) (where $a = -5$, $b = -3$) has the form :

$$W_{15}^{(-5,-3)}(z) = C_1(7z^2 + 4z + 1) + C_2(91 - 182z + 455z^2 - 455z^3 + 91z^4 + 273z^5 - 221z^6 - \\ -65z^7 + (\frac{455}{3})z^8 - (\frac{91}{3})z^9 - (\frac{1547}{33})z^{10} + (\frac{273}{11})z^{11} + (\frac{35}{11})z^{12} - 5z^{13} + z^{14}); \tag{29}$$

IOP Conf. Series: Journal of Physics: Conf. Series **937** (2017) 012009 doi:10.1088/1742-6596/937/1/012009

These results can be summarized as follows. Varing the values of the variables C_1 and C_2 in a polynomial (4) we get that configurations pass from one to another.

Table 1. The results of the search for parameter values a, b, n in the case of infinite number of nonequivalent stationary equilibrium of point vortices. n_1 - the degree of the smallest polynomial is a solution of equation (8), l - the number of pairs of coefficients with an infinite number of nonequivalent stationary equilibrium, n_2 - the degree of the largest polynomial is a solution of equation (8)

n_1	l	a	b	n_2	the general solution of equation (8)	number of configurations
2	1	-5	-3	15	$C_1 w_2^{(-5,-3)} + C_2 w_{15}^{(-5,-3)}$	∞
3	2	-9	-4	24	$C_1 w_3^{(-9,-4)} + C_2 w_{24}^{(-9,-4)}$	∞
		-6	-5	20	$C_1 w_3^{(-6,-5)} + C_2 w_{20}^{(-6,-4)}$	∞
4	2	-14	-5	35	$C_1 w_4^{(-14,-5)} + C_2 w_{35}^{(-14,-5)}$	∞
		-9	-6	27	$C_1 w_4^{(-9,-6)} + C_2 w_{27}^{(-9,-6)}$	∞
5	2	-20	-6	48	$C_1 w_5^{(-20,-6)} + C_2 w_{48}^{(-20,-6)}$	∞
		-10	-8	32	$C_1 w_5^{(-10,-8)} + C_2 w_{32}^{(-10,-8)}$	∞
6	2	-27	-7	63	$C_1 w_6^{(-27,-7)} + C_2 w_{63}^{(-27,-7)}$	∞
		-13	-9	39	$C_1 w_6^{(-13,-9)} + C_2 w_{39}^{(-13,-9)}$	∞
7	3	-35	-8	80	$C_1 w_7^{(-35,-8)} + C_2 w_{80}^{(-35,-8)}$	∞
		-21	-9	54	$C_1 w_7^{(-21,-9)} + C_2 w_{54}^{(-21,-9)}$	∞
		-14	-11	44	$C_1 w_7^{(-14,-11)} + C_2 w_{44}^{(-14,-11)}$	∞
8	4	-44	-9	99	$C_1 w_3^{(-44,-9)} + C_2 w_{99}^{(-44,-9)}$	∞
		-26	-10	65	$C_1 w_3^{(-26,-10)} + C_2 w_{65}^{(-26,-10)}$	∞
		-20	-11	55	$C_1 w_3^{(-20,-11)} + C_2 w_{55}^{(-20,-11)}$	∞
		-17	-12	51	$C_1 w_3^{(-17,-12)} + C_2 w_{51}^{(-17,-12)}$	∞

5. Conclusion

In this article we have studied the problem of finding any stationary configurations of point vortices with circulations $\Gamma_1, \Gamma_2, \Gamma_3$ in the plane. We considered the case when stationary equilibrium of point vortices consists of n point vortices with circulations $\Gamma_2 = a\Gamma_1$ and another one with circulations $\Gamma_3 = b\Gamma_1$, where a and b are integers. The values of circulations have been found for which the number of nonequivalent stationary equilibrium of point vortices is infinite.

IOP Conf. Series: Journal of Physics: Conf. Series **937** (2017) 012009 doi:10.1088/1742-6596/937/1/012009

Figure 1. Plots of vortex positions in stationary equilibrium. Circles denote vortices with circulation Γ_1, squares denote vortices with circulation Γ_2 and rhombus denote vortices with circulation Γ_3 *a)* z(n), n=63, $C_1 = 10$, $C_2 = 1$, *b)* z(n), n=63, $C_1 = 10^11$, $C_2 = 10^{11}$, *c)* z(n), n=63, $C_1 = 10$, $C_2 = 10^{12}$, *d)* z(n), n=63, $C_1 = 10$, $C_2 = 10^{14}$

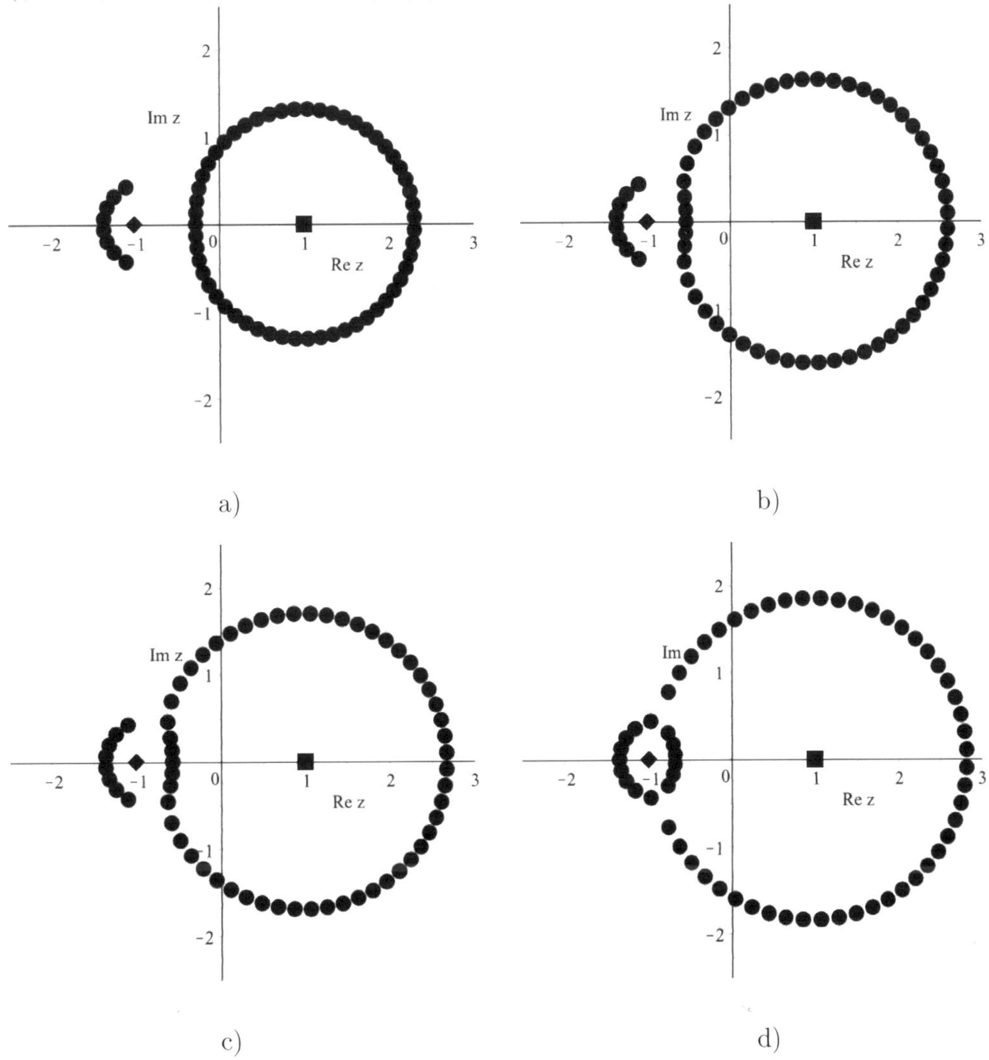

Some examples are given in Tab.1. The obtained equation is investigated. In particular, the existence of a polynomial solution is proved (Theorem 1).

References

[1] Tkachenko V K 1964 *Thesis* (Moscow: Institute of Physical Problems)
[2] Aref H 2007 *Vortex Crystals* (Journal of Mathematical Physics) p 79
[3] O'Neil K A 2006 *Physica D. Minimal polynomial systems for point vortex equilibria* **219** 69-79.
[4] Demina M V and Kudryashov N A 2012 *Journal of Physics A: Mathematical and Theoretical. Vortices and Polynomials: Non-uniqueness of the Adler-Moser polynomials for the Tkachenko equation* **45** 19
[5] Demina M V and Kudryashov N A 2016 *Regular and Chaotic Dynamics. Multi-particle dinamical systems and polynomials* **21** 3 351-66

[6] Demina M V and Kudryashov N A 2014 *Theoretical and Computational Fluid Dynamics. Rotation, collapse, and scattering of point vortices* **28** 3 357-68

[7] Helmholtz H 2002 *Fundamentals of the vortex theory. Translation from German under the editorship of SA Chaplygin* (Moscow-Izhevsk: Institute of Computer Research) (In Russian)

[8] Borisov A V and Mamaev I S 2005 *Mathematical methods of dynamics of vortex structures* (Moscow-Izhevsk: Institute for Computer Studies) (In Russian)

[9] Demina M V and Kudryashov N A 2012 *Regular and Chaotic Dynamics. Point Vortices and Classical Orthogonal Polynomials* **17** 5 371-84

[10] Smirnov V I 2010 *Course of Higher Mathematics* (BXV-Petersburg vol 3) (Saint Petersburg) (In Russian)

IOP Conf. Series: Journal of Physics: Conf. Series **937** (2017) 012010 doi:10.1088/1742-6596/937/1/012010

$J/\psi \longrightarrow e^+e^-$ decay selection criteria in the CBM experiment

O Yu Derenovskaya[1], V V Ivanov[1,2]

[1]Joint Institute for Nuclear Research, Dubna, RUSSIA

[2]National Research Nuclear University "MEPhI", Moscow, Russia

E-mail: odenisova@jinr.ru

Abstract. Currently the CBM experiment is being developed in GSI (Darmstadt, Germany) at the FAIR accelerator complex in an international collaboration with JINR (Dubna, Russia). Measurements of the $J/\psi \longrightarrow e^+e^-$ decay are one of the key objectives of the CBM experiment, the registration of such decays is planned to be conducted in real-time experiment. In the current paper the criteria, that allow for effective selection of signal decay events for AuAu collisions at 10 AGeV, are discussed in detail. This energy corresponds to the first phase of the CBM experiment.

1. Introduction

The CBM (Compressed Baryonic Matter) experiment is heavy-ion experiment dedicated to the investigation of the properties of highly compressed baryonic matter as is produced in nucleus-nucleus collisions at the Facility for Antiproton and Ion Research (FAIR) in Darmstadt, Germany [1].

The CBM experimental setup for studying dielectron decays is shown in Fig. 1.

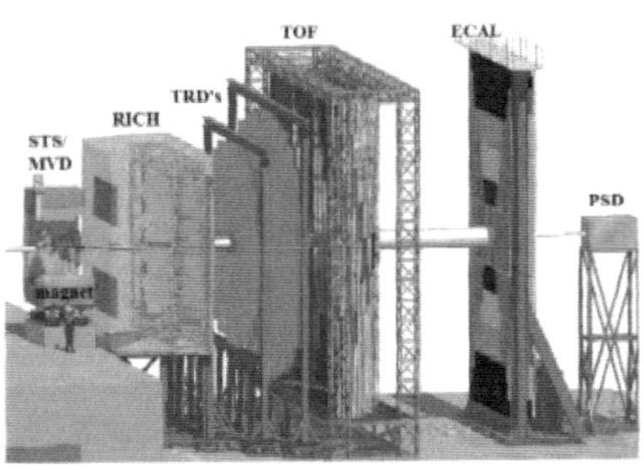

Figure 1. General view of the CBM setup for studying dielectron decays.

The main tracking detector of CBM STS (Silicon Tracking System) is located behind the target between the poles of the superconducting dipole magnet. The STS detector is intended to reconstruct trajectories and momenta of charged particles, as well as reconstruct primary and secondary vertices. The electron/positron identification system includes Ring Imaging CHerenkov (RICH) and Transition Radiation Detectors (TRD). The TRD is also used to reconstruct the trajectories of charged particles registered by the detector. The detector for Time-Of-Flight (TOF) measurement is intended for hadron identification. The Electromagnetic

IOP Conf. Series: Journal of Physics: Conf. Series **937** (2017) 012010 doi:10.1088/1742-6596/937/1/012010

CALorimeter (ECAL) serves to identify photons. The Projectile Spectator Detector (PSD) calorimeter is used for determining the reaction plane.

The study of charmonium, which is produced in nuclear-nuclear collisions at FAIR energies is one of the key objectives of the CBM experiment. Main difficulty of $J/\psi \to e^+e^-$ reconstruction is the extremely low yield of J/ψ mesons with a low probability (about 6%) of decay into the dielectron channel in the environment of high multiplicity of secondary particles (from 100 to 1000 particles per nucleus-nucleus collision) due to high interaction rate (up to 10^7 collisions per second). The important and relevant task for CBM collaboration is a fast and efficient selection of the signal events for $J/\psi \to e^+e^-$ reconstruction in the real time experiment.

The current paper presents criteria that provide an effective selection of signal events for Au+Au collisions at 10 AGeV. This energy corresponds to the first phase of the CBM experiment [2].

2. The J/ψ reconstruction technique

The CBM experimental setup is under development. Therefore, the current study was based on the Monte-Carlo simulated data. Signal decay events and background were simulated separately. The PLUTO [3] generator was used for the signal J/ψ decay, and the UrQMD [4] generator for the simulation of the background AuAu events. The so prepared sets of particles were transported through the CBM setup with help of GEANT3 [5] in the CBMROOT framework.

The selection criteria for J/ψ reconstruction are given below:

a) transverse momentum of the charged particles is more than 1 GeV/c;
b) number of counts in the TRD detector is greater than 2;
c) particles are identified as electrons in the TRD;
d) limit on the deviation of particle trajectory in the TRD $|d_{t_x}|$ is less than 0.035;
e) particles are identified as electrons in the TOF;
f) particles are linked to any of the Cherenkov rings in the RICH detector.

These criteria are applied in order of importance one after another, giving as a result a sample of selected events. The KF Particle Finder [6] is used to reconstruct the signal event topology as the last step.

3. The transverse momentum cut

The reconstruction of trajectories and momenta of charged particles using STS detector is performed as the first step. The cellular automaton algorithm is used for track reconstruction [7]. The Kalman filter is applied to determine the parameters of reconstructed tracks including spatial coordinates and momentum [8].

Fig. 2 shows the distribution of transverse momentum in the STS: signal events are shown in red, background events are blue. Shaded area corresponds to the selected cases, that is such which have a transverse momentum more than 1 GeV/c.

The major part of signal particles have a large transverse momentum p_t. This is consistent with the fact that kinematic criterion on transverse momentum works especially well when the total mass of daughter particles is much lower than the parent particle mass. Since $2m_e \ll m_{J/\psi}$ (here m_e is the electron mass and $m_{J/\psi}$ is the J/ψ meson mass) the limit of $p_t > 1$ can significantly suppress the background with a minimum loss of signal events (shaded area on Fig. 2) [9].

4. The number of counts in the TRD detector

In parallel with track reconstruction in the STS it is planned to use the trd data for event selection. The main purpose of the TRD detector is to exclude the pion contribution from the

IOP Conf. Series: Journal of Physics: Conf. Series **937** (2017) 012010 doi:10.1088/1742-6596/937/1/012010

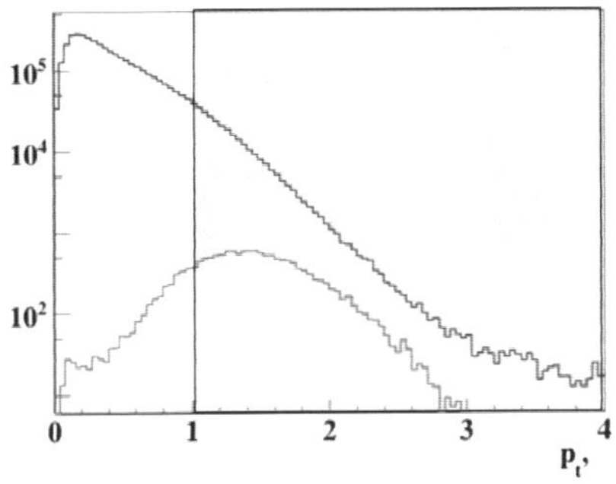

Figure 2. The distribution of transverse momentum in the STS: signal events are red, background events are blue, shaded area corresponds to the selected cases.

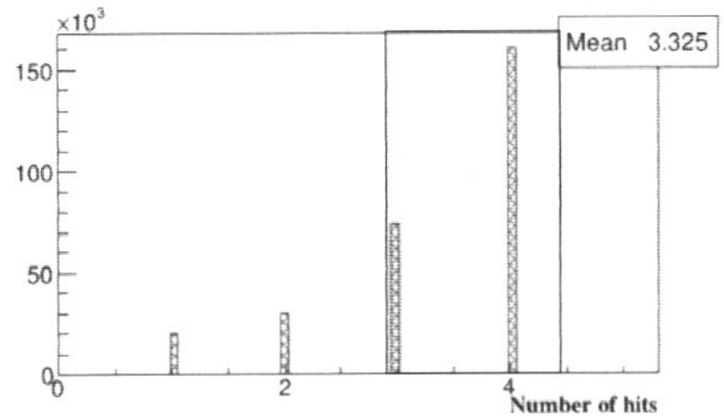

Figure 3. The distribution of the number of hits in the TRD-track, shaded area corresponds to the events that pass the further selection.

particle sample. The procedure of pion suppression includes several stages: (1) a reconstruction of trajectories of the particles registered by the TRD coordinate detectors, and (2) particle identification based on the energy losses only in those TRD modules that made a contribution to the reconstructed tracks.

The TRD consists of four stations, each of them registers the intersection of the particle with the detector plane (hit) and its energy loss. A track following technique and Kalman filter are applied for track reconstruction in the TRD [10]. Each track is associated with a set of measurements of the particle energy losses. With the help of mathematical methods, one should determine to which distribution (electrons or pions) these losses are related.

Fig. 3 shows the distribution of the number of hits in the TRD tracks. In order to perform an electron identification the information about the particle energy losses at least at three points is needed. Therefore, for the further analysis only such particles which have the number of hits in the track more than or equal to three are considered.

5. Electron identification in the TRD detector

Fig. 4 shows the distributions of the total energy losses for pions and electrons in the first TRD station. One can see that these distributions highly overlap. Therefore, it is impossible to distinguish between electrons and pions using measurements from one station.

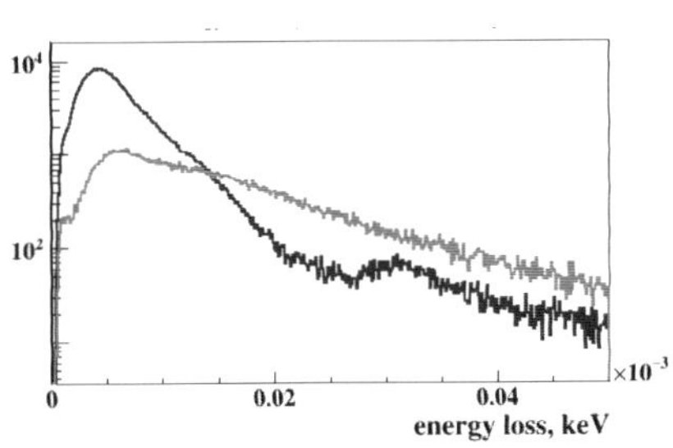

Figure 4. The distributions of the total energy losses for pions (blue line) and electrons (red line) in the first TRD station.

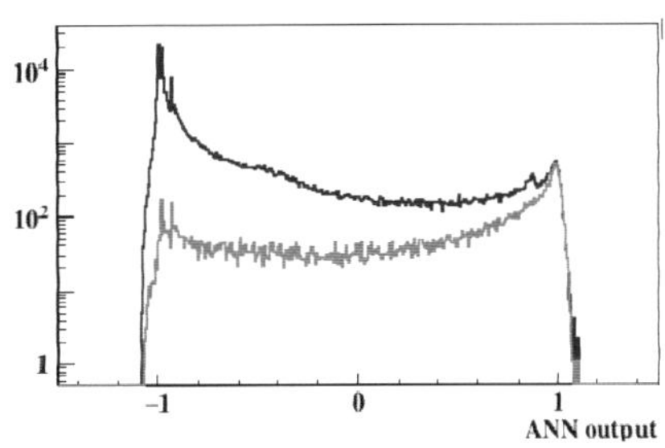

Figure 5. The distribution of the output values of the neural network for pions (blue line) and electrons (red line).

IOP Conf. Series: Journal of Physics: Conf. Series **937** (2017) 012010 doi:10.1088/1742-6596/937/1/012010

The possibility of electron and pion identification using an artificial neuron network, a multilayer perceptron (MLP), has been investigated [11, 12]. A three-layered perceptron implemented in the CBMROOT environment is currently employed in the CBM experiment for particle identification using the TRD [13]. Fig. 5 presents the distribution of the output values of the neural network for pions (blue line) and electrons (red line), which shows that the MLP allows reliable electron identification.

6. The limit on the particle deviation in the TRD

Additionally, to suppress the combinatorial background the limit on the deviation of the particle trajectory under the influence of a magnetic field is considered. Since the main magnetic field component is directed along OY axis, the magnetic field deflects electrons and positrons to the different directions in XOZ projection.

The deviation angle d_{t_x} is calculated as follows:

a) the target centre is connected by the straight line with the x-coordinate of the first hit of the corresponding TRD-track,

b) all the hits of the TRD track are connected by the straight line in the plane XOZ,

c) the angle between a) and b) lines are calculated.

Fig. 6 presents the deviation angle distribution for electrons (the blue line) and positrons (the green line) in case of background (a) and signal (b) events respectively. If we impose the limit on the deviation angle of the track $d_{t_x} < 0.035$, it is possible to further suppress the background.

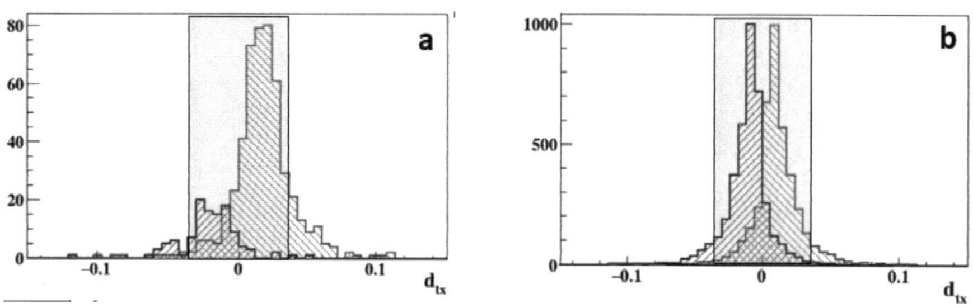

Figure 6. The deviation angle distribution for electrons (the blue line) and positrons (the green line) in case of background (a) and signal (b) events respectively.

Beside that the deviation of the particle trajectory in the TRD can be used to determine the charge sign of the electrons or positrons.

7. Electron identification with TOF detector

In addition to the TRD detector TOF is used to identify high energy electrons and exclude hadrons. ToF can measure the time of flight, which is needed for a certain particle to fly from the target to the ToF detector plane. If you know the particle momentum p and the trajectory length, you can calculate its mass m. The squared mass m^2 of charged particle dependence of their momentum p is used for particle identification in the TOF (Fig. 7) [12].

For the background there is a large contribution from protons (Fig. 7,a). In order to remove the protons from the sample we apply the method which is used in KF Particle Finder [6]. The particles which lie in the corridor marked with red lines are selected for the further analysis (Fig. 7,c).

IOP Conf. Series: Journal of Physics: Conf. Series **937** (2017) 012010 doi:10.1088/1742-6596/937/1/012010

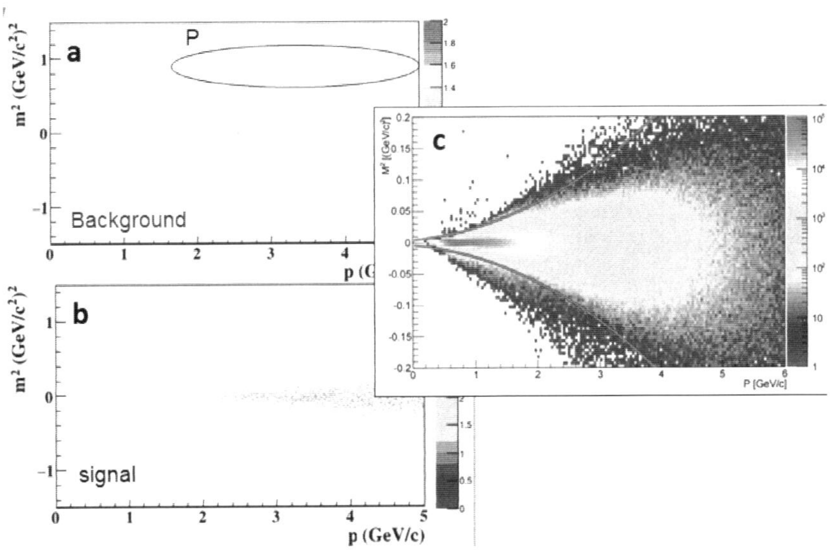

Figure 7. The distribution of the squared mass of charged particle dependence of their momentum for background (a) and signal (b) events respectively; c) the boundaries using for the electron selection (red lines).

8. Electron identification with RICH detector

The selected tracks are extrapolated to the RICH detector [14]. The primary function of this detector is electron identification and pion suppression in the momenta range from 0.5 to 10 GeV/c. The principle of the RICH operation is based on the detection of the Cherenkov radiation emitted during the passage of a relativistic charged particle through the detector radiator.

Fig. 8 shows the result of registration of the charged particles in one AuAu-collision in the RICH detector. Dots show places of registration of the Cherenkov radiation by the photomultipliers on the RICH plane and the reconstructed rings.

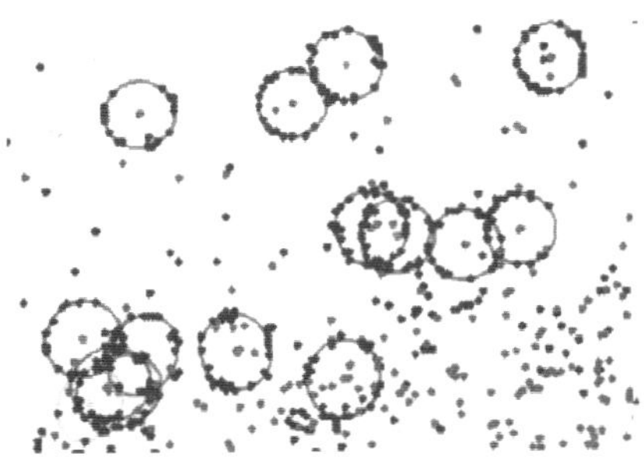

Figure 8. The result of registration of the one AuAu-collision in the RICH detector.

Further, we consider only those particles which are linked to any of the reconstructed Cherenkov ring.

9. The construction of the invariant mass spectrum

The KF Particle Finder was used to reconstruct the signal event topology [6]. This package is designed for the search and reconstruction of short-lived particles by the products of their decay. The set of J/ψ candidates are formed by combining all selected (using the criteria described

IOP Conf. Series: Journal of Physics: Conf. Series **937** (2017) 012010 doi:10.1088/1742-6596/937/1/012010

above) electrons with all selected positrons. In order to reduce the contribution of background, only particles emerged from the target are considered.

The observed number of signal events $N(s)$ may be estimated using the following formula:

$$N(s) = N(i) \cdot M \cdot BR \cdot Eff,$$

where $N(i)$ is the number of central AuAu collisions at 25 AGeV ($\sim 10^{12}$ events); M is the probability of J/ψ production in a central collision ($M = 5 \times 10^{-6}$ for the mentioned collisions) [15]; BR is the branching ratio of the J/ψ decay into dielectrons (6%) [16]; Eff is the efficiency of the signal registration by the CBM setup. This normalization has been used to construct the J/ψ distribution.

The combinatorial background was formed as a result of modeling the central UrQMD events for the mentioned above collisions. Since it is extremely time-consuming to generate 10^{12} collisions, the event-mixing technique was used to obtain the sufficient statistics. This method involves combining each background electron after applying all the criteria described above from the considered event with all positrons from other events. This approach allows to increase the statistics corresponding to the background spectrum quadratically.

The resulting invariant-mass spectrum of J/ψ mesons in AuAu collisions at 10 AGeV energy is shown in Fig. 9. The spectrum has been obtained by summing two distributions: signal and background. The reconstruction efficiency is about 11.6%. The calculated signal to the background ratio is about 0.14.

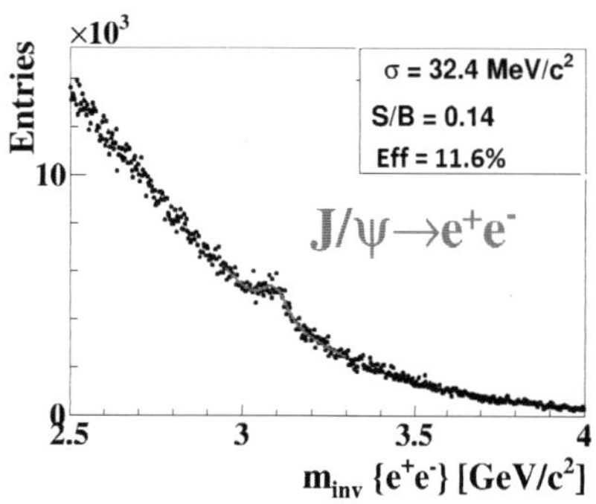

Figure 9. The reconstructed invariant-mass spectrum of J/ψ for central AuAu collisions at 10 AGeV.

10. Results

Table 1. The loss of signal events and the corresponding background suppression factor in each stage.

Criteria	Signal loss (%)	background suppression
a	16	92
b	18	26
c	28.5	99.5
d	7.5	16
e	1.1	54
f	1.6	61.5
Total	55	99.9

IOP Conf. Series: Journal of Physics: Conf. Series **937** (2017) 012010 doi:10.1088/1742-6596/937/1/012010

Table 1 shows the loss of signal events and the corresponding background suppression in each stage independently. This means that all the coefficients for each individual criterion was calculated to a sample formed as a result of applying the preceding criterion. First column are criteria: a) transverse momentum more than 1 GeV/c, b) number of counts in the TRD detector is greater than 2, c) particles are identified as electrons in the TRD, d) limit on the deviation of particle trajectory in the TRD $|d_{t_x}|$ is less than 0.035, e) particles are identified as electrons in the TOF, f) particles are linked to any of the Cherenkov rings in the RICH. The total values of the signal losses and the background suppression are in the last line.

The developed criteria lead to almost complete rejection of the background contribution, while selecting major part of the signal events.

References

[1] *The CBM Physics Book*, Lecture Notes in Physics, Vol. 814, ed B Friman and P Senger et al
[2] The CBM Collaboration 2012 *CBM Report 2012-01* GSI Darmstadt
[3] http://www-hades.gsi.de/computing/pluto/html/PlutoIndex.html
[4] Bleicher M, Zabrodin E, Spieles C et al 1999 *J.Phys.* G **25** 1859
[5] *GEANT - Detector Description and Simulation Tool* 1995 CERN Program Library, Long Write-up, **W5013**
[6] Zyzak M 2016 *Ph.D. thesis* Darmstadt
[7] Kulakov I S, Baginyan S A, Kisel P I, Ivanov V V 2013 *PIPAN, Letters* **10 No. 2** 162 170
[8] Zyzak V Ivanov V V Kisel P I Ablyazimov 2014 *Bulletin of PFUR Series Mathematics InformationSciences Physics* **2** 191 196
[9] Derenovskaya O Yu and Vassiliev I 2013 *PIPAN, Letters* **10** No 7(184) e.g. 1163 74
[10] Lebedev A et al 2010 *J. Phys.: Conf. Series* **219** 32 48
[11] Akishina T P Derenovskaya O Yu and Ivanov V V 2010 *Bulletin of PFUR Series Mathematics InformationSciences Physics* **1** 94 103
[12] Derenovskaya O Yu and Ivanov V V 2014 *PIPAN, Letters* **11** No 4(188) 862 885
[13] http://root.cern.ch/root/html/TMultiLayerPerceptron.html
[14] *Ring Imaging Cherenkov (RICH) Detector. Technical Design Report for the CBM* 2013 GSI Darmstadt
[15] Steinheimer J Botvina A and Bleicher M *arXiv:1605.03439 [nucl-th]*
[16] Nakamura K et al 2010 *J. Phys.* **G** 37

Lyapunov stability analysis for the generalized Kapitza pendulum

O V Druzhinina[1], L A Sevastianov[2], S A Vasilyev[2], D G Vasilyeva[2]
[1] Federal Research Center Computer "Science and Control" of Russian Academy of Sciences (FRC CSC RAS), 44/2 Vavilov St, Moscow, 119333, Russian Federation and V.A. Trapeznikov Institute of Control Sciences of Russian Academy of Sciences (ICS RAS), 65 Profsoyuznaya St, Moscow, 117997, Russian Federation
[2] Peoples Friendship University of Russia (RUDN University), 6 Miklukho-Maklaya St, Moscow, 117198, Russian Federation

E-mail: vasilyev_sa@rudn.university

Abstract. In this work generalization of Kapitza pendulum whose suspension point moves in the vertical and horizontal planes is made. Lyapunov stability analysis of the motion for this pendulum subjected to excitation of periodic driving forces and stochastic driving forces that act in the vertical and horizontal planes has been studied. The numerical study of the random motion for generalized Kapitza pendulum under stochastic driving forces has made. It is shown the existence of stable quasi-periodic motion for this pendulum.

1. Introduction

Kapitza pendulum is a rigid pendulum in which the pivot point vibrates in a vertical direction, up and down. The unique feature of the Kapitza pendulum is that the vibrating suspension can cause it to balance stably in an inverted position. A pendulum with vibrating point is a classical problem of perturbation theory. The phenomenon of stabilisation of the upper vertical position of the pendulum by fast vertical vibrations of the suspension point was discovered by A. Stephenson [13, 14]. P.L.Kapitsa developed a method of separation of slow and fast motions for the pendulum [5, 6]. Different aspects of this problem were discussed in many publications [1, 10, 11, 12]. In papers [3, 7, 9] the authors built various models of stochastic systems and considered their dynamics.

Without any doubt that time-varying systems, which are also referred to as non-autononuous systems and non-stationary systems, are more difficult to handle than time-invariant systems [15, 16]. For linear time-invariant system, it is known that there are only two kinds of stability concepts, namely, Lyapunov stability and asymptotic stability, which are totally determined by the eigenvalue set of the system matrix. However, for linear time-varying system, there are more stability concepts such as non-uniformly asymptotic stability, uniformly asymptotic stability, non-uniformly exponential stability and uniformly exponential stability. Moreover, differently from linear time-invariant systems, the stability of linear time-varying systems cannot be linked to the eigenvalue set of the system matrices directly [16]. Therefore, compared with time-invariant systems, study on the analysis and design of time-varying systems is very challenging and has been greatly retarded and only relatively less papers were available in the literature [2, 4, 8].

IOP Conf. Series: Journal of Physics: Conf. Series **937** (2017) 012011 doi:10.1088/1742-6596/937/1/012011

In this work generalization of Kapitza pendulum whose suspension point moves in the vertical and horizontal planes is made. Lyapunov stability analysis of the motion for this pendulum subjected to excitation of periodic driving forces and stochastic driving forces that act in the vertical and horizontal planes has been studied. The numerical study of the random motion for generalized Kapitza pendulum under stochastic driving forces has made. It is shown the existence of stable quasi-periodic motion of this pendulum.

2. Lyapunov stability analysis of the generalized Kapitza pendulum motion under periodic driving forces

2.1. Model of the generalized Kapitza pendulum under periodic driving forces

Let l and m be length of the massless rod and mass of the bob for this pendulum. Let $x(t)$ and $y(t)$ be horizontal and vertical Cartesian coordinates of the suspension point. Denote by $\theta(t)$ the angle between the rod of the pendulum and the vertical. In this case we can written the coordinates of bob in the form:

$$\begin{cases} x(t) &= l\,sin\,\theta(t) + a\,sin\,\omega_1\,t, \\ y(t) &= l\,cos\,\theta(t) + b\,sin\,\omega_2\,t, \end{cases} \tag{1}$$

where ω_1 and ω_2 are the frequencies of the harmonic vertical and horizontal forced oscillations of the suspension, a and b are amplitudes of the forced oscillations along the axes x and y. If in (1) we take the time derivative of $x(t)$ and $y(t)$ we can write

$$\begin{cases} v_x(t) &= l\,cos\,\theta(t)\,\dot{\theta}(t) + a\,\omega_1\,cos\,\omega_1\,t, \\ v_y(t) &= -l\,sin\,\theta(t)\,\dot{\theta}(t) + b\,\omega_2\,cos\,\omega_2\,t, \end{cases} \tag{2}$$

where $v_x(t) = \dot{x}(t)$, $v_y(t) = \dot{y}(t)$ are velocities along each of the axes.

Then the kinetic $K(t)$ and potential $V(t)$ energies of the bob are

$$\begin{cases} K(t) &= 0.5\,m\,\left(\dot{x}^2(t) + \dot{y}^2(t)\right), \\ V(t) &= mg\,y(t), \end{cases} \tag{3}$$

$$\begin{cases} K(t) &= 0.5\,m\left[\left(l\,cos\,\theta(t)\,\dot{\theta}(t) + a\,\omega_1\,cos\,\omega_1\,t\right)^2 + \left(-l\,sin\,\theta(t)\,\dot{\theta}(t) + b\,\omega_2\,cos\,\omega_2\,t\right)^2\right], \\ V(t) &= mg\,\left(l\,cos\,\theta(t) + b\,sin\,\omega_2\,t\right), \end{cases} \tag{4}$$

where m is mass of the bob and g is acceleration of free fall.

The total energy of the system is given by the sum of the kinetic and potential energies

$$E(t) = K(t) + V(t) = 0.5m^{-1}\left(p_x^2(t) + p_y^2(t)\right) + mg\,y(t), \tag{5}$$

where $p_x = mv_x$ and $p_y = mv_y$ are the components of momentum along the x and y axes.

Lagrangian of the system has the form:

$$L(t) = K(t) - V(t) = \frac{m}{2}\left(\dot{x}^2(t) + \dot{y}^2(t)\right) - mgl\,cos\,\theta(t). \tag{6}$$

The Euler - Lagrange equation for the phase of the pendulum as follows

$$\frac{d}{dt}\frac{\partial L(t)}{\partial \dot{\theta}} - \frac{\partial L(t)}{\partial \theta} = 0, \tag{7}$$

and the equation of the pendulum motion has the form:

$$\ddot{\theta}(t) - \frac{a\omega_1^2\,sin\,\omega_1 t}{l}cos\,\theta + \frac{b\omega_2^2\,sin\,\omega_2 t - g}{l}sin\,\theta = 0. \tag{8}$$

We can rewrite this equation in the form of the system of first order differential equations

$$\begin{cases} \dot{\theta}(t) &= \varphi(t), \\ \dot{\varphi}(t) &= l^{-1}\, a\omega_1^2\, sin\,\omega_1 t\, cos\,\theta(t) - l^{-1}\,(b\omega_2^2\, sin\,\omega_2 t - g) sin\,\theta(t)\,, \end{cases} \tag{9}$$

where φ is an auxiliary variable.

2.2. Stability analysis of motion under periodic driving forces
In this section Lyapunov stability analysis of the motion for this pendulum subjected to excitation of periodic driving forces that act in the vertical and horizontal planes has been studied.

Let θ_0 and φ_0 be parameters of the bob balance when equilibrium conditions can be obtained. Let $\Delta\theta$ and $\Delta\varphi$ be small deviations from the equilibrium state. In this case, the motion of the pendulum can be described as follows

$$\begin{cases} \theta &= \theta_0 + \Delta\theta(t), \\ \varphi &= \varphi_0 + \Delta\varphi(t). \end{cases} \tag{10}$$

Parameters θ_0 and φ_0 can be found from

$$\begin{cases} \varphi_0 = 0, \\ a\omega_1^2\, sin\,\omega_1 t\, cos\,\theta_0 - (b\omega_2^2\, sin\,\omega_2 t - g) sin\,\theta_0 = 0. \end{cases} \tag{11}$$

We can get the coordinates of the equilibrium position for the pendulum in the form:

$$\begin{cases} \varphi_0 = 0, \\ tan\,\theta_0 = a\omega_1^2\, sin\,\omega_1 t/(b\omega_2^2\, sin\,\omega_2 t - g). \end{cases} \tag{12}$$

Using the substitution variables (10) and linearization we can write the system

$$\begin{cases} \Delta\dot{\theta} = \Delta\varphi(t), \\ \Delta\dot{\varphi} = -l^{-1}\left(\sqrt{(a\omega_1^2\, sin\,\omega_1 t)^2 + (b\omega_2^2\, sin\,\omega_2 t - g)^2}\right)\Delta\theta(t)\,. \end{cases} \tag{13}$$

The characteristic polynomial of the linear system (13) has the form

$$D(p) = p^2 + l^{-1}\sqrt{(a\omega_1^2\, sin\,\omega_1 t)^2 + (b\omega_2^2\, sin\,\omega_2 t - g)^2}. \tag{14}$$

The necessary and sufficient condition for stability of generalized Kapitza pendulum motion is given by $D(p) > 0$. This condition of stability takes place for any $\omega_1 0$, ω_2 and $t \geq 0$.

3. Lyapunov stability analysis of the generalized Kapitza pendulum motion under stochastic driving forces

3.1. Model of the generalized Kapitza pendulum under stochastic driving forces
Let l and m be length of the massless rod and mass of the bob for this pendulum. Let $x(t)$ and $y(t)$ be horizontal and vertical Cartesian coordinates of the suspension point. Denote by $\theta(t)$ the angle between the rod of the pendulum and the vertical. In this case we can written the coordinates of bob in the form:

$$\begin{cases} x(t) &= l\, sin\,\theta(t) + a(t;\omega), \\ y(t) &= l\, cos\,\theta(t) + b(t;\omega), \end{cases} \tag{15}$$

IOP Conf. Series: Journal of Physics: Conf. Series **937** (2017) 012011 doi:10.1088/1742-6596/937/1/012011

where $a(t;\omega)$ and $b(t;\omega)$ are random functions that associated with acting upon the pendulum of stochastic forces that drive in the vertical and horizontal planes oscillations along x, y axes and $(t;\omega) \in \mathcal{T} \times \Omega$ $((\Omega; \boldsymbol{F}; \mathcal{P})$ is an abstract probability space.

Then the kinetic and potential energies of the bob are

$$\begin{cases} K(t) &= \frac{m}{2}\left[\left(l\,cos\,\theta(t)\,\dot{\theta}(t) + \dot{a}(t;\omega)\right)^2 + \left(-l\,sin\,\theta(t)\,\dot{\theta}(t) + \dot{b}(t;\omega)\right)^2\right], \\ V(t) &= mg\,(l\,cos\,\theta(t) + b(t;\omega))\,. \end{cases} \tag{16}$$

The total energy of the system is given by the sum of the kinetic and potential energies

$$E(t) = K(t) + V(t) = \frac{m}{2}\left(\dot{x}^2(t) + \dot{y}^2(t)\right) + mg\,(l\,cos\,\theta(t) + b(t;\omega))\,, \tag{17}$$

and then Lagrangian of the system has the form:

$$L(t) = K(t) - V(t) = \frac{m}{2}\left(\dot{x}^2(t) + \dot{y}^2(t)\right) - mg\,(l\,cos\,\theta(t) + b(t;\omega))\,. \tag{18}$$

The equation of the pendulum motion has the form:

$$\ddot{\theta}(t) + \frac{\ddot{a}(t;\omega)}{l} cos\,\theta\,(t) - \frac{\ddot{b}(t;\omega) + g}{l} sin\,\theta(t) = 0. \tag{19}$$

We can rewrite this equation in the form of the system differential equations first order

$$\begin{cases} \dot{\theta}(t) &= \varphi(t), \\ \dot{\varphi}(t) &= -l^{-1}\,\ddot{a}(t;\omega)\,cos\,\theta(t) + l^{-1}\,(\ddot{b}(t;\omega) + g) sin\,\theta(t)\,, \end{cases} \tag{20}$$

where φ is an auxiliary variable.

4. Stability analysis of motion under stochastic driving forces

In this section using Lyapunov stability analysis of the motion for this pendulum subjected to excitation of stochastic driving forces that random act in the vertical and horizontal planes has been studied.

Let θ_0 and φ_0 be parameters of the bob balance when equilibrium conditions can be obtained. Let $\Delta\theta$ and $\Delta\varphi$ be small deviations from the equilibrium state. In this case, the motion of the pendulum can be described as follows

$$\begin{cases} \theta &= \theta_0 + \Delta\theta(t), \\ \varphi &= \varphi_0 + \Delta\varphi(t). \end{cases} \tag{21}$$

Parameters θ_0 and φ_0 can be found from

$$\begin{cases} \varphi_0 = 0, \\ \ddot{a}(t;\omega)\,cos\,\theta_0 - (\ddot{b}(t;\omega) + g) sin\,\theta_0 = 0. \end{cases} \tag{22}$$

If the mathematical expectation $M_a = M\left[\ddot{a}(t)\right]$ and $M_b = M\left[\ddot{b}(t)\right]$ exist we can get the coordinates of the equilibrium position for the pendulum

$$\begin{cases} \varphi_0 = 0, \\ tan\,\theta_0 = M_a / \left(M_b + g\right)\,. \end{cases} \tag{23}$$

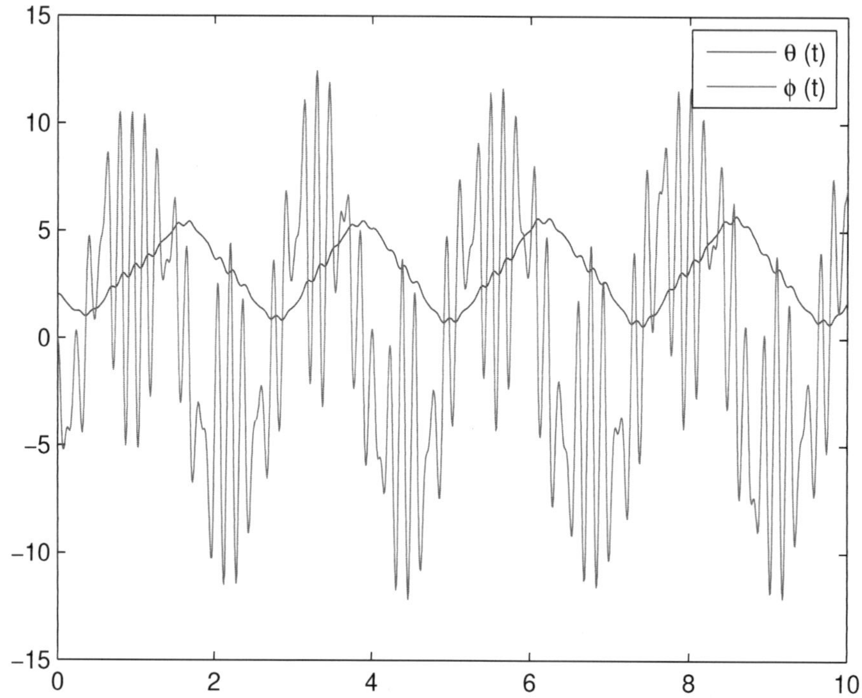

Figure 1. The motion of generalized Kapitza pendulum.

Using the substitution variables (21) and linearization we can write the system

$$\begin{cases} \Delta\dot{\theta} = \Delta\varphi(t), \\ \Delta\dot{\varphi} = -l^{-1}\left(\sqrt{M_a^2 + (M_b + g)^2}\right)\Delta\theta(t)\,. \end{cases} \tag{24}$$

The characteristic polynomial of the linear system (24) has the form

$$D(p) = p^2 + l^{-1}\sqrt{M_a^2 + (M_b + g)^2}. \tag{25}$$

The necessary and sufficient condition for average stability of generalized Kapitza pendulum motion is given by $D(p) > 0$. This condition of stability takes place for any M_a, M_b and $t \geq 0$.

5. Numerical study of the generalized Kapitza pendulum motion

In this section the numerical study of the random motion for generalized Kapitza pendulum under stochastic driving forces is presented. Suppose that random functions $a(t)$ and $b(t)$ can be represented in the form $a(t) = A(t;\omega)\, sin\, \nu_1 t$, $b(t) = B(t;\omega)\, sin\, \nu_2 t$ where $A(t;\omega)$ and $B(t;\omega)$ are random functions of amplitudes, ν_1 and ν_2 are the frequencies of the harmonic vertical and horizontal forced oscillations of the pendulum suspension. We can rewrite the system differential equations (20) in the form

$$\begin{cases} \dot{\theta} = \varphi(t), \\ \dot{\varphi} = -l^{-1}\left(\ddot{A}(t)\, sin\, \nu_1 t + 2\dot{A}(t)\, \nu_1\, cos\, \nu_1 t - A(t)\, \nu_1^2\, sin\, \nu_1 t\right) cos\, \theta(t) + \\ +l^{-1}\left(\ddot{B}(t)\, sin\, \nu_2 t + 2\dot{B}(t)\, \nu_2\, cos\, \nu_2 t - B(t)\nu_2^2\, sin\, \nu_2 t + g\right) sin\, \theta(t)\,, \end{cases} \quad (26)$$

where $\dot{A}, \dot{B}$ are random velocities and $\ddot{A}, \ddot{B}$ are random accelerations of the pendulum suspension long each of the axes.

The numerical example is presented in the figure (see Fig. 1) where $l = 0.5\, m.$ is length of the massless rod, $\nu_1 = 40\, sec.^{-1}$, $\nu_2 = 10\, sec.^{-1}$ are the frequencies of the harmonic vertical and horizontal forced oscillations of the pendulum suspension, $g = 10\, m./sec.^2$ is acceleration of free fall. The random functions of amplitudes $A(t)$ and $B(t)$ were taken in the form

$$\begin{cases} A(t) & = & a_0 + \epsilon_t, \\ B(t) & = & b_0 + \eta_t\,, \end{cases} \quad (27)$$

where parameters $a_0 = 0.1$, $b_0 = 0.05$, ϵ_t $N(0, \sigma_x^2)$ and η_t $N(0, \sigma_y^2)$ are independent normally distributed random variables (in the numerical example $\sigma_x = 0.01$ and $\sigma_y = 0.001$). In this numerical example it is shown the existence of stable quasi-periodic motion of this pendulum.

6. Conclusions

In this work generalization of Kapitza pendulum whose suspension point moving in the vertical and horizontal planes is made. Lyapunov stability analysis of the motion for this pendulum subjected to excitation of periodic driving forces and stochastic driving forces that random act in the vertical and horizontal planes has been studied. The numerical study of the random motion for generalized Kapitza pendulum under stochastic driving forces has made. It is shown the existence of stable quasi-periodic motion of this pendulum. The vertical and horizontal motion simulation of the pendulum can be used for analysis of buildings and structure stability during earthquake.

7. Acknowledgments

The publication was prepared with the support of the RUDN University Program 5-100 and partially funded by RFBF grants No 15-07-08795, No 16-07-00556.

References

[1] Bardin B S and Markeyev A P 1995 *On the stability of equilibrium of a pendulum with vertical oscillations of its suspension point* (J. Appl. Math. Mech. vol 59) pp 879–886
[2] Du H, Qian C, Yang S and Li S 2013 *Recursive design of finite-time convergent observers for a class of time-varying nonlinear systems* (Automatica vol 49, No 2) pp 601-609
[3] Gaidamaka Y, Sopin E and Talanova M 2016 *Approach to the analysis of probability measures of cloud computing systems with dynamic scaling* (Communications in Computer and Information Science vol 601) pp 121–131
[4] Jiang C, Teo K L and Duan G 2012 *A suboptimal feedback control for nonlinear time-varying systems with continuous inequality constraints* (Automatica vol 48, No 4) pp 660–665
[5] Kapitsa P L 1951 *Dynamic stability of a pendulum with oscillating point of suspension* (Sov. Phys. JETP (in Russian) vol 21) pp 588–597
[6] Kapitsa P L 1951 *A pendulum with vibrating point of suspension* Usp. Phys. Nauk (in Russian) vol 44) pp 7–20
[7] Korolkova A V, Eferina E G, Laneev E B, Gudkova I A, Sevastianov L A and Kulyabov D S 2016 *Stochastization of one-step processes in the occupations number representation* (Proceedings - 30th European Conference on Modelling and Simulation, ECMS) pp 698 — 704
[8] Mazenc F, Andrieu V and Malisoff M 2015 *Design of continuousdiscrete observers for time-varying non-linear systems* (Automatica vol 57) pp 135–144

[9] Samouylov K, Naumov V, Sopin E, Gudkova I and Shorgin S 2016 *Sojourn time analysis for processor sharing loss system with unreliable server* (Lecture Notes in Computer Science, including subseries Lecture Notes in Artificial Intelligence and Lecture Notes in Bioinformatics vol 9845) pp 284—297
[10] Levi M 1988 *Geometry of Kapitsa's potential* (Nonlinearity vol 11) pp 1365–1368
[11] Neishtadt A I and Shenga K 2016 *Bifurcations of phase portraits of pendulum with vibrating suspension point* https://arxiv.org/abs/1605.09448
[12] Ovseyevich A I 2006 *The stability of an inverted pendulum when there are rapid random oscillations of the suspension point* (J. Appl. Math. Mech. vol 70) pp 761–768
[13] Stephenson A 1908 *On induced stability* (Philosophical Magazine Series vol 6 No 15) pp 233–236
[14] Stephenson A 1908 *On a new type of dynamical stability* (Mem. Proc. Manch. Lit. Phil. Soc. vol 52 No 8) pp 1–10
[15] Zhang X, Liu L and Feng G 2015 *Leaderfollower consensus of time-varying nonlinear multi-agent systems* (Automatica vol 52) pp 8–14
[16] Zhou B, Cai G, and Duan G 2013 *Stabilization of time-varying linear systems via Lyapunov differential equations* (International Journal of Control vol 86, No 2) pp 332–347
[17] Zhou B 2015 *Global stabilization of periodic linear systems by bounded controls with applications to space-craft magnetic attitude control* (Automatica vol 60) pp 145–154

Spontaneous evaporation of the acetone drop

S Z Dunin and O V Nagornov

National Research Nuclear University MEPhI (Moscow Engineering Physics Institute), 31 Kashirskoe shosse, 115409 Moscow, Russia

E-mail: nagornov@yandex.ru

Abstract. The exact solution of the problem on evaporation of the acetone sessile drop at solid substrate is derived. The solution allows us to analyze the temperature field behavior at various thermodynamic parameters. The Marangoni forces are calculated in explicit form. The solution can be used to check numerical approaches for similar problems.

1. Introduction

The dissolved particles in liquid drop placed at the solid substrate form various structures at the substrate surface after evaporation of the drop. The prediction and control of these structure are important problems in many technological processes [1, 2, 3]. Analysis of the drop evaporation have been done in many papers. However, there are unsolved problem connected with behavior of the temperature near the contact line of the drop in framework of the isothermal model resulting in analytical solution [4]. The numerical approaches using the fitting parameters for the evaporative flux allowed for conclusion to take into account non-isothermal temperature field at the drop surface [5]. The experimental studies [6, 7] pointed out strong influence of thermodynamic parameters of liquid and substrate on the evaporation process. Non-isothermal temperature along the drop surface reduce to inhomogeneous gradient of the surface tension and liquid motion inside drop [8, 9].

Complexity and interdependence of the physical processes inside drop and their mathematical descriptions, phase transitioning, energy and mass balance, points of singularity and thermocappilar effects force researchers to simplify their models and use numerical methods. In several researches finite elements method (FEM) is used to define boundary conditions. In this approach, due to non-uniform temperature on drop surface the Marangoni force arises.

In this paper we derive analytical solution for the temperature and concentration fields, and derive the exact solutions for specific parameters of the system. The analysis is based on our previous approach [10].

2. Mathematical model

We consider a drop of partially wetting liquid resting on flat solid layer. We assume drop with pinned contact line, so during evaporation process the contact angle decreases to keep the same wetted area.The initial parameters of the drop are radius R, height h_0 and contact angle θ_c, respectively. In small drops due to Bond number $Bo = \rho g R h/\sigma = 0.03 \div 0.4$ the surface tension effects dominate gravitational effects and drops can be considered as spherical cap shape. We don't take into account the last phase of evaporation, where wetted area rapidly decreases. It is a subject for further investigations. We also consider the Peclet number to be small $Pe << 1$, so the thermal convection is negligible and the thermal diffusivity time to be much less than life time of the drop $t << R^2 k_{L,s}^{-1} = t_{L,s}$, where R is radius of the drop, k_s and k_L are thermal conductivities of substrate and liquid, respectively. So, temperature field in substrate and liquid satisfies Laplace's equation. In gas phase we consider no thermal convection (i.e. $Pe << 1$), so concentration field also satisfies Laplace's equation.

Thermal properties of liquid and substrate are taken into account by boundary conditions. We assume the heat flux from substrate is the main reason of evaporation. The heat inside the drop reaches its surface, where the phase transition occurs. Such problem formulation is closer to problems of the current static [6]. In this geometry the problem can be solved in toroidal coordinates (α, θ), where $\theta = 0$ and $\theta = \theta_c$ represent the drop base and surface respectively. We consider no normal mass and thermal fluxes through the substrate-gas boundary, and the temperature and thermal flux continuity on the liquid-substrate boundary. Thus, the system of governing equations can be written in the following form:

$$\Delta T_L = 0, \quad \Delta T_s = 0, \Delta c = 0, \tag{1}$$

$$c(0,\pi) = c_\infty, \quad T_s(0,-\pi) = T_\infty,$$

$$T_L(\alpha,0) = T_s(\alpha,0), \nabla_n T_L(\alpha,0) = k_R \nabla_n T_s(\alpha,0), \;\; \nabla_n T_s(\alpha,\pi) = 0, \tag{2}$$

$$c(\alpha,\theta_c) - c_{s\infty} = c'_{sT}[T_L(\alpha,\theta_c) - T_\infty], \quad \Gamma \nabla_n T_L(\alpha,\theta_c) = \nabla_n T_s(\alpha,\theta_c) \tag{3}$$

$$\Gamma = \frac{c'_{sT} LD}{k_L}$$

where T_L and T_s are temperature in liquid and substrate, respectively, $k_R = k_s / k_L$, k_s and k_L are the heat conductivity of substrate and liquid, respectively, T_∞ and c_∞ are ambient temperature and vapor concentration, respectively, and θ_c is the contact angle.

3. The problem solution

The general solution of Laplace's equation for temperature and concentration can be expressed as:

$$T_L(\alpha,\theta) - T_\infty = \{2(ch\alpha + \cos\theta)\}^{\frac{1}{2}} \int_0^\infty d\tau \, P_{-\frac{1}{2}+i\tau}(ch\alpha) A_s \frac{ch\tau\theta_c}{ch\tau\pi} \frac{ch\tau\pi ch\tau\theta + k_R sh\tau\pi sh\tau\theta}{ch\tau\pi ch\tau\theta_c + k_R sh\tau\pi sh\tau\theta_c} \; 0 \le \theta \le \theta_c$$

$$T_s(\alpha,\theta) - T_\infty = \{2(ch\alpha + \cos\theta)\}^{\frac{1}{2}} \int_0^\infty d\tau \, P_{-\frac{1}{2}+i\tau}(ch\alpha) A_s \frac{ch\tau\theta_c}{ch\tau\pi} \frac{ch\tau(\pi+\theta)}{ch\tau\pi ch\tau\theta_c + k_R sh\tau\pi sh\tau\theta_s} \; -\pi \le \theta \le 0$$

$$c(\alpha,\theta) - c_\infty = \{2(ch\alpha + \cos\theta)\}^{\frac{1}{2}} \int_0^\infty d\tau \, P_{-\frac{1}{2}+i\tau}(ch\alpha) B_D \frac{ch\tau\theta_c}{ch\tau\pi} \frac{ch\tau(\pi-\theta)}{ch\tau(\pi-\theta_c)} \theta_c \le \theta \le \pi, \tag{4}$$

$$B_D(\tau) - \Delta c = c'_{sT} A_s,$$

where A_s, B_D are nuclei of the integral equations that can be determined by the boundary conditions, $P_{-1/2+i\tau}(ch\alpha)$ is the Legendre function of the first kind. Subsituting temperatures and concentration (eqs. 4) in the boundary conditions (eq. 3), one can obtain:

$$\frac{\sin\theta_c}{2}\{(1-\Gamma)[c(\alpha,\theta_c) - c_{s\infty}] - \Gamma]\} = \sqrt{2}(ch\alpha + \cos\theta_c)^{\frac{3}{2}} \cdot$$

$$\cdot \int_0^\infty d\tau \, P_{-\frac{1}{2}+i\tau}(ch\alpha) \frac{ch\tau\theta_c}{ch\tau\pi} \tau[(B_D(\tau) - \Delta c) \frac{ch\tau\pi sh\tau\theta + k_R sh\tau\pi ch\tau\theta}{ch\tau\pi ch\tau\theta_c + k_R sh\tau\pi sh\tau\theta_c} + \Gamma \cdot B_D(\tau) \frac{sh\tau(\pi-\theta_c)}{ch\tau(\pi-\theta_c)}] \tag{5}$$

At $\Gamma = 0$ the temperature at the surface drop is constant. It is so-called the basic model at $B_D = \Delta c = c_{s\infty} - c_\infty$ and $c(\alpha, \theta_c) = c_{s\infty}$.

At $\Gamma = 1$ the exact solution can be found out for eq. (5). It takes place for materials similar to acetone ($\Gamma = 1.02$). For this configuration the heat flux goes from the substrate to liquid drop. In this case the nuclei are determined by equations:

$$B_D(\tau) = \Delta c \frac{k_R ch\tau(\pi - \theta_c)}{(1+k_R)ch\tau\pi ch\tau\theta_c}, \quad B_D(\tau) - \Delta c = c'_{sT} A_s \tag{6}$$

The temperature and concentration can be written as follows:

$$T_L(\alpha,\theta) - T_\infty = -\frac{1}{1+k_R}\frac{\Delta c}{c'_{sT}}\{2(ch\alpha + \cos\theta)\}^{\frac{1}{2}}\int_0^\infty d\tau\, P_{-\frac{1}{2}+i\tau}(ch\alpha)\frac{ch\tau\pi ch\tau\theta + k_R sh\tau\pi sh\tau\theta}{ch^2\tau\pi}$$

$$c(\alpha,\theta) - c_\infty = \Delta c\frac{k_R}{(1+k_R)}\{2(ch\alpha + \cos\theta)\}^{\frac{1}{2}}\int_0^\infty d\tau\, P_{-\frac{1}{2}+i\tau}(ch\alpha)\frac{ch\tau(\pi - \theta)}{ch^2\tau\pi} \tag{7}$$

4. Results

The solutions depend on k_R. For the substrates of high thermal conductivity (great values of k_R) the concentration coincides with one for the basic model. The concentration depends strongly on the material properties of substrate in case of low values of the thermal conductivity.

The case of $\Gamma = 1$ corresponds to the acetone drop. This case can take place at specific pressure and type of surrounding gas.

The solution can presented in the following form:

$$T_L(\alpha,\theta) - T_\infty = -\frac{1}{1+k_R}\frac{\Delta c}{c'_{sT}}[1 + k_R - \frac{2}{\pi}k_R arctg\frac{\sqrt{2}\sin\frac{\theta}{2}}{(ch\alpha + \cos\theta)\}^{\frac{1}{2}}}], 0 \le \theta \le \theta$$

$$c(\alpha,\theta) - c_\infty = \Delta c\frac{k_R}{(1+k_R)}[1 - \frac{2}{\pi}arctg\frac{\sqrt{2}\sin\frac{\theta}{2}}{(ch\alpha + \cos\theta)\}^{\frac{1}{2}}}] \quad \theta_c \le \theta \le \pi \tag{8}$$

The temperaturte changes monotonically along the drop surface. The maximum temperature is achieved at the edge drop.

The Marangoni force can be calculated from eq. (8):

$$F_M(\alpha,\theta_c) = \sigma\frac{ch\alpha + \cos\theta_c}{R}\frac{dT}{d\alpha} = \frac{sh\alpha}{1+k_R}\frac{\sin\frac{\theta_c}{2}}{\sqrt{2}\pi}\sigma\frac{\sqrt{ch\alpha + \cos\theta_c}}{R(ch\alpha + 1)}k_R\frac{\Delta c}{c'_{sT}}.$$

Next study will take into account the motion in liquid drop.

References

[1] Jing J P et al. 1998 Automated high resolution optical mapping using arrayed, fluid fixed DNA molecules *Proc. Natl. Acad. Sci.USA* **95** 8046–8051

[2] Sekitani T, Noguchi Y, Zschieschang U, Klauk H and Someya T 2008 Organic transistors manufactured using inkjet technology with subfemtoliter accuracy *Proc. Natl. Acad. Sci. USA* **105** 4976–4980

[3] Sirringhaus H, Kawase T, Friend R H, Shimoda T, Inbasekaran M, Wu W, Woo E P 2000 High-resolution inkjet printing of all-polymer transistor circuits *Science* **290** 2123–2126

[4] Deegan R D, Bakajin O, Dupont T F, Huber G, Nagel S R and Witten T A 2000 Contact line deposits in an evaporating drop *Phys. Rev. E* **62** 756–765

[5] Larson R G 2014 Transport and deposition patterns in drying sessile droplets *AIChE J.* **60** 1538–1571

[6] David S, Sefiane K and Tadrist L 2007 Experimental investigation of the effect of thermal properties of the substrate in the wetting and evaporation of sessile drops *Colloids and Surfaces A: Physicochem. Eng. Aspects* **298** 108–114
[7] Dunn G J, Wilson S K, Duffy B R and Sefiane K 2009 Evaporation of a thin droplet on a thin substrate with a high thermal resistance *Phys. Fluids* **21** 052101
[8] Zhang K, Ma L, Xu X, Luo J, Guo D 2014 Temperature distribution along the surface of evaporating droplets *Phys. Rev. E* **89** 032404
[9] Barash L Yu 2016 Marangoni convection in an evaporating droplet: Analytical and numerical descriptions *International Journal of Heat and Mass Transfer* **102** 445–454
[10] Dunin S Z, Nagornov O V and Trifonenkov V P 2017 Model of spontaneous evaporating droplet on solid horizontal substrate *Journal of Physics: Conference Series* **788** 012048

WEB-server for search of a periodicity in amino acid and nucleotide sequences

F E Frenkel[1], K G Skryabin[1] and E V Korotkov[1,2]

[1]Institute of Bioengineering, Research Center of Biotechnology of the Russian Academy of Sciences, Leninsky Ave. 33, bld. 2, 119071, Moscow, Russia.
[2]National Research Nuclear University "MEPhI", Kashirskoe shosse, 31. Moscow 115409, Russia.

genekorotkov@gmail.com

Abstract. A new web server (http://victoria.biengi.ac.ru/splinter/login.php) was designed and developed to search for periodicity in nucleotide and amino acid sequences. The web server operation is based upon a new mathematical method of searching for multiple alignments, which is founded on the position weight matrices optimization, as well as on implementation of the two-dimensional dynamic programming. This approach allows the construction of multiple alignments of the indistinctly similar amino acid and nucleotide sequences that accumulated more than 1.5 substitutions per a single amino acid or a nucleotide without performing the sequences paired comparisons. The article examines the principles of the web server operation and two examples of studying amino acid and nucleotide sequences, as well as information that could be obtained using the web server.

1. Introduction

At present, extensive sequencing of diverse genomes and determination of amino acid sequences for a variety of proteins are carried out [1]. All this leads to the fact that development of a complete mathematical model for regulating the cell genetic activity is becoming more and more realistic. Creation of such a model in some sense transfers a significant portion of bioengineering into the application area, since it will not be required to carry out molecular and biological experiments in order to create new species of both plants and animals. All this could be performed using mathematical simulation accompanied by subsequent implementation of the results of such simulation in the form of new genomes. However, there are significant difficulties along this path. The point is that at first it is required to find all the genes in the eukaryotes genomes, determine (annotate) the biological function thereof and distinguish the entire regulatory elements (promoters, enhancers and binding sites for various proteins), as well as to classify the regulatory elements. A purely experimental solution to this problem could be possible, but it would take relatively much time and would be rather expensive. Therefore, mathematical and algorithmic methods are used to solve this problem. The purpose of these operations is aimed at finding all regulatory sequences, determining the functions thereof and annotating all the genes. However, mathematical methods developed up till now are far from solving these problems. We are able to reliably annotate less than half of all the known genes; and, as far as the regulatory sequences are concerned, situation in this sense is even less optimistic. In order to search and annotate various biological sequences, position weight matrices (PWM) or Hidden Markov Models (HMM) are commonly used. Creation of PWM and HMM for the increasingly accurate identification of regulatory sequences and protein annotations rests on the problem of constructing a multiple alignment, since PWM and HMM are usually constructed using the multiple alignment of amino acid or nucleotide sequences. This determined the increasing interest of researchers to the problem of finding a multiple alignment.

The search for tandem repeats in the DNA amino acid and nucleotide sequences appears to be also one of the applications of the multiple alignment methods. Only in this case, sequences for multiple alignments are arranged in tandem in a singular amino acid or nucleotide sequence. A large number of mathematical algorithms and methods for finding tandem repeats using the multiple alignments were developed [2,3]. The problem, however, is to learn how to construct multiple alignments for less and less similar to each other sequences (for example, with the number of random replacements of more than 1.5 per a singular amino acid or nucleotide). In this case, it is possible to find more and more diffused tandem repeats, where such repeats could not be identified at present time [4]. As of today, the developed multiple alignment methods used to search for tandem repeats employ sequences pair comparison to construct multiple alignments [2]. This is connected to the fact that the solution of the problem of creating multiple alignments (number of sequences counting to several dozen and having the length of more than 100 amino acids or nucleotides) using the dynamic programming methods appears to be an NP complete problem and requires computer resources for its solution unattainable at the present moment. The use of paired alignment or the use of seeds for multiple alignments or similarities between words (k-tube) does not allow finding the multiple alignment the poorly similar amino acid or nucleotide sequences. For multiple alignments of the poorly similar sequences, a situation is quite realistic that the alignment could be statistically significant only for a large multiplicity of sequences (for example, more than 20 sequences) and appears to be insignificant for any pair of sequences, which makes it impossible to construct a guide tree. In this case, all the developed methods of searching and constructing multiple alignments would not be able to create it for such sequences. Therefore, the task of creating highly sensitive and highly specific PWM and HMM using the modern approaches could not be solved.

2. Mathematical method

Previously, we proposed an approach for multiple alignment and corresponding PWM, which was devoid of the aforementioned shortcomings [5,6]. This method was implemented to search for tandem repeats. Briefly, the essence of this approach is as follows. We are looking for tandem repeats in the *S* sequence; and for this purpose we first generate a multiplicity of random matrices of the $k \times n$ size, where n is the length of a period, and k is the size of the alphabet for the *S* sequence. Afterwards, these matrices are used to construct a local alignment of the studied *S* sequence with respect to each of the generated random matrices [6]. Before starting the alignment construction, the matrices were transformed in such a way that the distribution of the *F* similarity function for random sequences was the same for all the random matrices used [6]. Then, a random matrix was selected that possessed a local alignment characterized by the highest value of the *F* similarity function with the studied *S* sequence. Then this random matrix was optimized in order to achieve the maximum value of the *max F* similarity function; and we obtain the optimized *M* matrix. To ensure this optimization, a genetic algorithm [6] or a special optimization procedure is used [5]. After this, using the random mixing of the *S* sequence a multiplicity of random sequences was created. For each sequence out of this multiplicity, the value of *maxF* was calculated, which allowed us to determine for *maxF* the $\overline{maxF}$ average value and the *D(maxF)* dispersion, as well as the $Z = \overline{maxF} / \sqrt{D(maxF)}$. The algorithm was applied for the periods of the *n* length from 2 to 100; and for each length of a period we calculated the corresponding *Z* value. As a result of applying the algorithm, we receive the dependence of *Z* on *n*, which we will denote as *Z(n)*. It should be noted that in this study we used the method of dynamic programming, which allowed us to find local alignments. This means that the boundaries of a region, where *maxF* was obtained, could differ from the beginning and end of the sequence studied. This also means that the *Z(n)*values for different *n* could be obtained for different fragments of the sequence studied. Therefore, we would separately provide the boundaries of fragments, for which significant *Z(n)*values were obtained.

3. Examples of sequence analysis using Web-server

The algorithm was implemented in the form of software on the http://victoria.biengi.ac.ru/splinter/login.php web site. Let us consider several examples of the amino acid and nucleotide sequences analysis on this web server. The first sequence encodes the CEP250 centrosome-associated protein (Q9BV73 in Swiss-prot). For this sequence, the server created the *Z(n)* spectrum, which is shown in Fig. This Figure demonstrates that the periodicity possesses a period equal to 7 amino acids in the sequence; and, what is more, it was identified with insertions and deletions of amino acids. Using all the previously developed algorithms and software programs (such as REP([7], PTRStalker [8], TREKS [9] and many others) it turned to be impossible to identify this period. Table 1 shows the alignment fragment; and the optimized weight matrix created by the server is presented in Table 2. It could be seen from Table 1 that a period of 7 amino acids in length could only be obtained with insertions and deletions of symbols.

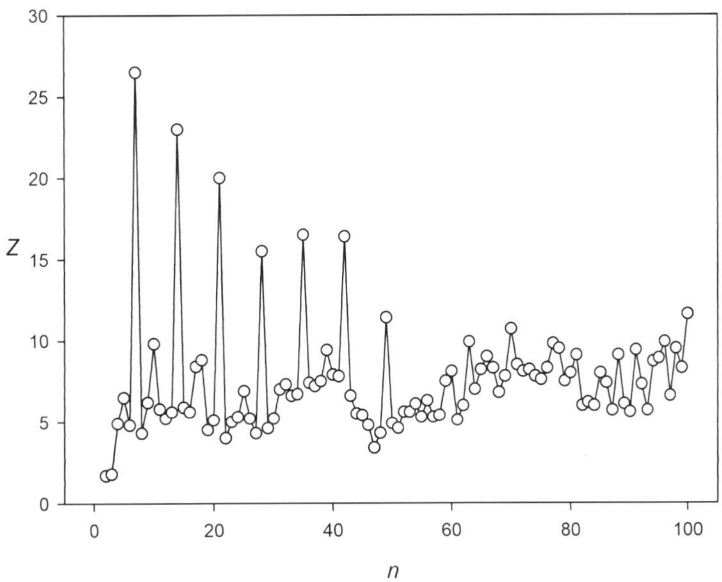

Figure 1.Z(n) dependence for the CEP250 centrosome-associated protein sequence (Q9BV73 in Swiss-prot).

Table 2 demonstrates that for each position of the period there appears a certain multiplicity of amino acids, which possesses some positive weight and a certain number of amino acids having the negative weight. This suggests that an extremely indistinct periodicity with the length of 7 symbols, which could be associated with formation of the α helices in the given protein, was identified. The presence of a period with the length of 7 amino acids in the α helical structures was demonstrated previously [10].

Table 1. Alignment fragment for *n*=7 in the Q9BV73 sequence. Entire alignment could be obtained by analyzing the Q9BV73 sequence at the http://victoria.biengi.ac.ru/splinter/login.php web site.

```
6712345671234567123456712345671234567123456712345*671234567123456712*345671234567
LEEQVLALQQQMAENQAASWRKLKNSQEAQQRQATLVRKLQAKVLQYRSWCQELEKRLEATGGPIPQRWENVEEPNLDELLVRLEEEQ

1234567123456712345671234567123456712345671234567123*45671234567123456712345671234567123
QRCESLAEVNTQLRLHMEKADVVNKALREDVEKLTVDWSRARDELMRKESQWQMEQEFFKGYLKGEHGRLLSLWREVVTFRRHFLEMK

456712345671234567123456712345671234567123456712345**6712345671234567123456712345671234 5
SATDRDLMELKAEHVRLSGSLLTCCLRLTVGAQSREPNGSGRMDGREPAQLLLLLAKTQELEKEAHERSQELIQLKSQGDLEKAELQD

67123456712345671234567123456712345671234567123456*7123456712345671234567123456712345671
RVTELSALLTQSQKQNEDYEKMIKALRETVEILETNHTELMEHEASLSRNAQEEKLSLQQVIKDITQVMVEEGDNIAQGSGHENSLEL

23456*712345671234567123456712345671
DSSIFSQFDYQDADKALTLVRSVLTRRRQAVQDLRQ
```

Table 2. Optimized *M* position-specific matrix, which was obtained by aligning the Q9BV73 sequence at the http://victoria.biengi.ac.ru/splinter/login.php web site.

	1	2	3	4	5	6	7
K	1.9	-0.4	2.0	-0.4	0.3	-1.8	-0.4
N	-2.1	-1.4	1.8	1.1	-1.4	1.1	1.1
I	-2.3	0.1	2.3	-1.7	-1.7	0.7	-0.5
M	-0.8	-1.4	2.0	2.3	-2.7	-0.8	-3.4
T	-0.5	-0.0	-2.5	0.8	1.2	0.4	1.7
R	-1.1	0.3	-0.8	0.3	0.0	-2.2	2.7
S	2.0	0.9	-1.4	1.6	2.1	-3.3	-1.0
L	-5.5	-8.0	3.6	-6.5	-4.4	3.8	-6.7
Y	-1.6	-1.6	1.8	-1.6	-0.2	1.8	-0.2
F	0.7	0.7	-0.5	-0.5	-0.5	0.7	-0.5
C	-1.1	-1.1	-1.1	-1.1	0.9	1.8	0.9
W	-1.4	-1.4	1.9	1.4	-2.3	1.7	-1.4
P	-0.6	1.2	-1.5	-0.6	2.2	-2.4	-0.6
H	-1.3	-0.2	-1.8	-1.3	1.7	-0.7	2.1
Q	2.7	2.6	-6.4	2.3	-2.3	-3.6	3.0
V	-2.0	-0.8	2.5	-0.0	-0.8	-0.4	-1.2
A	2.3	1.8	-1.9	1.0	2.3	0.2	-4.2
D	-1.5	2.4	-4.1	-0.2	2.3	-3.2	0.2
E	2.4	3.1	-8.3	2.9	2.7	-7.9	2.7
G	2.5	-2.8	1.6	-1.0	-0.1	-1.9	-1.5

IOP Conf. Series: Journal of Physics: Conf. Series **937** (2017) 012013 doi:10.1088/1742-6596/937/1/012013

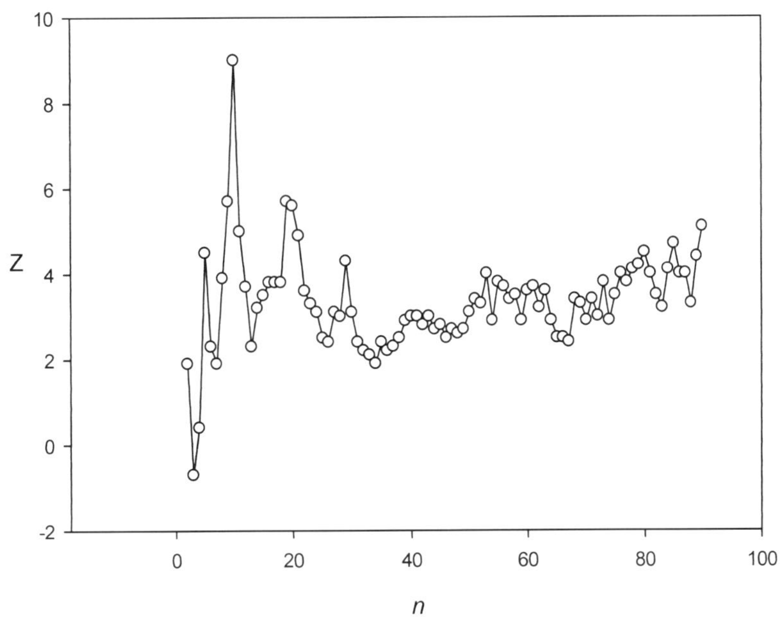

Figure 2. *Z(n)* dependence for nucleotide sequence taken from the chromosome 1 A.thaliana genome (14259442-14259804 b.p.) [11].

Table 3. Nucleotide sequence taken from the chromosome 1 A.thaliana genome (14259442-14259804 b.p.) alignment for n=10.

```
56789a12**3456789a123456789a123456789a123456789a123456789a1*
TTCGGTTAACCAATCGGTTA***TCGGTTCTATTCG***ATTTT*GGTTATTTTGGGTTT

**23456789a123456*789a123456789a123456789a123456789a1**23456
AGATTTTTATATCCATTTCGGTTCGGTTTGGTCCGGTTCCGGTCTTTCGGTTTTAAAATT

789a123456789a123****456789a123456789a123456789a123456789a12
AGTATCCATTCGGGTATTATGCTTCGGGTATTTTTGGATAT*TTCGGTTATTTTTGAGTA

3456789a123456789a123456789a123456789a123456789a123456789a12
TTTTCAGATATTTTTGGATATTTTCTGTTATTTTCGAGTAGTTTTAGTATTTTATATATA

3456789a123456789a1234567*89a12345678*9a123456789a123456789a
TATTCAAGTATTTTGGATTATTTGGTCTATAAATAGGTAATATTTTAAAATATATAATTG

123456789a123456789a123456789a123456789a123456789a123456789a
TAT*TTCGGATATTTTCAAGTAC*TCAAAATATTTCGGATCAGTATCGATTCGTTTTTGT

123456789a12
TTTTTTTGGTTA
```

IOP Conf. Series: Journal of Physics: Conf. Series **937** (2017) 012013 doi:10.1088/1742-6596/937/1/012013

Table 4. Optimized *M* position-specific matrix, which was obtained by aligning the sequence taken from the chromosome 1 A.thaliana genome (14259442-14259804 b.p.) for *n*=10.

	1	2	3	4	5	6	7	8	9	10
A	-5.7	11.4	-5.0	-1.1	-5.0	-4.4	-3.7	-1.1	0.2	0.9
T	3.7	-9.4	-0.2	-1.5	4.2	2.4	-6.3	-8.9	-7.6	-3.3
C	-2.2	2.7	-0.3	-3.2	-4.1	-1.2	11.4	-2.2	-4.1	-4.1
G	-5.6	-5.6	-2.0	-4.2	-5.6	-4.9	-1.3	10.9	8.0	0.1

Let us also consider, as an example, the study of the periodicity in the nucleotide sequence developed by the method. For this purpose, we once again examined the sequence taken from the chromosome 1 A.thaliana genome (14259442-14259804 b.p., NC_003070.1) [11]. Analysis of this sequence demonstrates that there appears a period equal to 10 nucleotides inside it (Fig. 2). This period could not be identified using the entire previously developed software programs, such as TRF [12], T-REKS [9], Mreps [13], ATR hunter [14], TRStalker [3], Repfind [15], as well as the other developed methods and algorithms. This also applies to the spectral approaches developed in certain woks [16–18].

Periodicity of 10-11 symbols (as well as multiple periods) could be associated with the DNA packaging [19]. We believe that sections, which periodicity with insertions and deletions constitutes 9-11 symbols, may be associated with the nucleus binding regions or with the formation of chromatin loops [20]. Periodic sections with the length of period 2 could be referred to micro- and minisatellite sequences [21]. Classical micro- and minisatellite with insertions and deletions correspond to the $Z>12.0$ values; at the same time, the number of substitutions constitutes less than 0.5 per a single nucleotide. That means that micro- and minisatellites found at the $6.0<Z<12.0$ level appear to be the ancient copies of sequences that later were subjected to significant evolutionary changes. Thus, the given server could be used to search for indistinct micro- and minisatellites with the length of more than 100 symbols. These sequences are of great importance in searching for new VNTR, SSR or STR [22] and creating the DNA markers. Similar results were also obtained for the D.melanogaster, C.elegans and H.sapiens genomes, which are presented in the database at the http://victoria.biengi.ac.ru/cgi-bin/indelper/index.cgi web site [4].

Thus, the http://victoria.biengi.ac.ru/splinter/login.php web server is able to identify the latent periodicity (where the average number of substitutions is equal to 1.5 per symbol or more) with insertions and deletions of symbols; and in accordance with this parameter, the given web resource has no analogues at the present moment.

Acknowledgement
This work was supported by a grant from the Russian Science Foundation (14-24-00175).

References

[1] Heather J M and Chain B 2016 The sequence of sequencers: The history of sequencing DNA. *Genomics* **107** 1–8
[2] Pellegrini M 2015 Tandem Repeats in Proteins: Prediction Algorithms and Biological Role. *Front. Bioeng. Biotechnol.* **3** 143
[3] Pellegrini M, Renda M E and Vecchio A 2010 TRStalker: an efficient heuristic for finding fuzzy tandem repeats. *Bioinformatics* **26** i358-66
[4] Frenkel F E, Korotkova M A and Korotkov E V. 2017 Database of Periodic DNA Regions in Major Genomes *Biomed Res. Int.* **2017** 1–9
[5] Korotkov E V and Korotkova M A 2015 [Developing the Mathematical Method in order to Search for Latent Periodicity in Protein Amino Acid Sequences with Deletions and Insertions]. *Biofizika* **60** 1057–68
[6] Pugacheva V.M., Korotkov A.E and Korotkov E.V. 2016 Search of latent periodicity in amino acid sequences by means of genetic algorithm and dynamic programming *Stat. Appl.*

Genet. Mol. Biol. **15**

[7] Andrade M a, Ponting C P, Gibson T J and Bork P 2000 Homology-based method for identification of protein repeats using statistical significance estimates. *J. Mol. Biol.* **298** 521–37

[8] Pellegrini M 2012 Using phylogenetic profiles to predict functional relationships ed J Helden, A Toussaint and D Thieffry *Methods Mol. Biol.* **804** 167–77

[9] Jorda J and Kajava A V. 2009 T-REKS: identification of Tandem REpeats in sequences with a K-meanS based algorithm *Bioinformatics* **25** 2632–8

[10] Trifonov E N 1998 3-, 10.5-, 200- and 400-base periodicities in genome sequences *Phys. A Stat. Mech. its Appl.* **249** 511–6

[11] Theologis A, Ecker J R, Palm C J, Federspiel N A, Kaul S, White O, Alonso J, Altafi H, Araujo R, Bowman C L, Brooks S Y, Buehler E, Chan A, Chao Q, Chen H, Cheuk R F, Chin C W, Chung M K, Conn L, Conway A B, Conway A R, Creasy T H, Dewar K, Dunn P, Etgu P, Feldblyum T V., Feng J, Fong B, Fujii C Y, Gill J E, Goldsmith A D, Haas B, Hansen N F, Hughes B, Huizar L, Hunter J L, Jenkins J, Johnson-Hopson C, Khan S, Khaykin E, Kim C J, Koo H L, Kremenetskaia I, Kurtz D B, Kwan A, Lam B, Langin-Hooper S, Lee A, Lee J M, Lenz C A, Li J H, Li Y, Lin X, Liu S X, Liu Z A, Luros J S, Maiti R, Marziali A, Militscher J, Miranda M, Nguyen M, Nierman W C, Osborne B I, Pai G, Peterson J, Pham P K, Rizzo M, Rooney T, Rowley D, Sakano H, Salzberg S L, Schwartz J R, Shinn P, Southwick A M, Sun H, Tallon L J, Tambunga G, Toriumi M J, Town C D, Utterback T, Van Aken S, Vaysberg M, Vysotskaia V S, Walker M, Wu D, Yu G, Fraser C M, Venter J C and Davis R W 2000 Sequence and analysis of chromosome 1 of the plant Arabidopsis thaliana *Nature* **408** 816–20

[12] Benson G 1999 Tandem repeats finder: a program to analyze DNA sequences. *Nucleic Acids Res.* **27** 573–80

[13] Kolpakov R, Bana G and Kucherov G 2003 mreps: Efficient and flexible detection of tandem repeats in DNA. *Nucleic Acids Res.* **31** 3672–8

[14] Wexler Y, Yakhini Z, Kashi Y and Geiger D 2005 Finding approximate tandem repeats in genomic sequences. *J. Comput. Biol.* **12** 928–42

[15] Betley J N, Frith M C, Graber J H, Choo S and Deshler J O 2002 A ubiquitous and conserved signal for RNA localization in chordates. *Curr. Biol.* **12** 1756–61

[16] Kravatskaya G I, Kravatsky Y V, Chechetkin V R and Tumanyan V G 2011 Coexistence of different base periodicities in prokaryotic genomes as related to DNA curvature, supercoiling, and transcription. *Genomics* **98** 223–31

[17] Lobzin V V. and Chechetkin V R 2000 Order and correlations in genomic DNA sequences. The spectral approach *Uspekhi Fiz. Nauk* **170** 57

[18] Sosa D, Miramontes P, Li W, Mireles V, Bobadilla J R and José M V 2013 Periodic distribution of a putative nucleosome positioning motif in human, nonhuman primates, and archaea: mutual information analysis. *Int. J. Genomics* **2013** 963956

[19] Herzel H, Weiss O and Trifonov E N 1999 10-11 bp periodicities in complete genomes reflect protein structure and DNA folding. *Bioinformatics* **15** 187–93

[20] Kadauke S and Blobel G A 2009 Chromatin loops in gene regulation. *Biochim. Biophys. Acta* **1789** 17–25

[21] Richard G-F, Kerrest A and Dujon B 2008 Comparative Genomics and Molecular Dynamics of DNA Repeats in Eukaryotes *Microbiol. Mol. Biol. Rev.* **72** 686–727

[22] Ulloa M 2005 Molecular Markers Learning Modules, Vol. 1 and 2/ Volume 1. Using Molecular Marker Technology in Studies on Plant Genetic Diversity *Crop Sci.*

The Influence of Plasma Inhomogeneity and Incident Wave Frequency on the Nonlinear Spatial Absorption of Alphven Waves In Dissipative Plasma

M B Gavrikov[1] **and A A Taiurskii**[1]
[1]Keldysh Institute of Applied Mathematics of the Russian Academy of Sciences, Miusskaya sq., 4, Moscow, 125047, Russia

E-mail: nadya_p@cognitive.ru, tayurskiy2001@mail.ru

Abstract. We consider the nonlinear absorption of an Alfven wave falling on the boundary of incompressible plasma occupying a half-space. This absorption is due to dissipative effects of such factors as magnetic and hydrodynamic viscosities, thermal conductivities of electrons and ions, bremsstrahlung of electrons, temperature relaxation. We examine the dependence of the absorption on the spatial inhomogeneity of plasma density and the frequency of the incident Alfven wave. It is shown that the parameters of an Alfven wave penetrating dissipative plasma stabilize in time and satisfy a certain boundary value problem for a system of ordinary differential equations on a half-line. A numerical solution of this boundary value problem allows us to find some important relationships characterizing the absorption of Alfven waves.

1. Introduction

For a plane Alfven wave falling on the boundary of homogeneous incompressible plasma occupying a half-space, the nonlinear spatial absorption due to dissipative effects was previously considered in [1], the nonlinearity of the absorption process being primarily determined by the dependence of the transport coefficients on the temperatures of electrons and ions. That investigation was based on the equations of electromagnetic hydrodynamics (EMHD) of plasma [2,3] taking into account its electron-ion structure. The EMHD equations are written out in section 2 of the present paper. In section 3, we indicate some exact solutions of the EMHD equations. These solutions, called plane Alfven waves, differ from the Alfven waves of the classical MHD, but turn into the latter in the long-wave limit. It has been shown in [1] that: (i) with bremsstrahlung of electrons taken into account, the parameters of the absorbed Alfven wave in dissipative plasma stabilize in time to a quasi-stationary regime; (ii) the Alfven wave penetrates plasma to a finite depth; (iii) the parameters of the stabilized Alfven wave can be found by solving a certain boundary value problem for a system of ordinary differential equations on a half-line. In the present paper, we examine the dependence of the said absorption on the inhomogeneous density of the incompressible plasma and the frequency ω of the incident Alfven wave. Our interest is mainly focused on two types of inhomogeneities with the Gaussian distribution $\rho(x)/\rho_0 = 1 + A\exp\left[-(x-x_0)^2/D\right]$, $D>0$, $x_0>0$, namely: (i) the case $A>0$ corresponding to a hump; (ii) the case $-1<A<0$ corresponding to a hollow. In section 4, we formulate an initial boundary value problem describing the decay of an Alfven wave in nonhomogeneous plasma, as well as a boundary value problem on a half-line for the parameters of the time-stabilized absorbed Alfven wave. Section 5 contains the results of numerical analysis of plasma parameters in the solar corona, which allows us to study the process of its anomalous heating by Alfven waves arising in the photosphere [4]. It is shown that the basic relations characterizing the absorption in homogeneous dissipative plasma still hold in the nonhomogeneous case. Moreover, for the depth to which the Alfven wave penetrates dissipative plasma and the maximal temperatures of electrons and ions, we find the dependence on the parameters A, D, ω, where A is the height of the hump, D is its width, and ω is the frequency of the incident wave.

2. Equations of Electromagnetic Hydrodynamics (EMHD) of Plasma

To study the dynamics of two-fluid plasma, we use the equations of electromagnetic hydrodynamics (EMHD). The EMHD equations of incompressible plasma have the form

$$\mathrm{div}\mathbf{U}=0,\quad \frac{\partial\rho}{\partial t}+\mathbf{U}\cdot\nabla\rho=0,\quad \frac{\partial\rho\mathbf{U}}{\partial t}+\mathrm{Div}\Pi=\mathrm{Div}\mathrm{P},$$
$$\mathbf{E}+\frac{c^2\lambda_+\lambda_-}{4\pi\rho}\mathrm{rotrot}\mathbf{E}=\frac{\mathbf{j}}{\sigma}-\frac{1}{c}[\mathbf{U},\mathbf{H}]+\frac{1}{\rho}\mathrm{Div}\mathrm{W}, \tag{1}$$
$$\frac{1}{c}\frac{\partial\mathbf{H}}{\partial t}+\mathrm{rot}\mathbf{E}=0,\quad \mathrm{div}\mathbf{E}=0,\quad \mathbf{j}=\frac{c}{4\pi}\mathrm{rot}\mathbf{H},\quad \mathrm{div}\mathbf{H}=0,$$

where Π is the momentum flux tensor, P is the viscous stress tensor, W is the "hall terms" tensor,

$$\Pi=\Pi^{\mathrm{h}}+\Pi^{\mathrm{p}}+\Pi^{\mathrm{c}},\quad \mathrm{P}=\Pi_*^{(\mathrm{c})}+\Pi^{(\mathrm{U})},$$
$$\mathrm{W}=(\lambda_--\lambda_+)(\Pi^{\mathrm{p}}+\Pi^{\mathrm{c}})+(\lambda_-p_+-\lambda_+p_-)\mathrm{I}_3+\lambda_+\lambda_-(\mathbf{jU}+\mathbf{Uj})-\Pi_*^{(\mathrm{U})}-\Pi^{(\mathrm{c})}, \tag{2}$$
$$\Pi^{\mathrm{h}}=\rho\mathbf{UU}+p_\Sigma\mathrm{I}_3,\ \Pi^{\mathrm{p}}=\frac{H^2}{8\pi}\mathrm{I}_3-\frac{\mathbf{HH}}{4\pi},\ \Pi^{\mathrm{c}}=\lambda_+\lambda_-\frac{\mathbf{jj}}{\rho}.$$

Here and in what follows, the subscripts $\pm$ refer to ions and electrons, $\lambda_\pm=m_\pm/e_\pm$, $\lambda_\Sigma=\lambda_++\lambda_-$, $p_\Sigma=p_++p_-$, $m_\Sigma=m_++m_-$, $\rho=\rho_++\rho_-$, $\mathbf{U}=(\rho_+\mathbf{v}_++\rho_-\mathbf{v}_-)/\rho$, I_3 is the three-dimensional identity tensor, σ is the conductivity of plasma. The viscous stress tensors (with plasma incompressibility taken into account) have the form

$$\Pi^{(\mathrm{U})}=2\mu_\Sigma\mathrm{D}^{\mathrm{U}},\ \Pi_*^{(\mathrm{U})}=2\mu_*\mathrm{D}^{\mathrm{U}},\ \Pi^{(\mathrm{c})}=2\mu^*\mathrm{D}^{\mathrm{c}}-(2/3)\mu^*\mathbf{j}\cdot\nabla(1/\rho)\mathrm{I}_3,$$
$$\Pi_*^{(\mathrm{c})}=2\mu_*\mathrm{D}^{\mathrm{c}}-(2/3)\mu_*\mathbf{j}\cdot\nabla(1/\rho)\mathrm{I}_3,\ \Pi_\pm=2\mu_\pm\mathrm{D}_\pm\mp(2/3)\mu_\pm\lambda_\pm\mathbf{j}\cdot\nabla(1/\rho)\mathrm{I}_3, \tag{3}$$

where $\mathrm{D}^{\mathrm{U}}=\mathrm{def}\mathbf{U}$, $\mathrm{D}^{\mathrm{c}}=\mathrm{def}(\mathbf{j}/\rho)$, $\mathrm{D}_\pm=\mathrm{def}\mathbf{v}_\pm$ are strain tensors, $\mu_\Sigma=\mu_++\mu_-$, $\mu_*=\lambda_-\mu_+-\lambda_+\mu_-$, $\mu^*=\lambda_-^2\mu_++\lambda_+^2\mu_-$, $\mu_\pm$ are hydrodynamic viscosities of electrons and ions. Here, the second viscosities of electrons and ions are assumed to be equal to zero. For plane flows considered below, we have $\mathbf{j}\cdot\nabla(1/\rho)=0$ and the expressions (3) become much simpler. Taking into account the dependence (see below) of the plasma conductivity σ, the hydrodynamic viscosities $\mu_\pm$, and the other transport coefficients on the temperatures $T_\pm$ of electrons and ions, we should supplement system (1) – (3) with the equations for the temperatures [5],

$$\rho_\pm c_{\mathrm{p}}^\pm\left[\frac{\partial T_\pm}{\partial t}+\mathrm{div}T_\pm\mathbf{v}_\pm\right]=\mathrm{div}(\chi_\pm\nabla T_\pm)+\mathrm{tr}(\Pi_\pm\mathrm{D}_\pm)+\frac{m_\mp}{m_\Sigma}\frac{j^2}{\sigma}\pm b(T_--T_+)-p_{\mathrm{T}}^\pm, \tag{4}$$

where $c_{\mathrm{p}}^\pm=\mathrm{k}/((\gamma-1)m_\pm)$ are heat capacities for constant pressure, k is the Boltzmann constant, $\chi_\pm$ are thermal conductivities of electrons and ions, $Q=\pm b(T_--T_+)$ is the heat transmitted to each other by plasma components in elastic collisions, $p_{\mathrm{T}}^\pm$ are the losses in bremsstrahlung of electrons and ions (the losses of ions are assumed to be equal to zero in what follows). Here, we should also take into account the expressions of hydrodynamic parameters of electrons and ions in terms of ρ, $\mathbf{U}$, $\mathbf{j}$:

$$\mathbf{v}_\pm=\mathbf{U}\pm(\lambda_\mp/\rho)\mathbf{j},\ \rho_\pm=(\lambda_\pm/\lambda_\Sigma)\rho. \tag{5}$$

In view of the dependence of σ, $\mu_\pm,\chi_\pm$, b, $p_{\mathrm{T}}^\pm$ on the other plasma parameters, especially, the temperatures $T_\pm$ (see below), equations (1) – (5) form a closed determined system of equations for ρ, $p_\pm$, $T_\pm$, $\mathbf{U}$, $\mathbf{H}$, $\mathbf{E}$.

The transport coefficients $\mu_\pm, \chi_\pm, \sigma, b$ are obtained from approximate solutions of kinetic equations [2] and, in what follows, are taken equal to [6–8]

$$\mu_+ = 3.44\cdot 10^{-18}(m_+/m_-)^{1/2}T_+^{5/2},\quad \mu_- = 1.857\cdot 10^{-18}T_-^{5/2},\quad \sigma = 0.906\cdot 10^7 T_-^{3/2},$$

$$b = 9.044\cdot 10^{38}(m_-/m_+)^3\rho^2 T_-^{-3/2},\ \chi_- = 3.657\cdot 10^{-5}T_-^{5/2},\ \chi_+ = 6.436\cdot 10^{-5}(m_-/m_+)^{1/2}T_+^{5/2},$$

$$p_T^+ = 0,\quad p_T^- = 8.044\cdot 10^{32}\,n_+ n_- T_-^{1/2}\ \text{erg}/(\text{s}\cdot\text{cm}^3),$$

where $T_\pm$ are measured in K. In the calculation results given below, the electron viscosity is assumed to be equal to zero. More detailed formulas for the dissipative coefficients can be found in [9].

3. Alfven Waves in EMHD

In the dissipation-free planar case, the EMHD equations admit the following exact solutions [9]:

$$U_\perp = u(t)\exp(\mathrm{i}\kappa x),\ H_\perp = h(t)\exp(\mathrm{i}\kappa x),\ E_\perp = e(t)\exp(\mathrm{i}\kappa x),\ T_\pm = \text{const},\ \rho = \text{const},\ \mathrm{U}_x = 0, \tag{6}$$

called *planar Alfven waves*. In (6), we have $\kappa > 0$ and use the complex notaion: $U_\perp = U_y + \mathrm{i}U_z$, $H_\perp = H_y + \mathrm{i}H_z$, $E_\perp = E_y + \mathrm{i}E_z$. Here, $H_x = \text{const}$, and $e(t)$ can be explicitly expressed through $u(t)$ and $h(t)$:

$$e(t) = \left(\frac{iH_x}{c}u(t) + \frac{i\kappa\Lambda \mathrm{v}_\mathrm{A}}{\omega_\mathrm{p}}h(t)\right)\Big/\left(1+r^2\right),\quad r = \frac{\kappa c}{\omega_\mathrm{p}},\quad \Lambda = \sqrt{\frac{\lambda_+}{\lambda_-}} - \sqrt{\frac{\lambda_-}{\lambda_+}},\quad \mathrm{v}_\mathrm{A} = \frac{H_x}{\sqrt{4\pi\rho}}. \tag{7}$$

The functions $u(t)$, $h(t)$ have the form

$$u(t) = C_1 e^{i\omega_+ t} + C_2 e^{i\omega_- t},\quad h(t) = (4\pi\rho)^{1/2}(\kappa \mathrm{v}_\mathrm{A})^{-1}\{C_1\omega_+ e^{i\omega_+ t} + C_2\omega_- e^{i\omega_- t}\}. \tag{8}$$

Here $\omega_\mathrm{p} = (4\pi\rho)^{1/2}(\lambda_+\lambda_-)^{-1/2}$ is plasma frequency, C_1, C_2 are arbitrary complex constants. The transverse component of the current density $j_\perp = j_y + \mathrm{i}j_z$ varies in the same way, $j_\perp = j(t)\exp(\mathrm{i}\kappa x)$, and $j_x = 0$, where $j(t) = -\kappa c/(4\pi)h(t)$. Finally,

$$\omega_\pm = \omega_\pm(\kappa) = \frac{\kappa \mathrm{v}_\mathrm{A}}{2}\left\{\frac{r\Lambda}{1+r^2} \pm \left[\frac{r^2\Lambda^2}{(1+r^2)^2} + \frac{4}{1+r^2}\right]^{1/2}\right\}. \tag{9}$$

It is convenient to invert formula (9):

$$\kappa = \kappa(\omega) = \frac{\omega_\mathrm{p}}{c}\cdot|\omega|\left(\frac{\mathrm{v}_\mathrm{A}^2\omega_\mathrm{p}^2}{c^2} + \frac{\omega_\mathrm{p}\mathrm{v}_\mathrm{A}}{c}\Lambda\omega - \omega^2\right)^{-1/2}, \tag{10}$$

where $-\omega_\mathrm{c}^+ < \omega < \omega_\mathrm{c}^-$ for $H_x > 0$, and $-\omega_\mathrm{c}^- < \omega < \omega_\mathrm{c}^+$ for $H_x < 0$.

4. Statement of the Spatial Absorption Problem

Consider a planar Alfven wave running from left to right in the domain $x \le 0$ and arriving at the dissipative plasma boundary $x = 0$, the plasma filling up the half-space $x \ge 0$. Further propagation of the Alfven wave in the domain $x \ge 0$ is accompanied by its absorption, which is the subject of this study. From the results of the preceding section, it follows that for the longitudinal magnetic field $H_x < 0$, the frequency ω of the incident Alfven wave varies within the interval $-\omega_\mathrm{c}^- = H_x/(\lambda_- c) < \omega < 0$, and if $H_x > 0$, then it varies on the interval $-\omega_\mathrm{c}^+ = -H_x/(\lambda_+ c) < \omega < 0$. Below, we limit ourselves to the more interesting first case. The plasma in the region $x \ge 0$ is assumed to be magnetized, stationary, and isothermal, with a given density distribution $\rho(x)$. Thus, at the initial instant in the domain $x \ge 0$, we have

$$U_\perp\big|_{t=0} = 0,\quad U_x\big|_{t=0} = 0,\quad H_\perp\big|_{t=0} = 0,\quad T_\pm\big|_{t=0} = T^0,\quad \rho\big|_{t=0} = \rho(\mathrm{x}), \tag{11}$$

where $\rho(x)$ tends to a finite limit $\rho > 0$ as $x \to +\infty$, with the constant ρ and the longitudinal magnetic field $H_x = \text{const}$ being the same as those in the domain $x \le 0$ from which the Alfven wave comes. The values of the dissipative plasma parameters on the boundary $x = 0$ coincide with the values of the Alfven wave parameters on the same boundary, and therefore, in view of (6) – (10), we have the following boundary conditions for $x = 0$:

$$U_\perp\big|_{x=0} = U_0 \exp(\mathrm{i}\omega t), \quad H_\perp\big|_{x=0} = U_0 \frac{\sqrt{4\pi\rho}}{\kappa \mathrm{v}_\mathrm{A}} \omega \exp(\mathrm{i}\omega t),$$

$$E_\perp\big|_{x=0} = \frac{\mathrm{i}U_0}{1+(\kappa c/\omega_\mathrm{p})^2}\left(\frac{H_x}{c} + \Lambda\sqrt{\lambda_+\lambda_-}\,\omega\right), \quad j_\perp\big|_{x=0} = -U_0 \frac{\omega c \rho}{H_x}\exp(\mathrm{i}\omega t). \tag{12}$$

At infinity, the dissipative plasma parameters coincide with the unperturbed plasma parameters for $t = 0$:

$$U_\perp\big|_{x=+\infty} = 0, \quad H_\perp\big|_{x=+\infty} = 0, \quad T_\pm\big|_{x=+\infty} = T^0, \quad E_\perp\big|_{x=+\infty} = 0, \quad j_\perp\big|_{x=+\infty} = 0. \tag{13}$$

Let us write system (1) – (5) in dimensionless form in the case of plane symmetry ($\partial/\partial y = \partial/\partial z = 0$), using the complex notation: $U = U_\perp = U_y + \mathrm{i}U_z$, $H = H_\perp = H_y + \mathrm{i}H_z$, $E = E_\perp = E_y + \mathrm{i}E_z$, $j = j_\perp = j_y + \mathrm{i}j_z$. Taking into account the relation $j_x = 0$, we have

$$\rho\frac{\partial U}{\partial t} - \frac{\partial}{\partial x}\left(H_x H + \mu_\Sigma \frac{\partial U}{\partial x} + \xi\mu_* \frac{\partial}{\partial x}\frac{j}{\rho}\right) = 0,$$

$$\frac{\partial H}{\partial t} + \mathrm{i}\frac{\partial E}{\partial x} = 0, \qquad j = \mathrm{i}\frac{\partial H}{\partial x}, \quad \rho = \rho(x),$$

$$E - \frac{\xi^2}{\rho}\frac{\partial^2 E}{\partial x^2} = \xi\zeta\frac{j}{\sigma} + \mathrm{i}H_x U + \frac{\xi}{\rho}\frac{\partial}{\partial x}\left[\Lambda H_x H - \mu_* \frac{\partial U}{\partial x} - \xi\mu^* \frac{\partial}{\partial x}\frac{j}{\rho}\right], \tag{14}$$

$$\frac{\partial T_\pm}{\partial t} = \frac{2Z_\pm(\gamma-1)}{\rho}\left\{\frac{\partial}{\partial x}\left(\chi_\pm \frac{\partial T_\pm}{\partial x}\right) + \frac{m_\mp}{m_\Sigma}\frac{\xi\zeta}{\sigma}|j|^2 \pm \frac{\zeta}{\xi}C_0\rho^2\frac{T_+ - T_-}{T_-^{3/2}} + \right.$$

$$\left. + \frac{T_\pm^{5/2}}{R_\pm}\left[\left|\frac{\partial U}{\partial x}\right|^2 + \frac{\lambda_\mp}{\lambda_\pm}\xi^2\left|\frac{\partial}{\partial x}\frac{j}{\rho}\right|^2 \pm 2\xi\sqrt{\frac{\lambda_\mp}{\lambda_\pm}}\,\mathrm{Re}\left(\frac{\partial U}{\partial x}\overline{\frac{\partial}{\partial x}\frac{j}{\rho}}\right)\right]\right\} - \frac{(1\mp 1)}{2}\frac{\xi_\mathrm{T}}{\xi}(\gamma-1)Z^2\rho T_-^{1/2},$$

where $\mu_\Sigma = \mu_+ + \mu_-$, $\mu_* = (\lambda_-/\lambda_+)^{1/2}\mu_+ - (\lambda_+/\lambda_-)^{1/2}\mu_-$, $\mu^* = (\lambda_-/\lambda_+)\mu_+ + (\lambda_+/\lambda_-)\mu_-$, $\mu_\pm = T_\pm^{5/2}/R_\pm$, $\sigma = \sigma_0 T_-^{3/2}$, $\chi_\pm = C_\pm(\xi/\zeta)T_\pm^{5/2}$, and σ_0, C_0, $R_\pm$, $C_\pm$ have the form

$$\sigma_0 = 2{,}59,\; C_0 = \frac{m_-}{m_+},\; R_+ = 1.108\frac{\zeta}{\xi}\left(\frac{m_-}{m_+}\right)^{1/2},\; R_- = 2.053\frac{\zeta}{\xi},\; C_- = 2.11\frac{m_+}{m_-},\; C_+ = 3.7\left(\frac{m_+}{m_-}\right)^{1/2}.$$

Finally, $Z_+ = Z$, $Z_- = 1$, and ξ, ζ, ξ_T are similarity numbers which, for $Z = 1$, have the form

$$\xi = \frac{\ell_\mathrm{c}}{L_0} = \frac{c\sqrt{\lambda_+\lambda_-}}{\sqrt{4\pi\rho_0}\,L_0}, \quad \zeta = 15\frac{(4\pi\rho_0)^{5/2}}{H_0^4}\frac{ce^3}{m_+^2}, \quad \xi_\mathrm{T} = 21.825\frac{\rho_0^{3/2}}{H_0^2}\frac{10^7 e^5}{c^2\hbar m_+ m_-},$$

where $\ell_\mathrm{c} = c/\omega_\mathrm{p}$ is skin length, L_0, ρ_0, H_0, etc., are the respective characteristic scales of length, density, magnetic field strength, etc. Moreover, when passing to dimensionless parameters, it has been assumed that $t_0 = L_0/\mathrm{v}_0$, $\mathrm{v}_0 = \mathrm{v}_\mathrm{A} = H_0/\sqrt{4\pi\rho_0}$, $E_0 = \mathrm{v}_0 H_0/c$, $j_0 = cH_0/(4\pi L_0)$, $T_0 = v_\mathrm{A}^2\lambda_\Sigma e/(2\mathrm{k})$.

Our aim is to describe steady quasi-stationary states of the absorbed wave of frequency ω. We seek these states as solutions of system (14) having the form $T_\pm = T_\pm(x)$, $U = U(x)\exp(\mathrm{i}\omega t)$, $H = H(x)\exp(\mathrm{i}\omega t)$, $E = E(x)\exp(\mathrm{i}\omega t)$, $j = j(x)\exp(\mathrm{i}\omega t)$. Substituting these expressions into (14) and dividing the resulting relations by $\exp(\mathrm{i}\omega t)$, we obtain the following system of ordinary differential equations for the amplitudes $T_\pm(x)$, $U(x)$, $E(x)$:

$$\mathrm{i}\rho\omega U + \frac{H_x}{\omega}E'' - \left(\mu_\Sigma \dot{U}'\right)' - \frac{\mathrm{i}\xi}{\omega}\left(\mu_*\left(\frac{E''}{\rho}\right)'\right)' = 0,$$

$$E - \left(\frac{\xi^2}{\rho} - \frac{\mathrm{i}\zeta\xi}{\omega\sigma} - \frac{\xi\Lambda H_x}{\omega\rho}\right)E'' - \mathrm{i}H_x U + \frac{\xi}{\rho}\left(\mu_* U'\right)' - \frac{\mathrm{i}\xi^2}{\omega\rho}\left(\mu_*\left(\frac{E''}{\rho}\right)'\right)' = 0, \tag{15}$$

$$\left(\chi_\pm T_\pm'\right)' + \frac{m_\mp}{m_\Sigma}\frac{\xi\zeta}{\omega^2\sigma}\left|E''\right|^2 \pm \frac{\zeta}{\xi}C_0\rho^2\frac{T_- - T_+}{T_-^{3/2}} +$$

$$+\frac{T_\pm^{3/2}}{R_\pm}\left[\left|U'\right|^2 + \frac{\lambda_\mp}{\lambda_\pm}\frac{\xi^2}{\omega^2}\left|\left(\frac{E''}{\rho}\right)'\right|^2 \pm 2\frac{\xi}{\omega}\sqrt{\frac{\lambda_\mp}{\lambda_\pm}}\mathrm{Re}\left(iU'\left(\frac{E''}{\rho}\right)'\right)\right] - \frac{(1\mp1)}{4}\frac{\xi_T}{\xi}Z^2\rho T_-^{1/2} = 0,$$

where $\rho(x)$ is a given function and the prime symbol indicates differentiation in x. Taking into account the above expressions of μ_Σ, μ_*, μ^*, σ in terms of $T_\pm$, we see that (15) is a nonlinear system of ordinary differential equations on $[0,+\infty)$ whose solutions allow us to calculate the functions $H(x) = -E'/\omega$ and $j(x) = -iE''/\omega$. System (15) is supplemented by the following dimensionless boundary conditions at the points $x = 0$ and $x = +\infty$:

$$T_\pm(0) = T^0,\quad U(0) = U_0,\quad E(0) = \frac{\mathrm{i}U_0}{1+\kappa^2\xi^2}\left(H_x + \xi\Lambda\omega\right),\quad E'(0) = -\frac{\rho U_0\omega^2}{\kappa H_x},$$

$$T_\pm(+\infty) = T^0,\quad U(+\infty) = 0,\quad E(+\infty) = 0,\quad E'(+\infty) = 0, \tag{16}$$

where $\kappa(\omega)$ is obtained from (10) in dimensionless form:

$$\kappa = \kappa(\omega) = |\omega|\rho^{1/2}\left(H_x^2 + H_x\Lambda\xi\omega - \xi^2\omega^2\right)^{-1/2}.$$

5. Analysis of Calculation Results

In order to find a solution of the boundary value problem (15), (16), we use the stabilization method [9] for $\xi = 1$, $\xi_T = 3$, $\zeta = 3\cdot10^3$, which corresponds to the typical parameters of the solar corona: $\rho_0 = 10^{-12}\,\mathrm{g/cm^3}$, $H_0 = 1\,\mathrm{G}$, $Z = 1$. We take $L_0 = \ell_c = 1\,\mathrm{cm}$ as the characteristic length, where $\ell_c = c/\omega_p$ is the skin length. Then we obtain the following characteristic values of the other parameters: $\mathrm{v_A} = H_0/\sqrt{4\pi\rho_0} = 2.8\cdot10^5\,\mathrm{cm/s}$ (velocity), $t_0 = L_0/\mathrm{v_A} = c\sqrt{\lambda_+\lambda_-}/H_0 = 3.45\cdot10^{-6}\,\mathrm{s}$, $T_0 = \mathrm{v_A^2}\lambda_\Sigma e/(2\mathrm{k}) = 10^3\ \mathrm{K}$, $r_\mathrm{D} = (\mathrm{k}T_0)^{1/2}/\left(2\pi^{1/2}\rho_0 n_0^{1/2}\right) \cong 2\cdot10^{-4}\,\mathrm{cm}$ (Debye radius), $n_0 = \rho_0/m_+$. In particular, we have $r_\mathrm{D} << L_0$, which is much smaller than the grid step $h = 10^{-2} \div 10^{-3}\,\mathrm{cm}$ of the difference scheme used in the stabilization method. Moreover, it is assumed that the incident Alfven wave has fixed amplitude $|U_0| = 0.1$ and $H_x = -1$. The wave frequency ω varies on the interval $-61 \cong -(\lambda_+/\lambda_-)^{1/2} < \omega < 0$. In the basic case $\rho(x) \equiv 1$, $\omega = -30$, it follows from [5,6] that the heating depth is equal to $d \cong 12.6$ and the maximal temperatures of electrons and ions have the values $T_- = 3.35$, $T_+ = 2.77$.

Suppose now that in the domain $x \geq 0$ the dissipative plasma density has a hump: $\rho(x) = 1 + A\exp\left[-(x-x_0)^2/D\right]$, $x_0 = 7$, $D = 3$, $A = 50$. Figure 1 shows that, in contrast to the basic case, the penetration depth of the Alfven wave has decreased about 3 times ($d \cong 4.5$), and the maximal temperatures of ions and electrons $T_+ = 2.23$, $T_- = 2.9$ have become smaller by about 15%.

Suppose now that the density function has a hollow: $\rho(x) = 1 + A\exp\left[-(x-x_0)^2/D\right]$, $x_0 = 7$, $D = 3$, $A = -0.9$. Figure 2 shows that now, in contrast to the basic case, the penetration depth has increased

IOP Conf. Series: Journal of Physics: Conf. Series **937** (2017) 012014 doi:10.1088/1742-6596/937/1/012014

by about 20% ($d \cong 15.1$), and the maximal temperatures $T_+ = 3.55$, $T_- = 2.74$ have also become larger. These results can be explained by the role of the bremsstrahlung of electrons whose density is equal to $p_{\mathrm{T}} \sim \rho^2\sqrt{T_-}$. In the case of the hump-shaped inhomogeneity, the bremsstrahlung substantially increases on the interval $\left[x_0 - \sqrt{2D}, x_0 + \sqrt{2D}\right] = [4.55, 9.45]$ and, according to the law of conservation of energy (see [5,6]), the thermal energy of electrons and ions in the domain $x \geq 0$ occupied by dissipative plasma causes a decrease of the penetration depth and the maximal values of the temperatures $T_\pm$. The same behavior can still be expected (figure 1 and figure 2 confirm this), if bremsstrahlung grows (drops) only in some part of the half-space $x \geq 0$ occupied by dissipative plasma. The above considerations can also explain why in the case of the hollow-shaped density, the heating of electrons increases both with respect to the depth and the maximal temperatures.

The above analysis suggests that with a decrease of the hump height A, the penetration depth of the Alfven wave, as well as the maximal temperatures of electrons and ions, increase. This is confirmed by the results obtained for inhomogeneous densities of the form $\rho(x) = 1 + A\exp\left[-(x-8)^2 / D\right]$ with various A (see table 1).

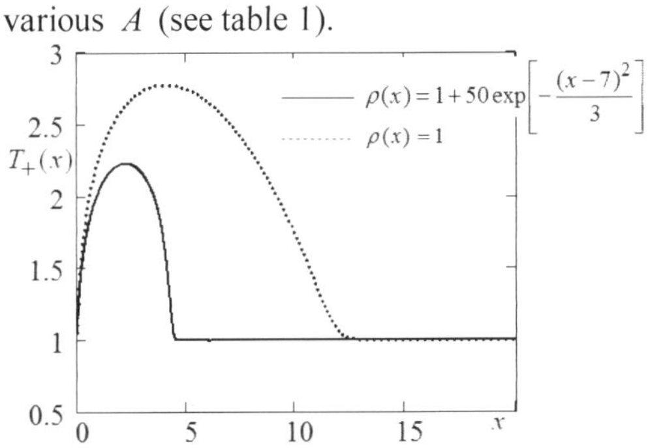

Figure 1. The influence of a hump-shaped density on the penetration depth of an Alfven wave. The temperature of ions for $\omega = -30$.

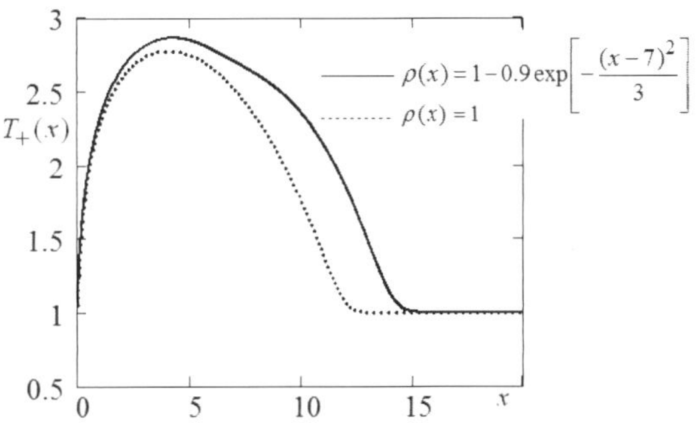

Figure 2. The influence of a hollow-shaped density on the penetration depth of an Alfven wave. The temperature of ions for $\omega = -30$.

Table 1. The penetration depth d and the maximal temperatures versus the hump height A; $\omega = -30$, $\rho(x) = 1 + A\exp\left[-(x-8)^2\right]$.

A	100	50	25	10	5	2.5	2	1.5	1	0.5	0.1	0
d	6.4	6.5	6.7	7.2	7.5	7.9	8.1	8.5	9.4	11.2	12.3	12.5
T_+	2.5	2.52	2.54	2.57	2.59	2.62	2.6	2.6	2.66	2.71	2.76	2.77
T_-	3.23	3.24	3.25	3.26	3.26	3.27	3.26	3.28	3.29	3.31.	3.34	3.35

We omit the graphs of $T_\pm(x)$ for the cases represented in table 1, since they are all of the same type, like those in figure 1 and figure 2, and their qualitative behavior is determined by the values of d, $T_\pm$ (note that the heating depths for electrons and ions, d_{e} and d_{i}, slightly differ, but $d_{\mathrm{i}} \geq d_{\mathrm{e}}$, $d_{\mathrm{i}} \cong d_{\mathrm{e}}$; in table 1, $d = \max(d_{\mathrm{i}}, d_{\mathrm{e}}) = d_{\mathrm{i}}$). As a compensation, we offer the graphs of the temperatures for $\omega = -25$, $A = 50, 25, 2, 0$ in figure 3.

The above considerations allow us to make some predictions. For instance, if D is decreased, then the penetration depth of an Alfven wave, as well as the maximal temperatures of ions and electrons, should increase, because of the contraction of the domain $\left[x_0 - \sqrt{2D}, x_0 + \sqrt{2D}\right]$, in which the bremsstrahlung grows, and therefore, the losses in bremsstrahlung become smaller. This conjecture is supported by calculations. Thus, for $A = 2$, $x_0 = 8$, $D = 0.5$, we obtain $d = 8.4$, $T_+ = 2.65$, $T_- = 3.28$, and the above prediction based on table 1 is true.

Now, consider the influence of the frequency ω of an Alfven wave on its absorption. Figure 4 shows the profiles of $T_+(x)$ in the case of hump-shaped density: $\rho(x) = 1 + A\exp\left[-(x-x_0)^2 / D\right]$, $x_0 = 8$,

$A=50$, $D=1$, and $\omega=-50$, $\omega=-35$, $\omega=-9$. Supplementing these results with more detailed data from table 2, we conclude that the maximal penetration depth of an Alfven wave and the maximal temperatures of electrons and ions are observed at midrange frequencies ($\omega\cong-25$) near the value $-0.5(\lambda_+/\lambda_-)^{1/2}$, the midpoint of the interval $(-(\lambda_+/\lambda_-)^{1/2},0)$ of admissible frequencies. For high frequencies, $\omega\cong-(\lambda_+/\lambda_-)^{1/2}$, or low frequencies, $\omega\cong0$, the Alfven wave practically does not heat dissipative plasma, which is penetrated only by the decaying electromagnetic field. These results are explained by the sharp drop of the electric field at the left boundary point $x=0$. According to the boundary conditions (16), for $\omega=-5$ (low frequencies), the absolute value of $E|_{x=0}$ is more than 3 times smaller than that of $E|_{x=0}$ for $\omega=-30$ (middle frequencies) and almost 3 times smaller than that for $\omega=-55$ (high frequencies). This, together with equations (14), leads to an approximately three-fold drop of the electric (and therefore, magnetic) field and the current density inside the domain occupied by dissipative plasma. A special role is played by the drop of the current density inside the domain $x\geq0$, which, for low frequencies, $\omega\cong-5$, is strengthened by a six-fold current density drop at the boundary $x=0$. For high frequencies, $\omega=-55$, the current density drop inside the domain $x\geq0$ is partially compensated by a two-fold increase of the current density at the boundary $x=0$. In any case, the almost three-fold drop of the current density inside the domain $x\geq0$ leads to an approximately 10-fold decrease of the Joule heat of plasma, which explains the sharp (almost by an order of magnitude) drop of the temperatures of electrons and ions, as well as practically no heating of dissipative plasma by the Alfven wave. Note that in the extreme case $\omega=0$, the current density on the boundary vanishes, $|H|\sim0.1$, and the absolute value of $E|_{x=0}$ drops almost 1000 times in comparison with $E|_{x=0}$ for $\omega=-30$. In the other extreme case $\omega\cong-(\lambda_+/\lambda_-)^{1/2}$, we have $H=0$, $E=0$ on the boundary, and the absolute value of $j|_{x=0}$ is about 2 times greater than that of $j|_{x=0}$ for $\omega=-30$.

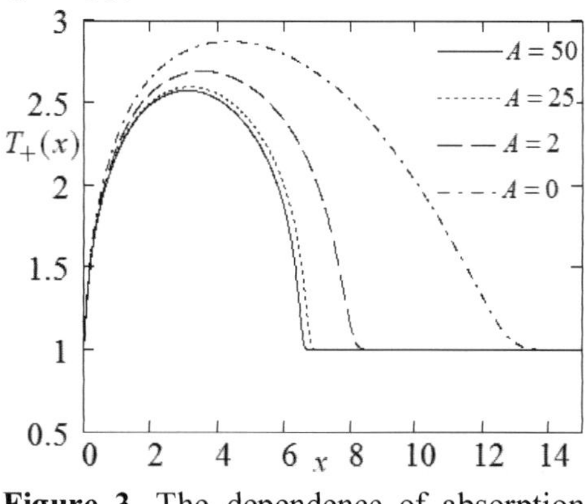

Figure 3. The dependence of absorption of an Alfven wave on the height A of the density hump. Ion temperature distribution for $\omega=-25$, $\rho(x)=1+A\exp\left[-(x-8)^2\right]$.

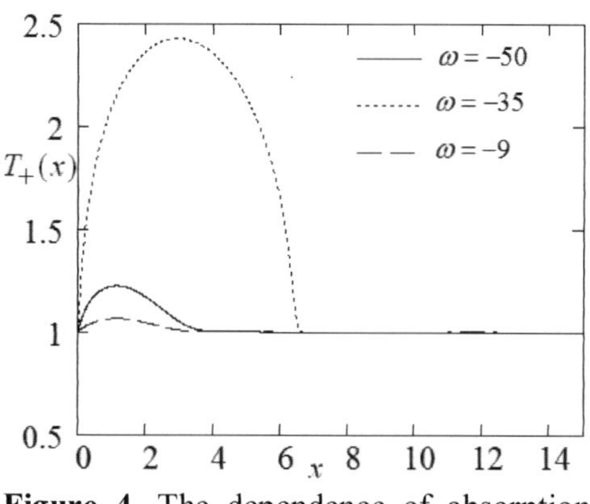

Figure 4. The dependence of absorption of an Alfven wave on its frequency ω. Ion temperature distribution for $\rho(x)=1+50\exp\left[-(x-8)^2\right]$.

Table 2. The dependence of the Alfven wave penetration depth d and maximal temperatures of electrons and ions on the frequency ω; $\rho(x)=1+50\exp\left[-(x-8)^2\right]$.

$-\omega$	60	55	50	45	35	30	25	20	10	9	8	1
d	0.96	1.3	3.7	6.3	6.5	6.6	6.6	6.6	6.3	2.9	0	0
T_+	1.04	1.05	1.22	1.94	2.43	2.52	2.57	2.57	1.91	1.06	1	1
T_-	1.2	1.22	1.27	2.15	3.01	3.24	3.34	3.31	2.07	1.07	1	1

6. Conclusion

Our main statement is that the inhomogeneity of plasma density and the incident wave frequency have a substantial effect on the penetration depth of an Alfven wave, as well as the heating of electrons and ions of dissipative plasma. More precisely, hump-shaped density distribution hinders the passage of an Alfven wave, decreasing its penetration depth and the temperatures of electrons and ions of dissipative plasma, while a hollow-shaped density distribution increases the penetration depth and the temperatures of electrons and ions. On the other hand, Alfven waves of high ($\omega \sim \omega_c^-$) and low ($\omega \sim 0$) frequencies practically have no heating effect on plasma and the absorption of the wave reduces to the decay of the electromagnetic field. The greatest heating effect on plasma, both in relation to the penetration depth and maximal temperatures of electrons and ions, is observed for midrange frequencies ($\omega \sim \omega_c^- / 2$).

Note that in the nonlinear heat equation for T_-, at the initial time instant, when $T_- \sim 1$, the terms corresponding to bremsstrahlung and temperature relaxation have the same order which is much smaller ($\sim 10^3$ times) than the Joule heat. With the growth of T_-, the value of the Joule heat, $\sim T_-^{-3/2}$, as well as the relaxation term $\sim T_-^{-1/2}$, experience a fast decrease, while the magnitude of bremsstrahlung $\sim T_-^{1/2}$ is increasing. As a result, in a short time these three terms acquire the same order of magnitude and none of these can be neglected! It should be added that the strong nonlinearity of the thermal conductivity coefficient $\chi_- \sim T_-^{5/2}$ substantially reduces the role of the Joule heat. Note also that the above results have been obtained under the assumption that $\mu_- = 0$, since the theoretical value of electron viscosity given in section 2 appears to be greater than the real one at least by 2 to 3 orders of magnitude. Absorption of Alfven waves for real values of electron viscosity μ_- requires further investigation.

7. Acknowledgments

This work has been supported by the Russian Science Foundation (Project 16-11-10278).

References

[1] Taiurskii A A and Gavrikov M B 2016 *AIP Conf. Proc.* **1773** 090003-1–14.
[2] Braginskii S I 1965 Transport processes in Plasma In *Reviews of Plasma Physics* vol 1 ed Leontovich M A (New York: Consultants Bureau) p 205.
[3] Gavrikov M B 2006 The basic equations for two–fluid electromagnetic hydrodynamics Part I *Preprint* 59 (Inst. Appl. Math. the Russian Academy of Science) 28 p.
[4] McIntosh S W, Pontien B P, Carlsson M, Hansteen V, Boerner P and Goossens M 2011 *Nature* **475** pp 478-480.
[5] Landau L D and Lifshitz E M 1987 *Fluid Mechanics* vol 6 (2nd ed Butterworth-Heinemann).
[6] Spitzer L *Physics of Fully Ionized Gases* 1962 (New York: 2nd ed Interscience).
[7] Chapman S and Cowling T G 1952 *The Mathematical Theory of Non-uniform Gases* (Cambridge University Press).
[8] Grim G 1983 Radiation Processes in Plasma [in Russian] In *Basics of Plasma Phisics* vol 1 (Energoatomizdat).
[9] Gavrikov M B and Taiurskii A A 2017 Spatial Nonlinear Alfven Wave Absorption by Dissipative Plasma *Herald of the Bauman Moscow State Tech. Univ. Nat. Sci.* **2** pp 40–59.

Traveling-wave solutions in continuous chains of unidirectionally coupled oscillators

S D Glyzin[1,2], A Yu Kolesov[1] and N Kh Rozov[3]

[1] Faculty of Mathematics, Yarostavl State University, Sovetskaya str., 14, Yaroslavl, 150000, Russia
[2] Scientific Center in Chernogolovka RAS, Lesnaya str., 9, Chernogolovka, Moscow region, 142432, Russia
[3] Faculty of Mechanics and Mathematics, Moscow State University, Leninskiye Gory, Main Building, Moscow, 119991, Russia

E-mail: glyzin@uniyar.ac.ru; kolesov@uniyar.ac.ru; fpo.mgu@mail.ru

Abstract. Proposed is a mathematical model of a continuous annular chain of unidirectionally coupled generators given by certain nonlinear advection-type hyperbolic boundary value problem. Such problems are constructed by a limit transition from annular chains of unidirectionally coupled ordinary differential equations with an unbounded increase in the number of links. It is shown that any preassigned finite number of stable periodic motions of the traveling-wave type can coexist in the model.

1. Introduction

The mathematical model of an annular chain of unidirectionally coupled generators may be given by a system of ordinary differential equations of the form

$$\dot{u}_j = F(u_j) + mD(u_{j+1} - u_j), \quad j = 1, \ldots, m, \quad u_{m+1} = u_1. \tag{1}$$

where $m \geq 2$, the dot means differentiation with respect to t, $u \in \mathbb{R}^n$, with $n \geq 2$, the vector function $F(u)$ belongs to the class $C^\infty(\mathbb{R}^n; \mathbb{R}^n)$ and D is an $n \times n$ square matrix. It is typically assumed that the partial system

$$\dot{u} = F(u) \tag{2}$$

corresponding to chain (1) has an equilibrium or a cycle.

Investigating chains of form (1) is relevant in view of various applications to radiophysics [1] – [5] and neurodynamics [6] – [8]. In particular, it is interesting to study the behavior of attractors of such chains as the number m of links increases without bound. Here, we study this problem in the case where partial system (2) has a stable zeroth equilibrium and the coupling matrix D is small in a suitable sense. For $m \gg 1$, it is totally appropriate to pass from discrete chain (1) to the corresponding continuous model. The transition procedure consists in approximating the variable j/m by a continuous index $x \in [0, 1] (\text{mod } 1)$ and replacing the term $m(u_{j+1} - u_j)$ in (1) with the derivative $\partial u/\partial x$. As a result, we obtain the boundary value problem

$$\frac{\partial u}{\partial t} = F(u) + D\frac{\partial u}{\partial x}, \quad u(t, x+1) \equiv u(t, x), \tag{3}$$

which is a mathematical model of a continuous annular chain of unidirectionally coupled generators.

We consider problem (3) as an evolutionary equation in a Banach space E consisting of vector functions $u(x)$ of class $W_2^1([0,1];\mathbb{R}^n)$ that are periodic with period 1. We define the norm in E as

$$\|u\|_E = \left(\int_0^1 \|u(x)\|^2 dx + \int_0^1 \|u'(x)\|^2 dx\right)^{1/2}.$$

Here and below, the same symbol $\|*\|$ denotes the Euclidean norms in the spaces $\mathbb{R}^n$ and $\mathbb{C}^n$ as well as the induced matrix norms corresponding to these norms. Next, we introduce the closed unbounded operator $L = Dd/dx : E \to E$ with the domain of definition $E(L)$ dense in E. The following statement is valid.

Lemma 1. *The operator L is a generator of the semigroup* $\exp(Lt), t \geq 0$

$$\exp(Lt)u(x) = \sum_{k\in\mathbb{Z}} \exp(2\pi ikDt)u_k \exp(2\pi ikx), \tag{4}$$

where

$$u(x) = \sum_{k\in\mathbb{Z}} u_k \exp(2\pi ikx), \quad u_k = \int_0^1 u(x)\exp(-2\pi ikx)dx, \quad k \in \mathbb{Z} \tag{5}$$

of class C_0 of linear bounded operators in E if and only if the whole spectrum of the matrix D lies on the real axis and there exists a basis of its eigenvectors in the space $\mathbb{R}^n$.

This lemma indicates that as $m \to \infty$, the chain (1) by no means always generates a well-posed boundary value problem (3). This naturally provokes constraints on the matrix D. Everywhere in what follows, we assume that all its eigenvalues are real and simple. We are then obviously in the framework of the applicability of Lemma 1; moreover, semigroup (4), (5) also exists for small perturbations of D. Hence, for the vector function $v(t) = u(t,\cdot) \in E$, we have the well-defined abstract semilinear differential equation

$$\dot{v} = Lv + \mathscr{F}(v), \quad \mathscr{F}\big(v(t)\big) \stackrel{\text{def}}{=} F\big(u(t,\cdot)\big). \tag{6}$$

In addition, by virtue of the well-known regularity theorem (see [9]), the composition operator $\mathscr{F} : E \to E$ appearing in (6) is infinitely Frechet differentiable and bounded on any bounded subset $\Omega \subset E$ together with any finite number of its derivatives.

The existence of semigroup (4) allows defining the notion of a generalized solution of Eq. (6) from E with an arbitrary initial condition

$$v(0) = v_0 \tag{7}$$

Namely, a generalized solution is a continuous E-valued vector function $v(t)$ that satisfies the following integral equation

$$v(t) = \exp(Lt)v_0 + \int_0^t \exp\big(L(t-s)\big)\mathscr{F}\big(v(s)\big)ds. \tag{8}$$

on some interval $0 \leq t \leq \bar{t}$ is a generalized solution.

The general theory of abstract semilinear differential equations can be found, for example, in the monographs [10], [11]. The facts presented there indicate that the properties of these equations are largely analogous to the properties of ordinary differential equations. In particular, the following several lemmas are valid.

Lemma 2. *For any bounded set $\Omega \subset E$ we can find a $t_0 = t_0(\Omega) > 0$, such that each generalized solution $v = v(t, v_0)$, $v(0, v_0) = v_0$ of Eq. (6) with the initial condition $v_0 \in \Omega$ is uniquely defined on the interval $0 \leq t \leq t_0$ and is continuous with respect to the totality of variables $(t, v_0) \in [0, t_0] \times \Omega$. If $v_0 \in \Omega \cap E(L)$ then this solution is classical for all $t \in [0, t_0]$.*

Lemma 3. *Let $v = v_*(t)$ be an arbitrary (generally speaking, generalized) solution to equation (6) on some interval $0 \leq t \leq \bar{t}$. Then there exists a sufficiently small $r_0 > 0$ such that any generalized solution $v = v(t, v_0)$ to this equation with an initial condition $v(0, v_0) = v_0 \in B$, $B = \{v \in E : ||v - v_*(0)||_E \leq r_0\}$, is defined on the interval $0 \leq t \leq \bar{t}$ and is jointly continuous with respect to the variables $(t, v_0) \in [0, \bar{t}\,] \times B$. If, in addition, $v_0 \in E(L)$, then the corresponding solution $v(t, v_0)$ turns out to be classical for all $[0, \bar{t}\,]$.*

Lemma 4. *For any generalized solution $v = v_*(t)$, defined on an interval $0 \leq t \leq \bar{t}$, there exists a sequence of classical solutions $v = v_k(t)$, $k \geq 1$, $t \in [0, \bar{t}\,]$, such that*

$$\lim_{k \to \infty} \max_{0 \leq t \leq \bar{t}} ||v_*(t) - v_k(t)||_E = 0.$$

Now we define the so-called maximal existence semi-interval $[0, t_{\max})$, where

$$t_{\max} = \sup \{ \; t_0 : \;\; v(t) \qquad t \in [0, t_0\,] \; \}. \tag{9}$$

According to Lemma 2, the set of values of t_0 involved in finding the least upper bound in (9) is certainly nonempty. If $t_{\max} < +\infty$, then the following statement applies.

Lemma 5. *We assume that for some generalized solution $v(t)$ of Eq. (6), the quantity in (9) is finite. Then the limit equality*

$$\lim_{t \to t_{\max} - 0} ||v(t)||_E = +\infty.$$

holds.

The lemmas presented here can be established in the same way as in [12], where similar results are proved. In what follows, under some additional constraints imposed on the matrix D and the nonlinearity $F(u)$, we will study the attractors of the nonlinear semiflow generated by generalized solutions to equation (6) in the phase space E.

2. Main results

Let us specify the statement of the problem. To this end, we consider the following boundary value problem similar to (6):

$$\frac{\partial u}{\partial t} = F(u, \varepsilon) + \sqrt{\varepsilon}(D_0 + \sqrt{\varepsilon} D_1)\frac{\partial u}{\partial x}, \quad u(t, x+1) \equiv u(t, x), \tag{10}$$

where $u \in \mathbb{R}^n, n \geq 3$; $\varepsilon > 0$ is a small parameter, and the vector function $F(u, \varepsilon)$ is jointly infinitely differentiable with respect to the variables $(u, \varepsilon) \in \mathbb{R}^n \times [0, \varepsilon_0]$, $\varepsilon_0 > 0$, and is such that $F(0, \varepsilon) \equiv 0$. Below we will need the Taylor expansion of the vector function $F(u, \varepsilon)$ at the point $(u, \varepsilon) = (0, 0)$,

$$F(u, \varepsilon) = (A_0 + \varepsilon A_1)u + F_2(u, u) + F_3(u, u, u) + O(\varepsilon^2 ||u|| + \varepsilon ||u||^2 + ||u||^4), \tag{11}$$

where A_0 and A_1 are $n \times n$ square matrices and F_2, F_3 are $\mathbb{R}^n$-valued quadratic and cubic symmetric forms.

Condition 1. The matrix A_0 in (11) has a simple pair of purely imaginary eigenvalues $\lambda = \pm i\omega_0$, $\omega_0 > 0$, and the corresponding eigenvectors a and a are normalized by the conditions

IOP Conf. Series: Journal of Physics: Conf. Series **937** (2017) 012015 doi:10.1088/1742-6596/937/1/012015

$(a, b) = 1$ and $(\overline{a}, b) = 0$ (here and below, the bar denotes complex conjugation, $(*, *)$ is the Euclidean inner product in $\mathbb{C}^n$, and the vector b is such that $A_0^* b = -i\omega_0 b$). The other eigenvalues of the matrix A_0 are assumed to lie in the left complex half-plane $\{\lambda : \operatorname{Re}\lambda < 0\}$.

Condition 2. The inequalities $\operatorname{Re}(A_1 a, b) > 0$ and $\operatorname{Re} d < 0$, hold, where

$$d = 2(F_2(\chi_0, a), b) + 2(F_2(\chi_2, \overline{a}), b) + 3(F_3(a, a, \overline{a}), b), \tag{12}$$

$$\chi_0 = -2A_0^{-1} F_2(a, \overline{a}), \quad \chi_2 = [2i\omega_0 I - A_0]^{-1} F_2(a, a), \tag{13}$$

and I stands for the identity matrix.

The Conditions 1 and 2, taken together, guarantee that a classical AndronovHopf bifurcation take place in the partial system

$$\dot{u} = F(u, \varepsilon) \tag{14}$$

Indeed, it follows from the results of [9], [10] that under these conditions, for any $0 < \varepsilon \ll 1$, system (14) has a stable cycle with amplitude of order $\sqrt{\varepsilon}$ and with frequency close to ω_0. This cycle is simultaneously a solution to the boundary value problem (10) that does not depend on the space variable x. Such a cycle will be called a spatially homogeneous (or simply homogeneous) cycle. The following constraint concerns the coupling matrices D_0 and D_1 that appear in (10).

Condition 3. The matrix D_1 is arbitrary and D_0 has simple eigenvalues $\lambda_k \in \mathbb{R}$, $k = 1, \ldots, n$, and corresponding eigenvectors e_k, $k = 1, \ldots, n$. These eigenvectors are assumed to be normalized by the conditions $(e_k, g_k) = 1$, where $D_0^* g_k = \lambda_k g_k$, $k = 1, \ldots, n$.

Now we linearize the boundary value problem (9) at the zero equilibrium state and apply the Fourier method with respect to the system of functions $\exp(2\pi kix)$, $k \in \mathbb{Z}$ to the resulting linear boundary value problem

$$\frac{\partial u}{\partial t} = (A_0 + \varepsilon A_1)u + \sqrt{\varepsilon}(D_0 + \sqrt{\varepsilon} D_1)\frac{\partial u}{\partial x}, \quad u(t, x+1) \equiv u(t, x).$$

As a result, we have the following countable system of ordinary differential equations for the u_k:

$$\dot{u}_k = \big(A_0 + \varepsilon A_1 + iz(D_0 + \sqrt{\varepsilon} D_1)\big)u_k, \quad z = 2\pi k\sqrt{\varepsilon}.$$

The stability of these equations is defined by the spectral properties of the matrix

$$\mathscr{D}(z) = A_0 + izD_0, \tag{15}$$

where we assume that the parameter z taking discrete values $2\pi k\sqrt{\varepsilon}$ varies continuously on the half-axis $z \geq 0$ (the case of $z < 0$ does not need separate consideration because $\mathscr{D}(-z) \equiv \overline{\mathscr{D}}(z)$).

Let us Conditions 1 and 3 are hold. Than in view of Condition 1 matrix (15) has eigenvalues $\lambda_1(z)$ and $\lambda_2(z)$ such that $\lambda_1(z) = \overline{\lambda}_2(-z)$ and $\lambda_1(0) = i\omega_0$, for all sufficiently small values of the parameter z. Next, its easy to see that $\lambda_1(z)$ at the point z = 0 has the Taylor expansion

$$\lambda_1(z) = i\omega_0 + \lambda_{1,1} z + \lambda_{1,2} z^2 + \ldots, \quad \lambda_{1,1} = i(D_0 a, b), \quad \lambda_{1,2} = i(D_0 a_1, b), \tag{16}$$

where a_1 is a solution to the linear system

$$(A_0 - i\omega_0 I)a_1 = i(D_0 a, b)a - iD_0 a, \tag{17}$$

where $(a_1, b) = 0$.

According to Condition 3, for $z \gg 1$, the eigenvalues $\lambda_k(z)$, $k = 1, \ldots, n$, of the matrix (15) satisfy the asymptotic relation

$$\lambda_k(z) = iz\lambda_k + (A_0 e_k, g_k) + O(1/z), \quad k = 1, \ldots, n, \tag{18}$$

where λ_k are the eigenvalues of the matrix D_0.

Condition 4. We assume that

$$\operatorname{Re}\lambda_{1,1} = 0, \quad \operatorname{Re}\lambda_{1,2} < 0, \quad (A_0 e_k, g_k) < 0, \quad k = 1, \ldots, n, \tag{19}$$

and the matrix (15) is a Hurwitz matrix for any $z > 0$.

Notice that for small z and for $z \gg 1$, the Hurwitz property of the matrix $\mathscr{D}(z)$ follows from Conditions 1 and 3 and relations (19). In the case of $z \sim 1$, this property needs to be postulated. Note also that the requirement $\operatorname{Re}\lambda_{1,1} = 0$ in (19) describes a certain singularity. This clarifies the role of the matrix D_1 in (10): it characterizes a deviation from the singularity.

We address the question of the existence and stability of periodic solutions of the traveling-wave type in boundary value problem (10) under Conditions 1–4. For this, we make the following substitution

$$u = \sqrt{\varepsilon}\, u_0(t, \tau, s, x) + \varepsilon u_1(t, \tau, s, x) + \varepsilon^{3/2} u_2(t, \tau, s, x) + \ldots \tag{20}$$

Here $\tau = \sqrt{\varepsilon}\, t$, $s = \varepsilon t$,

$$u_0 = \xi(\tau, s, x) a \exp(i\omega_0 t) + \overline{\xi}(\tau, s, x)\overline{a} \exp(-i\omega_0 t), \tag{21}$$

$\xi(\tau, s, x)$, $\xi(\tau, s, x+1) \equiv \xi(\tau, s, x)$ is a complex amplitude that (it will be specified later), and the functions $u_k(t, \tau, s, x)$, $k = 1, 2$ are some trigonometric polynomials in the variable $\omega_0 t$ with vector coefficients depending on τ, s, and x.

To find the functions $u_k(t, \tau, s, x)$, $k = 1, 2$, we substitute relations (20) and (21) together with the Taylor expansion (11) into (10) and successively equate the coefficients of ε and $\varepsilon^{3/2}$ on the left- and right-hand sides of the expression obtained. As a result, we arrive at linear inhomogeneous equations of the form

$$\frac{\partial u_k}{\partial t} = A_0 u_k + \gamma_k(t, \tau, s, x), \quad k = 1, 2, \tag{22}$$

where τ, s, x takes as parameters.

For $k = 1$, the inhomogeneity in (22) is defined by the equality

$$\begin{aligned} \gamma_1 = \gamma_{1,0}(\tau, s, x) + \gamma_{1,1}(\tau, s, x)\exp(i\omega_0 t) + \overline{\gamma}_{1,1}(\tau, s, x)\exp(-i\omega_0 t) + \\ + \gamma_{1,2}(\tau, s, x)\exp(2i\omega_0 t) + \overline{\gamma}_{1,2}(\tau, s, x)\exp(-2i\omega_0 t), \end{aligned} \tag{23}$$

where

$$\gamma_{1,0} = 2|\xi|^2 F_2(a, \overline{a}), \quad \gamma_{1,1} = D_0 a \frac{\partial \xi}{\partial x} - a \frac{\partial \xi}{\partial \tau}, \quad \gamma_{1,2} = \xi^2 F_2(a, a). \tag{24}$$

We seek the function $u_1(t, \tau, s, x)$ just as in (23), as a sum of zero, first, and second harmonics with coefficients $\beta_{1,r}$, $r = 0, 1, 2$. In this way, we obtain the following linear inhomogeneous systems for the indicated coefficients:

$$(ir\omega_0 I - A_0)\beta_{1,r} = \gamma_{1,r}, \quad r = 0, 1, 2. \tag{25}$$

In view of Condition 1 imposed on the spectrum of the matrix A_0, these systems are uniquely solvable for $r = 0, 2$. The solvability condition for this system for $r = 1$ is

$$(\gamma_{1,1}, b) = 0. \tag{26}$$

Now we take into account the explicit formula for $\gamma_{1,1}$ (see (24)) and the fact that the quantity $(D_0 a, b)$ is real by Condition 4. Then we obtain for the amplitude $\xi(\tau, s, x)$ the equation

$$\frac{\partial \xi}{\partial \tau} = \kappa \frac{\partial \xi}{\partial x}, \qquad \kappa = (D_0 a, b) \in \mathbb{R}.$$

This implies that

$$\xi = \xi(s, y), \qquad y = \kappa\tau + x. \tag{27}$$

Then we substitute relations (27) into equalities (24). As a result, we obtain the formula

$$u_1 = -i\, a_1 \frac{\partial \xi}{\partial y} \exp(i\omega_0 t) + i\, \bar{a}_1 \frac{\partial \bar{\xi}}{\partial y} \exp(-i\omega_0 t) + \chi_0 |\xi|^2 + \chi_2\, \xi^2 \exp(2i\omega_0 t) + \bar{\chi}_2\, \bar{\xi}^2 \exp(-2i\omega_0 t), \tag{28}$$

where a_1 is a solution of the system (17), and χ_0 and χ_2 are the vectors (13).

For $k = 2$ we obtain the equality

$$\gamma_2 = A_1 u_0 + 2F_2(u_0, u_1) + F_3(u_0, u_0, u_0) + D_0 \frac{\partial u_1}{\partial x} + D_1 \frac{\partial u_0}{\partial x} - \frac{\partial u_1}{\partial \tau} - \frac{\partial u_0}{\partial s}.$$

Relations (21) and (28) imply that the inhomogeneity γ_2 is a linear combination of the harmonics $\exp(\pm r i\omega_0 t)$, $r = 0, 1, 2, 3$. The function u_2 is sought in the same form. In this way, we obtain nondegenerate linear inhomogeneous systems for the coefficients of the harmonics with numbers $r = 0, 2, 3$ in the decomposition of this function. In the case of $r = 1$, a solvability condition similar to (26) arises and we see that the amplitude $\xi(s, y)$ is a solution to the boundary value problem

$$\frac{\partial \xi}{\partial s} = -i(D_0 a_1, b) \frac{\partial^2 \xi}{\partial y^2} + (D_1 a, b) \frac{\partial \xi}{\partial y} + (A_1 a, b)\xi + d\, |\xi|^2 \xi, \quad \xi(s, y+1) \equiv \xi(s, y), \tag{29}$$

where d is the parameter (12). It is important that in view of the inequality $\mathrm{Re}\left[-i(D_0 a_1, b)\right] > 0$ (see Condition 4), this problem turns out to be parabolic.

Following the tradition, we will call the boundary value problem (29) the quasinormal form of the original problem (10). Note that for $D_0 = D_1 = 0$ the equation in (29) is a truncated normal form of (14) on a stable two-dimensional invariant manifold. If $D_0 \neq 0$, then the boundary value problem (10) does not have an analogous invariant manifold, and the derivation of problem (29) becomes a purely formal procedure. Indeed, when deriving this problem, we discarded the terms that, although being small in order of magnitude compared with ε, are not subordinate to the other terms. In spite of all that has been said above, the quasinormal form (29) nevertheless contains some information on the attractors of the original problem (10). Before formulating relevant rigorous results, we give two definitions.

A *traveling-wave-type self-similar cycle of problem* (29) is a periodic solution of the form

$$\xi(s, y) = \xi_0 \exp\left[2\pi i(\sigma s + p y)\right], \tag{30}$$

where $\xi_0, \sigma = \mathrm{const} \in \mathbb{R}$, $p \in \mathbb{Z}$.

A *two-dimensional self-similar torus of problem* (29) is a quasiperiodic solution of the problem of the form

$$\xi(s, y) = \xi_0(\sigma_1 s + y) \exp(2\pi i \sigma_2 s), \tag{31}$$

where $\sigma_1, \sigma_2 = \mathrm{const} \in \mathbb{R}$ and $\xi_0(y)$ is a 1-periodic complex-valued function. We assume that the solution (31) does not reduce to the solution (30).

The following two standard assertions concern the correspondence between the solutions (30) and (31) of the quasinormal form and analogous steady-state regimes of problem (10).

Theorem 1. *Suppose that the quasinormal form (29) has a self-similar cycle of the form (30) that is exponentially orbitally stable or dichotomic (in the metric of the phase space $(\mathrm{Re}\,\xi, \mathrm{Im}\,\xi) \in \mathrm{W}_2^1 \times \mathrm{W}_2^1$, where W_2^1 is the Sobolev space of 1-periodic functions). Then, for any positive integer l, there exists a sufficiently small $\varepsilon_l > 0$ such that for all $0 < \varepsilon \le \varepsilon_l$ the original boundary value problem (10) admits a traveling-wave-type cycle with the same stability properties that is defined by the equalities*

$$u = \sqrt{\varepsilon}\,\xi_0\big[\exp\big(2\pi i(\theta + px)\big)a + \exp\big(-2\pi i(\theta + px)\big)\overline{a}\big] + \varepsilon\mathscr{U}(\theta + px, \varepsilon),$$
$$\frac{d\theta}{dt} = \frac{\omega_0}{2\pi} + \sqrt{\varepsilon}\,p\kappa + \varepsilon\sigma + \varepsilon^{3/2}\Delta(\varepsilon). \tag{32}$$

Here ξ_0, σ and p are the constants from (30), κ is the constant from (27), $\Delta(\varepsilon)$ is a bounded function of ε, and the vector function $\mathscr{U}(\theta, \varepsilon) \in \mathbb{R}^n$, $\mathscr{U}(\theta + 1, \varepsilon) \equiv \mathscr{U}(\theta, \varepsilon)$, is continuous and bounded on the set $(\theta, \varepsilon) \in \mathbb{R} \times (0, \varepsilon_l]$ together with its partial derivatives with respect to θ up to the order l inclusive.

Theorem 2. *For every self-similar torus (31) of the quasinormal form (29) that is exponentially orbitally stable or dichotomic, the original problem (10) has a two-dimensional invariant torus with the same stability properties for all sufficiently small $\varepsilon > 0$. This torus admits a parametric representation analogous to (32):*

$$u = \sqrt{\varepsilon}\left[\xi_0(\theta_1 + x)\exp(2\pi i\theta_2)a + \overline{\xi}_0(\theta_1 + x)\exp(-2\pi i\theta_2)\overline{a}\right] + \varepsilon\mathscr{U}(\theta_1 + x, \theta_2, \varepsilon),$$
$$\frac{d\theta_1}{dt} = \sqrt{\varepsilon}\,\kappa + \varepsilon\sigma_1 + \varepsilon^{3/2}\Delta_1(\theta_2, \varepsilon), \quad \frac{d\theta_2}{dt} = \frac{\omega_0}{2\pi} + \varepsilon\sigma_2 + \varepsilon^{3/2}\Delta_2(\theta_2, \varepsilon), \tag{33}$$

where the functions $\mathscr{U}$ and Δ_j, $j = 1, 2$ depend periodically (with period 1) on the phase variables θ_1 and θ_2 and are continuous and bounded in $(\theta_1, \theta_2, \varepsilon)$ together with any fixed number of their derivatives with respect to (θ_1, θ_2).

Note that the special dependence on x in (33) and the absence of the argument θ_1 in the functions Δ_j, $j = 1, 2$, are associated with the invariance of this torus with respect to the rotational symmetry, i.e., with respect to the changes $x \to x + c$ and $\theta_1 \to \theta_1 - c$, where c is an arbitrary real number.

The theorems formulated above can be proved by the methods described in [13] and the monograph [14].

3. Attractors of the quasinormal form. Numerical results

We reduce the boundary value problem (29) to a canonical form. To this end, taking into account the inequalities $\mathrm{Re}\,(A_1 a, b) > 0$ and $\mathrm{Re}\,d < 0$ (see Condition 2), we successively make the following changes in it

$$\xi\exp(-i\,\mathrm{Im}\,(A_1 a, b)s) \to \xi, \quad y + \mathrm{Re}\,(D_1 a, b)s \to y, \quad \mathrm{Re}\,(A_1 a, b)s \to s,$$
$$\xi = \sqrt{-\mathrm{Re}\,(A_1 a, b)/\mathrm{Re}\,d} \cdot w, \quad s \to t, \quad y \to x.$$

Then we obtain a model boundary value problem

$$\frac{\partial w}{\partial t} = \nu\left((1 - ic_1)\frac{\partial^2 w}{\partial x^2} + ic_2\frac{\partial w}{\partial x}\right) + w - (1 + ic_3)|w|^2 w, \quad w(t, x + 1) \equiv w(t, x), \tag{34}$$

where

$$\nu = \frac{\mathrm{Re}\,[-i(D_0 a_1, b)]}{\mathrm{Re}\,(A_1 a, b)} > 0, \quad c_1 = -\frac{\mathrm{Im}\,[-i(D_0 a_1, b)]}{\mathrm{Re}\,[-i(D_0 a_1, b)]} \in \mathbb{R},$$
$$c_2 = \frac{\mathrm{Im}\,(D_1 a, b)}{\mathrm{Re}\,[-i(D_0 a_1, b)]} \in \mathbb{R}, \quad c_3 = \frac{\mathrm{Im}\,d}{\mathrm{Re}\,d} \in \mathbb{R}. \tag{35}$$

The problem (34) always has the homogeneous cycle

$$w = \exp(-ic_3 t). \tag{36}$$

This cycle is a particular case of a traveling wave; i.e., it is contained in the family of cycles

$$w = w_p \exp(i\sigma_p t + 2\pi p i x), \tag{37}$$

where $w_p, \sigma_p = \text{const} \in \mathbb{R}$, $p \in \mathbb{Z}$. Conditions for the existence and stability of such cycles are given in the following assertion.

Theorem 3. *Suppose that*

$$1 - 2\pi\nu p(2\pi p + c_2) > 0. \tag{38}$$

for some $p \in \mathbb{Z}$. *Then the boundary value problem* (34) *admits a* p*-th traveling wave* (37) *whose amplitude and frequency are defined by the formulas*

$$w_p = \sqrt{1 - 2\pi\nu p(2\pi p + c_2)}, \quad \sigma_p = 4\pi^2 p^2 \nu\, c_1 - c_3 w_p^2. \tag{39}$$

The cycle (37), (39) *is exponentially orbitally stable if the infinite series of inequalities*

$$\begin{gathered}(w_p^2 + 4\nu\pi^2 k^2)^2 \left[2w_p^2(1 - c_1 c_3) + 4\nu\pi^2 k^2(1 + c_1^2) - \nu(4\pi p + c_2)^2\right] > \\ > \nu(4\pi p + c_2)^2(-w_p^2 c_3 + 4\nu\pi^2 k^2 c_1)^2, \quad k \in \mathbb{N}\end{gathered} \tag{40}$$

holds, and is unstable if at least one of these inequalities is strictly violated.

The analysis of conditions (38) and (40) for $0 < \nu \ll 1$ and for fixed c_1, c_2, and c_3 shows that in this case the boundary value problem (34) has traveling waves (37), (39) with numbers $p = 0, \pm 1, \ldots, \pm p_0$, where p_0 is any predetermined fixed positive integer. All these cycles are simultaneously stable (unstable) if the inequality $c_1 c_3 - 1 < 0$ (> 0) holds. This and Theorem 1 imply that for $c_1 c_3 < 1$, as the quantity $\operatorname{Re}(A_1 a, b)$ increases (or, equivalently (see (35)), as the parameter ν decreases), the original boundary value problem (10) exhibits a buffer phenomenon, i.e., an unbounded accumulation of stable cycles of the form (32).

Two conclusions hold, firstly, the homogeneous cycle (36) is stable (unstable) for $|c_2| - 2\pi < 0$ (> 0) in case, when $\nu \gg 1$ and $c_j \sim 1$, $j = 1, 2, 3$, and secondly, all traveling waves (37), (39) provided by Theorem 3 are unstable in the case of $\nu, c_1, c_3 \sim 1$, $c_1 c_3 > 1$ and $|c_2| \gg 1$. Thus, the most interesting range of the parameters c1, c2, and c3 is specified by the inequalities

$$c_1 c_3 > 1, \quad |c_2| > 2\pi. \tag{41}$$

Indeed, under these conditions, the boundary value problem (34) has nontrivial dynamics both for $\nu \gg 1$ (its homogeneous cycle (36) is certainly unstable) and for $\nu \ll 1$ (all existing traveling waves are unstable). Below we will show by numerical analysis that for fixed c_j, $j = 1, 2, 3$, satisfying inequalities (41) and for $\nu \ll 1$ problem (34) admits a chaotic attractor A_ν whose Lyapunov dimension $d_L(A_\nu)$ (calculated by the well-known KaplanYorke formula [15], algorithm from [16]) tends to infinity as $\nu \to 0$. In a similar way, an indefinite growth of the dimension of the chaotic attractor is observed for $\nu, c_1, c_3 \sim 1$, and $c_1 c_3 > 1$ as $|c_2|$ increases.

To describe the corresponding numerical experiments, we introduce points $x = j/N$, $j = 1, \ldots, N$, where N is an arbitrary fixed positive integer, and replace the partial derivatives with respect to x in (34) at these points with the corresponding symmetric differences. As a result, we arrive at the following finite-dimensional model for the variables $w_j(t) = w(t, x)|_{x=j/N}$:

$$\begin{gathered}\dot{w}_j = \nu\left[(1 - ic_1)N^2(w_{j+1} - 2w_j + w_{j-1}) + ic_2 N(w_{j+1} - w_{j-1})/2\right] + w_j - \\ -(1 + ic_3)|w_j|^2 w_j, \quad j = 1, \ldots, N, \quad w_0(t) = w_N(t), \quad w_{N+1}(t) = w_1(t).\end{gathered} \tag{42}$$

IOP Conf. Series: Journal of Physics: Conf. Series **937** (2017) 012015 doi:10.1088/1742-6596/937/1/012015

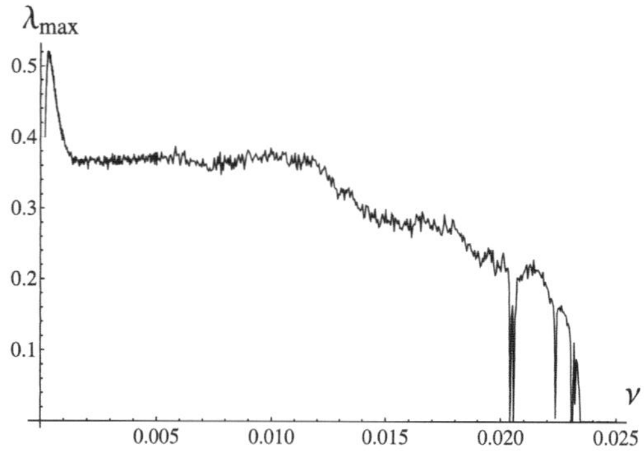

Figure 1.

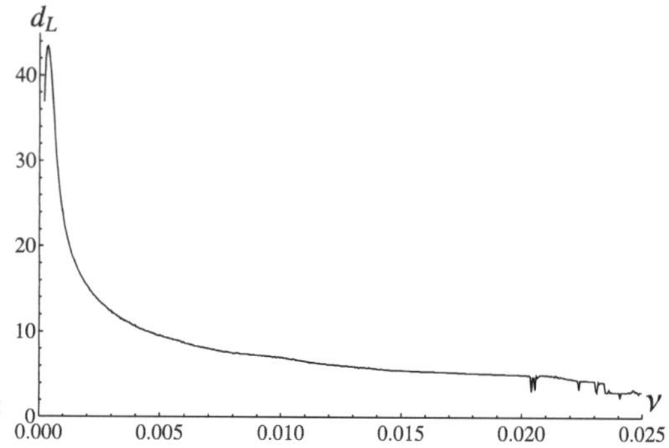

Figure 2.

For the parameter values

$$c_1 = 1, \quad c_2 = 8, \quad c_3 = 4, \tag{43}$$

which satisfy inequalities (41), a chaotic attractor $A_\nu(N)$ arises in system (42) as ν decreases. A clear picture of this process is given by the graphs of the leading Lyapunov exponent $\lambda_{\max}\big(A_\nu(N)\big)$ and the Lyapunov dimension $d_L\big(A_\nu(N)\big)$ of this attractor versus ν. For $N = 30$, the corresponding graphs drawn point-by-point with a step of 0.00005 on the interval $0 \le \nu \le 0.025$ are shown in Figs. 1 and 2. A characteristic feature of the dependence of $d_L\big(A_\nu(N)\big)$ on ν is the presence of a maximum at small ν. This is associated with the fact that for a fixed N, as $\nu \to 0$, the attractors in the difference model (42) become simpler (stable traveling waves arise that have no analogs in the distributed model (34)). Therefore, for too small values of ν, the difference model ceases to adequately describe the dynamics of problem (34). Note in addition that as N increases, the above-mentioned maximum shifts to zero and the value of $d_L\big(A_\nu(N)\big)$ at the maximum point indefinitely increases.

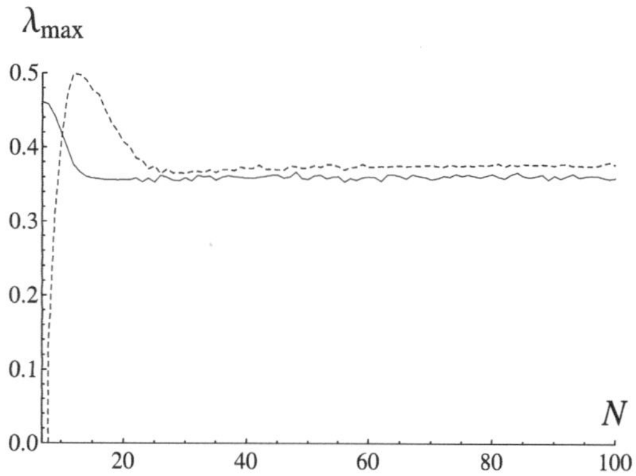

Figure 3.

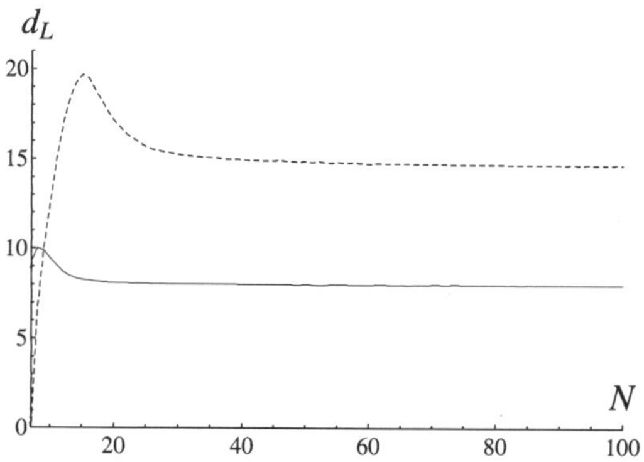

Figure 4.

At the next stage, we verify that as N increases, the attractor $A_\nu(N)$ of the difference model (42) converges in a some sense to the attractor A_ν of the boundary value problem (34). Indeed, numerical calculations for fixed ν show that the quantities $\lambda_{\max}\big(A_\nu(N)\big)$ and $d_L\big(A_\nu(N)\big)$ stabilize as N increases. Figures 3 and 4 illustrate the dependence of these quantities on N for

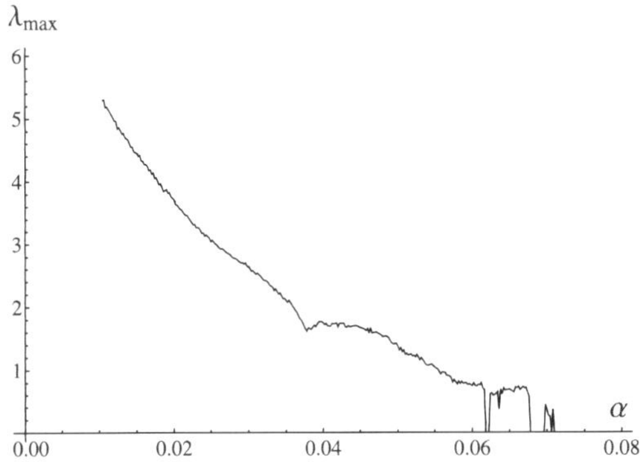

Figure 5.

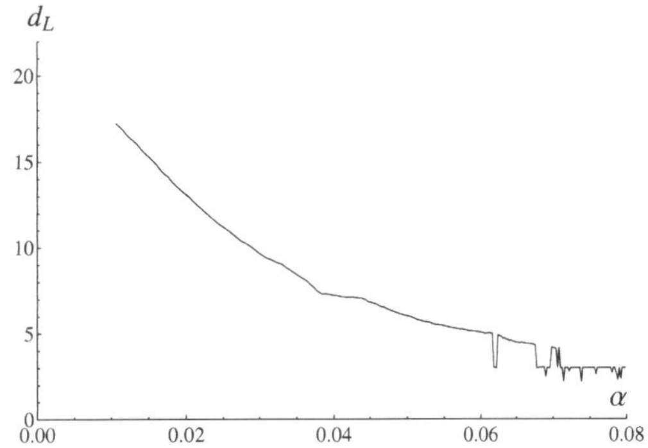

Figure 6.

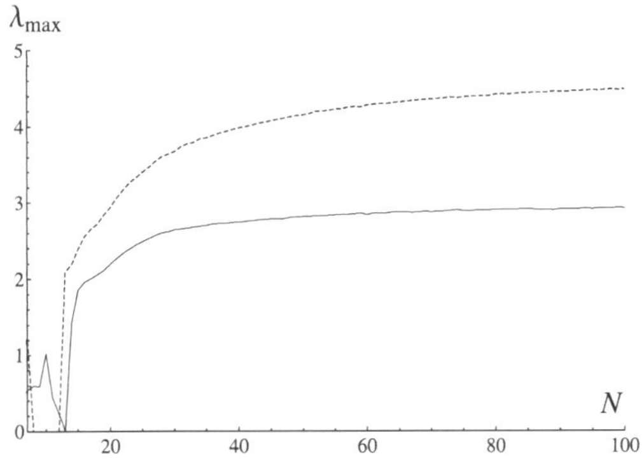

Figure 7.

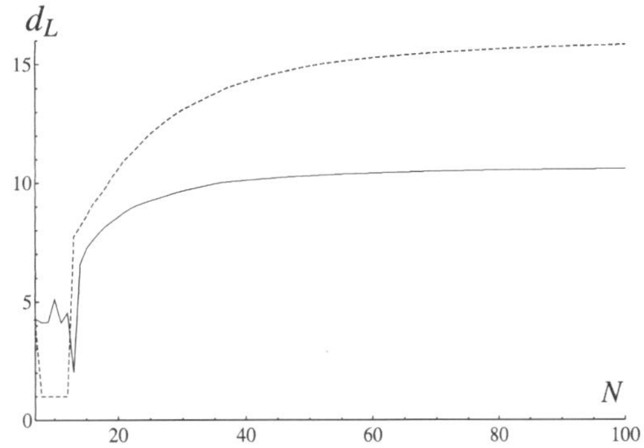

Figure 8.

$5 \leq N \leq 100$ when $\nu = 0.007$ and $\nu = 0.002$ (solid lines correspond to the case of $\nu = 0.007$, and the dashed lines, to the case of $\nu = 0.002$). Here it is important that smaller values of the parameter ν correspond to greater values of d_L. Thus, we have every reason to expect that for the set of parameters (43) the dimension $d_L(A_\nu)$ of the chaotic attractor A_ν of the boundary value problem (34) indefinitely increases as $\nu \to 0$.

A similar conclusion on the indefinite growth of the dimension of the chaotic attractor of the boundary value problem (34) is valid for $c_1 = 1$, $c_3 = 4$, and $\nu = 1$ as the quantity $\alpha = 1/c_2 > 0$ decreases. To verify this, for the indicated values of the parameters we present the results of numerical analysis of the system

$$\begin{gathered} \dot{w}_j = i\nu\, N(w_{j+1} - w_{j-1})/2 - (1 + ic_3)|w_j|^2 w_j + \\ +\alpha\big[\nu\,(1 - ic_1)N^2(w_{j+1} - 2w_j + w_{j-1}) + w_j\big], \quad j = 1, \ldots, N, \end{gathered} \tag{44}$$

which is obtained from (42) after the substitutions $w_j/\sqrt{c_2} \to w_j$ and $c_2 t \to t$.

Just as in the previous case, for small α system (44) admits a chaotic attractor $A_\alpha(N)$. For $N = 30$, the dependence of the leading Lyapunov exponent λ_{max} and the Lyapunov dimension d_L of this attractor on α is demonstrated in Figs. 5 and 6. These graphs are drawn point-by-point with a step of 0.0001 on the interval $0.01 \leq \alpha \leq 0.08$. The interval $0 \leq \alpha \leq 0.01$ has not been considered because the difference model (44) ceases to be adequate to the continuous model (34) for fixed N and very small α (chaos is destroyed, and stable traveling waves whose analogs in the continuous model are unstable arise). Next, Figs. 7 and 8 demonstrate the dependence of $\lambda_{\max}$ and d_L on N for $5 \le N \le 100$ when $\alpha = 0.03$ and $\alpha = 0.02$ (solid and dashed lines, respectively). The information presented supports the conjecture that $d_L(A_\alpha) \to +\infty$ as $\alpha \to 0$, where $A_\alpha = \lim A_\alpha(N), N \to \infty$ is the attractor of the original boundary value problem (34) for $c_1 = 1$, $c_3 = 4$, $\nu = 1$, and $\alpha = 1/c_2$.

To conclude the description of the results of numerical calculation, we note that for the parameter values (43) and for all $\nu \ge 0.1$ problem (34) has a stable self-similar two-dimensional torus of the form (31). Thus, we see that the applicability domain of Theorem 2 is certainly nonempty.

4. Conclusions

The reason why the order of smallness of the coupling matrix (10) is chosen to be $\sqrt{\varepsilon}$ is associated with the spectral properties of the matrix (15). Such a coupling order has allowed us to apply the known method of quasinormal forms to the boundary value problem (10) and obtain rigorous results on the correspondence between special steady-state regimes of the boundary value problems (29) and (10). A similar correspondence between the chaotic attractors of these problems has not been established; however, there is every reason to expect that such a correspondence also takes place. Based on the analytic and numerical analysis performed above within the framework of this hypothesis, we get a sufficiently clear picture of the local dynamics of the boundary value problem (10) in the neighborhood of the zero equilibrium state. The main features of this dynamics are the following two alternatives that take place for an appropriate choice of the parameters: either the buffer phenomenon occurs or a chaotic attractor of an arbitrarily high Lyapunov dimension arises (see also [16]–[19]).

Acknowledgments

This work was supported by the Russian Science Foundation (project nos. 14-21-00158)

References

[1] Glyzin S D, Kolesov A Yu, and Rozov N Kh 2006 *Comput. Math. Math. Phys.* **46** 1724–1736
[2] Mishchenko E F, Sadovnichii V A, Kolesov A Yu, and Rozov N Kh 2012 *Multifaceted Chaos* (Moscow: Fizmatlit, Russian)
[3] Kapitaniak T and Chua L O 1994 *Internat. J. Bifurcation Chaos Appl. Sci. Engrg.* **4** 477–482
[4] Mariño I.P, Pérez-Muñuzuri V, Pérez-Villar V, Sánchez E, and Matías M A 1999 *Phys. D* **128** 224–235
[5] Perlikowski P, Yanchuk S, Wolfrum M, Stefanski A, Mosiolek P, Kapitaniak T 2010 *Chaos* **20** 013111
[6] Glyzin S D, Kolesov A Yu, and Rozov N Kh 2013 *Theor. Math. Phys.* **175** 499–517
[7] Glyzin S D, Kolesov A Yu, and Rozov N Kh 2013 *Izv. Math.* **77** 271–312
[8] Glyzin S D, Kolesov A Yu, and Rozov N Kh 2013 *Differ. Equ.* **49** 1193–1210.
[9] Hassard B D, Kazarinoff N D, and Wan Y H 1981 *Theory and Applications of Hopf Bifurcation* (London Math. Soc. Lect. Note Ser., Vol. 41, Cambridge: Cambridge Univ. Press)
[10] Kolesov A Yu, and Rozov N Kh 2004 *Invariant Tori of Nonlinear Wave Equations* (Moscow: Fizmatlit, Russian)
[11] Henry D 1981 *Geometric Theory of Semilinear Parabolic Equations* (Lect. Notes Math., Vol. 840, Berlin: Springer)
[12] Glyzin S D, Kolesov A Yu, and Rozov N Kh 2009 *Theor. Math. Phys.* **158** 246–261
[13] Kolesov Yu S 1994 *Sb. Math.* **78** 367– 78
[14] Mishchenko E F, Sadovnichii V A, Kolesov A Yu, and Rozov N Kh 2005 *Autowave Processes in Nonlinear Media with Diffusion* (Moscow: Fizmatlit, Russian)
[15] Frederickson P, Kaplan J L, Yorke E D, and Yorke J A 1983 *J. Diff. Eqns.* **49 (2)** 185–207
[16] Glyzin D S, Glyzin S D, Kolesov A Yu, and Rozov N Kh 2005 *Differ. Equ.* **41** 284–289
[17] Glyzin S D, Kolesov A Yu, and Rozov N Kh 2010 *Comput. Math. Math. Phys.* **50** 816–830
[18] Glyzin S D, Kolesov A Yu, and Rozov N Kh 2011 *Comput. Math. Math. Phys.* **51** 1307–1324
[19] Glyzin S D 2013 *Automatic Control and Computer Sciences* **47** 452–469.

IOP Conf. Series: Journal of Physics: Conf. Series **937** (2017) 012016 doi:10.1088/1742-6596/937/1/012016

Multiplicity of chaotic attractors in a model of lasers with variable feedback delay

E V Grigorieva[1,3] **and S A Kaschenko**[2,3]

[1]Belarus State Economic University, 220070, Partizanski av. 26, Minsk, Belarus
[2]Yaroslavl State University, 150000 Sovetskaya av. 14, Yaroslavl, Russia
[3]National Research Nuclear University MIFI, Moscow, Russia

E-mail: grigorieva@tut.by, kasch@uniyar.ac.ru

Abstract. We demonstrate coexistence of chaotic attractors induced by periodic modulation of the delay time in the feedback circuit which controls pumping rate in a laser. Depending on the n-number of pulses in the delay interval, the dynamics of the pulse regime is described by the dynamics of the $(2n + 2)$ -dimensional map. It is possible to switch chaotic regimes of different structures by a single impulse perturbation of pumping in a suitable phase of the oscillations.

1. Introduction

Delayed feedback (FB) is widely used to control the dynamics of nonlinear optical and electro-optical systems [1]. Recently, modifications of standard FB schemes have been actively studied, in particular, high-frequency periodic modulation of the delay time in the FB scheme was discussed to stabilize the state of unstable equilibrium [2,3]. On the other hand, the variation of the delay time with a frequency comparable with the relaxation one can be used to obtain oscillating regimes with given characteristics. This may be of interest for methods of optical processing and information coding, optical vibrometry and in other applications, see, for example, [5].

In this paper, we show the coexistence of irregular pulse regimes caused by periodic modulation of the delay time in the FB circuit. Such regimes can be classified as slowly oscillating with inter-spikes intervals longer than the delay time or fast oscillating with inter-spikes intervals shorter than the delay.

Of particular importance can be the task of purposeful switching of coexisting attractors. Previously, quick switching periodic cycles by external impulse was suggested for a loss-driven CO_2 laser [6]. With a suitable choice of the switching pulse characteristics the system may be found on stable periodic cycles, as well as on unstable cycles. It was found that there is an optimal time of application of impact, which corresponds to the minimum duration of the transition process to the target cycle. The theoretical explanation was given on the base of the 3-dimensional non-autonomous model [7].

In the present paper, we study coexisting chaotic attractors with basins that may have fractal boundaries and are embedded in the infinite-dimensional phase space of a system with retarded argument. Thus, the possibility of a single-cycle transition is not obvious. Nevertheless, we note the conditions that can ensure successful switching by a quick change of the system position within the phase space. To this end we derive asymptotically the finite-dimensional maps responsible for the dynamics of spikes. With an impulse-perturbation of the pumping rate in a suitable phase of the oscillation, it is possible to achieve switching chaotic regimes of different structures.

2. Model of laser dynamics

Relaxation oscillations in a semiconductor laser with an optoelectronic FB can be studied on the base of the model proposed in [8], which we supplements by periodic modulation of the delay time,

$$\frac{du}{dt} = vu(y-1), \ \frac{dy}{dt} = q + \gamma u(g(t)) - y - yu, \tag{1}$$

where q is the pumping rate; v is the ratio of the rate of decay photons in the cavity to the rate of relaxation of populations; γ is the feedback factor (FB level); t is current time normalized at the time of relaxation of the population inversion. The optoelectronic FB is represented by the term $\gamma u(g(t))$, where γ is the feedback coefficient, the delayed argument has the form $g(t) = t - (\tau_0 + B\cos\Phi(t))$, the modulator phase $\Phi(t) = \omega t + \varphi$, τ_0 is the constant time of the radiation transformation in the FB circuit, B and ω is the amplitude and frequency of the delay modulation, respectively, φ is the initial phase of the modulating signal.

Here we will consider positive FB with $\gamma > 0$. The delay modulation amplitude and frequency should be limited by the inequalities which ensures positive values of the delay $\tau(t) > 0$ and positive derivation $g'(t) > 0$ (ensures pulsed structure of the solution). Further we study the system with the assumptions

$$B < \tau_0, \ B\omega < 1. \tag{2}$$

For class B lasers including semiconductor lasers, some solid-state lasers, CO_2 gas lasers typical values $v \sim 10^3$ are large while the other parameters are of the order of unit. Considering $v \gg 1$ as a large parameter, an asymptotic approach for spiking can be done. The method of investigation is following. The phase space of system (1) is the direct product of the Banach space $C_{[-\tau,0]}$ of continues functions by the number line R^1, i.e., the values of the functions from $C_{[-\tau,0]}$ and the value $y(0) \in R^1$ are given as initial conditions. In this space, we shall distinguish a (fairly wide) set $S(\xi)$ dependent on the vector parameter ξ and consider the solutions with initial conditions from this set. It is possible to construct uniform asymptotic approximations of all such solutions and show that after a certain time these solutions again fall within $S(\xi)$. Thus, the operator of the shifting along the trajectories, which makes a function from S corresponds to a function also from S, is naturally determined. The properties of this operator are mainly determined by the finite-dimensional map $\bar{\xi} = f(\xi)$. To a fixed point of the map there corresponds the fixed point of the operator, and to the later point there corresponds a periodic solution of the same stability in the original system. Examples of the investigation of similar systems were given in our previous works [9].

Solutions in form of spikes can be classified as slowly oscillating (SO) solutions with inter-spikes intervals longer than the delay time, fast oscillating (FO^n) solutions with inter-spikes intervals shorter than the delay, mixed slowly and fast (MSF) oscillating solutions. For each type of solutions we will find asymptotically (at $v \to \infty$) the maps responsible for the dynamics of spikes.

3. Slowly oscillating solution

For SO solutions any interval between spikes greater than the time delay. Fix a time point when the radiation spike begins as the starting point, so that radiation intensity $u(0) = 1$ whereas before that, in the delay interval, the radiation intensity is of noise level.

The set of initial conditions can be define as follows,

$$S(c,\varphi)=\{y(0)=c,\ \Phi(0)=\varphi,\ u(s)=\psi(s),\ s\in[-\tau_0-B,0]\} \tag{3}$$

where $c\in(1,q]$, $\varphi\in[0,2\pi]$, and the function $\psi(s)\in S_0$ is chosen from the set S_0 of the functions with the properties,

$$S_0=\{\psi(s):0\leq\psi(s)\ll 1,\ \psi(0)=1,\ \int_{-\tau_0-B}^{0}\psi(s)ds\leq v^{-1/2}\}.$$

Note, the parameter $c>1$ that provides $\dot{u}(0)>0$. The form of the functions $\psi(s)$ is not specified, hence, the set $S(c,\varphi)$ is wide enough.

Now we integrate the system step-by-step dividing the evolution path into segments $[t_i,t_{i+1}]$, where the asymptotic estimates (under $v\to\infty$) for functions $u(t-\tau)$ and $u(t)$ are known. The solution values at the end of the interval are used as the initial conditions for the next interval. In this way we obtain an analytical estimate of the SO solution up to the time $t=t_2$, when the next radiation pulse begins.

The main result is that the problem of further constructing the solution for $t>t_2$ has returned to the original problem with the initial conditions $u(t_2+s)=\bar{\psi}(s)\in S_0$, $s\in[-(\tau_0+B),0]$ and $y(t_2)=\bar{c}+o(1)$, $\Phi(t_2)=\bar{\varphi}$, which belong to set S given by Eq.(3) with replacing (c,φ) by $(\bar{c},\bar{\varphi})$, namely,

$$\begin{aligned}\bar{c}&=q+(c-p-q)e^{-T}+\frac{\gamma p}{g'(\tilde{t}_0)}e^{-(T-\tilde{t}_0)},\\ \bar{\varphi}&=\omega T+\varphi,\ mod\ 2\pi,\end{aligned} \tag{4}$$

where $\tilde{t}_0(\varphi)$ is the positive root of the equation $\tau_0+B\cos(\omega\tilde{t}_0+\varphi)=\tilde{t}_0$; $p(c)$ is the positive root of the equation $c-p=ce^{-p}$; and $T(c,\varphi)$ is the root of the equation

$$(q-1)T+(c-p-q)(1-e^{-T})+\gamma\frac{p}{g'(\tilde{t}_0)}(1-e^{-(T-\tilde{t}_0)})=0. \tag{5}$$

Note, the value of p characterizes the pulse energy and the value of T characterizes the interpulse interval. Integrating the system, we have supposed that inequality

$$a(\tilde{t}_0)<0 \tag{6}$$

where $a(t)=(q-1)t+(c-p-q)(1-e^{-t})$ is fulfilled in order to get the $T>\tau_0+B$.

Let us introduce the trajectory shift operator $\Pi\big(c,\varphi,\psi(s)\big)=\big(y(t_2),\Phi(t_2),u(s+t_2)\big)$, $s\in[-\tau_0-B,0]$, which associates (by means of solutions) any element (c,φ) from the set S with another element $(\bar{c},\bar{\varphi})$ from the same set S. Evolution of the operator and, in turn, evolution of slowly oscillating pulsed solution is determined by the iterations of the two-dimensional map (5). In particular, to the stable fixed point (c_0,φ_0) there corresponds stable periodic slowly oscillating pulsations of the period $T_0=T(c_0,\varphi_0)$ in the original system and $T_0>\tau_0+B$. If map (5) has another attractor and inequality (6) is fulfilled for each iteration of the map, then the attractor corresponds to pulsed solution of SO structure in system (1). To bifurcations of the mapping attractor there correspond bifurcations of pulsed regime. Inequality (6) provides SO structure of such solutions and, in this way, restricts the domain in parameters space and in phase space where SO solutions can realized.

By computing map (5) with $q=1.5$, $\gamma=0.3$, $\tau_0=0.3$ and $\omega=29.8$ we find that at sufficiently small modulation amplitudes, $0<B<0.012$, there is a stable mapping point which corresponds to a stable cycle with a period longer than the delay. At $0.012<B<0.04$, cycles

IOP Conf. Series: Journal of Physics: Conf. Series **937** (2017) 012016 doi:10.1088/1742-6596/937/1/012016

and chaotic attractors (quasiperiodic and chaotic SO oscillations) are observed. At $B > 0.04$ that corresponds to a violation of the condition $B\omega < 1$, a transition takes place to oscillations a with interpulse intervals longer and shorter than the delay time.

Thus, there are sufficiently wide domains of modulation parameters at which chaotic spikes of special SO-sructure may be observed.

4. Fast oscillating solutions

Inequality (6) determines the parameters with which the intervals between neighboring pulses are greater than the delay. Violation of this condition leads to occurrence of $n \geq 1$ pulses on any interval of delay length. We call such regimes fast oscillating (FO^n) modes, since the time intervals between pulses are shorter than the delay time in the FB circuit.

Consider FO^1 regimes with one pulse on the delay interval. The set of initial conditions for Eqs.(1) can be given as follows,

$$S(c, \varphi, \xi, p_1) = \{y(0) = c,\ \Phi(0) = \varphi,\ u(s) = \psi_1(s),\ s \in [-\tau_0 - B, 0]\}, \tag{7}$$

where $c \in (1, q]$, $\varphi \in [0, 2\pi]$, $\psi_1(s) \in S_1(\xi, p_1)$, and

$$S_1(\xi, p_1) = \left\{ \psi_1(s) \in C_{[-\tau_0 - B, 0]} :\ \psi_1(0) = 1,\ \int_{-\tau+\xi}^{-\tau+\xi+\delta_1} \psi_1(s)ds = p_1, \right.$$
$$\left. \int_{-\tau}^{-\tau+\xi} \psi_1(s)ds + \int_{-\tau+\xi+\delta_1}^{0} \psi_1(s)ds < v^{-1/2} \right\}.$$

The parameter ξ determines the moment of pulse onset on the delay interval so that the value $T = (\tau_0 + B\cos\varphi - \xi) > 0$ is the interval between pulses, the parameter $p_1 > 0$ is the energy of the pulse in the delay interval. The $\psi_1(s)$ values are asymptotically small in the intervals between pulses, the pulse shape is not specified, for example, a pulse can be square (non-smooth). It is only necessary that the pulse width δ_1 should be sufficiently short, $\delta_1 \to 0$, with $v \to \infty$. Note also, the condition $c > 1$ ensures the positive derivative $u'(0) > 0$ at the initial moment, which we determine at the moment of a new pulse onset. Thus the set $S(c, \varphi, \xi, p_1)$, depending on four parameters, is rather wide.

The main result is that in time $t_2 = T + o(1)$ we find the system in the state analogous to the initial one with replacing (c, ξ, p_1, φ) by $(\bar{c}, \bar{\xi}, \bar{p}_1, \bar{\varphi})$, where

$$\begin{aligned} \bar{c} &= q + (c - p - q)e^{-T} + \frac{\gamma p_1}{g'(\tilde{\xi})} e^{\tilde{\xi} - T}, \\ \bar{p}_1 &= p, \\ \bar{\xi} &= \tau_0 + B\cos\varphi - T, \\ \bar{\varphi} &= \omega T + \varphi,\ mod\ 2\pi, \end{aligned} \tag{8}$$

with $\tilde{\xi} = \tilde{\xi}(\xi, \varphi)$ is the root of the equation $\tilde{\xi} - B\cos(\omega\tilde{\xi} + \varphi) = \xi - B\cos\varphi$, $p = p(c)$ is the root of the equation $c - p = c\exp(-p)$ and $T = T(c, \varphi, p_1, \xi)$ being positive root of the equation

$$(q - 1)T + (c - p - q)(1 - e^{-T}) + \gamma \frac{p_1}{g'(\tilde{\xi})}(1 - e^{\tilde{\xi}_1 - T}) = 0.$$

The obtained 4D-map describes the dynamics of FO^1-solutions if the inequalities

$$a(\tilde{\xi}) < 0,\ T(c, \varphi, p_1, \xi) < \tau_0 + B \tag{9}$$

are valid for each iteration of the map. Conditions (9) provide the reproduction of FO^1-structure of pulsed solution.

Numerical simulations show that the conditions can be valid for map's attractors only in the case of $\gamma > 0$ (positive feedback). With modulation amplitude B increases, chaotic attractor can form, hence, one can expect chaotic spiking of special FO^1-structure in the original infinite-dimensional delayed system.

Inequalities (9) restrict the domain in the parameter space of the system, where solutions of this type can be realized. It appears that the parameter domain for FO^1-solution may intersect with the domain of SO-solutions, hence the multiplicity of different periodic spike regimes takes place.

It is possible to construct analogously $(2n+2)$-dimensional map responsible for FO^n solutions representing $n > 1$ spikes within the delay interval and find the parameter domain where such solutions coexist. Below we numerically demonstrate the coexistence of chaotic spike regimes in the case of the modulated delay.

5. Numerical simulation

In order to verify the above theoretical conclusions we provide numerical simulations of the original delayed system with $v = 10^3$ and $q = 1.5$ that are typical for a semiconductor laser. Note, v takes sufficiently large value. Other parameters were chosen so that both maps (5) and (8) have the chaotic attractor: FB level $\gamma = 0.3$, the delay $\tau_0 = 0.6$, modulation amplitude of the delay $B = 0.032$, modulation frequency $\omega = 29.814$ was chosen comparable with natural relaxation frequency. Preliminary calculation of maps is really necessary, because the parameter regions where various solutions are realized are relatively narrow, and also depend on the initial conditions.

In accordance with set (3) we choose the initial function $u(s)$ in the form $u(s) = 0$, $s \in [-(\tau_0 + B), 0)$, $u(0) = 1$. The initial value of the modulation phase is $\Phi(0) = 3.6$. The initial value of the inversion $y(0) = 1.25$ was chosen so that map (3) has the SO chaotic attractor. At $y(0) = 1.15$ map (6) has the FO^1 chaotic attractor and at $y(0) = 1.08$ there is FO^2 chaotic attractor. In order to specify structure of irregular spiking, we show the dependence of inter-

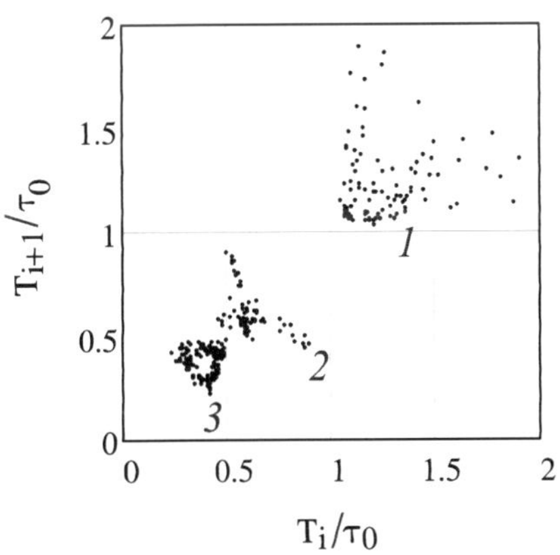

Figure 1. Coexistence of SO (1), FO^1 (2) and FO^2 (3) spiking which are obtained under different initial conditions for the delayed system (1).

pulse intervals T_{i+1} on T_i that correlates with a projection of the maps obtained. In Fig.2 coexistence of three chaotic attractors of different structure is demonstrated. One can see that all points of the map corresponding to SO spikes are located in domain $T_i > \tau_0$. In domain $\tau_0/2 < T_i < \tau_0$ the map corresponds to FO^1 chaotic attractor, and in domain $\tau_0/3 < T_i < \tau_0/3$ the map corresponds to FO^2 chaotic attractor with two pulses on any delay interval.

5.1. *Switching to the desired attractor*

Here we study the effects of an additional short-time square-shape control signal in the pumping circuit . In Eqs. (1) we set

$$q(t) = q + \rho,\ t \in [t_x, t_x + \Theta],$$

where ρ is the force amplitude, t_x is the moment of the signal application, Θ is its duration of the order or shorter than radiation pulse width, $\eta = \rho\Theta$ is the pulse energy which assumed to be not an asymptotically small value.

In order to switch on the target attractor one has to move the phase trajectory into its attractive basin. The phase space of the delayed system (1) is infinite-dimensional. Therefore, we can not fully describe the attractive basins of coexisting solutions. But, using asymptotic conditions (6) and (9) for existence of the solution of each type, it is possible to propose perturbations guaranteeing access to the target attractor.

In order to move from an SO solution to a FO solution, it is necessary to violate condition (6). To do this, we must apply a perturbation of the corresponding energy η at the time t_x in the interval less than τ_0, following the radiation pulse, as shown in Fig. 2a.

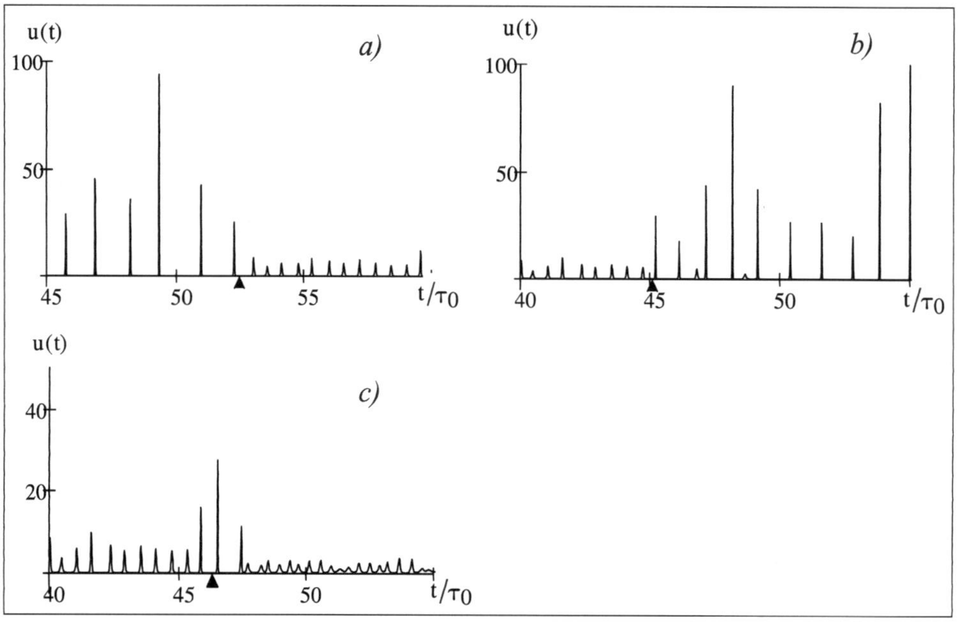

Figure 2. Switching induced by pump-perturbation signal: a) from irregular SO spiking to FO^1 spiking, b) from irregular FO^1 spiking to SO spiking, c) from irregular FO^1 spiking to FO^2 spiking. The moment of application of the control signal is marked with a black triangle.

In order to move from an FO^1-solution to a SO-solution, it is necessary to violate condition (9), namely to get $T_i > \tau_0/2$. To do this, we must apply a perturbation of the corresponding energy η at the time t_x synchronized with the radiation pulse, as shown in Fig. 2b. At last, if the perturbation restricts condition (9), namely to get $T_i < \tau_0/3$, FO^1-solution can switch to FO^2-solution, as shown in Fig. 2c.

6. Conclusion

In conclusion, we have analytically described pulsed solutions of various structure in the laser diode with variable delayed FB. Doing so, the problem on dynamics of the original infinite-dimensional system has been reduced to the problem on the dynamics of nonlinear finite-dimensional maps. The advantages of the proposed asymptotic method are as follows: 1) the initial conditions (attractive basin) are determined for the desired regime; 2) the domain of parameters can be found for the desired regime; 3) the dependencies of characteristics (amplitude and inter-spike interval) can be obtained from the control parameters; 4) bifurcations of spikes can be followed.

A hierarchy of spiking regimes has been proposed on the base of pattern complexity, i.e. on the number n of the pulses within the delay interval. We distinguish between slowly oscillating solutions determined by the 2D-map, fast oscillating solutions determined by the $(2n+2)$D-maps and mixed solutions determined by the 2D-map. Since the dynamics of such a map completely determines pulse dynamics, including chaotic pulsing, we expect that the dimension of the corresponding attractor in the infinite-dimensional phase space would be respectively limited. With increasing amplitude of modulation of the delay time, we observed intermittency of slowly and fast oscillating solutions and merge of attractors. Such a scenario leading to annihilation one of the coexisting states can be proposed for controlled monostability of a chaotic attractor.

The obtained maps can be used for finding the parameters of the system at which chaotic pulsing with special characteristics are realized. In the case of negative FB, we demonstrate a chaotic spikes following in strictly alternating intervals of less than and more than the delay time. In the case of positive FB, we demonstrate possible coexistence of slowly and fast oscillating solutions (multistability of spiking) at the same parameters. These results are promising for developing methods for dynamic control, in particular, for the fast switching of attractors by a suitable external shock.

Acknowledgments
This study is supported by State assignment, project No 1.10160.2017/5.1.

References

[1] Dahms T, Hövel P and Schöll E 2008 *Phys. Rev.* E **78** 056213
[2] Gjurchinovski A, Jüngling T, Urumov V and Schöll E 2013 *Phys. Rev.* E **88** 032912
[3] Grigorieva E V and Kaschenko S A 2015 *Physica* D **291** 1
[4] Otsuka K, Kazutaka A, Ko Y and Fukazawa T 2002 *Opt.Lett.* D **27** 1339
[5] Won-Ho Kye, Muhan Choi, Chil-Min Kim and Young-Jai Park 2005 *Phys. Rev.* E **71** 045202
[6] Chizhevsky V N and Glorieux P 1995 *Phys. Rev.* E **51** R2701
[7] Chizhevsky V N, Grigorieva E V and Kashchenko S A 1997 *Opt. Commun.* **133** 189
[8] Giacomelli G, Calzavara M and Arecchi F T 1989 *Opt. Commun.* **74** 97
[9] Grigorieva E V and Kaschenko S A 2017 *Asymptotic representation of relaxation oscillations in lasers* (Switzerland: Springer)

IOP Conf. Series: Journal of Physics: Conf. Series **937** (2017) 012017 doi:10.1088/1742-6596/937/1/012017

Simulation of complex magnesium alloy texture using the axial component fit method with central normal distributions

T M Ivanova[1,3] and V N Serebryany[2,4]

[1]Department of High Mathematics, National Research Nuclear University, 31 Kashirskoe Shosse, Moscow 115409, Russia

[2]Laboratory of Crystal Structural Research, Institute of Metallurgy and Material Science, RAS, 49 Leninsky Prospect, Moscow 119991, Russia

E-mail: [3]ivatatiana@gmail.com, [4]vns@imet.ac.ru

Abstract. The component fit method in quantitative texture analysis assumes that the texture of the polycrystalline sample can be represented by a superposition of weighted standard distributions those are characterized by position in the orientation space, shape and sharpness of the scattering. The components of the peak and axial shapes are usually used. It is known that an axial texture develops in materials subjected to direct pressing. In this paper we considered the possibility of modelling a texture of a magnesium sample subjected to equal-channel angular pressing with axial components only. The results obtained make it possible to conclude that ECAP is also a process leading to the appearance of an axial texture in magnesium alloys.

1. Introduction

The subject of the study of quantitative texture analysis (QTA) is the experimental determination and interpretation of the statistical distribution of crystallites in polycrystalline materials. The knowledge of the preferential orientations of crystallites is necessary in particular to find out the influence of the technological process on the occurrence of preferential orientations [1]. The most commonly used in QTA values are functions describing crystallite distributions (ODF) and pole density functions - pole figures (PF) describing the distribution of crystallographic planes. ODF is defined on the rotation group of a three-dimensional space $SO(3)$ and pole figures are defined on a unit sphere S^2. The main problem of the QTA is simulation of an ODF $f(g)$ in a polycrystalline sample based on a set of experimentally obtained pole figures $\left\{P_{\vec{h}_\lambda}(\vec{y})\right\}$. Among the various methods for solving this problem [1] - [5], the method of texture components [2] - [4], [6] has the advantage in the case of samples with low symmetry. Two types of components are usually considered. These are peak and axial ones. The first reaches its maximum at a certain point in the orientation space and has the form of a more or less wide bell. Averaging the peak component over all possible rotations about a fixed axis $\vec{n}$ generates an axial component that reaches a maximum on a certain circle in the rotation space.

2. The component fit method

The problem of an ODF simulation based on a set of pole figures is ill-posed because of the non-uniqueness of its solution [5] and the instability to measurement errors on pole figures [6]. The

IOP Conf. Series: Journal of Physics: Conf. Series **937** (2017) 012017 doi:10.1088/1742-6596/937/1/012017

uniqueness of the solution can be achieved on the class of functions that are the sum of fixed number of components with anchored positions in the orientation space. In our model, the ODF $f(g), g \in SO(3)$ is represented by the sum of K texture components of with peak or axial scattering:

$$f^{model}\left(g\right) = c_0 + \sum_{k=1}^{K} c_k f\left(g, g_{0k}, \vec{n}_k, \varepsilon_k\right),$$

where c_k is the weight of corresponding component, g_{0k} is its maximum position, $\vec{n}_k$ is an axis of axial component, ε_k is scattering parameter, c_0 is the volume fraction of the non-textured component . These parameters are to be determined. Considering the pole figures as integral projections of the ODF on the unit sphere:

$$P_{\vec{h}}(\vec{y}) = \frac{1}{4\pi}\int_0^{2\pi}\left\{f\left(\left\{\vec{h},\phi\right\}^{-1}\left\{\vec{y},0\right\}\right) + f\left(\left\{-\vec{h},\phi\right\}^{-1}\left\{\vec{y},0\right\}\right)\right\}d\phi,$$

we get that any pole figure will be represented by the projections of the components of the PDF with the same parameters:

$$P^{model}\left(\vec{h}, \vec{y}\right) = c_0 + \sum_{k=1}^{K} c_k P\left(\vec{h}, \vec{y}, g_{0k}, \vec{n}_k, \varepsilon_k\right),$$

$\vec{h}, \vec{y}$ are directions in the crystallite and the sample, respectively. We note that the normalization factors of the pole figures N_λ are also unknown parameters of the problem, since it is not possible to normalize pole figures in the process of measurement because of incompleteness of the data. In this work unknown model parameters are searched by minimizing the square of the weighted residual taken at all measured points on all pole figures:

$$r^2 = \sum_{\lambda=1}^{\Lambda}\sum_{i=1}^{I}\sum_{j=1}^{J} W\left(\lambda, i, j\right)\left\{N_\lambda P_{\vec{h}_\lambda}(\vec{y}_{ij}) - P^{model}\left(\vec{h}_\lambda, \vec{y}_{ij}\right)\right\}^2 \to \min.$$

This problem of non-linear optimization is solved numerically. To achieve the stability of the solution, regularization is applied [7] - [9]. The desired ODF is constructed from the received parameters of the model. The adequacy of the model is estimated using the RP-value:

$$RP(\Delta) = \frac{1}{N\left(\Delta\right)}\sum_{ij}^{N(\Delta)}\left|\frac{P_{h_\lambda}(\vec{y}_{ij}) - P^{model}(h_\lambda, \vec{y}_{ij})}{P_{h_\lambda}(\vec{y}_{ij})}\right| \cdot 100\%,$$

where the summation is carried out over all points at which the experimentally measured magnitude of the pole density is not less than the parameter Δ, and $N\left(\Delta\right)$ is the number of these points [1], [6].

3. Simulation of complex magnesium alloy texture

We consider a sample of magnesium alloy Mg-4.5% Nd subjected to equal-channel angular pressing (ECAP). The crystal lattice of magnesium has a hexagonal symmetry. The symmetry of the sample is monoclinic. The presence of a low symmetry determined the choice of component fit method for the ODF simulation. For this sample, six incomplete non-normalized pole figures were obtained by X-ray diffraction. There were visually determined individual maxima and rings of increased pole density on experimental pole figures. The individual maxima were supposed to be modeled by peak components. However, no reasonable number of peak components made it possible to adequately describe the available PFs, therefore, the possibility of pole density approximation by superposition of only the axial components was considered. It turned out that with a certain overlapping of each other, the axial components also create peaks of pole density of the type, as in the available experimental PFs. In the process of approximation it is established that all pole figures can be adequately described by a model including seven axial components and a background component. Figure 1 shows the positions of the axial components on the stereographic projections of pole figures [10-13], [10-11], [10-10].

IOP Conf. Series: Journal of Physics: Conf. Series **937** (2017) 012017 doi:10.1088/1742-6596/937/1/012017

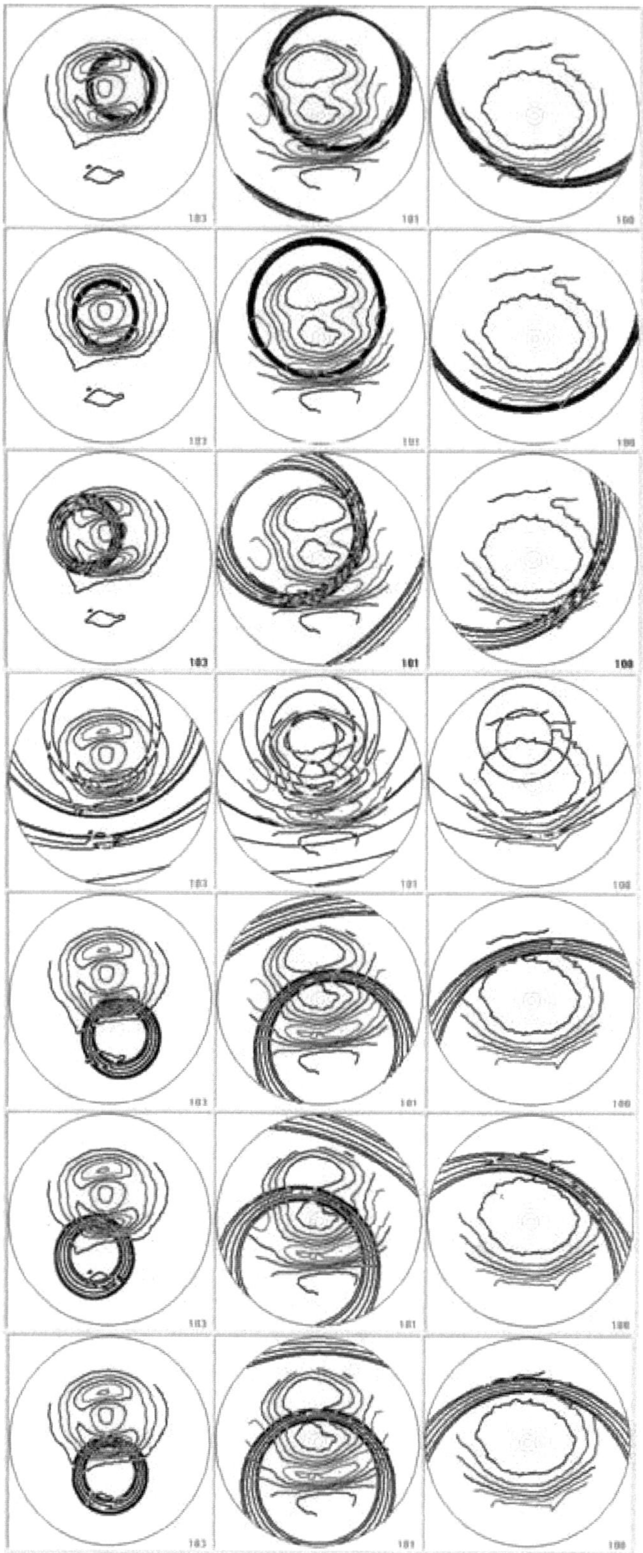

Figure 1. The positions of the seven axial texture components of the magnesium sample on the pole figures [10-13], [10-11], [10-10] are listed in descending order of the component weights.

The components found have the following parameters:

$$c_1 = 29.8,\ \varepsilon_1 = 0.010, \vec{n}_1 = \{259.2^o, 23.0^o\}, g_{01} = \{259.2^o, 23.9^o, 9.9^o\},$$

$$c_2 = 18.7,\ \varepsilon_2 = 0.012, \vec{n}_2 = \{295.2^o, 30.1^o\}, g_{02} = \{295.2^o, 32.3^o, 14.9^o\},$$

$$c_3 = 13.3,\ \varepsilon_3 = 0.007, \vec{n}_3 = \{223.2^o, 32.6^o\}, g_{03} = \{223.2^o, 39.5^o, 14.9^o\},$$

$$c_4 = 8.7,\ \varepsilon_4 = 0.024, \vec{n}_4 = \{259.2^o, 43.7^o\}, g_{04} = \{79.2^o, 37.0^o, 9.8^o\},$$

$$c_5 = 6.3,\ \varepsilon_5 = 0.014, \vec{n}_5 = \{74.7^o, 36.5^o\}, g_{05} = \{74.7^o, 43.4^o, 14.9^o\},$$

$$c_6 = 4.9,\ \varepsilon_6 = 0.010, \vec{n}_6 = \{114.3^o, 37.3^o\}, g_{06} = \{114.3^o, 44.1^o, 14.9^o\},$$

$$c_7 = 3.4,\ \varepsilon_7 = 0.008, \vec{n}_7 = \{90.0^o, 32.7^o\}, g_{07} = \{90.0^o, 35.9^o, 14.9^o\},\ c_0 = 14.9.$$

For the model found we obtain the values $RP(0.5) \approx 14\%$, $RP(1.0) \approx 10\%$. Figure 2. shows the experimental and model pole figures [0004] and [10-11] in the form of the level lines of the pole density on stereographic projections.

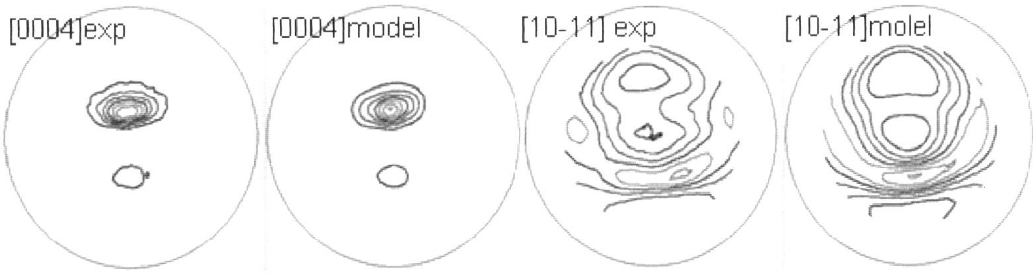

Figure 2. Experimental and model pole figures [0004] and [10-11].

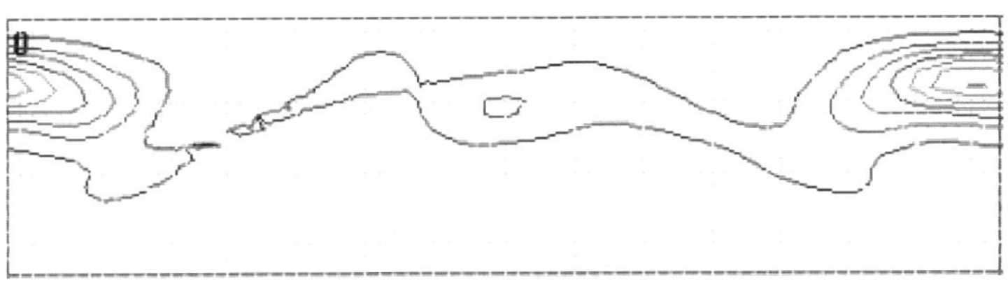

Figure 3. The isolines of the orientation distribution function at zero section.

The model ODF with axial components with central normal distribution (Legendre polynomials) is constructed on the basis of the parameters found:

$$f(g) = c_0 + \sum_{k=1}^{7} c_k \sum_{l=0}^{\infty} (2l+1)\exp\left(-l(l+1)\varepsilon_k^2\right) P_l\left(g\vec{n}_k, g_{0k}\vec{n}_k\right).$$

The ODF isolines for the sample of the magnesium alloy at zero section are shown in figure 3.

4. Conclusion

It is shown that texture development of magnesium alloys in the process of equal-channel angular pressing proceeds according to the principle of axial components formation. The complex magnesium alloy Mg-4.5% Nd texture in the sample subjected to ECAP can be adequately described by a superposition of seven axial components with a central normal distribution.

IOP Conf. Series: Journal of Physics: Conf. Series **937** (2017) 012017 doi:10.1088/1742-6596/937/1/012017

References

[1] Bunge H J 1982 *Texture Analysis in Material Science* (Butterworths:London)
[2] Bucharova T I and Savyolova T I 1993 Application of normal distribution on SO(3) and S2 for orientation distribution function approximation *Text. Microstruct.* **23** 161–176
[3] Nikolayev D I and Savyolova T I 1997 Normal distribution on the rotation group SO(3) , *Text. Microstruct.* **29** 201–233
[4] Helming K and Eschner T 1990 A new approach to texture analysis of multiphase materials using a texture component *Model Cryst. Res. Technol.* **25**, K203–K208 (1990).
[5] Matthies S 1979 On the reproducibility of the orientation distribution function from pole figures (ghost phenomena) *phys. stat. sol.* (b) **92** N2 K135–137
[6] Ivanova T M and Savelova T I 2006 Robust method of approximating the orientation distribution function by canonical normal distributions *Phys. Met. Metallog.* **101** N2 114–118
[7] Kahaner D, Moler C, Nash S 1989 *Numerical Methods and Software* (Englewood Cliffs, NJ: Prentice Hall)
[8] Golub G and Van Loan Ch 1996 *Matrix Computations* (Johns Hopkins University Press, Baltimore)
[9] Dennis J E and Schnabel R B 1983 *Numerical Methods for Unconstrained Optimization and Nonlinear Equations* (Englewood Cliffs, NJ: Prentice Hall)

IOP Conf. Series: Journal of Physics: Conf. Series **937** (2017) 012018 doi:10.1088/1742-6596/937/1/012018

On inverse problem of determination the right-hand side term in higher order degenerate parabolic equation with integral observation in time

V L Kamynin and T I Bukharova

Department of Mathematics, National Research Nuclear University MEPhI (Moscow Engineering Physics Institute), 31 Kashirskoe shosse, 115409 Moscow, Russia

E-mail: VLKamynin@mephi.ru, TIBukharova@mephi.ru

Abstract. One studies the unique solvability of the inverse problem of source determination in higher order degenerate parabolic equation with integral observation. The explicit estimates are obtained.

1. Introduction

We study the existence and uniqueness of generalized solution $\{u(t,x),p(x)\}$ of the inverse problem for parabolic equation

$$D_t u+(-1)^m a(t,x) D_x^{2m} u = p(x) g(t,x)+r(t,x), \quad (t,x)\in Q; \tag{1}$$

with initial and boundary conditions

$$u(0,x)=u_0(x), \ x\in[0,l]; \ D_x^j u(t,0)=D_x^j u(t,l)=0, \ j=0,1,\ldots,m-1, \ t\in[0,T], \tag{2}$$

and the additional condition of integral observation

$$\int_0^T u(t,x)\chi(t)\,\mathrm{d}t=\varphi(x), \quad x\in[0,l]; \tag{3}$$

here $Q=[0,T]\times[0,l]$, where T and l are some numbers,. $D_t u=\dfrac{\partial u}{\partial t}$, $D_x^j u=\dfrac{\partial^j u}{\partial x^j}$, $j=1,2,\ldots,2m$, $D_x^0 u\equiv u$.

In the setting of the inverse problem (1)–(3) the leading coefficient $a(t,x)$ of the equation (1) is assumed only nonnegative so the equation may turn out to be degenerate.

The interest to inverse problems for degenerate parabolic equation is arises, for example, in connection with the problem in modelling the price formation for options in financial markets [1-3, etc.]. Earlier the closed inverse problem investigated by the authors for the case when the additional condition is posed in the form of integral observation with respect to spatial variable [4], see also [5,6] for the case of second order degenerate parabolic equation.

In this paper we use Lebesgue, Hölder and Sobolev spaces with corresponding norms in usual sense (see [7]).

We assume that the functions occurring in the input data of the problem (1)–(3) are measurable and satisfy the following conditions:

$$0\le a(t,x)\le a_1,\ \frac{1}{a}\in L_q(Q),\ q>1,\ \frac{D^j a}{a}\in L_\infty(Q), j=1,2,...,m-1;\ \left\|\frac{1}{a}\right\|_{L_q(Q)}\le a_2; \tag{A}$$

$$\frac{g}{a},\frac{r}{a}\in L_1(0,T;L_\infty(0,l)),\ \frac{g^2}{a},\frac{r^2}{a}\in L_1(Q),\ \chi(t)\in L_\infty(0,T),\ D_t\left(\frac{\chi}{a}\right)\in L_1(0,T;L_\infty(0,l)),$$

$$\left\|\frac{r}{a}\right\|_{L_1(0,T;L_\infty(0,l))}\le K_r,\ \left\|\frac{g^2}{a}\right\|_{L_1(Q)}\le K_g^*,\ \left\|\frac{r^2}{a}\right\|_{L_1(Q)}\le K_r^*,\ \|x\|_{L_\infty(0,T)}\le K_\chi,\ \left|\frac{\chi(T)}{a(T,x)}\right|\le a_3, \tag{B}$$

$$\left|\frac{\chi(0)}{a(0,x)}\right|\le a_4,\ \left\|D_t\left(\frac{\chi}{a}\right)\right\|_{L_1(0,T;L_\infty(0,l))}\le K_{a,\chi},\ \left|\int_0^T \frac{g(t,x)}{a(t,x)}\chi(t)dt\right|\ge g_0>0;$$

$$u_0(x)\in \overset{0}{W}{}_2^m(0,l),\ \|D_x^m u_0\|_{L_2(0,l)}\le M_0,\ \|u_0(x)\|_{L_\infty(0l)}\le M_1; \tag{C}$$

$$\varphi(x)\in W_\infty^{2m}(0,l),\ D_x^j\varphi(0)=D_x^j\varphi(l),\ j=0,1,...,m-1,\ \|D_x^{2m}\varphi\|_{L_\infty(0,l)}\le K_\varphi^*. \tag{D}$$

Here $a_1,\ a_2,\ g_0, K_g^*,\ K_{a,\chi},\ K_\chi=\text{const}>0,\ a_3,a_4,K_r,K_r^*,\ K_\varphi^*,M_0,\ M_1=\text{const}\ge 0.$

In what follows we shall use the notations

$$G(x)=\int_0^T\frac{g(t,x)}{a(t,x)}\chi(t)dt,\ R(x)=\int_0^T\frac{r(t,x)}{a(t,x)}\chi(t)dt,\ q^*=\frac{2q}{q+1}. \tag{4}$$

Definition 1. By generalized solution of the inverse problem (1)–(3) we mean a pair of functions $\{u(t,x),p(x)\}$ such that

1) $u(t,x)\in C^{0,\beta}(Q)\cap L_\infty\left(0,T;\overset{0}{W}{}_2^m(0,l)\right)\cap W_s^{1,2m}(Q),\ s>1,\ \beta\in(0,1),\ p(x)\in L_\infty(0,l);$

2) this pair satisfies equation (1) almost everywhere in Q;

3) $u(t,x)$ satisfies (2), (3) in classical sense.

2. A priori estimates and solvability of the direct problem

Suppose that $p(x)\in L_\infty(0,l)$ is a known function and consider the direct problem (1)–(2) with this function $p(x)$ in the right-hand side of equation (1).

Using the ideas and results of [7] we prove the following theorem

Theorem 1. Let conditions (A)–(C) are satisfied and $p(x)\in L_\infty(0,l)$. Let q^* be a constant defined in (4). Then there exists a unique solution $u(t,x)$ of direct problem (1)–(2) and the following estimates

$$\sup_{0\le t\le T}\|D_x^m u(t,\cdot)\|_{L_2(0,l)}^2\le\|D_x^m u_0\|_{L_2(0,l)}^2+\left\|\frac{(pg+r)^2}{a}\right\|_{L_1(Q)}\le M_0^2+2\|p\|_{L_\infty(0,l)}^2K_g^*+2K_r^*, \tag{5}$$

$$\|D_x^{2m}u\|_{L_{q^*}(Q)}^2\le a_2\left(M_0^2+2\|p\|_{L_\infty(0,l)}^2K_g^*+2K_r^*\right), \tag{6}$$

$$\|D_t u\|_{L_2(Q)}^2\le c_1, \quad (7$$

$$|u(t_1,x_1)-u(t_2,x_2)|\le c_2|x_1-x_2|^{1/2}+c_3|t_1-t_2|^{1/6} \tag{8}$$

hold, where $c_1,\ c_2,\ c_3=\text{const}>0$ and these constants depend on $m,l,T,a_1,a_2,M_0,K_g^*,\ K_r^*$.

3. Unique solvability of the inverse problem

Let us assume that input data of the inverse problem (1)–(3) satisfy the conditions (A)–(D) and $q^* > 1$ is defined in (4). We derive the operator equation for the unknown coefficient $p(x)$. For this purpose we multiply the equation (1) by $\chi(t)$, divide it by $a(t,x)$ and integrate the resulting relation over the segment $[0,T]$. Taking into account conditions (2), (3) and notation (4) after integration by parts we obtain the relation

$$p(x)=\frac{1}{G(x)}\left[\frac{\chi(T)}{a(T,x)}u(T,x)-\int_0^T D_t\left(\frac{\chi}{a}\right)u\ \mathrm{d}t+(-1)^m D_x^{2m}\varphi-\frac{\chi(0)}{a(0,x)}u_0(x)-R(x)\right]. \quad (9)$$

Let us introduce the operator $\mathrm{A}: L_\infty(0,l)\to L_\infty(0,l)$ by the formula

$$\mathrm{A}(p)=\frac{1}{G(x)}\left[\frac{\chi(T)}{a(T,x)}u(T,x)-\int_0^T D_t\left(\frac{\chi}{a}\right)u\ \mathrm{d}t+(-1)^m D_x^{2m}\varphi-\frac{\chi(0)}{a(0,x)}u_0(x)-R(x)\right], \quad (10)$$

where $u(t,x)$ is a solution of the direct problem (1), (2) with given function $p(x)$ in the right-hand side of (1). We note that such a solution exists and is unique in view of Theorem 1. Then the relation (9) can be written as

$$p=\mathrm{A}(p). \quad (11)$$

By standard arguments (see, for example, [5, 6]) we prove

Lemma 1. Let conditions (A)–(D) hold. Then the pair of functions $\{u(t,x),p(x)\}$ is a generalized solution of the problem (1)–(3) if and only if this pair of functions satisfies the relations (1), (2), (11).

Now we prove the main result

Theorem 2. Let conditions (A)–(D) hold. Suppose that

$$\kappa\equiv\frac{1}{g_0}\left(a_3+K_{a,\chi}\right)l^{m-1/2}\sqrt{K_g^*}<1. \quad (12)$$

Then there exists a unique generalized solution $\{u(t,x),p(x)\}$ of the inverse problem (1)–(3) with $u(t,x)\in W_{q^*}^{1,2}(Q)$, which satisfies the estimate

$$\|p\|_{L_\infty(0,l)}\le\frac{1}{(1-\kappa)g_0}\left[\left(a_3+K_{a,\chi}\right)l^{m-1/2}\left(M_0^2+K_r^*\right)^{1/2}+K_\varphi^*+a_4M_1+K_\chi K_r\right] \quad (13)$$

and the function $u(t,x)$ satisfies the estimates (5)–(8).

Proof. Suppose that $p^{(1)}(x),p^{(2)}(x)\in L_\infty(0,l)$ and $u^{(1)}(t,x),u^{(2)}(t,x)$ are solutions of the direct problem (1), (2) with $p^{(1)}(x)$ and $p^{(2)}(x)$ in the right-hand side of equation (1) respectively. Set $v(t,x)=u^{(1)}(t,x)-u^{(2)}(t,x)$, $y(x)=p^{(1)}(x)-p^{(2)}(x)$. Then the pair $\{v(t,x),y(x)\}$ satisfies the relations

$$D_tv+(-1)^m a(t,x)D_x^{2m}v=y(x)g(t,x), \quad (14)$$

$$v(0,x)=0,\quad D_x^jv(t,0)=D_x^jv(t,l)=0,\ j=0,1,\dots,m-1. \quad (15)$$

Let us show that A defined by the relation (10) is a contraction operator in $L_\infty(0,l)$. Indeed, in view of definition of the operator A we have

$$\left|\mathrm{A}\left(p^{(1)}\right)-\mathrm{A}\left(p^{(2)}\right)\right|=\left|\frac{1}{G(x)}\left[\frac{\chi(T)}{a(T,x)}v(T,x)-\int_0^T D_t\left(\frac{\chi}{a}\right)v\,\mathrm{d}t\right]\right|\le\frac{1}{g_0}\left(a_3+K_{a,\chi}\right)\|v\|_{L_\infty(Q)}. \quad (16)$$

By inequality (5) applied to the function $v(t,x)$ with regard to the relations (14), (15) we have

$$\|v\|_{L_\infty(Q)}\le l\|D_x v\|_{L_\infty(Q)}\le l^{m-1}\left\|D_x^{m-1}v\right\|_{L_\infty(Q)}\le l^{m-1/2}\left\|D_x^m v\right\|_{L_2(Q)}\le l^{m-1/2}\sup_x\left|y^2(x)\right|^{1/2}\sqrt{K_g^*}. \quad (17)$$

Substituting (17) into (16) we obtain

$$\left\|\mathrm{A}\left(p^{(1)}\right)-\mathrm{A}\left(p^{(2)}\right)\right\|_{L_\infty(0,l)}\le\frac{l^{m-1/2}}{g_0}\left(a_3+K_{a,\chi}\right)\sqrt{K_g^*}\left\|p^{(1)}-p^{(2)}\right\|_{L_\infty(Q)}\equiv\kappa\left\|p^{(1)}-p^{(2)}\right\|_{L_\infty(Q)}. \quad (18)$$

Thus by virtue of (12) operator A is a contraction operator in $L_\infty(0,l)$. Therefore, in view of Lemma 1 there exists a unique generalized solution $\{v(t,x), y(x)\}$ of the inverse problem (1)–(3) in the rectangle Q, and also $u(t,x)\in W_{q^*}^{1,2}(Q)$ by virtue of Theorem 1.

Now let us prove the estimate (13). Denote by $z^0(t,x)$ the solution of direct problem (1), (2) with $p(x)\equiv 0$ in the equation (1). Then by virtue of (5) we have

$$\sup_{0\le t\le T}\left\|D_x^m z^0(t,\cdot)\right\|_{L_2(0,l)}^2\le M_0^2+K_r^*. \quad (19)$$

Let $p(x)\equiv 0$ be the zeroth approximation of the solution of equation (11) when applying the method of iterations. Then

$$\mathrm{A}(0)=\frac{1}{G(x)}\left[\frac{\chi(T)}{a(T,x)}z^0(T,x)-\int_0^T D_t\left(\frac{\chi}{a}\right)z^0\,\mathrm{d}t+(-1)^m D_x^{2m}\varphi-\frac{\chi(0)}{a(0,x)}u_0(x)-R(x)\right],$$

Whence in view of conditions (B)–(D) and estimate (19)

$$\begin{aligned}\|\mathrm{A}(0)\|_{L_\infty(0,l)}&\le\frac{1}{g_0}\left[\left(a_3+K_{a,\chi}\right)\left\|z^0\right\|_{L_\infty(Q)}+K_\varphi^*+a_4M_1K_\chi K_r\right]\le\\&\le\frac{1}{g_0}\left[\left(a_3+K_{a,\chi}\right)l^{m-1/2}\left(M_0^2+K_r^*\right)^{1/2}+K_\varphi^*+a_4M_1+K_\chi K_r\right].\end{aligned} \quad (20)$$

Let $p_1(t)$ be the first approximation of the solution of equation (11) when applying the method of iterations. Then, using the well-known estimate of the n-th approximation of the solution in the iterative process (see, for example, [8, p.43]) and also the estimate (18) we obtain

$$\|p_1-p\|_{L_\infty(0,l)}\le\frac{\kappa}{1-\kappa}\|\mathrm{A}(0)\|_{L_\infty(0,l)},$$

hence, in view of (20)

$$\|p\|_{L_\infty(0,l)}\le\frac{1}{1-\kappa}\|\mathrm{A}(0)\|_{L_\infty(0,l)}\le\frac{1}{(1-\kappa)g_0}\left[\left(a_3+K_{a,\chi}\right)l^{m-1/2}\left(M_0^2+K_r^*\right)^{1/2}+K_\varphi^*+a_4M_1+K_\chi K_r\right].$$

This estimate coincides with (13). The estimates (5)–(8) are proved in Theorem 1. Theorem 2 is proved.

This work was supported by the Program of Competitiveness increase of the National Research Nuclear University MEPhI (Moscow Engineering Physics Institute); contract No.02.a03.21.0005, 27.08.2013.

References

[1] Bouchouev I and Isakov V 1999 Uniqueness, stability and numerical methods for the inverse problem that arises in financial markets *Inverse problems* **15** N3 95–116
[2] Jiang Lishang, Chen Qihong, Wang Lijun and Zhang J E 2003 A new well-posed algorithm to

recover implied local volatility *Quantitative Finance* **3** N6 451–457
[3] Jiang Lishang and Tao Yourshan 2001 Identifying the volatility of underlying assets from option prices *Inverse problems* **17** N1 137–155
[4] Kamynin V L and Bukharova T I 2017 Inverse problem of determination of the right-hand side term in the degenerate higher-order parabolic equation on a plane *Numerical Analyses and its Aplications NAA 2016 Lecture Notes in Computer Science* **10187 Springer** 391–397
[5] Kamynin V L 2015 On the solvability of the inverse problem for determining the right-hand side of a degenerate parabolic equation with integral observation *Math. Notes* **98** N5 765–777
[6] Kamynin V L 2017 Inverse problem of determination of the right-hand side term in the degenerate parabolic equation with unbounded coefficients *Comp. Math. and Math. Physics* **57** N5 832–841
[7] Kruzhkov S N 1979 Quasilinear parabolic equations and systems with two independent variables *Trudy Sem. im. Petrovskogo I G* **5** 217–272
[8] Lyusternik L A and Sobolev V I 1982 *Kratkii kurs Functional'nogo Analiza* (Moskva: Vysh. Shkola)

About Global Stable of Solutions of Logistic Equation with Delay

S A Kaschenko[1,2,3] **and D O Loginov**[1,4]

[1]Yaroslavl State University, 8/10 Kirova str., Yaroslavl, 150000, Russian Federation
[2]National Research Nuclear University MEPhI, 31 Kashira Highway, Moscow, 115409, Russian Federation

E-mail: [3]kasch@uniyar.ac.ru

E-mail: [4]dimonl@inbox.ru

Abstract. The article is devoted to the definition of all the arguments for which all positive solutions of logistic equation with delay tend to zero for $t \to \infty$. The authors have proved the acquainted Wright's conjecture on evaluation of a multitude of such arguments. An approach that enables subsequent refinement of this evaluation has been developed.

1. Introduction

Logistic equation with delay

$$\dot{u} = r[1 - u(t - T)]u \tag{1}$$

can be seen in many applied problems (e.g. [1, 2, 3]). Considerable literature is concerned with the analysis of its solution. In the context of the problem the coefficients r and T are positive. Initial data space for the equation (1) is the space $C_{[-T,0]}$. As the analysis of only nonnegative solutions of the equation (1) is of interest, then only nonnegative initial functions from $C_{[-T,0]}$ are revised. The solution with nonnegative initial function posed for $t = t_0$ is nonnegative for all $t > t_0$. Hereafter the term "solution" is used only for nonnegative solutions of the equation (1).

The number of arguments in the equation (1) can be reduced by applying normalizing time substitution $t \to Tt$. As a result we get the following equation

$$\dot{u} = \lambda[1 - u(t - 1)]u, \qquad \lambda = rT. \tag{2}$$

In this equation it is opportune to make a substitution $u = 1 + x$. Then we get the following equation

$$\dot{x} = -\lambda x(t - 1)[1 + x], \qquad x \geq -1. \tag{3}$$

Let us enumerate acquainted and necessary for future properties of the equation (3) solutions:

(i) Every solution of the equation (3) at sufficiently large t complies with the following inequation

$$x(t) \leq \exp(\lambda) - 1. \tag{4}$$

(ii) Equilibrium point stability $x_0 \equiv 0$ is defined by the position of characteristical quasipolynomial roots (of an equation linear on x_0 of the equation).

$$\mu = -\lambda \exp(-\mu). \tag{5}$$

On condition that

$$0 < \lambda < \frac{\pi}{2} \tag{6}$$

all roots of the equality (5) have negative real components, therefore the equilibrium point of x_0 in the equation (3) is asymptotically stable. The remarkable thing is that the equation (5) for $\lambda = \pi/2$ has a couple of purely imaginary roots and the rest its roots have negative real components. Nonlinear analysis of the equation (3) in small neighbourhood of x_0 cited in works of a lot of authors (e.g. [1, 5]) has shown that zero equilibrium point in the equation (3) for $\lambda = \pi/2$ is also asymptotically stable. Furthermore for $\lambda > \pi/2$ this equilibrium point is instable and stable cycle is bifurcated from it.

The following article deals with the matter of equilibrium point global stability $u_0 \equiv 1$ in the equation (2) or, what is the same, equilibrium point $x_0 \equiv 1$ in the equation (3). This means that the problem of finding such values of argument λ that all solutions of the equation (2) with positive initial functions (or all solutions of the equation (3) with initial functions complying with inequation $x > -1$) tend to u_0 (accordingly to x_0) for $t \to \infty$ is set. In works [1, 2, 3, 4, 5, 6, 7, 8] it was shown that global stability region includes all values of the argument λ, complying with inequation $0 < \lambda \leq 3/2$. In the work [1] deals with a hypothesis that there is a more strong result positing that a multitude of such values of λ is posed by the following inequation

$$0 < \lambda < 37/24. \tag{7}$$

The authors of the following work think that for argumentation of this result using the method of the work [1] you need to overcome "immeasurable" difficulties. In the following paragraphs evaluation of the inequation (7) will be proved and the procedure for continual improvement of the inequation (7) evaluation will be developed. The possibility of getting more accurate evaluations in comparison with the inequation (4) for solutions of the equation (2) is opened up as the condition of the inequation (7) is not fulfilled. Subject to the above mentioned the inequation $\lambda \leq \pi/2$ for values of λ from global stability region is trivially performed.

We also note that a numerical study of equation (3) showed that for $0 < \lambda < \pi/2$ all solutions of this equation tend to zero for $t \to \infty$. However as can be seen analytically it is clear from the following paragraphs that only for the condition $0 < \lambda < 37/24$. In the paper [9], an opinion was expressed that global stability includes $\lambda \in [1.5, 1.5706]$. But no arguments confirming it was not given.

2. Main Structure

First we note that for every solution of $x(t)$ (of the equation (3)) and for every whole $n > 1$ there is such t_0, that for $t > t_n$ the function $x(t)$ is n times continuously differentiable. Moreover from the inequation (4) the possibility of finding the values of coefficients of a_n and b_n opens up for for which

$$-b_n \leq x^{(n)}(t) \leq a_n \qquad (n = 1, 2, \cdots). \tag{8}$$

For example, for $n = 0$ we get $b_0 = -1$, $a_0 = \exp(\lambda) - 1$.

Let us assume that solution $x(t)$ of the equation (3) for some $\alpha \in (0, 1)$ and $M \in (0, \exp(\lambda) - 1)$ complies with the following inequations

$$-\alpha \leq x(t) \leq M \qquad (t \geq t_0). \tag{9}$$

Then for every n from the equation (3) and the inequation (9) we can get explicit formulae for coefficients $a_n = a_n(\alpha, M)$ and $b_n = b_n(\alpha, M)$ that occur in the inequation (8).

Let us arbitrary settle the number $m \geq 1$ and label by $C(m, \alpha, M)$ and $S(m, \alpha, M)$ extremums of composed functions

$$C(m, \alpha, M) = \inf \int_0^1 x(s)ds \tag{10}$$

and

$$S(m, \alpha, M) = \sup \int_0^1 x(s)ds, \tag{11}$$

where upper and lower bounds are taken by all m times continuously differential solutions of the equation (3) complying with the inequation (9) for $n = 1, \cdots, m$ and for which $x(1) = 0$.

Let us consider that the hypothesis that solution of the equation (3) is equal to zero may not be fulfilled, that is some solution can be of fixed sign (for $t \geq t_0$). But then this solution, as it appears from the equation (3), steadily tends to zero for $t \to \infty$ and so such solutions are not of interest. In this regard it is useful to highlight that these steadily tending to zero solutions can exist only on condition that $0 < \lambda < \exp(-1)$.

2.1. Transition to Two-dimensional Projectivity

Let us set

$$M_1 = \exp(-\lambda C(m, \alpha, M)) - 1, \tag{12}$$

$$\alpha_1 = 1 - \exp(-\lambda S(m, \alpha, M)). \tag{13}$$

Therefore there occurs the following two-dimensional projectivity

$$(\alpha, M) \overset{\Phi(\alpha,M)}{\longrightarrow} (\alpha_1, M_1) \overset{\Phi(\alpha_1,M_1)}{\longrightarrow} (\alpha_2, M_2) \overset{\Phi(\alpha_2,M_2)}{\longrightarrow} \cdots \tag{14}$$

To make all solutions of the equation (3) complying with the inequation $x \geq -1$ tend to zero, it is sufficient that for all $\alpha \in (0, 1)$ and $M \in (0, \exp(\lambda) - 1)$ the following inequations are performed

$$\alpha_1 < \alpha, \qquad M_1 < M. \tag{15}$$

Remark 2.1 *Let us consider the iterations (α_n, M_n) $(n = 2, 3, \cdots)$ of $\Phi(\alpha, M)$ transformation:*

$$(\alpha_{n+1}, M_{n+1}) = \Phi(\alpha_n, M_n). \tag{16}$$

Let us assume that for some λ we have

$$\alpha_0(\lambda) = \lim_{n\to\infty} \sup_{k>n} \alpha_k, \qquad M_0(\lambda) = \lim_{n\to\infty} \sup_{k>n} M_k.$$

Then for solutions of the equation (3) the following equations are performed

$$\lim_{t\to\infty} \inf_{\tau\geq t} x(\tau) \geq -\alpha_0(\lambda), \qquad \lim_{t\to\infty} \sup_{\tau\geq t} x(\tau) \leq M_0(\lambda). \tag{17}$$

IOP Conf. Series: Journal of Physics: Conf. Series **937** (2017) 012019 doi:10.1088/1742-6596/937/1/012019

2.2. Reduction of Two-dimensional Projectivity of the Equations (12), (13) to Two Projectivities of the First Order

Here we suppose that $m \leq 2$.

Let us assume that for some solution of $x(t)$ for all $t \geq t_0$ the following inequation is performed

$$x(t) \geq -\alpha \qquad (0 < \alpha < 1). \tag{18}$$

First let us note that for $m = 1$ the expression $C(1, \alpha, M)$ does not depend from M. Let us assume that $C(\alpha) = C(1, \alpha, M) = \cdots$.

Let

$$M_1 = \exp(-\lambda C(\alpha)) - 1.$$

Let us consider the expression $C(m, \alpha, M_1)$ and assume that

$$M_2 = \exp(-\lambda C(m, \alpha, M_1)) - 1.$$

Then let us pose $M_3, \cdots$ from the formula

$$M_{n+1} = \exp(-\lambda C(m, \alpha, M_n)) - 1 \qquad (n = 2, 3, \cdots). \tag{19}$$

Let us note that there occur inequations

$$M_{n+1} \leq M_n.$$

We denote by $M(\ alpha)$ the limit value of the sequence M_n:

$$M(\alpha) = \lim_{n\to\infty} M_n.$$

Herewith $M(\alpha)$ is a stable fixed-point value of the equation (19) projectivity.

On the next stage let us consider the equations

$$\alpha_1 = 1 - \exp(-\lambda S(\alpha, M(\alpha)))$$

and

$$\alpha_{n+1} = 1 - exp(-\lambda S(\alpha_n(\alpha), M(\alpha_n(\alpha))) \qquad (n = 1, 2, \cdots). \tag{20}$$

Then we give the conclusive result.

Theorem 2.2 *Let us assume that for all $\alpha \in (0, 1)$ the following limit equality is performed*

$$\lim_{n\to\infty} \alpha_n(\alpha) = 0.$$

Than all solutions of the equation (3) tend to zero for $t \to \infty$.

Remark 2.3 *For the validity of the Theorem 2.2 statement it is sufficient that for all $\alpha \in (0, 1)$ the inequation $\alpha_1 < \alpha$ is performed.*

The next two section are concerned with the study of case when $m = 1$. As will be shown, in these case the two-dimensional projectivity of the equations (12), (13) reduces to one-dimensional one. The case when $m = 1$ the importance of getting prior evaluations for the solutions is shown. In this regard at first, the results in case of rather rough evaluations are given and then it is shown that more accurate evaluations enable to prove sufficient extension of global stability region. Finally in Section 4 for $m = 2$ based on the Theorem 2.2 of the use of more accurate evaluations evaluation of the inequation (7).

3. The Case when $m = 1$

Let $m = 1$. It is reasonable to consider only the case when $\lambda \geq 1$.

Let us show that at a accurate prior evaluation solutions it is possible to extend the range of such values of λ at which there is zero equilibrium point global stability in the equation (3).

Theorem 3.1 *Let*

$$0 < \lambda \leq \frac{3}{2}.$$

Then all solutions of the equation (3) tend to zero for $t \to \infty$.

Proof. The equation (3) for $x \neq 0$ is equivalent to the following equation

$$(1+x)^{-1}\dot{x} = -\lambda x(t-1). \tag{21}$$

Then from evaluations of the inequation (9) we get that

$$-\lambda M \leq (1+x)^{-1}\dot{x} \leq \lambda\alpha. \tag{22}$$

Therefore we get the more accurate evaluations for $x(t)$

$$\begin{cases} -\lambda M \exp(-\lambda M(t-1)) \leq \dot{x} \leq \lambda\alpha\exp(\lambda\alpha(t-1)), \\ -1+\exp(-\lambda\alpha(1-t)) \leq x(t) \leq -1+\exp(\lambda M(1-t)). \end{cases} \tag{23}$$

It follows that

$$x_0(t) = \begin{cases} -\alpha, & t \in [0, t_0], \quad t_0 = 1 + (\lambda\alpha)^{-1}\ln(1-\alpha), \\ -1+\exp(-\lambda\alpha(1-t)), & t \in (t_0, 1], \end{cases}$$

$$x^0(t) = \begin{cases} M, & t \in [0, t^0], \quad t^0 = 1 - (\lambda M)^{-1}\ln(1+M), \\ -1+\exp(\lambda M(1-t)), & t \in (t^0, 1]. \end{cases}$$

For $C(1,\alpha,M)$ and $S(1,\alpha,M)$ the following equations are true

$$C(1,\alpha,M) = \int_0^1 x_0(s)ds = -\alpha + \frac{1}{\lambda} + (1-\alpha)(\lambda\alpha)^{-1}\ln(1-\alpha),$$

$$S(1,\alpha,M) = \int_0^1 x^0(s)ds = M + \frac{1}{\lambda} - (1+M)(\lambda M)^{-1}\ln(1+M).$$

And in this context the two-dimensional projectivity of the equations (12), (13) reduce to one-dimensional projectivity of the equality

$$f(\lambda,\beta) = 1 - \exp[-1 - \lambda m(\beta) + m^{-1}(\beta)(1+m(\beta))\ln(1+m(\beta))]$$

and

$$m(\beta) = -1 + \exp[-1 + \lambda\beta - \beta^{-1}(1-\beta)\ln(1-\beta)].$$

Let us note that $f'(0) = (\lambda - \frac{1}{2})^2$, so $f'(0) < 1$ for $\lambda < \frac{3}{2}$.

Thus, based on more accurate evaluations we succeeded to show that for $0 < \lambda < \frac{3}{2}$ all solutions of the equation (3) tend to zero for $t \to \infty$. Let us note that for $\lambda \leq \frac{\pi}{2}$, for every solution at larger t the following evaluation is performed

$$x(t) \geq -\beta_0 = -0.5254171515.$$

The theorem is proved.

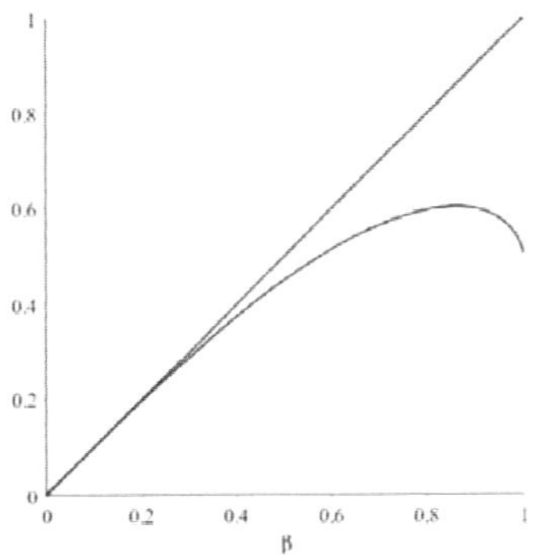

Figure 1. Graph of the function $f(\lambda, \beta)$ where $\lambda = \frac{3}{2}$

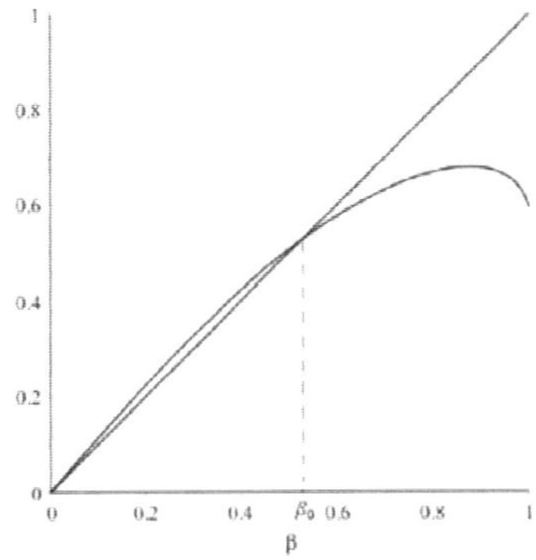

Figure 2. Graph of the function $f(\lambda, \beta)$ where $\lambda = \frac{\pi}{2}, \beta_0 = 0.5254171515$

4. The Case when $m = 2$

Now we formulate our main result.

Theorem 4.1 *Let $0 < \lambda < \frac{37}{24}$. Then all solutions of the equation (3), where certain functions corresponding the valuation $x(s) > -1$, tend to zero for $t \to \infty$.*

Proof. Let us arbitrary settle solution $x(t)$ of the equation (3). Let us assume that for all $t \geq t_0$, the following evaluations are performed

$$x(t) \geq -\alpha \qquad (1 > \alpha > 0). \tag{24}$$

Let us label by M the supremum of this solution, i.e.

$$\sup_{t > t_0} x(t) = M. \tag{25}$$

Then for the value M from the previous formulations comes the evaluation

$$M \leq \exp\left[-\lambda^2 \alpha \left(1 - \frac{1}{2\lambda}\right)\right] - 1 = M_1(\alpha, \lambda). \tag{26}$$

In this context we can take $a_2 = a_2(M_1)$ and with regard to (26)

$$a_2 = \lambda^2 M_1 (1 + 2M_1).$$

Let us consider the inequations (23) to improve the achieved evaluation. Then for $x_0(t)$ we get the following function

$$x_0(t) = \begin{cases} -\alpha, & -1 \leq t \leq t_1, \\ -\alpha + \frac{1}{2} a_2 (t - t_1)^2, & t_1 \leq t \leq t_2, \\ -1 + \exp(\lambda \alpha t), & t_2 < t \leq 0, \end{cases}$$

Here

$$t_2 = \frac{\ln\left(\frac{a_2 \Delta_0^2}{2} - \alpha + 1\right)}{\alpha \lambda}, \quad t_1 = t_2 - \Delta_0, \quad \Delta_0 = \frac{a_2 - \sqrt{a_2^2 + a_0}}{\alpha \lambda a_2},$$

$$a_0 = 2\alpha^2\lambda^2 a_2(\alpha - 1).$$

The projectivity $M_{n+1}(\alpha, M_n, \lambda)$ is expressed in the following way

$$M_{n+1} = \exp[-\lambda C(2, \alpha, M_n)] - 1. \tag{27}$$

Consequent iterations of the projectivity

$$M_1 \to M_2 \to \cdots .$$

Lead to the conclusion that the desired upper evaluation of $x(t)$ solution doesnt rank over the fixed-point value of (27) projectivity.

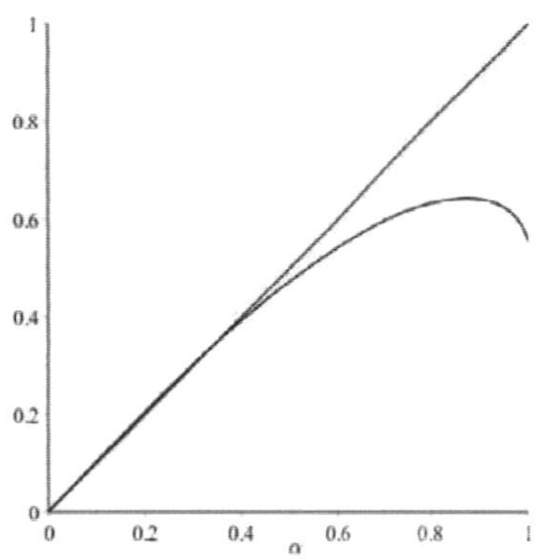

Figure 3. Graph of the function $f(\lambda, \beta)$ where $\lambda = \frac{37}{24}$

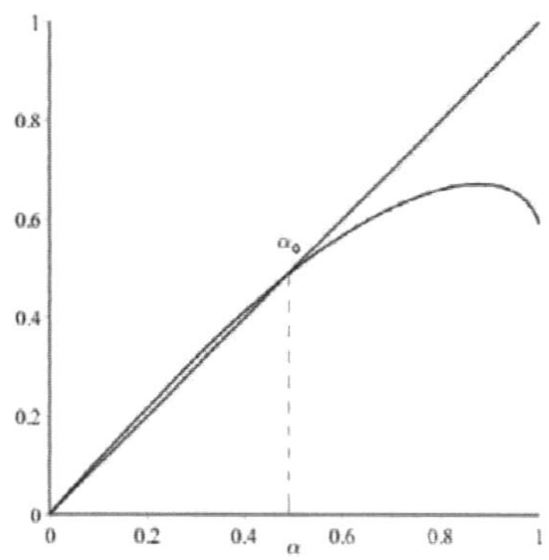

Figure 4. Graph of the function $f(\lambda, \beta)$ where $\lambda = \frac{\pi}{2}, \alpha_0 = 0.4809292426$

After definition of the fixed point $M(\alpha)$, the function $x^0(t)$ on which the desired value $S(2, \alpha, M(\alpha))$ is realized, is written as

$$x^0(t) = \begin{cases} M(\alpha), & 0 \le t \le \tau_1, \\ M(\alpha) + \frac{1}{2}b_2(t - \tau_1)^2, & \tau_1 < t \le \tau_2, \\ -1 + \exp(-\lambda M(\alpha)(t-1)), & \tau_2 < t \le 1. \end{cases}$$

Here

$$\tau_2 = 1 - \frac{\ln\left(\frac{b_2\Delta_1^2}{2} + M(\alpha) + 1\right)}{M(\alpha)\lambda}, \quad \tau_1 = \tau_2 - \Delta_1, \quad \Delta_1 = -\frac{b_2 + \sqrt{b_2^2 - b_0}}{M(\alpha)\lambda b_2},$$

$$b_0 = 2M(\alpha)^2\lambda^2 b_2(M(\alpha) + 1), \quad b_2 = -\lambda^2\alpha(1 + M(\alpha))^2.$$

Then for equations $C(2, \alpha, M(\alpha))$ and $S(2, \alpha, M(\alpha))$ we have the formulas

$$C(2, \alpha, M(\alpha)) = \int_{-1}^{0} x_0(s)ds = [6a_2^2\alpha^2\lambda^2(1-\alpha)\ln\left(\frac{\frac{1}{2}(a_2 - \sqrt{a_2^2 + a_0})^2}{(\alpha^2\lambda^2 a_2) - \alpha + 1}\right) +$$

$$+3\sqrt{a_2^2 + a_0 a_2^2} - 2a_2^3 + (-6\alpha^4\lambda^3 + 6\alpha^3\lambda^2)a_2^2 - (a_2^2 + a_0)^{\frac{3}{2}}] \cdot (6a_2^2\alpha^3\lambda^3)^{-1},$$

$$S(2,\alpha,M(\alpha)) = \int_0^1 x^0(s)ds = [6M^2(\alpha)\lambda^2(M^2(\alpha)+M(\alpha)\lambda - M(\alpha)-2)b_2^2 -$$
$$-6b_2^2M^2(\alpha)\lambda^2(M(\alpha)+1)\ln\left(\frac{\frac{1}{2}(-b_2-\sqrt{b_2^2-b_0})^2}{M^2(\alpha)\lambda^2 b_2} + M(\alpha)+1\right) -$$
$$-9\sqrt{b_2^2-b_0}b_2^2 - 10b_2^3 + 6b_2b_0 - (b_2^2-b_0)^{\frac{3}{2}}]\cdot(6b_2^2M^3(\alpha)\lambda^3)^{-1}.$$

Hence, that if $0<\lambda<\frac{37}{24}$ then all orbits of the mapping (27) tend to zero for $n\to\infty$. The theorem is proved.

Intimate, that $f'_\alpha(\lambda,0)=(\lambda-\frac{13}{24})^2$, thus $f'_\alpha(\lambda,0)<1$ for $\lambda<\frac{37}{24}$.

5. Conclusions

Wright's conjecture [1] that under condition of the inequation (7) zero equilibrium point of the equation (3) is globally stable, has been proved. An approach of subsequent refinement of corresponding evaluations has been developed.

References

[1] E M Wright 1955 *A non-linear difference-differential equation* (Journal für die reine und angewandte Mathematik (Crelles Journal)) Vol 194 pp 66 – 87

[2] Y Kuang 1993 *Delay differential equations with applications in population dynamics* (New York: In: Mathematics in Science and Engineering Vol. 191, Academic Press)

[3] S A Gourley, J So and J Wu 2004 *Nonlocality of Reaction-Diffusion Equations Induced by Delay: Biological Modeling and Nonlinear Dynamics* (J. Math. Sci.) Vol. 124 pp 5119 – 5153

[4] S Kakutani and L Markus 1958 *On the non-linear difference differential equation* $y'(t)=[A+By(t-\gamma)]y(t)$ (Princeton: Contributions Theory Nonlinear Oscillation) Vol 4 pp 411 – 418

[5] E V Grigorieva and S A Kashchenko 1993 *Regular and chaotic pulsations in laser diode with delayed feedback* (Int.J.Bifur.& Chaos) Vol. 3 No. 3 pp 1515 – 1528

[6] S A Kashchenko 1985 *K voprosu ob otsenke v prostranstve parametrov oblasti globalnoy ustoychivosti uravneniya Hatchinsona (in Russian) [On the problem of estimating in the parameter space the domain of global stability of the Hutchinson equation]* (Yaroslavl: Nelineynye kolebaniya v zadachakh ekologii YarGU) pp 55 – 62

[7] S A Kashchenko 2012 *Asimptotika resheniy obobcshennogo urovneniya Hatchinsona (in Russian) [Asymptotics of the Solutions of the Generalized Hutchinson Equation]* (Yaroslavl: Modelirovanie i analiz informacionnih sistem) Vol. 19 No. 13 pp 32 – 61.

[8] S A Kashchenko 2013 *Asymptotics of the Solutions of the Generalized Hutchinson Equation* (Automatic Control and Computer Sciences) Vol. 47 No. 7 pp 470 – 494

[9] B Bánhelyi, T Csendes, T Krisztin and A Neumaier 2014 *Global attractivity of the zero solution for Wright's equation* (SIAM Journal on Applied Dynamical Systems) Vol. 13 No. 1 pp 537 – 563

Acknowledgments

This study is supported by State assignment, project No 1.10160.2017/5.1

Relaxation periodic solutions of one singular perturbed system with delay

A A Kashchenko

Mathematical Faculty, Yaroslavl State University, Sovetskaya str. 14, Yaroslavl 150003, Russia

E-mail: a.kashchenko@uniyar.ac.ru

Abstract. In this paper, we consider a singularly perturbed system of two differential equations with delay, simulating two coupled oscillators with a nonlinear compactly supported feedback. We reduce studying nonlocal dynamics of initial system to studying dynamics of special finite-dimensional mappings: rough stable (unstable) cycles of these mappings correspond to exponentially orbitally stable (unstable) relaxation solutions of initial problem. We show that dynamics of initial model depends on coupling coefficient crucially. Multistability is proved.

1. Introduction

In this paper, we study nonlocal dynamics of a system of two delay differential equations

$$\begin{cases} \dot{u}_0 + u_0 = \lambda F(u_0(t-T)) + \gamma(u_1 - u_0), \\ \dot{u}_1 + u_1 = \lambda F(u_1(t-T)) + \gamma(u_0 - u_1), \end{cases} \quad (1)$$

which simulates two coupled oscillators with nonlinear feedback. Here parameter λ, delay time T, and coupling parameter γ are positive. We suppose that $F(u)$ is some nonlinear function, whose values are concentrated in a bounded domain: for some $p > 0$ we have equality

$$F(u) = \begin{cases} f(u), & |u| < p, \\ 0, & |u| \geq p. \end{cases}$$

Equations with this type of nonlinearity arise in electrical engineering [1] and in radiophysics [2].

Here function $f(u)$ is smooth and satisfies the following conditions:

$$\begin{cases} uf(u) > 0 & \text{if} \quad 0 < |u| < p, \\ f(-p) = f(0) = f(p) = 0, \\ f(u) = const_1(p-u)^2 & \text{if} \quad 0 < p - u \ll 1, \\ f(u) = const_2(p+u)^2 & \text{if} \quad 0 < p + u \ll 1. \end{cases}$$

We suppose that parameter λ is sufficiently large ($\lambda \gg 1$) and coupling parameter γ is sufficiently small ($0 < \gamma \ll 1$). We consider three cases: $\gamma = O((\ln \lambda)^{-1})$; $\gamma = O(\lambda^{-\alpha}(\ln \lambda)^{-1})$, where $0 < \alpha \leq 1/2$, and $\gamma = O(\lambda^{-\alpha})$, where $1/2 < \alpha < 1$ as $\lambda \to +\infty$.

We study nonlocal dynamics of system (1) with special method of large parameter [3, 4, 5, 6, 7]. In phase space $C_{[-T,0]}(R^2)$ of system (1), we choose sets $S(x)$ depending on parameter x. Then we investigate the asymptotic behavior as $\lambda \to +\infty$ of all solutions of (1) with initial-value conditions from $S(x)$. We obtain that all considered solutions get into a set $S(\bar{x})$ after a certain time. We get representation $\bar{x} = \psi(x) + o(1)$ as $\lambda \to +\infty$. Then the first return operator $\Pi : \Pi S(x) \subset S(\bar{x})$ is constructed. Its iterations are principally determined by dynamics of mapping $\bar{x} = \psi(x)$. Theorem of correspondence is proved: rough stable (unstable) cycle of mapping $\bar{x} = \psi(x)$ corresponds to exponentially orbitally stable (unstable) relaxation periodic solution of initial problem.

2. Dynamics of system (1) under condition $\gamma = \gamma_1(\ln\lambda)^{-1}$

2.1. Building a mapping

Let

$$\gamma = \gamma_1(\ln\lambda)^{-1}, \quad \gamma_1 > 0. \tag{2}$$

Let us choose a set of initial conditions for system (1). Let x be some value such that $|x| \geq 1$. Let value k be equal either to 1 or -1. Let parameter m be equal either to 0 or 1. Fix some values x, k, and m. Let $S(x) \subset C_{[-T,0]}(R^2)$ be a set of continuous functions $u_m(s)$ and $u_{1-m}(s)$ ($s \in [-T, 0]$) such that

$$|u_m(s)| > p, \quad |u_{1-m}(s)| > p \quad \text{if} \quad s \in [-T, 0),$$
$$u_m(0) = kp, \quad u_{1-m}(0) = xp, \quad kx > 0. \tag{3}$$

We build asymptotics of all solutions of system (1) with initial conditions from set $S(x)$ (see [6]). We get that each solution of system (1) with initial conditions from set $S(x)$ belongs to a set $S(\bar{x})$ after time t_*. It means that

$$|u_{\bar{m}}(t)| > p, \quad |u_{1-\bar{m}}(t)| > p \quad \text{for all} \quad t \in [t_* - T, t_*),$$
$$u_{\bar{m}}(t_*) = kp, \quad u_{1-\bar{m}}(t_*) = \bar{x}p, \quad k\bar{x} > 0.$$

Consequently, at point $t = t_*$ we return to initial situation with replacement m by $\bar{m}$, x by $\bar{x}$.

By asymptotics of solutions we obtain representation $t_* = (1 + o(1))\ln\lambda$ as $\lambda \to +\infty$.

Let

$$R(x) = \begin{cases} (1+\exp(-2\gamma_1))(1-\exp(-2\gamma_1))^{-1} & \text{if} \quad |x|e^{-T} \geq 1, \\ \dfrac{g(2T)(1+\exp(-2\gamma_1)) + g(2T,x)(1-\exp(-2\gamma_1))}{g(2T)(1-\exp(-2\gamma_1)) + g(2T,x)(1+\exp(-2\gamma_1))} & \text{if} \quad |x|e^{-T} < 1, \end{cases}$$

where

$$g(t) = \int\limits_T^t \exp(s-t) f(kp\exp(T-s))ds,$$
$$g(t,x) = \int\limits_{T+\ln(x)}^t \exp(s-t) f(xp\exp(T-s))ds.$$

Then from asymptotics of solutions we obtain that

$$\bar{m} = \begin{cases} m & \text{if} \quad 0 < R(x) < 1, \\ 1-m & \text{if} \quad R(x) > 1; \end{cases} \tag{4}$$

$$\bar{x} = k\max\{R(x), 1/R(x)\} + o(1). \tag{5}$$

IOP Conf. Series: Journal of Physics: Conf. Series **937** (2017) 012020 doi:10.1088/1742-6596/937/1/012020

2.2. Theorem of correspondence

Let us consider the main part of mapping (5):

$$x_{n+1} = k \max\{R(x_n), 1/R(x_n)\}. \tag{6}$$

We have the following theorem.

Theorem 1. *Suppose that mapping (4), (6) has a cycle (m_i, x_i) of period n and absolute value of multiplier of cycle $x_1, x_2, \dots, x_n$ is not equal to 1. Let (2) hold. Then for all positive and sufficiently large λ system (1) has a periodic relaxation solution of period $T_*(\lambda) = n(1+o(1)) \ln \lambda$ as $\lambda \to +\infty$ of the same stability as the cycle (m_i, x_i).*

Proof of existence of relaxation periodic solutions of system (1) follows directly from constructed asymptotics and special method of large parameter. One can see in [5] detailed proof (for another model) of the fact that rough cycle of constructed mapping has the same stability as corresponding relaxation periodic solution of initial problem.

It follows from asymptotics of solutions that if $|x_i|e^{-T} < 1$, then amplitude of both components of solution is of order $O(\lambda)$ as $\lambda \to +\infty$ on this "step", and if $|x_i|e^{-T} \geq 1$, then amplitude of component u_{m_i} is of order $O(\lambda)$ and amplitude of component u_{1-m_i} is of order $O(\lambda/\ln \lambda)$ as $\lambda \to +\infty$. On Figure 1 typical view of inhomogeneous relaxation periodic solutions of system (1) is shown.

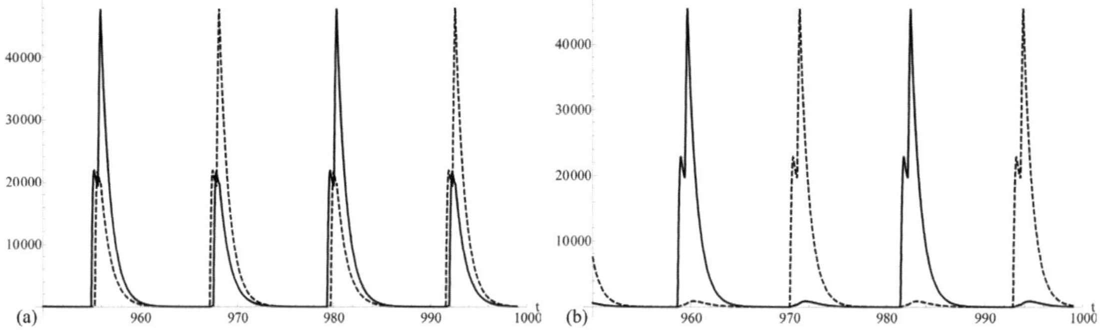

Figure 1. (a) Stable inhomogeneous periodic solution of system (1) with peaks of the same order (b) Stable inhomogeneous periodic solution of system (1) with peaks of different orders

2.3. Number of stable coexisting solutions of system (1)

The following result on dynamics of mapping (6) holds.

Lemma 1. *If x_* is a fixed point of mapping (6) and multiplier of (6) at $x = x_*$ is negative, then x_* is a stable fixed point of (6).*

For a given set of initial parameters (γ_1, T, p) and different functions $f(u)$ mapping (6) may have different number of stable fixed points. For example, for $\gamma_1 = 0.2$, $T = 1$, $p = 1$ and function $F(u)$, shown on Figure 2, there are twelve stable fixed points of mapping (6) (see Figure 3).

Stable fixed point $x = 1$ ($x = -1$) of mapping (6) corresponds to exponentially orbitally stable positive (negative) homogeneous periodic relaxation solution of system (1). One stable fixed point x (where $|x| > 1$) of mapping (6) corresponds to two exponentially orbitally stable inhomogeneous periodic relaxation solutions of system (1) (we may take $m = 0$ or $m = 1$ in (3) and get two different solutions). Thus if mapping (6) has n stable fixed points x_i where $|x_i| > 1$ $(i = 1, \dots, n)$, then for all sufficiently large $\lambda > 0$ system (1) has $2n$ coexisting exponentially

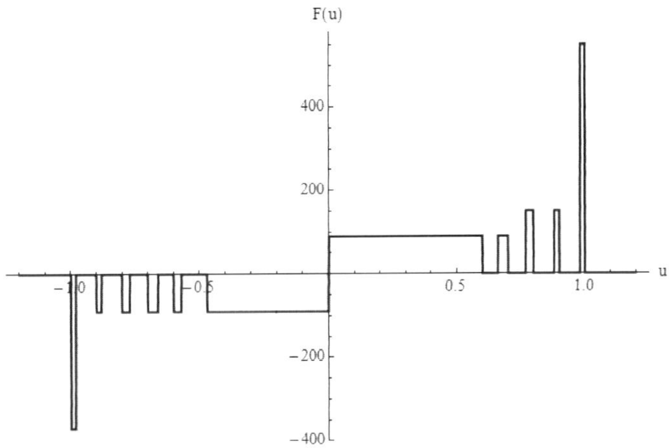

Figure 2. Example of function $F(u)$. Values of parameters: $p = 1$.

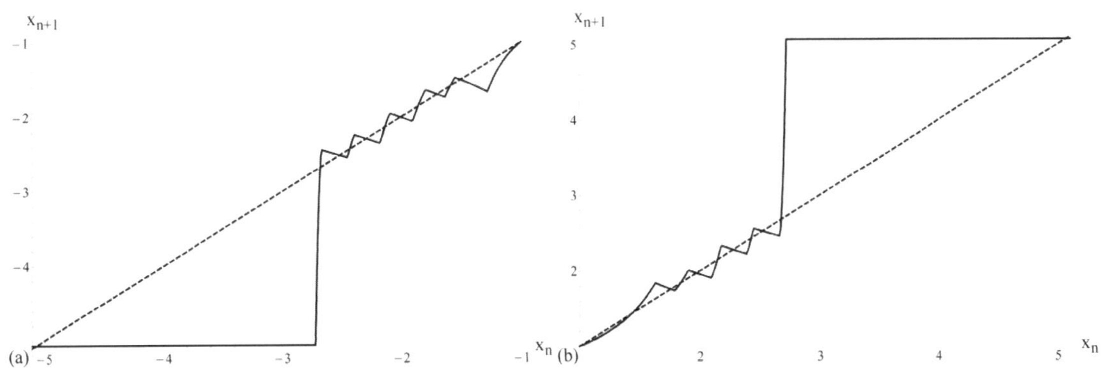

Figure 3. Example of mapping (6) and mapping $x_{n+1} = x_n$ (dashed line) for (a) $x_n \le -1$ (b) $x_n \ge 1$. Values of parameters: $\gamma_1 = 0.2$, $T = 1$, $p = 1$.

orbitally stable inhomogeneous periodic relaxation solutions. Note that all types of solutions: homogeneous, inhomogeneous with peaks of the same order and inhomogeneous with peaks of different orders at $\lambda \to +\infty$ may be stable simultaneously.

3. Dynamics of system (1) under condition $\gamma = \gamma_1 \lambda^{-\alpha} (\ln \lambda)^{-1}$

3.1. Building a mapping

Let

$$\gamma = \gamma_1 \lambda^{-\alpha} (\ln \lambda)^{-1}, \quad \gamma_1 > 0, \quad 0 < \alpha \le \frac{1}{2}. \tag{7}$$

Let us define a set of initial conditions for system (1). Let, as in the previous case, value k be equal either to 1 or -1 and parameter m be equal either to 0 or 1. Let x be some nonzero value ($x \neq 0$). Let β be some value from $(0, \alpha]$. Fix some values x, k, m, and β. Denote $S(x) \subset C_{[-T,0]}(R^2)$ as a set of continuous on $s \in [-T, 0]$ functions $u_m(s)$ and $u_{1-m}(s)$ such that

$$|u_m(s)| > p, \quad |u_{1-m}(s)| > p \quad \text{if} \quad s \in [-T, 0),$$

$$u_m(0) = kp, \quad u_{1-m}(0) = xp\lambda^{\beta}, \quad kx > 0.$$

Note that now value $u_{1-m}(0)$ is asymptotically large as $\lambda \to +\infty$ (in previous case it had the order of unity).

As in the previous case, we build asymptotics of solutions of system (1) with initial conditions from set $S(x)$ and get that each solution after time t_* belongs to a set $S(\bar{x})$.

In this case $t_* = (1-\alpha)(1+o(1))\ln\lambda$ as $\lambda \to +\infty$ and

$$u_m(t_*) = \bar{x}p\lambda^{\bar{\beta}}, \quad u_{1-m}(t_*) = kp, \quad k\bar{x} > 0.$$

Here

$$\bar{\beta} = \alpha, \tag{8}$$

$$\bar{x} = \begin{cases} \dfrac{g(2T)}{|xp\exp(-2T) + \frac{\gamma_1}{2}g(2T)|} + o(1) & \text{if} \quad \beta = \dfrac{1}{2}, \\ k[\gamma_1(1-\alpha)]^{-1} + o(1) & \text{if} \quad \beta < \frac{1}{2}. \end{cases} \tag{9}$$

This implies that at point $t = t_*$ we return to initial situation with swap u_m and u_{1-m} and replacing β, x by $\bar{\beta}$, $\bar{x}$ using (8), (9). Notice that beginning from the second "step" we have $\beta = \alpha$.

If $\beta = \alpha = 1/2$, then for the leading term of $\bar{x}$ we have the following mapping

$$x_{n+1} = \frac{g(2T)}{|x_n p\exp(-2T) + \frac{\gamma_1}{2}g(2T)|}, \quad (n = 1, 2, ...). \tag{10}$$

If $x_1 > 0$, then $k = 1$, $g(2T) > 0$ and sequence x_n tends to a stable positive fixed point. If $x_1 < 0$, then $k = -1$, $g(2T) < 0$ and sequence x_n tends to a stable negative fixed point.

If $0 < \alpha < 1/2$, then for the leading term of value $\bar{x}$ we obtain mapping

$$x_{n+1} = k[\gamma_1(1-\alpha)]^{-1}, \quad (n = 1, 2, ...). \tag{11}$$

Beginning from x_2 sequence x_n is constant. This mapping has one stable positive fixed point and one negative stable fixed point.

3.2. Number of stable coexisting solutions of system (1)

Each stable fixed point of mapping (10) (mapping (11)) corresponds to two exponentially orbitally stable relaxation periodic solutions of system (1). Thus we obtain the following result.

Theorem 2. *Let (7) hold. Then for all sufficiently large $\lambda > 0$ system (1) has four (two positive and two negative) exponentially orbitally stable relaxation periodic solutions with period $T_*(\lambda) = [2(1-\alpha) + o(1)]\ln\lambda$ as $\lambda \to +\infty$.*

It follows from asymptotics of solutions that if $0 < \alpha < 1/2$, then both components of solution have two peaks on period (see Figure 4 (a)). One peak is of order $O(\lambda)$ and another peak is of order $O(\lambda^{1-\alpha}(\ln\lambda)^{-1})$ as $\lambda \to +\infty$. But if $\alpha = 1/2$, then both components have a single peak on period (see Figure 4 (b)). This peak is of order $O(\lambda)$ as $\lambda \to +\infty$.

4. Dynamics of system (1) under condition $\gamma = \gamma_1\lambda^{-\alpha}$

Let

$$\gamma = \gamma_1\lambda^{-\alpha}, \quad \gamma_1 > 0, \quad \frac{1}{2} < \alpha < 1.$$

Let us define set $S(x) \subset C_{[-T,0]}(R^2)$ of initial conditions. Let, as in the previous cases, parameter k be equal either to 1 or -1 and parameter m be equal either to 0 or 1. Let parameter x be nonzero ($x \neq 0$). Let β satisfy inequality $0 < \beta < \alpha$. Fix some values of parameters k, x, m, and β. Denote as $S(x)$ a set of continuous on $s \in [-T, 0]$ functions $u_m(s)$, $u_{1-m}(s)$ such that

$$|u_m(s)| > p, \quad |u_{1-m}(s)| > p \quad \text{if} \quad s \in [-T, 0)$$

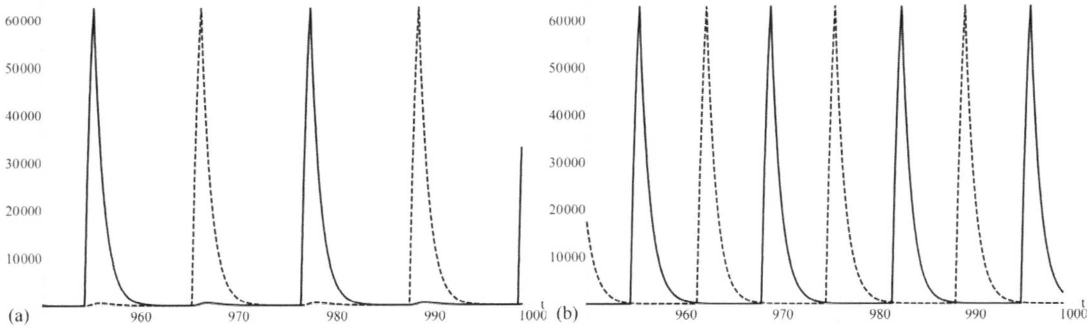

Figure 4. (a) Stable inhomogeneous periodic solution of system (1) (case $\alpha = 0.2$) (b) Stable inhomogeneous periodic solution of system (1) (case $\alpha = 0.5$)

and

$$u_m(0) = kp, \quad u_{1-m}(0) = xp\lambda^{\beta}, \quad kx > 0.$$

We build asymptotics of all solutions of system (1) with initial conditions from set $S(x)$ (see [7]). We obtain that after time t_* each considered solution belongs to a set $S(\bar{x})$. It means that

$$|u_m(t)| > p, \quad |u_{1-m}(t)| > p \quad \text{for all} \quad t \in [t_* - T, t_*)$$

and

$$u_m(t_*) = \bar{x}p\lambda^{\bar{\beta}}, \quad u_{1-m}(t_*) = kp, \quad k\bar{x} > 0.$$

From asymptotics of solutions we get representation

$$t_* = \begin{cases} (\beta + o(1)) \ln \lambda & \text{if} \quad 1 - \alpha < \beta < \alpha, \\ (1 - \alpha + o(1)) \ln \lambda & \text{if} \quad 0 < \beta \leq 1 - \alpha. \end{cases}$$

Quantities $\bar{\beta}$ and $\bar{x}$ can be expressed by formulas

$$\bar{\beta} = \begin{cases} 1 - \beta & \text{if} \quad 1 - \alpha < \beta < \alpha, \\ \alpha & \text{if} \quad 0 < \beta \leq 1 - \alpha; \end{cases} \tag{12}$$

$$\bar{x} = \begin{cases} \dfrac{g(2T)}{|x| p \exp(-2T)} & \text{if} \quad 1 - \alpha < \beta < \alpha, \\ k[\gamma_1(1-\alpha)\ln\lambda]^{-1} & \text{if} \quad 0 < \beta \leq 1 - \alpha \end{cases} \tag{13}$$

up to accuracy $o(1)$.

Thus at $t = t_*$ we return to initial situation with exchange u_m and u_{1-m} and replacing β and x by $\bar{\beta}$ and $\bar{x}$ respectively using (12), (13).

For β from interval $(1 - \alpha, \alpha)$ we obtain an iterative process with respect to parameters β and x

$$\beta_{n+1} = 1 - \beta_n, \quad x_{n+1} = \frac{g(2T)}{|x_n| p \exp(-2T)}, \quad n = 1, 2, \ldots. \tag{14}$$

It follows from (14) that $\beta_{n+2} = \beta_n$ and $x_{n+2} = x_n$. Thus all solutions of mapping (14) are non-rough cycles of period two. For initial system (1) for all sufficiently large $\lambda > 0$ one can assert existence of non-rough inhomogeneous relaxation periodic asymptotic (with respect to the residual) solutions of period $T_*(\lambda) = (1 + o(1)) \ln \lambda$, where the values of parameters β and

x with an accuracy up to $o(1)$ vary according to (14). Conclusions about existence of exact periodic solutions and their stability can not be made.

Note that each component of the non-rough asymptotic (with respect to the residual) solutions of system (1) once on period $T_*(\lambda)$ in a short time interval (of the order of $O(1)$) increases from a value of the order of unity to a value of the order $O(\lambda)$ and the remaining time it decreases exponentially to a value of the order of unity.

5. Conclusions

In this paper we study nonlocal dynamics of a model of two coupled oscillators with compactly supported feedback function. We show that studying dynamics of initial system of delay differential equations can be reduced to studying dynamics of finite-dimensional mappings: rough stable (unstable) cycles of these mappings correspond to exponentially orbitally stable (unstable) relaxation periodic solutions of initial problem. Dynamics of system (1) depends on the value of coupling coefficient crucially. If $\gamma = O(1)$, then the only stable solution of system (1) with initial conditions from set $S(x)$ is homogeneous relaxation solution; if $\gamma = O((\ln\lambda)^{-1})$, then homogeneous relaxation solution of system (1) may be stable or unstable and many stable inhomogeneous relaxation solutions may coexist; if $\gamma = O(\lambda^{-\alpha}(\ln\lambda)^{-1})$ (where $0 < \alpha \leq 1/2$), then four stable inhomogeneous relaxation periodic solutions coexist; and if $\gamma = O(\lambda^{-\alpha})$ (where $1/2 < \alpha < 1$), then system (1) has a two-parameter family of non-rough inhomogeneous relaxation periodic asymptotic (with respect to the residual) solutions.

Acknowledgments

This study is supported by State assignment, work No 1.6074.2017/П220.

References

[1] Kilias T, Kelber K, Mogel A and Schwarz W 1995 *International journal of electronics* **79** 737–753
[2] Dmitriev A S and Kislov V Y 1989 *Stochastic oscillations in radiophysics and electronics* (Nauka, Moscow)
[3] Grigorieva E V and Kaschenko S A 1993 *International Journal of Bifurcation and Chaos* **3** 1515–1528
[4] Kaschenko D, Kaschenko S and Schwarz W 2012 *International Journal of Bifurcation and Chaos* **22** 1250184
[5] Kaschenko S A 2015 *Automatic Control and Computer Sciences* **49** 547–581
[6] Kashchenko A A 2016 *Modeling and Analysis of Information Systems* **23** 841–849
[7] Kashchenko A A 2017 *Modeling and Analysis of Information Systems* **24** 649–654

IOP Conf. Series: Journal of Physics: Conf. Series **937** (2017) 012021 doi:10.1088/1742-6596/937/1/012021

The behavior of solutions of an equation with a large spatially distributed control

I Kashchenko
Yaroslavl State University, Yaroslavl, Russia
E-mail: ikashchenko@yandex.ru

Abstract. The paper is devoted to the dynamical properties of the scalar complex equation and system of two equations with spatially distributed parameters. Main assumption is that the coefficient of spatial distribution is sufficiently large. Using asymptotic methods we construct the families of special parabolic equations, which do not contain big and small parameters, which nonlocal dynamics determines the behaviour of solutions of the original equation.

1. Introduction

In this paper we study the behavior of solutions of the two problems with spatial distribution. First one is the complex equation

$$\dot{u} = \Phi(u) + Ke^{i\varphi}\Big(\int\limits_{-\infty}^{\infty} F(s)u(t, x+s)ds - u\Big) \tag{1}$$

with periodic boundary conditions

$$u(t, x + 2\pi) \equiv u(t, x). \tag{2}$$

The second problem is a system of two scalar equations Consider also system of equations

$$\begin{aligned} \dot{u}_1 &= \Phi_1(u_1) + K\left[\int\limits_{-\infty}^{\infty} F(s)u_2(t, x+s)ds - u_1\right], \\ \dot{u}_2 &= \Phi_2(u_2) + K\left[\int\limits_{-\infty}^{\infty} F(s)u_1(t, x+s)ds - u_2\right] \end{aligned} \tag{3}$$

with periodic boundary conditions

$$u_j(t, x + 2\pi) \equiv u_j(t, x) \quad (j = 1, 2). \tag{4}$$

Here $\Phi(u)$, $\Phi_1(u)$ and $\Phi_2(u)$ are sufficiently smooth nonlinear functions; $u = u(t, x)$ is complex value function, $u_1 = u(t, x)$ and $u_2(t, x)$ are real value functions. Parameter K is real and positive, and $\varphi \in (-\frac{\pi}{2}, \frac{\pi}{2})$. Normalization conditions

$$\int\limits_{-\infty}^{\infty} F(s)ds = 1.$$

are assumed to hold. The dynamics and applications of equations of such kind are studied in numerous works [1, 2, 3, 4, 5]. In particular, in the [4] the function $F(x)$ was defined as $F(x) = \exp(k|x|)$. In this paper, we assume that $F(x)$ is smooth and possibly asymmetric. More specifically,

$$F(x) = \frac{1}{\sqrt{\mu\pi}} \exp\left(-\frac{(x+h)^2}{\mu}\right), \qquad (\mu > 0).$$

The research technique is based on the special asymptotic method developed in [6, 7]. In this context, the parameter μ is assumed to be sufficiently small: $0 < \mu \ll 1$. Note that the corresponding parameter in [4] was approximately equal to 0.2. (The main analytical results in [4] were formulated assuming that the control parameter K is small.) In contrast, we consider the situation when this parameter is sufficiently large:

$$K = \varepsilon^{-1}, \quad 0 < \varepsilon \ll 1.$$

From the problem with two small parameters, it is convenient to pass to a single small parameter. In this context, we set

$$\mu = \kappa\varepsilon^{\alpha},$$

where κ and α are fixed positive parameters.

Under these conditions, we examine the behavior of all solutions to boundary value problems (1), (2) and (3), (4) with initial conditions from an arbitrary (bounded as $K \to \infty$, $\mu \to 0$) domain of the phase space $C_{[0,2\pi]}$ (for (1), (2)) and $C_{[0,2\pi]} \times C_{[0,2\pi]}$ (for 3), (4)).

It is convenient to divide the left- and right- hand sides of (1) and (3) by K and make the substitution

$$t = \varepsilon\tau.$$

As a result, we obtain from (1), (2) the problem

$$\frac{\partial u}{\partial \tau} = \varepsilon\Phi(u) + e^{i\varphi}[\int_{-\infty}^{\infty} F(s)u(\tau, x+s)ds - u], \qquad u(\tau, x+2\pi) \equiv u(\tau, x). \tag{5}$$

The characteristic equation for the linear part of problem (5) for $\varepsilon = 0$ has the form

$$\lambda_m = 1 - e^{i\varphi}\exp[-ihm - \kappa\varepsilon m^2] \qquad (m \in Z). \tag{6}$$

Dividing the left- and right- hand sides of (3) by K yields the system

$$\begin{aligned} \frac{\partial u_1}{\partial \tau} &= \varepsilon\Phi_1(u_1) + \int_{-\infty}^{\infty} F(s)u_2(\tau, x+s)ds - u_1, \\ \frac{\partial u_2}{\partial \tau} &= \varepsilon\Phi_2(u_2) + \int_{-\infty}^{\infty} F(s)u_1(\tau, x+s)ds - u_2, \\ & u_j(t, x+2\pi) \equiv u_j(t, x). \end{aligned} \tag{7}$$

The characteristic equation of system (7) with $\varepsilon = 0$ has the roots

$$\lambda_m = -1 + \exp(-\frac{1}{4}d^2k^2), \quad \tilde{\lambda}_m = -1 - \exp(-\frac{1}{4}d^2k^2) \quad (m \in Z).$$

In the next Section we study the case of symmetric (with respect to zero) $F(x)$. In Section 3, the deviation h is assumed to be asymptotically small; i.e., the function $F(x)$ is close to symmetric. In Section 4, the dynamics of problems (5) and (7) is analysed in the case where $F(x)$ is substantially asymmetric; i.e., h is not small. We show that the dynamic properties of these problems are mainly determined by the nonlocal dynamics of a specially constructed families parabolic boundary value problem.

2. Dynamics in the case of a symmetric function $F(s)$

Let $h = 0$ in this section.

1. First, we consider problem (5) in the case $\alpha = 1$. To formulate the main result introduce parabolic equation

$$\frac{\partial \xi}{\partial t} = \kappa e^{i\varphi} \frac{\partial^2 \xi}{\partial y^2} + \Phi(\xi) \tag{8}$$

with the periodic boundary conditions

$$\xi(t, y + 2\pi) \equiv \xi(t, y). \tag{9}$$

The following theorem holds [5, 8].

Theorem 1. *Let (8), (9) has a periodic solution $\xi_0(t, y)$. Then the boundary value problem (5) has an asymptotic (in the residual) solution in the form*

$$u_0(\tau, x) = \xi_0\left(\varepsilon\tau, x\right) + O(\varepsilon).$$

This theorem means that (8), (9) is analogue of normal form for problem (5).

Note that the condition of periodicity ξ_0 can be replaced by the condition that ξ_0 and its derivatives should be bounded for all $t \geq 0$, $y \in [0, 2\pi]$.

2. Consider problem (7) for $\alpha = 1$. The normal form in this situation is very similar [9]

$$\frac{\partial \xi}{\partial t} = \frac{1}{4}\kappa \frac{\partial^2 \xi}{\partial y^2} + \frac{1}{2}(\Phi_1(\xi) + \Phi_2(\xi)) \tag{10}$$

with boundary conditions (9).

The analogue of Theorem 1 takes place and solution of (7), is given by the formula

$$\begin{pmatrix} u_1(\tau, x) \\ u_2(\tau, x) \end{pmatrix} = \begin{pmatrix} \xi(\varepsilon\tau, x) \\ \xi(\varepsilon\tau, x) \end{pmatrix} + O(\varepsilon).$$

3. Consider problems (5) and (7) for $\alpha > 1$. In this situation instead of equations (8) and (10) we have families of equations

$$\frac{\partial \xi}{\partial t} = z^2 \kappa e^{i\varphi} \frac{\partial^2 \xi}{\partial y^2} + \Phi(\xi) \tag{11}$$

and

$$\frac{\partial \xi}{\partial t} = \frac{1}{4}\kappa z^2 \frac{\partial^2 \xi}{\partial y^2} + \frac{1}{2}(\Phi_1(\xi) + \Phi_2(\xi)) \tag{12}$$

with periodic boundary conditions (9). Here z is arbitrary real parameter. Let $\theta_z = \theta_z(\varepsilon)$ be value from $[0, 1)$ such that $z\varepsilon^{(1-\alpha)/2} + \theta_z$ is integer. The following two theorems hold [8, 9].

Theorem 2. *Let for some z (11), (9) has a periodic solution $\xi_0(t, y)$. Then the boundary value problem (5) has an asymptotic (in the residual) solution in the form*

$$u_0(\tau, x, \varepsilon) = \xi_0\left(\varepsilon\tau, (z\varepsilon^{(1-\alpha)/2} + \theta_z)x\right) + O(\varepsilon).$$

Theorem 3. *Let for some z (12), (9) has a periodic solution $\xi_0(t, y)$. Then the boundary value problem (7) has an asymptotic (in the residual) solution in the form*

$$\begin{pmatrix} u_1(\tau, x) \\ u_2(\tau, x) \end{pmatrix} = \begin{pmatrix} \xi(\varepsilon\tau, (z\varepsilon^{(1-\alpha)/2} + \theta_z)x) \\ \xi(\varepsilon\tau, (z\varepsilon^{(1-\alpha)/2} + \theta_z)x) \end{pmatrix} + O(\varepsilon).$$

Since z is arbitrary, problem (5) (and (7) too) has unbounded number of solutions with given asymptotic for sufficiently small ε.

3. Dynamics in the case of a small h

In this section, the parameter h is assumed to be sufficiently small:

$$h = \varepsilon^{\alpha/2} h_1. \tag{13}$$

1. Consider equation (5).

Assuming that $h \neq 0$, $\varphi \neq 0$, and ε is sufficiently small, characteristic equation (6) has an asymptotically large number of roots with a positive real part bounded away from zero as $\varepsilon \to 0$. Thus, problem (5) cannot have stable steady-state regimes in a bounded domain of the phase space $C_{[0,2\pi]}$. In this context, given a fixed φ_1, we assume that

$$\varphi = \varepsilon^{\alpha} \varphi_1.$$

Consider parabolic equation depending on the arbitrary parameter z:

$$\frac{\partial \xi}{\partial t} = (z^2 \kappa + \frac{1}{2} h_1^2) \frac{\partial^2 \xi}{\partial y^2} + \Phi(\xi). \tag{14}$$

with the periodic boundary conditions

$$\xi(t, y + 2\pi) \equiv \xi(t, y). \tag{15}$$

This system plays the role of normal form of (5) in the case (13).

Theorem 4. *Let the boundary value problem (14), (15) has a periodic solution $\xi_0(t, y)$. Then the problem (5) has an asymptotic (in the residual) solution*

$$u_0(\tau, x) = \xi_0 \Big(\varepsilon\tau, (z\varepsilon^{(1-\alpha)/2} + \theta_z)x + (\varepsilon^{\alpha/2} h_1 + \varepsilon^{\alpha} \varphi_1 h_1) t \Big) + O(\varepsilon^{\alpha/2}).$$

2. Now consider problem (7) . Let (13) holds, then the normal form is

$$\frac{\partial \xi}{\partial t} = (\frac{1}{4} \kappa z^2 + \frac{1}{2} h_1^2) \frac{\partial^2 \xi}{\partial y^2} + \frac{1}{2} (\Phi_1(\xi) + \Phi_2(\xi))$$

with periodic boundary conditions (15).

The analogue of Theorem 4 takes place and solution of (7), is given by the formula

$$\begin{pmatrix} u_1(\tau, x) \\ u_2(\tau, x) \end{pmatrix} = \begin{pmatrix} \xi(\varepsilon\tau, (z\varepsilon^{(1-\alpha)/2} + \theta_z)x - \varepsilon^{\alpha} h_1 \tau) \\ \xi(\varepsilon\tau, (z\varepsilon^{(1-\alpha)/2} + \theta_z)x - \varepsilon^{\alpha} h_1 \tau) \end{pmatrix} + o(1).$$

3. Assume now that h in (7) is small but is considerably larger than $\varepsilon^{\alpha/2}$. More precisely, for some fixed $h_1 > 0$ and $\beta \in (0, \alpha/2)$, let

$$h = h_1 \varepsilon^{\beta}.$$

The most interesting situations occurs in the case [9]

$$\frac{1}{2} \leq \beta \leq (\alpha - 1)/2. \tag{16}$$

We introduce some notation. Let z and v be arbitrary and fixed real values. Define δ and γ as $\delta = (\alpha - 1)/2 - \beta$ and $\gamma = \beta - 1/2$. Let $\theta_1 = \theta_1(\varepsilon) \in [0, 2)$ be such that the value of $z\varepsilon^{-\delta} + \theta_1$ is an odd integer, and let $\theta_2 = \theta_2(\varepsilon) \in [0, 1)$ and $\theta_3 = \theta_3(\varepsilon) \in (0, 1]$ be such that the values of $\pi h_1^{-1} \varepsilon^{-\beta} (z\varepsilon^{-\delta} + \theta_1) + \theta_2$ and $v\varepsilon^{-\gamma} + \theta_3$ are integers (for $\gamma = 0$, we set $v = 0$ and $\theta_3 = 1$).

In the considered case the normal form is the system of two parabolic equations in the two-dimensional domain

$$\begin{aligned}\frac{\partial\xi}{\partial t} &= \frac{\kappa\pi^2 z^2}{4h_1^2}\frac{\partial^2\xi}{\partial y^2} + \frac{1}{2}h_1^2 v^2\frac{\partial^2\xi}{\partial w^2} + R_1(\xi,\eta),\\ \frac{\partial\eta}{\partial t} &= \frac{\kappa\pi^2 z^2}{4h_1^2}\frac{\partial^2\eta}{\partial y^2} + \frac{1}{2}h_1^2 v^2\frac{\partial^2\eta}{\partial w^2} + R_2(\xi,\eta)\end{aligned} \tag{17}$$

with boundary conditions

$$\xi(t,y,w+2\pi) \equiv \xi(\tau,y,w) \equiv \xi(\tau,y+2\pi,w), \tag{18}$$

$$\eta(t,y,w+2\pi) \equiv \eta(\tau,y,w) \equiv -\eta(\tau,y+\pi,w). \tag{19}$$

Here

$$\begin{aligned}R_1(\xi,\eta) &= \frac{1}{4}[\Phi_1(\xi+\eta) + \Phi_1(\xi-\eta) + \Phi_2(\xi+\eta) + \Phi_2(\xi-\eta)],\\ R_2(\xi,\eta) &= \frac{1}{4}[\Phi_1(\xi+\eta) - \Phi_1(\xi-\eta) + \Phi_2(\xi+\eta) - \Phi_2(\xi-\eta)].\end{aligned}$$

Theorem 5. Let inequalities (16) hold, and let z and v be arbitrary fixed parameters. Suppose that the boundary value problem (17)–(19) has a periodic (by t) solution $\xi_0(t,y,w)$, $\eta_0(t,y,w)$. Then the problem (7) has an asymptotic (in the residual) solution

$$\begin{aligned}u_{10}(\tau,x) &= \xi_0(\varepsilon\tau,y,w) + \eta_0(\varepsilon\tau,y,w) + o(1),\\ u_{20}(\tau,x) &= \xi_0(\varepsilon\tau,y,w) + \eta_0(\varepsilon\tau,y,w) + o(1),\end{aligned}$$

where $y = x\left(\pi h_1^{-1}\varepsilon^{-\beta}(z\varepsilon^{-\delta}+\theta_1)+\theta_2\right)$, $w = x\,(v\varepsilon^{-\gamma}+\theta_3) - h_1\varepsilon^{\frac{1}{2}}\tau$.

4. The case of substantially asymmetric $F(x)$

Let h is not small but is close to a number that is rationally commensurable with π. This means that

$$h = \frac{\pi m_1}{m_2} - \varepsilon^\alpha h_1.$$

where m_1 and m_2 are relative prime integers.

1. First, consider problem (5).

Denote $N = m_2$ if m_1 is even and $N = 2m_2$ if m_1 is odd. As in Sections 1, 3, we obtain a boundary value problem similar to (14), (15):

$$\frac{\partial\xi}{\partial t} = N^2(z^2\kappa + \frac{1}{2}h_1^2)\frac{\partial^2\xi}{\partial y^2} + \Phi(\xi), \qquad \xi(t,y+2\pi) \equiv \xi(t,y). \tag{20}$$

The solutions of problems (5) and (20) are related by the formula [5, 8]

$$u(\tau,x) = \xi\left(\varepsilon\tau, N((z\varepsilon^{(1-\alpha)/2}+\theta_z)x - (\varepsilon^{\alpha/2}h_1 + \varepsilon^\alpha\varphi_1 h_1)\tau)\right) + O(\varepsilon^{\alpha/2}).$$

2. Results for problem (7) depends on algebraic properties of m_1. First, consider the case of even m_1. Then normal form is [9]

$$\frac{\partial\xi}{\partial t} = (\frac{1}{4}z^2\kappa + \frac{1}{2}h_1^2)m_2^2\frac{\partial^2\xi}{\partial y^2} + \frac{1}{2}(\Phi_1(\xi)+\Phi_2(\xi)), \quad \xi(t,y+2\pi) \equiv \xi(t,y), \tag{21}$$

and the solutions of problem (21) are related to the asymptotic (in the residual) solutions of problem (7) by the formula

$$\begin{pmatrix} u_1(\tau,x)\\ u_2(\tau,x)\end{pmatrix} = \begin{pmatrix}\xi(\varepsilon\tau,(z\varepsilon^{(1-\alpha)/2}+\theta_z)m_2x - m_2\varepsilon^{\alpha/2}h_1\tau)\\ \xi(\varepsilon\tau,(z\varepsilon^{(1-\alpha)/2}+\theta_z)m_2x - m_2\varepsilon^{\alpha/2}h_1\tau)\end{pmatrix} + o(1).$$

In the case of odd m_1, the situation is more complicated. Instead of (21), we have the two-dimensional boundary value problem

$$\frac{\partial \xi}{\partial t} = (\frac{1}{4}z^2\kappa + \frac{1}{2}h_1^2)m_2^2\frac{\partial^2 \xi}{\partial y^2} + R_1(\xi, \eta), \quad \xi(t, y + 2\pi) \equiv \xi(t, y),$$
$$\frac{\partial \eta}{\partial t} = (\frac{1}{4}z^2\kappa + \frac{1}{2}h_1^2)m_2^2\frac{\partial^2 \eta}{\partial y^2} + R_2(\xi, \eta), \quad \eta(t, y + \pi) \equiv -\eta(t, y)$$

and

$$\begin{pmatrix} u_1(\tau, x) \\ u_2(\tau, x) \end{pmatrix} = \begin{pmatrix} \xi(\varepsilon\tau, y) + \eta(\varepsilon\tau, y) \\ \xi(\varepsilon\tau, y) - \eta(\varepsilon\tau, y) \end{pmatrix} + o(1).$$

Here $y = (z\varepsilon^{((1-\alpha)/2} + \theta_z)m_2 x - m_2\varepsilon^{\alpha/2}h_1\tau$.

3. If

$$h = \frac{\pi m_1}{m_2} - \varepsilon^{\beta}h_1$$

and condition (16) holds then the dynamics of system (7) for small ε becomes even more complicated. Following the constructions in [9], we can obtain systems of $2m_2$ parabolic equations, which dynamics determines the behavior of solutions to problem (7). As an example, consider one of the simplest cases when

$$m_1 = 2, m_2 = 3.$$

We introduce some notation. Let z and v be arbitrary and fixed real values. Define δ and γ as $\delta = (\alpha - 1)/2 - \beta$ and $\gamma = \beta - 1/2$. Let $\theta_1 = \theta_1(\varepsilon) \in [0, 2)$ be such that the value of $z\varepsilon^{-\delta} + \theta_1$ is an odd integer, and let $\theta_2 = \theta_2(\varepsilon) \in [0, 1)$ and $\theta_3 = \theta_3(\varepsilon) \in (0, 1]$ be such that the values of $\frac{1}{3}\pi h_1^{-1}\varepsilon^{-\beta}(z\varepsilon^{-\delta} + \theta_1) + \theta_2$ and $v\varepsilon^{-\gamma} + \theta_3$ are integers. For $k = 0, 1$ and $j = 1, 2$ define

$$\Phi_{kj}(x_1, x_2) = \frac{1}{2\pi}\int_0^{2\pi} \Phi_j(x_1 + x_2\exp(iks) + \overline{x}_2\exp(-iks))\, ds$$

and let

$$M_{k1}(\xi_1, \eta_1, \xi_2, \eta_2) = \frac{1}{4}[\Phi_{k1}(\xi_1 + \eta_1, \xi_2 + \eta_2) + \Phi_{k1}(\xi_1 - \eta_1, \xi_2 - \eta_2) + \\ + \Phi_{k2}(\xi_1 + \eta_1, \xi_2 + \eta_2) + \Phi_{k2}(\xi_1 - \eta_1, \xi_2 - \eta_2)]$$

$$M_{k2}(\xi_1, \eta_1, \xi_2, \eta_2) = \frac{1}{4}[\Phi_{k1}(\xi_1 + \eta_1, \xi_2 + \eta_2) - \Phi_{k1}(\xi_1 - \eta_1, \xi_2 - \eta_2) + \\ + \Phi_{k2}(\xi_1 + \eta_1, \xi_2 + \eta_2) - \Phi_{k2}(\xi_1 - \eta_1, \xi_2 - \eta_2)].$$

Consider boundary value problem for two real variables ξ_1 and η_1 and two complex variables ξ_2 and η_2

$$\frac{\partial \xi_1}{\partial t} = \frac{z^2\kappa\pi^2}{36h_1^2}\frac{\partial^2 \xi_1}{\partial y^2} + \frac{1}{2}h_1^2v^2\frac{\partial^2 \xi_1}{\partial w^2} + M_{01}(\xi_1, \eta_1, \xi_2, \eta_2),$$

$$\frac{\partial \eta_1}{\partial t} = \frac{z^2\kappa\pi^2}{36h_1^2}\frac{\partial^2 \eta_1}{\partial y^2} + \frac{1}{2}h_1^2v^2\frac{\partial^2 \eta_1}{\partial w^2} + M_{02}(\xi_1, \eta_1, \xi_2, \eta_2),$$

$$\frac{\partial \xi_2}{\partial t} = \frac{z^2\kappa\pi^2}{36h_1^2}\frac{\partial^2 \xi_2}{\partial y^2} + \frac{1}{2}h_1^2v^2\frac{\partial^2 \xi_2}{\partial w^2} + M_{11}(\xi_1, \eta_1, \xi_2, \eta_2),$$

$$\frac{\partial \eta_2}{\partial t} = \frac{z^2\kappa\pi^2}{36h_1^2}\frac{\partial^2 \eta_2}{\partial y^2} + \frac{1}{2}h_1^2v^2\frac{\partial^2 \eta_2}{\partial w^2} + M_{12}(\xi_1, \eta_1, \xi_2, \eta_2),$$

$$\xi_j(t,y,w) = \xi_j(t,y+2\pi,w) = \xi_j(t,y,w+2\pi), \quad j=1,2,$$
$$\eta_j(t,y,w) = \eta_j(t,y+\pi,w) = \eta_j(t,y,w+2\pi), \quad j=1,2.$$

And solution of (7) can be found by formula

$$u_1(\tau,x) = \xi_1(\varepsilon\tau,y,3w) + \eta_1(\varepsilon\tau,y,3w) + [\xi_2(\varepsilon\tau,y,3w) + \eta_2(\varepsilon\tau,y,3w)]\exp(-i(2\pi(3h_1\varepsilon^{\delta})^{-1}))x+ \\ +[\overline{\xi}_2(\varepsilon\tau,y,3w) + \overline{\eta}_2(\varepsilon\tau,y,3w)]\exp(i(2\pi(3h_1\varepsilon^{\delta})^{-1}))x + o(1),$$

$$u_2(\tau,x) = \xi_1(\varepsilon\tau,y,3w) - \eta_1(\varepsilon\tau,y,3w) + [\xi_2(\varepsilon\tau,y,3w) - \eta_2(\varepsilon\tau,y,3w)]\exp(-i(2\pi(3h_1\varepsilon^{\delta})^{-1}))x+ \\ +[\overline{\xi}_2(\varepsilon\tau,y,3w) - \overline{\eta}_2(\varepsilon\tau,y,3w)]\exp(i(2\pi(3h_1\varepsilon^{\delta})^{-1}))x + o(1),$$

where $y = x(\pi(3h_1\varepsilon^{\beta})^{-1}(z\varepsilon^{-\delta}+\theta_1)+\theta_2)$, $w = x(v\varepsilon^{-\gamma}+\theta_3) - h_1\varepsilon^{\frac{1}{2}}\tau$.

5. Conclusions

Special nonlinear families of generally parabolic evolution equations (without small or large parameters) have been constructed to determine the leading terms of the asymptotic representations of solutions to the original boundary value problems. It is well known that equations of this type have rich and complex dynamics. The presence of the continual parameters (z and sometimes v) in these families suggests that systems (1) and (3) with a large parameter have multistability.

The construction of the leading terms in the asymptotic expansions of solutions to systems (1) and (3) gives the possibility of finding solutions with any prescribed degree of accuracy in ε. In a number of cases, based on the norm estimates derived in [10] for linear operators, the existence of exact solutions (with the constructed asymptotics) can be proved and their stability can be analyzed. The above results can be extended to more general systems of equations. For example, it is of interest to study new phenomena arising in the case of a two-dimensional domain (see [11]). Note that, for delay equations, we obtain evolution equations with a deviating spatial variable; i.e., the time delay passes into a spatial deviation.

Acknowledgments

This study is supported by Base part of State assignment, project No 1.5722.2017/8.9.

References

[1] Haken H 1983 *Synergetics; An Introducion, 3rd Revised and Enlarged Edition* (Berlin: Springer)
[2] Haken H 2002 *Brain dynamics; synchronization and activity patterns in pulse-coupled neural nets with delays and noise* (Berlin: Springer)
[3] Kuramoto Y 1984 *Chemical oscillations, waves and turbulence* (New-York: Springer)
[4] Kuramoto Y and Battogtokh D 2002 *Nonlinear phenomena in complex systems* **5** 380–385
[5] Kashchenko I and Kaschenko S 2016 *Communications in Nonlinear Science and Numerical Simulation* **34** 123–129
[6] Kashchenko I S and Kashchenko S A 2010 *Doklady Mathematics* **82** 850–853
[7] Kashchenko S A 1990 *U.S.S.R. Computational Mathematics and Mathematical Physics* **30** 186–197
[8] Kashchenko I S and Kashchenko S A 2011 *Doklady Mathematics* **83** 408–412 ISSN 1064-5624
[9] Kashchenko I S and Kashchenko S A 2012 *Doklady Mathematics* **85** 129–133 ISSN 1064-5624
[10] Kashchenko S A 1992 *Journal of Soviet Mathematics* **60** 1742–1764
[11] Kashchenko S A 1989 *Differential Equations* **25** 193–199

IOP Conf. Series: Journal of Physics: Conf. Series **937** (2017) 012022 doi:10.1088/1742-6596/937/1/012022

Calculation of Rayleigh type sums for zeros of the equation arising in spectral problem

A B Kostin and V B Sherstyukov[1]

National Research Nuclear University MEPhI (Moscow Engineering Physics Institute), 31 Kashirskoe shosse, 115409 Moscow, Russia

E-mail: abkostin@yandex.ru, shervb73@gmail.com

Abstract. For zeros of the equation (arising in the oblique derivative problem)

$$\mu\, J_n'(\mu)\cos\alpha + i\, n\, J_n(\mu)\sin\alpha = 0, \qquad \mu\in\mathbb{C},$$

with parameters $n\in\mathbb{Z}$, $\alpha\in[-\pi/2,\pi/2]$ and the Bessel function $J_n(\mu)$ special summation relationships are proved. The obtained results are consistent with the theory of well-known Rayleigh sums calculating by zeros of the Bessel function.

1. Introduction

In [1] the authors was studied the question on the complex location of eigenvalues $\lambda=\mu^2$ of following spectral problem for Laplace operator in disk $D=\{(x,\,y)\mid x^2+y^2<1\}$:

$$\Delta w + \mu^2 w = 0 \quad \text{in} \quad D, \qquad \frac{\partial w}{\partial l} = 0 \quad \text{on} \quad \partial D. \tag{1}$$

Here, Δ is the Laplace operator, $\mu^2\in\mathbb{C}$ is a spectral parameter, $\boldsymbol{l}$ is a direction forming a fixed angle $\alpha\in(-\pi/2,\pi/2)$ with outer normal $\boldsymbol{n}$ to ∂D. Our interest in this subjects have been stimulated by a well-known work [2] in which the absence of the basis property in $L_2(D)$ system of root functions of the problem (1) was proved. In [2] (see, e.g. [1]) have been shown that all eigenvalues of the spectral problem (1) are described by zeros of the equation

$$\mu\, J_n'(\mu)\cos\alpha + i\, n\, J_n(\mu)\sin\alpha = 0, \qquad n\in\mathbb{Z}, \tag{2}$$

with given $\alpha\in(-\pi/2,\pi/2)$ and the Bessel function $J_n(\mu)$ of variable $\mu\in\mathbb{C}$. For the aims of the work we expand change area of parameters n, α. We assume that pairs (n,α) belong to the set of permissible values

$$\mathcal{P}\equiv(\mathbb{Z}\times[-\pi/2,\pi/2])\setminus(\{0\}\times\{-\pi/2,\pi/2\}). \tag{3}$$

For $(n,\alpha)\in\{0\}\times\{-\pi/2,\pi/2\}$ the equation (2) is degenerate and its zeros fill the whole complex plane. Introducing the set (3) we except such combination of parameters. It is also obviously

[1] The work of the authors is partially supported by the program of increasing of competitiveness of the National Nuclear Research University MEPhI, project No. 02.a03.21.0005 from 27.08.2013

that $\mu = 0$ is a zero (2) for any values n and α in (3). The others zeros of equation (2) which we call non-trivial are of interest. For the permissible values of parameters when $\alpha = 0$ (or $n = 0$) and $\alpha = \pm\pi/2$ zeros of corresponding equations $\mu J_n'(\mu) = 0$ and $J_n(\mu) = 0$ are real and well studied (see [3, Ch. 15]). For the main case $0 < |\alpha| < \pi/2$ from the general theory of entire functions and from the results of [1] follows that for a fixed $n \neq 0$ the equation (2) has an infinite countable set of non-trivial zeros at that all of them are simple, don't lie on real and imaginary axes and are located on the complex plane symmetrically with respect to the point $\mu = 0$.

The report will present further results on the properties of zeros of the equation of form (2). To describe the the results we introduce the even entire function of exponential type

$$L(\mu; n, \alpha) \equiv \frac{\mu\, J_n'(\mu) \cos\alpha + i\, n\, J_n(\mu) \sin\alpha}{(\mu/2)^n}, \quad \mu \in \mathbb{C}. \tag{4}$$

For $\mu \neq 0$ the equation (2) is equivalent to the equation $L(\mu; n, \alpha) = 0$ with coincidence of the multiplicities of zeros. It was noted (see also section 2) that the function (4) has an infinite countable set of zeros. We denote the set by $\{\pm\, \mu_{n,k}(\alpha)\}_{k\in\mathbb{N}}$, where $\mu_{n,k}(\alpha)$ are located in half-plane $\operatorname{Re}\mu > 0$, rearranged on increasing of modules and $\mu_{n,k}(\alpha) \to \infty$ at $k \to \infty$. In view of the obvious chain of relationships

$$L(\mu; n, \alpha) = 0 \Leftrightarrow L(\overline{\mu}; n, -\alpha) = 0 \Leftrightarrow L(\overline{\mu}; -n, \alpha) = 0 \Leftrightarrow L(\mu; -n, -\alpha) = 0 \tag{5}$$

it is enough to study zeros $\mu \in \mathbb{C}$ of the equation

$$L(\mu; n, \alpha) = 0, \qquad n \in \mathbb{N},\ \alpha \in [0, \pi/2]. \tag{6}$$

Systematically investigating the problem on the location of zeros of the equation (6) on half-plane $\mathbb{C}$ of variable μ, we have got explicit formulas for sums of special structure containing zeros $\mu_{n,k}(\alpha)$. For example we established the following relationship

$$\sum_{k=1}^{\infty} \frac{1}{\mu_{n,k}^2(\alpha)} = \frac{n + 2\cos^2\alpha}{4n(n+1)} - i\,\frac{\sin 2\alpha}{4n(n+1)} \equiv \frac{n + 1 + \exp(-2\alpha\, i)}{4n(n+1)}, \qquad n \in \mathbb{N},\ \alpha \in [0, \pi/2]. \tag{7}$$

If $\alpha = \pi/2$ then the formula (7) coincides with the particular expression

$$\sigma^{(2)}(n) \equiv \sum_{k=1}^{\infty} \frac{1}{j_{n,k}^2} = \frac{1}{4(n+1)} \tag{8}$$

for the known Rayleigh sums (see [4], [3], [5], [6])

$$\sigma^{(2p)}(n) \equiv \sum_{k=1}^{\infty} \frac{1}{j_{n,k}^{2p}}, \qquad p \in \mathbb{N},$$

where $j_{n,k}$ are positive zeros of the Bessel function $J_n(\mu)$. If $\alpha = 0$ then the formula (7) again gives well-known expression (see [7], [8], [9])

$$\sum_{k=1}^{\infty} \frac{1}{j'^2_{n,k}} = \frac{n+2}{4n(n+1)},$$

where $j'_{n,k}$ are positive zeros of the derivative $J_n'(\mu)$ of the Bessel function.

The authors know a number of others studies (see, e.g. the bibliography in reviews [6], [10]), dedicated to close questions for the equation of form

$$A\,\mu\, J_\nu'(\mu) + B\, J_\nu(\mu) = 0,$$

but as a rule either with real parameters A, B, ν or without explicit indication of the range of parameters. In this connection, even the formula (7) appears to be a new.

In the main part of the work we will found the basic properties of zeros of the entire function (4) and then we will obtain some summation relationships like the formula (7).

IOP Conf. Series: Journal of Physics: Conf. Series **937** (2017) 012022 doi:10.1088/1742-6596/937/1/012022

2. The properties of zeros of the function $L(\mu; n, \alpha)$

Let us study in more detail the properties of zeros of the entire function $L(\mu; n, \alpha)$ denoting it for simplifying through $L(\mu)$ for fixed values $n \in \mathbb{N}$, $\alpha \in [0, \pi/2]$. We briefly explain why the function $L(\mu)$ has an infinite set of zeros. Using the expansion in the power series of the Bessel function $J_n(\mu)$ we write

$$L(\mu) = \sum_{k=0}^{\infty} \frac{(-1)^k}{k!} \, \frac{(2k+n)\cos\alpha + i\, n\, \sin\alpha}{(k+n)!} \left(\frac{\mu}{2}\right)^{2k}, \qquad \mu \in \mathbb{C}. \tag{9}$$

According to (9) and standard formulas for calculation of the order and the type of an entire function by the coefficients of the Taylor series (see [11, p. 6]) we obtain that $L(\mu)$ is a function of the order $\rho_L = 1$ and of the type $\sigma_L = 1$. We suppose that the entire function $L(\mu)$ has the finite number of zeros or doesn't have them at all. It is well known that then $L(\mu)$ can be represented in the form

$$L(\mu) = e^{a\mu} P(\mu), \qquad \mu \in \mathbb{C},$$

where $a \in \mathbb{C}$, $P(\mu)$ is a polynomial. As $L(-\mu) = L(\mu)$ then $a = 0$ that is $L(\mu) = P(\mu)$. But it contradicts to the expression (9). Since the zeros of an entire function are isolated, it follows that the set of zeros $L(\mu)$ is countable and doesn't have finite limit points (the only limit point is $\mu = \infty$). The parity $L(\mu)$ leads to the symmetry of the zeros with respect to the point $\mu = 0$.

The simplicity of zeros of the function $L(\mu)$ is an obvious consequence of the general fact about the simplicity of non-trivial zeros of the equation (2) that was proved in [16].

Now we show that all zeros of the equation (6) lie in some specific domain (depending on n, α), contained in the horizontal strip. We use the expansion of the Bessel function in the infinite product

$$J_n(\mu) = \frac{1}{n!} \left(\frac{\mu}{2}\right)^n \prod_{k=1}^{\infty} \left(1 - \frac{\mu^2}{j_{n,k}^2}\right), \qquad \mu \in \mathbb{C},$$

and we calculate its logarithmic derivative. We have

$$\frac{J_n'(\mu)}{J_n(\mu)} = \sum_{k=1}^{\infty} \frac{2\mu}{\mu^2 - \nu_{n,k}} + \frac{n}{\mu}, \tag{10}$$

where $\nu_{n,k} \equiv j_{n,k}^2$. Differentiating (10) we obtain the equality

$$\left(\frac{J_n'(\mu)}{J_n(\mu)}\right)' = 2\sum_{k=1}^{\infty} \frac{1}{\mu^2 - \nu_{n,k}} - 4\mu^2 \sum_{k=1}^{\infty} \frac{1}{(\mu^2 - \nu_{n,k})^2} - \frac{n}{\mu^2}. \tag{11}$$

On the other hand the Bessel equation

$$\mu^2 J_n''(\mu) + \mu\, J_n'(\mu) + (\mu^2 - n^2) J_n(\mu) = 0$$

gives the relationship

$$\left(\frac{J_n'(\mu)}{J_n(\mu)}\right)' = \frac{n^2 - \mu^2}{\mu^2} - \frac{J_n'(\mu)}{\mu J_n(\mu)} - \left(\frac{J_n'(\mu)}{J_n(\mu)}\right)^2. \tag{12}$$

From (10)–(12) we conclude

$$4\mu^2 \sum_{k=1}^{\infty} \frac{1}{(\mu^2 - \nu_{n,k})^2} = 4\mu^2 \left(\sum_{k=1}^{\infty} \frac{1}{\mu^2 - \nu_{n,k}}\right)^2 + 4(n+1) \sum_{k=1}^{\infty} \frac{1}{\mu^2 - \nu_{n,k}} + 1.$$

Replacing $\lambda = \mu^2$ and denoting

$$S^{(1)}(\lambda;\, n) \equiv \sum_{k=1}^{\infty} \frac{1}{\lambda - \nu_{n,\,k}}, \quad S^{(2)}(\lambda;\, n) \equiv \sum_{k=1}^{\infty} \frac{1}{(\lambda - \nu_{n,\,k})^2},$$

we obtain the identity

$$4\lambda\, S^{(2)}(\lambda;\, n) = 4\lambda \left(S^{(1)}(\lambda;\, n)\right)^2 + 4(n+1)\, S^{(1)}(\lambda;\, n) + 1. \tag{13}$$

If the parameters satisfy the conditions $n \in \mathbb{N}$, $\alpha \in [0, \pi/2)$, we rewrite the equation (2) in the equivalent form at $\mu \neq 0$:

$$\frac{J_n'(\mu)}{2\mu\, J_n(\mu)} + \frac{i\, n\, \tan\alpha}{2\mu^2} = 0. \tag{14}$$

Let $\mu = \mu_0$ be an any fixed non-trivial zero of the equation (2). Comparing (10) and (14) we see that

$$S^{(1)}(\lambda_0;\, n) = -\frac{n(1 + i\tan\alpha)}{2\lambda_0}, \tag{15}$$

where $\lambda_0 \equiv \mu_0^2 = \xi_0 + i\eta_0$. Substituting λ_0 in (13) and taking into account (15) after groups of transformations we obtain

$$4|\lambda_0|^4 \sum_{k=1}^{\infty} \frac{(\xi_0 - \nu_{n,\,k})^2 - \eta_0^2}{|\lambda_0 - \nu_{n,\,k}|^4} = \left(\xi_0 - \frac{n^2}{\cos^2\alpha} - 2n\right)(\xi_0^2 - \eta_0^2) + 2\xi_0\eta_0\,(\eta_0 - 2n\tan\alpha),$$

$$8\eta_0|\lambda_0|^4 \sum_{k=1}^{\infty} \frac{\xi_0 - \nu_{n,\,k}}{|\lambda_0 - \nu_{n,\,k}|^4} = 2\xi_0\eta_0 \left(\xi_0 - \frac{n^2}{\cos^2\alpha} - 2n\right) - (\eta_0 - 2n\tan\alpha)(\xi_0^2 - \eta_0^2).$$

Combining the two last formulas with the formula (15) we have the equality

$$8\eta_0^3|\lambda_0|^2 \sum_{k=1}^{\infty} \frac{\nu_{n,\,k}}{|\lambda_0 - \nu_{n,\,k}|^4} = (2n\tan\alpha - \eta_0)|\lambda_0|^2 + \xi_0\eta_0\, \frac{n^2}{\cos^2\alpha}.$$

By the results [1] the condition $\eta_0 \geqslant 0$ is satisfied. From here we have

$$(2n\tan\alpha - \eta_0)|\lambda_0|^2 + \xi_0\eta_0\, \frac{n^2}{\cos^2\alpha} \geqslant 0. \tag{16}$$

The inequality (16) means that for fixed $n \in \mathbb{N}$, $\alpha \in [0, \pi/2)$ the squares of zeros of the equation (6) on the plane $\mathbb{C}_\lambda$, where $\lambda = \mu^2 = \xi + i\eta = r(\cos\varphi + i\sin\varphi)$, are located in the domain Ω specified by the conditions

$$r\sin\varphi \leqslant 2n\tan\alpha + \frac{n^2}{2\cos^2\alpha}\sin 2\varphi, \qquad 0 \leqslant \varphi \leqslant \pi/2. \tag{17}$$

From [1] it follows that in the condition (17) one can take $0 \leqslant \varphi \leqslant \alpha$. For $\lambda = \xi + i\eta \in \Omega$ we obviously have

$$0 \leqslant \eta \leqslant 2n\tan\alpha + \frac{n^2}{2\cos^2\alpha}.$$

In other words all points are located in the half-strip

$$0 \leqslant \operatorname{Im}\lambda \leqslant 2n\tan\alpha + \frac{n^2}{2\cos^2\alpha}, \qquad \operatorname{Re}\lambda \geqslant 0$$

of the plane $\mathbb{C}_\lambda$. Further, from (17) at the same n, α we extract that all zeros $\mu_{n,k}(\alpha)$ on the plane $\mathbb{C}_\mu$ where $\mu = \rho(\cos\psi + i\sin\psi)$ lie in the domain

$$0 \leqslant \operatorname{Im}\mu \leqslant \sqrt{\left(n\tan\alpha + \frac{n^2}{4\cos^2\alpha}\sin 4\psi\right)\tan\psi}, \quad 0 \leqslant \psi \leqslant \alpha/2, \qquad \operatorname{Re}\mu \geqslant 0. \tag{18}$$

Taking into account the relationship (5) from (18) now we easily obtain the following properties of zeros of the equations (2) and (6).

(i) For given $n \in \mathbb{Z}$, $\alpha \in (-\pi/2, \pi/2)$ all zeros of the equation (2) lie in the strip

$$|\operatorname{Im}\mu| \leqslant \frac{\sqrt{|n|(|n| + 2\sin 2|\alpha|)}}{2\cos\alpha},$$

and for $n \in \mathbb{Z}\setminus\{0\}$, $\alpha = \pm\pi/2$ the equation (2) has only real zeros.

(ii) For fixed $n \in \mathbb{N}$, $\alpha \in (0, \pi/2)$ and $k \to \infty$ it's satisfied $\operatorname{Im}\mu_{n,k}(\alpha) \to 0$.

(iii) For fixed $(n, \alpha) \in \mathcal{P}$ and for any $h > 0$ outside the strip $|\operatorname{Im}\mu| \leqslant h$ it lies not more than the finite number of zeros of the equation (2).

3. The Rayleigh type sums

As in the previous section for convenience of exposition we choose and fix parameters in the equation (2) so that $n \in \mathbb{N}$, $\alpha \in [0, \pi/2]$. In fact, this allow us to work with the equation (6). At the end of the section we give the general result.

By analogy with the Rayleigh sums [4] we define more general sums of the Rayleigh type

$$\eta^{(r)}(n,\alpha) \equiv \frac{1}{2}\sum \frac{1}{(\pm\mu_{n,k}(\alpha))^r}, \qquad r \in \mathbb{N}, \tag{19}$$

where the summation is performed for all non-trivial zeros $\pm\mu_{n,k}(\alpha)$ of the equation (2). From the theory of entire functions it follows that the convergence exponent of the sequence of zeros of the function $L(\mu) = L(\mu; n, \alpha)$ doesn't exceed a value of the order $\rho_L = 1$ (in our case these characteristics coincide). Thus, the series in (19) converges absolutely for $r > 1$. For all odd exponents $r = 2m - 1 \in \mathbb{N}$ the sum $\eta^{(2m-1)}(n,\alpha)$ is equal to zero. For even exponents $r = 2m \in \mathbb{N}$ we have

$$\eta^{(2m)}(n,\alpha) = \sum_{k=1}^{\infty} \frac{1}{\mu_{n,k}^{2m}(\alpha)}, \qquad m \in \mathbb{N}. \tag{20}$$

We remind that through $\mu_{n,k}(\alpha)$ are denoted zeros of the equation (6) (non-trivial zeros of the equation (2)) located in the half-plane $\operatorname{Re}\mu > 0$.

We set a problem of calculating of the sums (20). Above it was noted that $L(\mu)$ is the even entire function of exponential type with simple zeros $\{\pm\mu_{n,k}(\alpha)\}_{k\in\mathbb{N}}$. From the expansion $L(\mu)$ in the power series (9) we have

$$L(0) = \frac{\cos\alpha + i\sin\alpha}{(n-1)!} = \frac{\exp(\alpha i)}{(n-1)!}.$$

Representing $L(\mu)$ as the canonical product

$$L(\mu) = \frac{\exp(\alpha i)}{(n-1)!}\prod_{k=1}^{\infty}\left(1 - \frac{\mu^2}{\mu_{n,k}^2(\alpha)}\right) \tag{21}$$

and selecting its logarithmic derivative we write the identity

$$\frac{L'(\mu)}{2\mu\, L(\mu)} = \sum_{k=1}^{\infty} \frac{1}{\mu^2 - \mu_{n,\,k}^2(\alpha)}. \tag{22}$$

We calculate of the value of the left hand side (22) at $\mu = 0$. Using the expansion (9) after standard transformations we find

$$\left.\frac{L'(\mu)}{2\mu\, L(\mu)}\right|_{\mu=0} = -\frac{(n+2)\cos\alpha + i\, n\,\sin\alpha}{4n(n+1)\,(\cos\alpha + i\,\sin\alpha)}.$$

We substitute $\mu = 0$ in (22) and one obtain the relationship

$$\eta^{(2)}(n,\alpha) \equiv \sum_{k=1}^{\infty} \frac{1}{\mu_{n,\,k}^2(\alpha)} = \frac{(n+2)\cos\alpha + i\, n\,\sin\alpha}{4n(n+1)\,(\cos\alpha + i\,\sin\alpha)} \equiv \frac{n+1+\exp(-2\alpha\, i)}{4n(n+1)},$$

coinciding with (7).

To calculate the sums of the Rayleigh type of higher orders it is convenient to make a substitute $t = \mu^2$ in (22) and to briefly write

$$\sum_{k=1}^{\infty} \frac{1}{t - \mu_{n,k}^2(\alpha)} = -\frac{1}{4}\,\frac{u(t)}{v(t)}, \tag{23}$$

where

$$u(t) \equiv \sum_{k=0}^{\infty} \frac{(-1)^{k+1}}{k!}\,\frac{(n+2k+2)\cos\alpha + i\, n\,\sin\alpha}{(n+k+1)!}\left(\frac{t}{4}\right)^k,$$

$$v(t) \equiv \sum_{k=0}^{\infty} \frac{(-1)^{k}}{k!}\,\frac{(n+2k)\cos\alpha + i\, n\,\sin\alpha}{(n+k)!}\left(\frac{t}{4}\right)^k.$$

Differentiating the necessary number of times for the equality (23) we obtain the following general formula for sums of the Rayleigh type

$$\eta^{(2m)}(n,\alpha) \equiv \sum_{k=1}^{\infty} \frac{1}{\mu_{n,\,k}^{2m}(\alpha)} = \frac{1}{4(m-1)!}\left.\left(\frac{u}{v}\right)^{(m-1)}\right|_{t=0}, \qquad m \in \mathbb{N}. \tag{24}$$

We exemplify to obtain the sums $\eta^{(2m)}(n,\alpha)$ at $m = 2, 3$. We give only the results of calculations.

For $m = 2$ we have

$$\eta^{(4)}(n,\alpha) = \frac{1}{4}\left\{\frac{u'(0)}{v(0)} - \frac{v'(0)u(0)}{v^2(0)}\right\},$$

whence

$$\eta^{(4)}(n,\alpha) \equiv \sum_{k=1}^{\infty} \frac{1}{\mu_{n,\,k}^4(\alpha)} = \frac{(n+1)(n+2) + 4(n+1)\exp(-2\alpha\, i) + (n+2)\exp(-4\alpha\, i)}{2^4\, n^2\,(n+1)^2\,(n+2)}.$$

In particular, for $\alpha = 0$ we have the formula

$$\sum_{k=1}^{\infty} \frac{1}{j'^{\,4}_{n,\,k}} = \frac{n^2+8n+8}{2^4\, n^2\,(n+1)^2\,(n+2)},$$

coinciding with the formula P. L. Kapitsa (see [8]). In the other extreme case $\alpha = \pi/2$ the obtained relationship is transformed in the well-known Rayleigh sum

$$\sum_{k=1}^{\infty} \frac{1}{j_{n,k}^4} = \frac{1}{2^4 (n+1)^2 (n+2)} \equiv \sigma^{(4)}(n).$$

For $m = 3$ we have

$$\eta^{(6)}(n,\alpha) = \frac{1}{8} \left\{ \frac{u''(0)}{v(0)} - 2\,\frac{v'(0)u'(0)}{v^2(0)} - \frac{v''(0)u(0)}{v^2(0)} + 2\,\frac{v'^2(0)u(0)}{v^3(0)} \right\},$$

whence

$$\eta^{(6)}(n,\alpha) \equiv \sum_{k=1}^{\infty} \frac{1}{\mu_{n,k}^6(\alpha)} =$$

$$= \frac{2(n+1)^2(n+3) + 3(n+1)(5n+6)\,e^{-2\alpha i} + 6(n+1)(n+3)\,e^{-4\alpha i} + (n+2)(n+3)\,e^{-6\alpha i}}{2^6\, n^3\,(n+1)^3\,(n+2)(n+3)}.$$

For $\alpha = 0$ it leads to the sum

$$\sum_{k=1}^{\infty} \frac{1}{j'^6_{n,k}} = \frac{n^3 + 16n^2 + 38n + 24}{2^5\, n^3\,(n+1)^3\,(n+2)(n+3)},$$

and for $\alpha = \pi/2$ it leads to the corresponding Rayleigh sum

$$\sum_{k=1}^{\infty} \frac{1}{j_{n,k}^6} = \frac{1}{2^5\,(n+1)^3\,(n+2)(n+3)} \equiv \sigma^{(6)}(n).$$

Let us give some expressions for special double sums. With fixed $n \in \mathbb{N}$, $\alpha \in [0, \pi/2)$ for any zero $\mu_{n,k}(\alpha)$ from (10), (14) we have

$$\sum_{m=1}^{\infty} \frac{1}{\mu_{n,k}^2(\alpha) - j_{n,m}^2} = -\,\frac{n(1 + i\,\tan\alpha)}{2\,\mu_{n,k}^2(\alpha)}.$$

Now summing over all k, we obtain the relationship

$$\sum_{k=1}^{\infty}\sum_{m=1}^{\infty} \frac{1}{j_{n,m}^2 - \mu_{n,k}^2(\alpha)} = \frac{n + 2 + i\,n\,\tan\alpha}{8(n+1)}, \qquad (25)$$

containing both zeros of the equation (2) and zeros of the Bessel function $J_n(\mu)$. In particular, for $\alpha = 0$ the equality (25) takes the form

$$\sum_{k=1}^{\infty}\sum_{m=1}^{\infty} \frac{1}{j_{n,m}^2 - j'^2_{n,k}} = \frac{n+2}{8(n+1)}.$$

The authors didn't find the two last formulas in the literature.

In conclusion we give a generalization of the formula (24) for the sums $\eta^{(2m)}(n,\alpha)$ on the permissible values of parameters $(n,\alpha) \in \mathcal{P}$. For $n \in \mathbb{N}$, $\alpha \in [0, \pi/2]$ and $m = 1, 2, 3$ the necessary formulas are given above and the method of calculating such sums for subsequent $m \in \mathbb{N}$ is also indicated. Let us analyze other cases.

(i) Let $n = 0$, $\alpha \in (-\pi/2, \pi/2)$. The equation (2) at $\mu \neq 0$ is equivalent to each from the equations

$$J_0'(\mu) = 0 \Leftrightarrow J_1(\mu) = 0,$$

and the required sums are $\eta^{(2m)}(0, \alpha) \equiv \sigma^{(2m)}(1)$.

(ii) Let $n \in \mathbb{N}$, $\alpha = -\pi/2$. The equation (2) is equivalent to the equation

$$J_n(\mu) = 0,$$

and the required sums are $\eta^{(2m)}(n, -\pi/2) \equiv \sigma^{(2m)}(n)$.

(iii) Let $n \in \{-1, -2, \dots\}$, $\alpha = \pm\pi/2$. The equation (2) is equivalent to the equation

$$J_p(\mu) = 0, \qquad p = -n \in \mathbb{N},$$

and the required sums are $\eta^{(2m)}(n, \pm\pi/2) \equiv \sigma^{(2m)}(|n|)$.

(iv) Let $n \in \{-1, -2, \dots\}$, $\alpha \in [0, \pi/2]$. The equation (2) is equivalent to the equation

$$\mu\, J_p'(\mu) \cos\alpha - i\, p\, J_p(\mu) \sin\alpha \;=\; 0, \qquad p = -n \in \mathbb{N}.$$

Zeros of the equation are the numbers $\overline{\mu}_{p,\,k}(\alpha)$ thus $\eta^{(2m)}(n, \alpha) = \overline{\eta}^{(2m)}(|n|, \alpha)$.

(v) Let $n \in \mathbb{N}$, $\alpha \in [-\pi/2, 0]$. After denoting $\beta = -\alpha \in [0, \pi/2]$ the equation (2) takes the form

$$\mu\, J_n'(\mu) \cos\beta - i\, n\, J_n(\mu) \sin\beta \;=\; 0.$$

Zeros of the equation are the numbers $\overline{\mu}_{n,\,k}(\beta)$ and therefore $\eta^{(2m)}(n, \alpha) = \overline{\eta}^{(2m)}(n, |\alpha|)$.

(vi) Let $n \in \{-1, -2, \dots\}$, $\alpha \in [-\pi/2, 0]$. We denote $p = -n \in \mathbb{N}$, $\beta = -\alpha \in [0, \pi/2]$. Then the equation (2) takes the form

$$\mu\, J_p'(\mu) \cos\beta + i\, p\, J_p(\mu) \sin\beta \;=\; 0,$$

zeros of the equation are the numbers $\mu_{p,\,k}(\beta)$ and therefore $\eta^{(2m)}(n, \alpha) = \eta^{(2m)}(|n|, |\alpha|)$.

Thereby, all possible cases are sorted and the method of calculating of the Rayleigh type sums for all values $(n, \alpha) \in \mathcal{P}$ is specified.

References

[1] Kostin A B and Sherstyukov V B On complex roots of an equation arising in the oblique derivative problem // IOP Conf. Series: Journal of Physics: Conf. Series 788 (2017) 012052 doi:10.1088/1742-6596/788/1/012052. P. 1–7

[2] Il'in V A and Moiseev E I 1994 On the Absence of the Basis Property of a System of Root Functions of a Problem with an Oblique Derivative *Differ. Uravn.* **30** 128–143

[3] Watson G N 1944 *A Treatise on the Theory of Bessel Functions* (Cambridge: Cambridge University Press)

[4] Rayleigh 1874 Note on the Numerical Calculation of the Roots of Fluctuating Functions *Proc. London Math. Soc.* **5** 119–124

[5] Kishore N 1963 The Rayleigh Function *Proc. Amer. Math. Soc.* **14** 527–533

[6] Kerimov M K 1999 The Rayleigh Function: Theory and Methods of Its Calculation *Comput. Math. Math. Phys.* **39** 1883–1925

[7] Lamb H 1884 Note on the Induction of Electric Currents in a Cylinder placed across the Lines of Magnetic Force *Proc. London Math. Soc.* **15** 270–274

[8] Kapitsa P L 1951 Computation of Negativ Even Power Sums of the Zeros of Bessel Functions *Dokl. Akad. Nauk SSSR* **77** 561–564

[9] Meiman N N 1956 On Recurrence Formulas for Power Sums of the Zeros of Bessel Functions *Dokl. Akad. Nauk SSSR* **108** 190–193

[10] Kerimov M K 2008 Overview Some New Results Concerning the Theory and Applications of the Rayleigh Special Function *Comput. Math. Math. Phys.* **48** 1454–1507

[11] Levin B Ya 1996 Lectures on Entire Functions. — Rhode Island: Amer. Math. Soc., Providens

IOP Conf. Series: Journal of Physics: Conf. Series **937** (2017) 012023 doi:10.1088/1742-6596/937/1/012023

Coordinated interaction of two hydraulic cylinders when moving large-sized objects

G V Kreinin[1], S Yu Misyurin[1,2], A V Lunev[1]

[1] Blagonravov Institute of Machine Science, RAS, Moscow, Russia
[2] National Nuclear Research University MEPhI, Moscow, Russia

ssmmrr@mail.ru

Abstract. The problem of the choice of parameters and the control scheme of the dynamics system for the coordinated displacement of a large mass object by two hydraulic piston type engines is considered. As a first stage, the problem is solved with respect to a system in which a heavy load of relatively large geometric dimensions is lifted or lowered in the progressive motion by two unidirectional hydraulic cylinders while maintaining the plane of the lifted object in a strictly horizontal position.

1. Introduction

At present, the problem associated with the uniform elevation of a heavy object by two or more drives is quite relevant. This problem occurs in various situations, where it is necessary to move the load evenly up or down with a minimum skew (see Figure 1: 3 - table, 1 - drives, 2 - base) [1, 2]. For example, in 3D Printing Tools the object (product), which is produced using "printing", is installed on the position table and moved in the vertical direction. Today quite large objects are "printed", for example, the objects with the size of 1500mm x 1500mm x 1500mm and with maximum weight up to 200 kg. There are several implementations where the position table is moved by means of two stepping engines with the transfer of the screw-nut (ball couple). Synchronization is carried out with the help of a control system, where a displacement sensor is an optical rotating encoder on each engine. The synchronization accuracy in the displacement of the engines is approximately 0.1 mm. But in many problems this accuracy is not enough and the power of the stepper engine may not be enough.

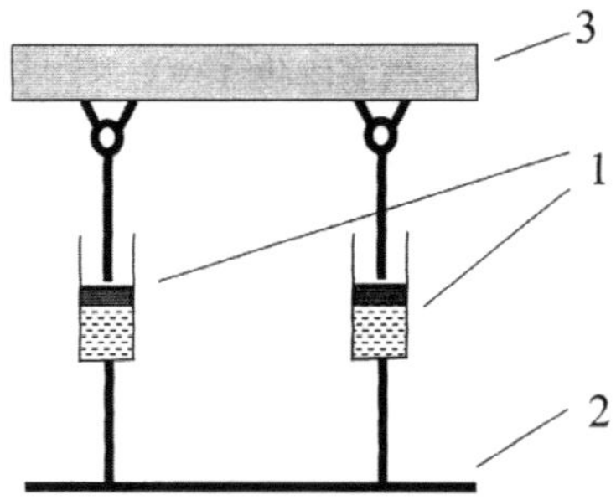

Figure 1.

IOP Conf. Series: Journal of Physics: Conf. Series **937** (2017) 012023 doi:10.1088/1742-6596/937/1/012023

2. Movement scheme

It is proposed to use hydraulic drives that are controlled by the redistribution of fluid flows between them, including (if it is possible) introducing reverse flows.

Figure 2 shows a diagram using two hydraulic drives. A distinctive feature of the system under consideration is the adopted control scheme. It uses three hydraulic switchgears (***β, β₁, β₂***). The first of them performs the function of regulating the overall speed of elevating (and lowering) the object; the other two support the horizontal position of the object within the specified accuracy in the presence of disturbing factors of different nature - changes in the resistance forces, pressure in the system, consumption characteristics of regulators, and others.

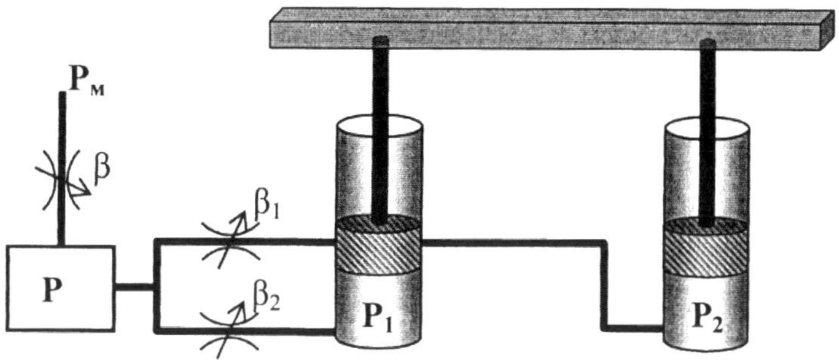

Figure 2.

The second feature of the system is the scheme of turning on hydraulic devices. The total flow of fluid consumed by both drives passes through the first device, where it is simultaneously divided into two identical flows. The pressure created in the system and the first and second hydraulic cylinders are $\boldsymbol{P}$, $\boldsymbol{P_1}$, $\boldsymbol{P_2}$, correspondingly.

3. System equations

The dimensionless equations of motion of the drives

$$c_1\ddot{\xi}_1 = \sigma_1 - \chi_{L1} - \kappa_1\dot{\xi}_1$$
$$c_2\ddot{\xi}_2 = \sigma_2 - \chi_{L2} - \kappa_2\dot{\xi}_2$$

The derivation and analysis of these equations is given in [3-5]. Here ξ and ξ are dimensionless variables characterizing the motion of the first and second drives.

The equation of pressure variation in the intermediate cavities

$$\dot{\sigma} = \frac{\varepsilon}{\xi_0}\Lambda\left[\beta\,\mathrm{sgn}(1-\sigma)\sqrt{|1-\sigma|} - A_1 - A_2\right]$$
$$A_1 = \beta_1\alpha_1\,\mathrm{sgn}(\sigma-\sigma_1)\sqrt{|\sigma-\sigma_1|},$$
$$A_2 = \beta_2\alpha_2\,\mathrm{sgn}(\sigma-\sigma_2)\sqrt{|\sigma-\sigma_2|}$$

$\Lambda = \frac{f_1}{f}\sqrt{\frac{2m}{Fs^*\rho}}$; here m – weight of the object being moved; f – effective area of the channel cross-section at the entrance to the system; F – piston area; $\alpha_1=f_1/f$; $\alpha_2=f_2/f$ here f_1 and f_1 – effective areas of feed channels for working cavities of drives; $\boldsymbol{\beta}$, $\boldsymbol{\beta_1}$, $\boldsymbol{\beta_2}$ – the relative areas of opening of the corresponding channels.

Equations of pressure changes in working cavities of drives

$$\dot{\sigma}_1 = \frac{\varepsilon}{\xi_{01} + \xi_1}[\beta_1 \alpha_1 \Lambda \operatorname{sgn}(\sigma - \sigma_1)\sqrt{|\sigma - \sigma_1|} - \dot{\xi}_1]$$

$$\dot{\sigma}_2 = \frac{\varepsilon}{\xi_{02} + \xi_2}[\beta_2 \alpha_2 \Lambda \operatorname{sgn}(\sigma - \sigma_2)\sqrt{|\sigma - \sigma_2|} - \dot{\xi}_2]$$

The control system equations $\beta = \lambda \gamma_1$; γ_1 – constant control signal at the first stage; $\beta_1 = \beta_0 + \lambda_2 \Delta\xi$; $\beta_2 = \beta_0 - \Delta\xi$;

Using the Painlev'e test one can study the analytical properties as was suggested in reference [6-8]

4. Conclusions

As a result of a numerical experiment, it was shown that it is possible to achieve a minimum deviation of the operation of the two drives to 0.01 mm. A sign of the deviation of the object from the horizontal position is the difference in the movements of the hydraulic actuator rods from some initial (basic) position at a given time, determined from the difference in the signals of the corresponding impulse-type position sensors. Additionally, a physical level sensor, whose positive property is independence from the current load position, can be used. Movement of the whole system is stable, deviations in the system are compensated by control. In the numerical experiment, it is proposed to carry out further multicriteria optimization as shown in [9].

This work was supported by the Russian Foundation for Basic Research, in the framework of a research project 16-29-04401 "Development of methods and algorithms for the synthesis of drive complexes (robots) working in different environments, taking into account the interaction of two or more robots.

This work was supported by the MEPhI Academic Excellence Project (agreement with the Ministry of Education and Science of the Russian Federation of August 27, 2013, project no. 02.a03.21.0005).

References

[1] Min-FuHsieh, Chin-JueiTung, Wu-SungYao etc. Servo design of a vertical axis drive using dual linear motors for high speed electric discharge machining. International Journal of Machine Tools and Manufacture, Volume 47, Issues 3–4, March 2007, Pages 546-554

[2] DirkVorberg, Hans-HenningSchulze. Linear Phase-Correction in Synchronization: Predictions, Parameter Estimation, and Simulations. Journal of Mathematical Psychology. Volume 46, Issue 1, February 2002, Pages 56-87.

[3] G.V.Kreinin and S.Yu.Misyurin. On Some General Laws in Drive Dynamics. ISSN 1052-6188, Journal of Machinery Manufacture and Reliability, 2008, Vol. 37, No.6, pp.546-551.

[4] G.V.Kreinin and S.Yu.Misyurin. Choice of the Law for a Position Control System // ISSN 1052-6188, Journal of Machinery Manufacture and Reliability, 2012, Vol. 41, No. 4, pp.331-336.

[5] S. Yu. Misyurin and G. V. Kreinin. Dynamics and design of a power unit with a hydraulic piston actuator. ISSN 1028_3358, Doklady Physics, 2016, Vol. 61, No. 7, pp. 354–359.

[6] N. A. Kudryashov. JOURNAL OF PHYSICS A-MATHEMATICAL AND GENERAL Volume: 35 Issue: 21 Pages: 4617-4632 Article Number: PII S0305-4470(02)33994-5 Published: MAY 31 2002.

[7] N. A. Kudryashov; A. S. Zakharchenko. APPLIED MATHEMATICS LETTERS Volume: 32 Pages: 53-56 Published: JUN 2014

[8] N. A. Kudryashov, D. I. Sinelshchikov, M. V. Demina. PHYSICS LETTERS A Volume: 375 Issue: 7 Pages: 1074-1079 Published: FEB 14 2011

[9] A. P. Nelyubin, T. P. Galkin, A. A. Galaev, D. D. Popov, S. Yu. Misyurin, V. V. Pilyugin. 2017 Usage of Visualization in the Solution of Multicriteria Choice Problems. Scientific Visualization vol. 9(5).

IOP Conf. Series: Journal of Physics: Conf. Series **937** (2017) 012024 doi:10.1088/1742-6596/937/1/012024

Numerical modelling of doped graphane superconducting and normal properties

N A Kudryashov, A A Kutukov

National Research Nuclear University MEPhI (Moscow Engineering Physics Institute), 31 Kashirskoye shosse, 115409, Moscow, Russia

E-mail: alexkutuk@gmail.com

Abstract. The normal properties of graphane with various degrees of doping are calculated with the help of the generalized Eliashberg theory. Within the theory of strong electron-phonon interaction, a superconducting order parameter of the doped graphane has been found. The critical temperature of superconductivity of graphane has been calculated as a function of the doping degree taking into account the renormalization of the electron mass, the chemical potential, and the density of electronic states.

1. Introduction

Graphene is the two-dimensional hexagonal lattice of carbon atoms. This material was discovered and obtained in 2004 [1]. As a result of the interaction of graphene with atomic hydrogen, a new substance called graphane with the chemical formula CH can be formed. The existence of graphane was theoretically predicted in 2006 [2] and the work devoted to experimental obtained graphane was published in 2009 [3]. Graphane is obtained by reversible hydrogenation of graphene and retains two-dimensional hexagonal lattice. In order to the graphane has a metallic conductivity and can transform into a superconducting state it must be doped. It is proposed in [4] to dope graphane similarly to B-doped diamond and study the electronic structure of the substance based on the density functional theory. In [4] using the Eliashberg function the electron-phonon interaction constant was calculated and the critical temperature of the doped graphane was estimated from the McMillan formula. In this paper we calculate the critical temperature of the doped graphane taking into account the renormalization of the density of electronic states, the electron mass, and the chemical potential.

2. Numerical study of normal and superconducting properties of doped graphane

To determine the critical temperature of superconductivity for a strong electron-phonon interaction, the generalized Eliashberg theory [5] is used. To calculate the normal state, a system of nonlinear integral equations (1)-(3) is used for the real and imaginary parts of the self-energy part of the electron Green's function and the renormalization of the density of electronic states [6]. Real part of the self-energy part of the electron Green's function

$$Re\Sigma(\omega) = -P\int_0^{+\infty} dz\alpha^2(z)F(z)\int_0^{+\infty} dz'\Big\{f(-z')\left(-\frac{N_n(-z')}{z'+z+\omega}+\frac{N_n(z')}{z'+z-\omega}\right) \\ +f(z')\left(-\frac{N_n(-z')}{z'-z+\omega}+\frac{N_n(z')}{z'-z-\omega}\right)\Big\}, \tag{1}$$

where $f(z') = \frac{1}{e^{z'/T}+1}$ is the Fermi distribution. Imaginary part of the self-energy part of the electron Green's function

$$Im\Sigma(\omega) = -\pi\int_0^{+\infty} dz\alpha^2(z)F(z)\Big\{[N_n(\omega-z)+N_n(\omega+z)]\,n_B(z) \\ +N_n(\omega-z)f(z-\omega)+N_n(\omega+z)f(z+\omega)\Big\}, \tag{2}$$

$n_B(z) = \frac{1}{e^{z'/T}-1}$ is the Bose distribution. The renormalized by the electron-phonon interaction electronic density of states

$$N_n(z') = -\frac{1}{\pi}\int_{-\mu}^{\infty} d\xi'\, N_0(\xi')\,\frac{Im\Sigma(z')}{[z'-\xi'-Re\Sigma(z')]^2+[Im\Sigma(z')]^2}. \tag{3}$$

Figure 1 shows the results of solving the system of equations (1)-(3) for normal state of 1% doped graphane. The frequency is expressed in dimensionless units corresponding to the limiting frequency of the phonon spectrum 0.158 eV.

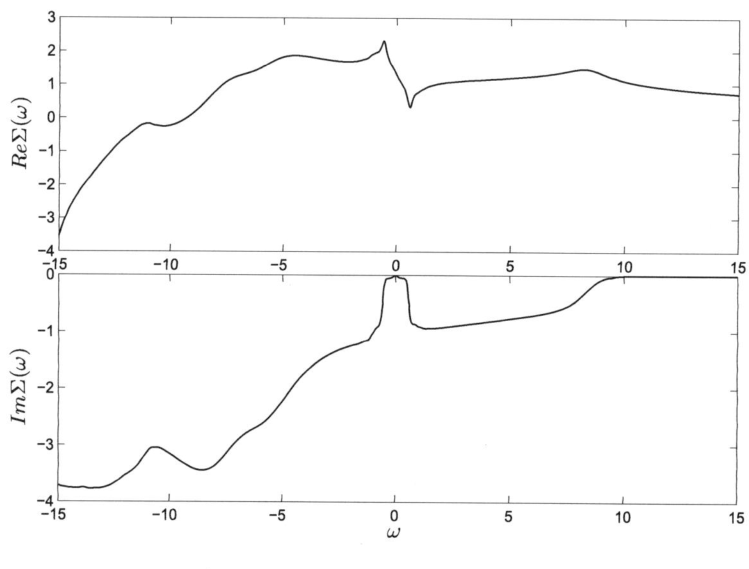

Figure 1. The real and the imaginary part of the self-energy part of the electron Green's function for 1% doped graphane and T=60K

The superconducting properties of doped graphane are described by a system of nonlinear integral equations for the superconducting order parameter [7], which are solved numerically by the iteration method. Figure 2 shows the results of solving the system of equations for superconducting state of 8% doped graphane for 25 and 30 iterations. The frequency is expressed in dimensionless units corresponding to the limiting frequency of the phonon spectrum 0.164 eV.

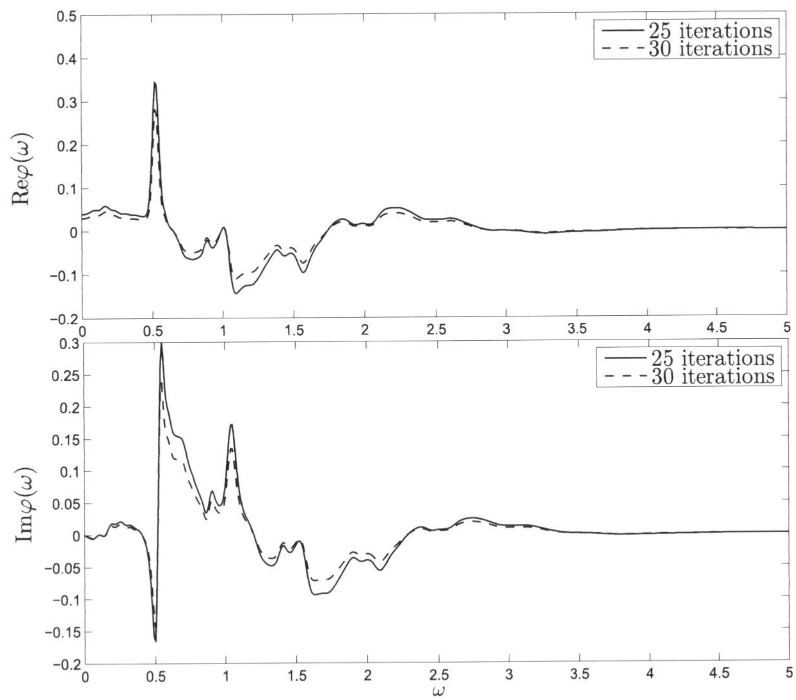

Figure 2. The real and the imaginary part of the order parameter for 8% doped graphane and T=90K

3. Conclusion

The critical temperature of superconductivity is found by determining the transition to the zero value of the order parameter. For 1% and 8% of the doped graphane the critical temperature of superconductivity T = 70 K and T = 90 K respectively was found. The calculations for intermediate degrees of doping indicate that the superconducting transition temperature increases with increasing degree of doping. When performing the calculations, we used the Eliashberg function and the density of electronic states found in [4].

Acknowledgments

This work was supported by the Ministry of Education and Science of the Russian Federation (base part of state task, project no. 1.9746.2017/BCh)

References

[1] Novoselov K S, Geim A K, Morozov S V, Jiang D, Zhang Y, Dubonos S V, Grigorieva I V, Firsov A A 2004 *Science* **306** (5696) 666-9.
[2] Sofo J O, Chaudhari A S, Barber G D 2006 Graphane: a two-dimensional hydrocarbon *arXiv:cond-mat/0606704*.
[3] Elias D C, Nair R R, Mohiuddin T M G, Morozov S V, Blake P, Halsall M P, Ferrari A C, Boukhvalov D W, Katsnelson M I, Geim A K, Novoselov K S 2009 *Science* **323** (5914) 610-3.
[4] Savini G, Ferrari A C, Giustino F 2010 *Phys. Rev. Lett.* **105** 037002-1-4.
[5] Mazur E A 2010 *Europhys. Lett.* **90** 47005-16.
[6] Kudryashov N A, Kutukov A A and Mazur E A 2016 *J. Exp. Theor. Phys* **123** (3) 481-8.
[7] Kudryashov N A, Kutukov A A and Mazur E A 2016 *JETP Letters* **104** (7) 460-5.

Equilibrium states of a variational formulation for the Ginzburg-Landau equation

A N Kulikov, D A Kulikov

Faculty of Mathematics, Yaroslavl State University, Sovetskaya str., 14, Yaroslavl, 150003, Russia

E-mail: kulikov_d_a@mail.ru

Abstract. Periodic boundary value problem for one of the versions of the complex Ginzburg-Landau equation, which is commonly called the variational Ginzburg-Landau equation are studied. Questions of existence and stability in the sense of Lyapunov, and also the local bifurcations problem of spatially nonhomogeneous equilibrium states are investigated. Three types of such solutions for the given problem are indicated. The exact formulas of the solutions for the first two types are suggested. Equilibrium states of the second type are expressed through elliptic functions. The third type of equilibrium states appears as a result of bifurcations of automodel equilibrium states, i.e., solutions of the first type in the case when the stability changes. It is shown that equilibrium states of the second and third types are unstable.

1. Introduction

The cubic complex equation can be written in the form

$$u_t = u - (1 + ic)u|u|^2 + (d_1 + id_2)u_{xx}, \tag{1.1}$$

where $u = u_1(t, x) + iu_2(t, x), c, d_1, d_2 \in R, d_1 > 0$ is commonly called a quantic time-dependent Ginzburg-Landau equation (QTDGL) [1-3]. More general forms of this equation and also particular variants of this equation are considered. For example, in applications to nonlinear optics and to some model problems of oscillations theory one considers its weakly dissipative form when $d_1 = 0$ [3-4].

In [1] the case when equation (1.1) has the form (1.1) is singled out in a separate case

$$u_t = (1 + ic)[u - u|u|^2 + du_{xx}]\,(d > 0). \tag{1.2}$$

For equation (1.2) in [1] it was suggested a special name - variational QTDGL (see [5]). In the given work it will be considered the following boundary value problem with $c = 0$

$$u_t = u - u|u|^2 + du_{xx}, \tag{1.3}$$

$$u(t, x + 2\pi) = u(t, x), \tag{1.4}$$

which was obtained in studying of phenomena in condensed media (Ψ^4 – Ginzburg-Landau's model). The questions of existence and stability of nontrivial equilibrium states will be further considered. For such solutions it will be obtained explicit or asymptotic formulas. We will understand the stability in the sense of the norm of the phase space of solutions for the boundary value problem (1.3), (1.4). We add the following initial condition to the boundary value problem (1.3), (1.4)

$$u(0,x) = f(x). \tag{1.5}$$

The mixed problem (1.3), (1.4), (1.5) is locally solvable if a complex-value function $f(x) \in H_2^2$ i.e., belongs to the space of periodic functions with period 2π, which have distributional derivatives. They and this function belong to $L_2(-\pi,\pi)$ (see [6-7]), i.e., $\int\limits_{-\pi}^{\pi} |f^{(k)}(x)|^2dx < \infty,\ k = 0,1,2.$

2. Automodel equilibrium states

The boundary value problem (1.3), (1.4) has the solutions of the form

$$u_n(x) = \eta_n \exp(inx),\ n = 0, \pm 1, \dots, \pm n_0. \tag{2.1}$$

where a natural number n_0 will be defined later. Together with solutions (2.1) there are exist the solutions $u_n(x+h) = \eta_n \exp(inh)\exp(inx), h \in R$. Therefore, considering the solutions (2.1) we suppose that $\eta_n \in R_+\ (\eta_n > 0)$, and a positive constant η_n should be found as the corresponding root of the equation $\eta - \eta^3 - d\eta n^3 = 0$. Except a zero root the last equation has a root $\eta_n = (1-dn^2)^{1/2}$ if $1 - dn^2 > 0$. Therefore, the given equilibrium states S_n exist, if $|n| \le n_0$, where $n_0 = [1/\sqrt{d}]$ or $n_0 = [1/\sqrt{d}] - 1$. The last choice of n_0 is realized, if $(1/\sqrt{d}) \in N$. We remind that as $[a]$ we denote an integer part of the number a.

To analyze the stability of $S_n\ (u_n(x) = \eta_n \exp(inx))$ we set $u_n(x) = \eta_n \exp(inx)(1 + w(t,x))$. Then, for $w(t,x) = w_1(t,x) + iw_2(t,x)$ we obtain the boundary value problem in the form

$$w_t = -\eta_n^2(w + \overline{w}) - \eta_n^2[2w\overline{w} + w^2 + w|w|^2] + dw_{xx} + 2dinw_x, \tag{2.2}$$

$$w(t, x+2\pi) = w(t,x). \tag{2.3}$$

The boundary value problem (2.2), (2.3) for a complex-value function $w(t,x) = w_1(t,x) + iw_2(t,x)$ may be rewritten in a real form for a vector-function $w(t,x) = colon(w_1, w_2)$

$$w_t = A_n w + F_2(w) + F_3(w),\ n = 0; \pm 1; \dots, \pm n_0. \tag{2.4}$$

$$w(t, x+2\pi) = w(t,x), \tag{2.5}$$

$$A_n w = \begin{pmatrix} -2\eta_n^2 w_1 + dw_{1xx} - 2dnw_{2x} \\ dw_{2xx} + 2dnw_{1x} \end{pmatrix},\ F_2(w) = -\eta_n^2 \begin{pmatrix} 3w_1^2 + w_2^2 \\ 2w_1w_2 \end{pmatrix},$$
$$F_3(w) = -\eta_n^2 \begin{pmatrix} w_1(w_1^2 + w_2^2) \\ w_2(w_1^2 + w_2^2) \end{pmatrix}.$$

A linearized form of the boundary value problem (2.4), (2.5) supposes the analysis of the equation $w_t = A_n w$ with the boundary conditions (2.5). We should found the eigenfunctions of a linear differential operator A_n in the following form $E_m(x,n) = h_m \exp(imx), h_m = colon(h_{1m}, h_{2m})$ for each $n\quad (|n| \le n_0)$.

Standard constructions allow reducing the question about finding the eigenvalues A_n to the analysis of the characteristic equation

$$det\begin{pmatrix} -2\eta_n^2 - dm^2 - \lambda & -imdn \\ imdn & -dm^2 - \lambda \end{pmatrix} = 0 \text{ or } \lambda^2 + P_n(m)\lambda + Q_n(m) = 0,$$

where $Q_n(m) = dm^2(dm^2 + 2\eta_n^2 - 4dn^2), P_n(m) = 2(dm^2 + \eta_n^2)$. It is clear that $P_n(m) > 0$. The inequality $Q_n(m) > 0$ holds for $m \neq 0$, if $d + 2 - 6dn^2 > 0$. It is understood that $Q_n(0) = 0$.

Lemma 1. Let $n = \pm 1, \ldots, \pm n_0$ and $d < d_n = 2/(6n^2 - 1)$. Then for the given d, the eigenvalues A_n, except $\lambda_0 = 0$, are located in the half-lane $Re\lambda \leq -\gamma_0(n) < 0$.

The proof is reduced to the verification of the inequalities $Q_n(m) > 0$. The analysis of characteristic equation shows that eigenvalues $\lambda_n(m) \in R$ and $\lim\limits_{m\to\infty} \lambda_n(m) = -\infty$.

Theorem 1. The boundary value problem (1.3), (1.4) has $2n_0 + 1$ the one-dimensional invariant manifolds $S_n(\alpha)$ which are formed by solutions $u_n(x + \alpha), \alpha \in R, n = 0, \pm 1, \ldots, n_0$.

Let $n \neq 0$. Then, a one-dimensional invariant manifold $S_n(\alpha)$ is attracting (local attractor), if $d \in (0; d_n)$, unstable (saddle), if $d \in (d_n, \infty)$.

For $n = 0$ the manifold $S_0(\alpha)$ is a local attractor for all $d > 0$. We underline that each of equilibrium states of the family $S_n(\alpha)$, which are existing with $d \in (0; 1/n^2)$ are stable with $d \in (0; d_n)$ and unstable with $d \subset (d_n; 1/n^2), d_n < 1/n^2$ for all considering $n\,(n \neq 0)$.

3. Equilibrium states of the second kind

Let $u(t, x) = u_1(t, x) + iu_2(t, x)$ and $u_2(t, x) = 0$. Then for the function $v(t, x) = u_1(t, x)$ we derived the following boundary value problem

$$v_t = v - v^3 + dv_{xx}, \tag{3.1}$$

$$v(t, x + 2\pi) = v(t, x). \tag{3.2}$$

The boundary value problem (3.1),(3.2) has a zero equilibrium state, but may have the nonzero equilibrium states. They should be found as nontrivial solutions $v(x)$ of the boundary value problem

$$dv'' + v - v^3 = 0, \tag{3.3}$$

$$v(t, x + 2\pi) = v(t, x). \tag{3.4}$$

The ordinary differential equation (3.3) has the solutions, which can be express through the elliptic functions

$$v(x) = psn(\delta x, k), \tag{3.5}$$

where $k \in (0; 1)$ is a real parameter (a module of an elliptic sine). We can assume that $p, \delta > 0$. The function (3.5) satisfies the equation (3.3), if $d\delta^2(1 + k^2) = 1$, and $2d\delta^2 k^2 = p^2$, i.e., $p = \sqrt{2}(k/\sqrt{1 + k^2}), \delta = (d(1 + k^2))^{-1/2}$. We should further choose $k \in (0; 1)$ so that the function (3.5) has a period 2π. It has a period $2\pi/m$, where $m = 1, 2, \ldots,$ if

$$\frac{1}{\sqrt{dm}} = G(k) = \frac{2}{\pi}(1 + k^2)^{1/2} \int\limits_0^{\pi/2} \frac{dy}{\sqrt{1 - k^2 \sin^2 y}}. \tag{3.6}$$

It is clear that $G(0) = 1, G'(k) > 0$ for all $k \in (0; 1), \lim\limits_{k\to 1} G(k) = \infty$. Furthermore, for each natural number m equation (3.6) has the solution $k_m \in (0; 1)$ if $\sqrt{dm} < 1$.

Lemma 2. Let $m < 1/\sqrt{d}$. For all such natural m the solution (3.5) exists, where $p_m = \dfrac{\sqrt{2}k_m}{\sqrt{1 + k_m^2}}, \delta_m = \dfrac{1}{\sqrt{d}\sqrt{1 + k_m^2}}$, and k_m is the corresponding root of the equation (3.6), $k_m \in (0; 1)$. The corresponding solution to the given m we denote as $v_m(x)$ and investigate the question about its stability. Let $u(t, x) = v_m(x) + w(t, x), w(t, x) = w_1(t, x) + iw_2(t, x)$. A complex-value function w satisfies the boundary value problem

$$w_t = w - v_m^2(2w + \overline{w}) - v_m(2w\overline{w} + w^2) - w^2\overline{w} + dw_{xx}, \tag{3.7}$$

$$w(t, x + 2\pi) = w(t, x) \tag{3.8}$$

for this problem it should be considered the question of stability of a zero equilibrium state. For that, we consider a linear variant of the boundary value problem (3.7), (3.8). We write the boundary value problem in a real form. For this, we set $w = colon(w_1, w_2)$. And as a result, we obtain a linear differential equation in R^2

$$w_t = B_m w, \tag{3.9}$$

where $w(t, x)$ satisfies the periodic boundary conditions, and $B_m w = \begin{pmatrix} w_1 + dw_{1xx} - 3v_m^2 w_1 \\ w_2 + dw_{2xx} - v_m^2 w_2 \end{pmatrix}$.

The given linear operator B_m has the eigenvalue $\lambda_{0m} = 1 - \dfrac{1}{1 + k_m^2} > 0$, corresponding to the eigenvector function $colon(0, cn(\delta_m x, k_m))$, where $cn(y, k)$ is an elliptic cosine, and the periodic boundary value problem for the partial differential equation (3.9) has the exponentially growing solution $w_1 = 0, w_2(t, x) = cn(\delta_m x, k_m)\exp(\lambda_{0m} t)$. The main boundary value problem (1.3), (1.4) with the solution $v_m(x)$ has a two-parametric family of analogous solutions

$$u = \exp(ih_2)v_m(x + h_1),\ h_1, h_2 \in R. \tag{3.10}$$

Thus, it was proved the statement.

Theorem 2. Let $dm_0^2 < 1, m_0 \in N$. Then, the boundary value problem (1.3), (1.4) has m_0 two-dimensional invariant manifolds $V_2(m)$ filled with the unstable equilibrium states (3.10), $m = 1, \ldots, m_0$.

4. Bifurcations of the automodel solutions

Consider the local bifurcations problem for the solutions

$$u_n(x) = \eta_n \exp(inx), \eta_n = (1 - dn^2)^{1/2}$$

in the case when the stability changes. Here, $n = \pm 1, \pm 2, \ldots$, and the case $n = 0$ we do not consider since the corresponding solution $u_0(x) = 1$ does not change the stability. We remind that the critical value is $d = d_n = 2/(6n^2 - 1)$.

In this section it will be shown that for $d < d_n$ (subcritical bifurcations) the two-dimensional invariant manifolds, which are formed by equilibrium states with more complex structure than $u_n(x + \alpha)$, bifurcate from the family of equilibrium states $S_n(\alpha)$. In a boundary value problem (1.3), (1.4) we set

$$d = d_n(\varepsilon) = d_n(1 + \varepsilon\frac{\gamma}{2}),\ u = u_n(x)(1 + w(t, x)),\ u_n(x) = \eta_n(\varepsilon)\exp(inx). \tag{4.1}$$

Here, $\varepsilon \in (0; \varepsilon_0), 0 < \varepsilon_0 << 1, \gamma = \pm 1$. The value γ will be chosen below. Finally, $\eta_n^2(\varepsilon) = 1 - dn^2 - d_n\varepsilon(\gamma/2)n^2$. As a result, for we obtain the boundary value problem

$$w_t = -\eta_n^2(\varepsilon)[w + \overline{w} + 2w\overline{w} + w^2 + w^2\overline{w}] + dw_{xx} + 2dinw_x, \tag{4.2}$$

$$w(t, x + 2\pi) = w(t, x). \tag{4.3}$$

We note that the substituting $u_n(x)$ instead of $u_n(x + h)$, where $h \in R$, leads to the same boundary value problem (4.2), (4.3) where the right part of equation does not depends on h. We may rewrite the boundary value problem (4.2), (4.3) for the complex-value functions $w = w_1(t, x) + iw_2(t, x)$ in a real form for $w = colon(w_1, w_2)$

$$w_t = A_n w + F_2(w) + F_3(w) + \varepsilon B_n w + \varepsilon F_4(w) + \varepsilon F_5(w), \tag{4.4}$$

$$w(t, x + 2\pi) = w(t, x), \tag{4.5}$$

$$A_n w = -\eta_n^2 \begin{pmatrix} 2w_1 \\ 0 \end{pmatrix} + d_n \begin{pmatrix} w_{1xx} \\ w_{2xx} \end{pmatrix} + 2nd_n \begin{pmatrix} -w_{2x} \\ w_{1x} \end{pmatrix}, \ B_n w = d_n n^2 \gamma \begin{pmatrix} w_1 \\ 0 \end{pmatrix} +$$
$$+ d_n \frac{\gamma}{2} \begin{pmatrix} w_{1xx} \\ w_{2xx} \end{pmatrix} + nd_n\gamma \begin{pmatrix} -w_{2x} \\ w_{1x} \end{pmatrix}, \ F_2(w) = -\eta_n^2 G_2(w), \ F_3(w) = -\eta_n^2 G_3(w),$$
$$F_4(w) = \frac{d_n}{2} n^2 \gamma G_2(w), \ F_5(w) = \frac{d_n}{2} n^2 \gamma G_3(w),$$
$$G_2(w) = \begin{pmatrix} 3w_1^2 + w_2^2 \\ 2w_1 w_2 \end{pmatrix}, \ G_3(w) = \begin{pmatrix} w_1(w_1^2 + w_2^2) \\ w_2(w_1^2 + w_2^2) \end{pmatrix}.$$

As the phase space of the solutions of the boundary value problem (4.4), (4.5) we assume $H_{2,2}^2$, i.e., the space of the two-dimensional periodic vector-functions $(f_1(x), f_2(x))$ with a period 2π , where $f_j(x) \in W_2^2[-\pi; \pi]$, i.e., belongs to the Sobolev space.

From the results of section 1, it follows that a linear differential operator A_n has a triple zero eigenvalue with the corresponding eigenfunctions

$$E_0(x) = \begin{pmatrix} 0 \\ 1 \end{pmatrix}, \ E_1 = \begin{pmatrix} \cos x \\ -2n \sin x \end{pmatrix}, \ E_2 = \begin{pmatrix} \sin x \\ 2n \cos x \end{pmatrix}.$$

Consider the subspace $H_* \subset H_{2,2}^2$ consisting the vector-functions $f(x) \in H_{2,2}^2$ which have as a first component an even function, and as the second one an odd function. This subspace is invariantly for the solutions of the boundary value problem (4.4), (4.5). In H_* the linear operator A_n has a simple zero eigenvalue with the corresponding eigenfunction $E_1(x)$ We consider in this space the question about local bifurcations in a neighborhood of a zero equilibrium state. From the theorem of existence and properties of the central invariant manifold $M_1(\varepsilon)$ (in the given case $dim M_1(\varepsilon) = 1$) it follows that the question about local bifurcations may be reduced to the analysis of an ordinary differential equation (normal form)

$$\dot{z} = \varepsilon \Psi_n(z) + o(\varepsilon), \tag{4.6}$$

where a form of the function $\Psi_n(z)$ will be defined below, $z = z(t)$. Briefly remind the algorithm allowing to form the right part of the equation (4.6). We will find the solutions belonging to $M_1(\varepsilon)$ in the following form (see [8-9])

$$w(x, z, \varepsilon) = \varepsilon^{1/2} Q_1(x, z) + \varepsilon Q_2(x, z) + \varepsilon^{3/2} Q_3(x, z) + o(\varepsilon^{3/2}), \tag{4.7}$$

where $z = z(t)$ are solutions of (4.6), $Q_1(x, z) = zE_1(x)$ vector-functions $Q_2(x, z), Q_3(x, z)$ as the functions of x for all z belong to H_*. These equalities satisfy

$$\int_{\pi}^{-\pi} (Q_{1k} Q_{11} + Q_{2k} Q_{21}) dx = 0, k = 2, 3, Q_k = Q_k(x, z) = colon(Q_{1k}(x, z), Q_{2k}(x, z)).$$

The equality (4.7) should be understood as the equation defining the integral (invariant) manifold $M_1(\varepsilon)$. Substituting the sum (4.7) in the boundary value problem (4.4), (4.5) with (4.6), equating powers of ε we obtain the two nonhomogeneous boundary value problems to define $Q_2(x, z), Q_3(x, z)$

$$A_n Q_2 + F_2(Q_1) = 0, \tag{4.8}$$
$$Q_2(x + 2\pi) = Q_2(x, z), \tag{4.9}$$
$$A_n Q_3 + F_3(Q_1) + \Phi_3 - \Psi_n(z) E_1(x) + BQ_1 = 0, \tag{4.10}$$

IOP Conf. Series: Journal of Physics: Conf. Series **937** (2017) 012025 doi:10.1088/1742-6596/937/1/012025

$$Q_3(x+2\pi) = Q_3(x,z). \tag{4.11}$$

Here, $\Psi_n(z)$ is "the main part of the right part of equation (4.6),

$$\Phi_3 = -2\eta_n^2 \begin{pmatrix} 3Q_{11}Q_{12} + Q_{21}Q_{22} \\ Q_{11}Q_{22} + Q_{21}Q_{12} \end{pmatrix}, Q_j = Q_j(x,z) = \begin{pmatrix} Q_{1j}(x,z) \\ Q_{2j}(x,z) \end{pmatrix}, j = 1, 2.$$

The nonhomogeneous boundary value problem (4.8), (4.9) is solvable in the considered class of functions H_*, and the corresponding solution has the following form

$$Q_2(x,z) = z^2 \begin{pmatrix} q_0 + q_1 \cos 2x \\ q_2 \sin 2x \end{pmatrix}, q_0 = -\frac{4n^2+3}{4}, q_1 = \frac{1-4n^2}{4}, q_2 = \frac{n(4n^2-1)}{2}.$$

Remark. Remind a well-known fact: the nonhomogeneous boundary value problem

$$A_n y = G(x), y(x+2\pi) = y(x),$$

where $y = (y_1(x), y_2(x)), G(x) = (G_1(x), G_2(x))$, where periodic functions $G_j(x) \in L_2(-\pi, \pi)$, $j = 1, 2$ with a period 2π have a solution from H_* if

$$\int_{-\pi}^{\pi} (G_1 Q_{11} + G_2 Q_{21}) dx = 0.$$

An analogue equality for the components of the vector-function $(y_1(x), y_2(x))$ singles out the solution from H_* .

The solvability condition applying to the nonhomogeneous boundary value problem (4.10), (4.11) allows to define $\Psi_n(z)$ It turned out that

$$\Psi_n(z) = \beta_n z + l_n z^3, \; l_n = 3\frac{4n^2-1}{6n^2-1}, \; \beta_n = \frac{\gamma}{4n^2+1}.$$

In particular, $l_n > 0$ for all n ($l_1 = 1.80, l_2 = 1.96, l_3 = 1.98, \lim_{n\to\infty} l_n = 2, \lim_{n\to\infty} \beta_n = 0$). We now consider the truncated normal form i.e., the differential equation

$$\dot{z} = \varepsilon[\beta_n z + l_n z^3].$$

For $\gamma = -1$ it has the two nonzero equilibrium points (for $\gamma = 1$ they do not exist)

$$z_1 = (\alpha_n)^{-1/2}, z_2 = -(\alpha_n)^{-1/2}, \alpha_n = l_n/|\beta_n|.$$

Both of these equilibrium states are unstable, and a zero equilibrium state of the differential equation (4.6) is asymptotically stable.

Thus, it was proved the statement.

Lemma 3. There exists $\varepsilon_0 > 0$ such that for all $\varepsilon \in (0; \varepsilon_0), d = d_n(1 - \varepsilon/2)$ the boundary value problem (4.4), (4.5) with an additional condition of belonging of solutions to the space H_* has the equilibrium states

$$w(x,z,\varepsilon) = \varepsilon^{1/2} Q_1(x, z_j) + \varepsilon Q_2(x, z_j) + o(\varepsilon)$$

where z_j is a coordinate of one of the nonzero equilibrium states of equation (4.6). These equilibrium states are unstable.

If $d = d_n(1 + \varepsilon/2), \varepsilon > 0$ then, the nonlinear equilibrium states are absent.

We also note that $w(x, z_2, \varepsilon) = w(x + \pi, z_1, \varepsilon)$ From the previous statements it follows the theorem for the main boundary value problem (1.3), (1.4) (see [8-9]).

Theorem 3. For the indicated ε and d within the formulation of lemma 3 for $d = d_n(1-\varepsilon/2)$ the boundary value problem (1.3), (1.4) has the two-parametric family of equilibrium states

$$\begin{aligned} u = \eta_n \exp(inx + ih_1)[1 + \left(\frac{\varepsilon}{\alpha_n}\right)^{1/2} (\cos(x + h_2) - 2ni \sin(x + h_2)) + \\ + \frac{\varepsilon}{\alpha_n}(q_0 + q_1 \cos(2x + 2h_2) + iq_2 \sin(2x + 2h_2)) + o(\varepsilon)], \end{aligned} \tag{4.12}$$

where $h_1, h_2 \in R$, and the constants $q_0 = q_0(n), q_1 = q_1(n), q_2 = q_2(n), \alpha_n$ were defined earlier.

The family of solutions (4.12) forms the two-dimensional integral manifold $V_2(\varepsilon)$. This manifold is not attracting (local attractor). It is filled with the unstable equilibrium states of the boundary value problem (1.3), (1.4).

5. Conclusion

In the present work it was shown that in the periodic boundary value problem for the variational Ginzburg-Landau equation can exist three types of the equilibrium states, which differ from a trivial. These equilibrium states can be divided into two classes: spatially homogeneous and spatially nonhomogeneous, i.e., substantially depending on a variable x. Nontrivial, but spatially homogeneous equilibrium states always exist and they are stable. A different situation exists for spatially nonhomogeneous equilibrium states, and conditions of existence and stability are obtained in the sections 1,2,3. We underline that for them the situation when they are unstable is more typical.

6. Acknowledgments

The results of the given work were presented on the 6th International conference "The problems of mathematical physics and mathematical modeling" (Moscow, NRNU MEPhI, 25-27 May 2017). Authors thank the participants of the section "Methods of mathematical physics" and professor N.A. Kudryashov for the helpful discussion of the report.

7. References

[1] Aranson I S and Kramer L 2002 *Reviews of modern physic* **74** 99
[2] Malinetskiy G G and Potapov A B and Podlazov A V 2006 *Nelineynaya dinamica. Podhody,resultaty, nadeshdy*(Moscow: Komkniga(in Russian))
[3] Drazin P G 2002 *Introduction to hydrodynamic stability*(Cambridge: univers. press)
[4] Kolesov A Yu and Kulikov A N Rozov N Ch 2006 *Doklady Mathematics* **73** 125
[5] Kulikov A N and Rudy A S 2003 *Chaos, Solitons and Fractals* **15** 75
[6] Krein S G 1977 *Lineyniye differentsialniye uravneniya v banachovom prostranstve*(Moscow: Nauka (in Russian))
[7] Sobolevskiy P E 1977 *Trudy MMO* **10** 297
[8] Kulikov A N and Kulikov D A 2010 *Differential equations* **46** 1299
[9] Kulikov A N and Kulikov D A 2013 *Journal of Mathematical Sciences* **46** 273

IOP Conf. Series: Journal of Physics: Conf. Series **937** (2017) 012026 doi:10.1088/1742-6596/937/1/012026

A posteriori error estimates in voice source recovery

A S Leonov[1] **and V N Sorokin**[2]

[1]National Research Nuclear University MEPhI (Moscow Engineering Physics Institute), Moscow, 115409 Russia

[2]Institute of Problems of Information Transfer, Russian Academy of Sciences, Bolshoi Karetnyi per., Moscow, 127994 Russia

E-mail: asleonov@mephi.ru vns@iitp.ru

Abstract. The inverse problem of voice source pulse recovery from a segment of a speech signal is under consideration. A special mathematical model is used for the solution that relates these quantities. A variational method of solving inverse problem of voice source recovery for a new parametric class of sources, that is for piecewise-linear sources (PWL-sources), is proposed. Also, a technique for a posteriori numerical error estimation for obtained solutions is presented. A computer study of the adequacy of adopted speech production model with PWL-sources is performed in solving the inverse problems for various types of voice signals, as well as corresponding study of a posteriori error estimates. Numerical experiments for speech signals show satisfactory properties of proposed a posteriori error estimates, which represent the upper bounds of possible errors in solving the inverse problem. The estimate of the most probable error in determining the source-pulse shapes is about 7-8% for the investigated speech material. It is noted that a posteriori error estimates can be used as a criterion of the quality for obtained voice source pulses in application to speaker recognition.

1. Introduction

The voice source pulse excites acoustic oscillations in the vocal tract. Mathematically, it is proportional to the time derivative of the volume velocity of the air flow in the glottis. Its characteristics are unique for each person. The individual features of voice source form give the basis for an approach to automated speaker recognition regardless of context, as well as for larynx pathology diagnostics and other applications. So, the inverse problem of finding parameters of voice source pulses from speech segments is of importance for applications.

Methods of solving this inverse problem are considered for a long time in many mathematical statements (see, e.g., a review in [1]). However, the question of error estimation for obtained pulse form remains poorly understood. The opinion about the accuracy of obtained solution is mainly formed intuitively. Therefore, it is not known quantitatively to what extent the found form of a particular source pulse can be considered reliable and whether it is possible to use this form, for example, in assessing pathology or in speaker recognition.

The reason is that the inverse problem of voice source recovery appears to be ill-posed and non-linear. So, its solution requires the use of special regularization methods (see, e.g., [2, 3]). Obtaining error estimates for solutions of such ill-posed inverse problems is a very complicated problem, which can be solved only by using significant a priori information about the desired solution. The information of this kind is not available in our task. Thus, the existing theory of *a priori errors estimates* (see, e.g. [3]) is practically not applicable to the inverse problem under

IOP Conf. Series: Journal of Physics: Conf. Series **937** (2017) 012026 doi:10.1088/1742-6596/937/1/012026

consideration. Therefore, we propose a method for voice source recovery with a quantitative *a posteriori error estimate* for real speech signals, based on the theoretical results presented in [4, 5].

The initial in our study is a special mathematical model of speech generation by parametrical vocal sources, proposed in [6] (Sec.2). As in most known approaches, it is assumed in the model that the resonant frequencies (formants) of the vocal tract are the same at the intervals of open and closed vocal slit. The model takes into account the experimental dependence of the formants damping on the frequency, the radiation of speech signal from the lips etc. In processing, formants are calculated from the speech segment corresponding to the closed vocal slit. Signal segmentation, i.e., estimation of the instants of the glottis opening and closure, is performed according to the technique from [7]. For the adopted model, the inverse problem of determining shapes of the voice source pulse from registered segments of speech is posed in a form of parametrical operator equation.

In Sec.3, we give a variational technique for solving the inverse problem for each pitch period. To this end, we postulate a new parametric voice source, so-called piecewise linear source (PWL-source). Further, we apply presented method to the processing of speech signals. In fact, sources and processes in the vocal tract differ from their description in the mathematical model. In this regard, we demonstrate a computer study of the adequacy of the proposed model (Sec.4). The procedure of error estimate based on the work [4] is described in Sec.5. We apply it to different types of speech signals (Sec.6). In particular, we adduce the results of corresponding calculations for the speech signals from the database CMU ARCTIC [8] (2 speakers, about 22,000 analyzed impulses) and the base of Russian numerals (56 speakers, the total number of analyzed impulses is more than 322,000). It was found that, in general, PWL-sources adequately describe speech processes in the vocal tract. The most possible value of a posteriori error estimate in PWL-source recovery is about 7-8%. Sec.7 contains the conclusions.

A posteriori estimate gives upper bound for possible errors in the solution to the inverse problem for the selected segment of speech. If it turns out that this bound is *sufficiently small*, then the calculated form of the voice source preserves to a greater extent the individual properties of the speaker. Such form can be used in a procedure of speaker identification. *Large* upper bounds of the error serve as an indicator that the individual features of the speaker can be significantly distorted in found voice sources. Such source pulses should be excluded from the recognition procedure. In general, knowledge of the upper bound of errors makes it possible to more realistically assess the prospects for not only the speaker identification by voice source, but also other applied problems using such a source.

2. Mathematical model "from voice source to speech signal"

We use a model for the speech production of vowels based on the Webster horn equation with viscous-friction losses and with standard impedance boundary conditions (see [6]). The model relates the speech signal $f(t)$ on a pitch period $t \in [0, T_0]$ with the voice source function $q(t)$ exciting vibrations in the vocal tract. The function $q(t) \in C[0, T_0]$ is associated with the volume velocity $v(t)$ of the air flow in the glottis by relation $q(t) = q_0 v'(t)$, $q_0 = \text{const} > 0$. We assume that there is no systematic distortion of the signal such as reverberation, and that the function $f(t)$ recorded by the microphone is related to the pressure $p(x,t)$ at the end $x = l$ of the vocal tract by the formula

$$f(t) = f_0 \left(\lambda p(l,t) + \frac{(1-\lambda)}{T_0} \int_0^t p(l,\tau) d\tau \right), \quad f_0 = \text{const} > 0; \ t \in [0, T_0].$$

This assumption is based on the theoretical reasonings on the radiation of a speech signal from lips to a microphone. The value $\lambda \in [0, 1]$ is determined by conditions for speech recording, the type of microphone, and other factors. When solving inverse problems, we will consider λ to be determined. Then the connection between the source function $q(t)$ and the registered function $f(t)$ is postulated in the form [6, 9]:

$$\begin{aligned} f(t) &= \int_0^t K(t-\tau)\left(\lambda q(\tau) + \frac{(1-\lambda)}{T_0}\int_0^\tau q(\eta)d\eta\right)d\tau + \\ &+ \sum_{n=1}^{N_0} e^{-(\mu+\sigma(f_n))t}\left(A_n\cos(2\pi f_n t) + \frac{B_n}{f_n}\sin(2\pi f_n t)\,t\right) + C_0 \end{aligned} \quad (1)$$

with

$$K(t) = K_0\sum_{n=1}^{N_0} e^{-(\mu+\sigma(f_n))t}\frac{\sin\omega_m t}{\omega_m\frac{d\Delta}{d\omega}(\omega_m)},\ \Delta(\omega) = (\omega_1-\omega)\prod_{m=2}^{N_0}\frac{\omega_m-\omega}{(m-1)^2},\ K_0 = \text{const.} \quad (2)$$

Here $\{f_m\}_{m=1}^{N_0}$ are resonant frequencies of the vocal tract, N_0 is the number of such frequencies taken into account and $\omega_m = 2\pi f_m$. The function $\sigma(f)$ represents the dependence of the damping decrement on the frequency and is taken from practical measurements. Parameters of the model are the values μ, λ, A_n, B_n, C_0, where $\mu > 0$ is the coefficient of viscous losses, the coefficients A_n, B_n, C_0 are determined by the initial conditions for the pressure in the vocal tract, and also by the numbers λ, μ. With known values μ, λ, A_n, B_n, C_0 and for a given source function $q(t)$, formulas (1) and (2) make it possible to calculate the theoretical analogue of the recorded speech signal. The formula (1) can be written in the operator form

$$f = B(\mu, \lambda, A_n, B_n, C_0)\,[q(t)], \quad (3)$$

where the operator B, depending on the parameters, describes the mathematical model (1), (2).

We state our *inverse problem of source recovery* as follows. Given the fundamental tone period T_0, vocal slit closure $t_{cl} \in (0, T_0)$, frequencies $\{f_m\}_{m=1}^{N_0}$ and the speech segment $f(t)$, $t \in [0, T_0]$, we wish to find from the operator equation (1) the unknown source function $q(t)$ and associated source function $v(t)$, and the parameters $\mu > 0$, $\lambda \in [0, 1]$ and A_n, B_n, C_0.

3. Method for solving the inverse problem

It is common to use special parametric classes of functions $q(t) = q(t, a)$ to solve the inverse problem in question. Such functions reflect the approximate behavior of the source pulse and require physically and physiologically grounded a priori constraints on the parameters a. We use in this work a piecewise linear model of the voice source. A typical plot of such a source and its parameters $a = (t_1, t_2, t_3, t_4, t_5, A_1, A_2, A_3)$ are shown in Fig.1. The value A_4, presented in Fig.1, depends on other parameters and is determined from the condition $v(t_{cl}) = \int_0^{t_{cl}} q(t)dt = 0$ as $A_4 = (A_1 t_2 + A_2(t_3 - t_1) + A_3(t_4 - t_2))\,/(t_{cl} - t_4)$. There are specific physiological constraints on the parameters a. For example, the inequality $0 < t_1 < t_2 < t_3 < t_4 < t_{cl}$ must hold. Without going into other details here, we denote the set of all these constraints as M_q. There are similar restrictions on the parameters $p = (\mu, \lambda, A_n, B_n, C_0)$. We shall denote them as M_p

Now we formulate the method for solving our inverse problem in variational form. To this end, we introduce the relative residual of the operator equation (3):

$$\Phi(q(t, a), \mu, \lambda, A_n, B_n, C_0) = \frac{\|B(\mu, \lambda, A_n, B_n, C_0)\,[q(t, a)] - f(t)\|}{\|f(t)\|}.$$

Here we use the norm

$$\|f(t)\|^2 = \int_0^{t_{cl}} \left\{f(t)e^{-\mu t}\right\}^2 dt$$

in the space $L_2[0, T_0]$. This corresponds to our desire to ensure the proximity of the problem data and their computed counterparts for the part of the pitch period where the source operates. The norm and the functional Φ actually specifies the closeness of the functions

$$U(t) = e^{-\mu t} f(t),\ U_{calc}(t) = e^{-\mu t} B\left(\mu, \lambda, A_n, B_n, C_0\right)[q(t, a)].$$

Exactly these functions will be presented below as plots of $U(t), U_{calc}(t)$ (see Fig.2).

We reduce our inverse problem to the solution of an optimization problem, namely we wish to find a function $q = q(t, \bar{a})$, $\bar{a} \in M_q$, and parameters $(\bar{\mu}, \bar{\lambda}, \bar{A_n}, \bar{B_n}, \bar{C_0}) \in M_p$ such that

$$\Phi(q(t, \bar{a}), \bar{\mu}, \bar{\lambda}, \bar{A_n}, \bar{B_n}, \bar{C_0}) = \min\left\{\Phi(q(t, a), \mu, \lambda, A_n, B_n, C_0) : a \in M_q, (\mu, \lambda, A_n, B_n, C_0) \in M_p\right\}. \tag{4}$$

In fact, problem (4) gives *quasisolutions* of the operator equation (3). Such problems can be ill-posed. The method for solving them are studied in detail in [10]. We note that the quantity $\Phi_{\min} = \Phi(q(t, \bar{a}), \bar{\mu}, \bar{\lambda}, \bar{A_n}, \bar{B_n}, \bar{C_0})$ found after solving the problem (4) shows to what extent the adopted mathematical model (1) is adequate to the data of the inverse problem. Problem (4) is a mathematical programming problem, and we solve it with the help of known stable numerical algorithms and program modules (see, for example, [2, 10]).

4. Model verification

For practical applications of the source recovery problem, it is important how our mathematical model (1) and forms of PWL-sources are adequate to the speech signals and experimantal forms of voice source pulses. Fig.2 shows an example of solving our inverse problem for a speaker whose source functions $v_{exp}(t)$ and their derivatives $q_{exp}(t)$ can be obtained independently from additional experimental data presented in [11]. This allows us to compare the velocity

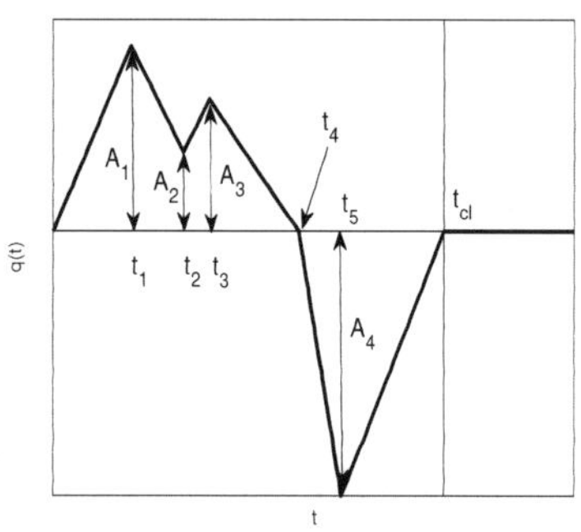

Figure 1. PWL-source and its parameters.

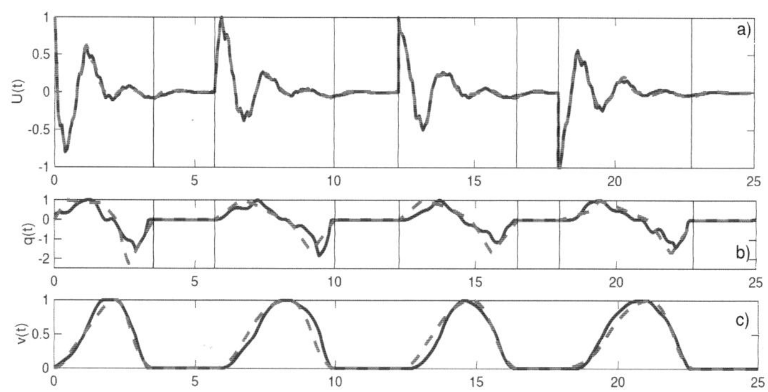

Figure 2. Experimental (——) and calculated (- - - -) functions: a) $U_{exp}(t)$ and $U_{calc}(t)$; b) $q_{exp}(t)$ and $q_{calc}(t)$; c) $v_{exp}(t)$ and $v_{calc}(t)$

pulses $v_{exp}(t)$ and their derivatives $q_{exp}(t)$ and functions $v_{calc}(t)$, $q_{calc}(t)$ obtained by solving the optimization problem (4). All these functions are normalized to a maximum. One can see that the pulses of the volume velocity are recovered satisfactorily. Also, we note the high proximity of the functions $U_{exp}(t)$ and $U_{calc}(t)$, which means the adequacy of the model used.

5. Method of a posteriori error estimation

The main difficulty in estimating the accuracy of voice source recovery is determined by numerical instability or ill-conditionality of the problem (4). It is for this reason that attempts to obtain theoretical a priori error estimates lead to estimates that can not be practically used. This is expressed in the fact that a priori estimates contain constants that either can not be found numerically, or can be estimated extremely roughly. As a result, attempts to express a priori error estimate as a number lead to overestimation of errors. In this connection, practical error estimates of the solutions to the problem (4) should be performed a posteriori, using, for example, the recently developed theory [4, 5]. Now we describe the technique of obtaining these a posteriori estimates for our problem.

After solving the problem (4), we have at our disposal quantities $(q(t,\tilde{a}),\tilde{\mu},\tilde{\lambda},\tilde{A}_n,\tilde{B}_n,\tilde{C}_0)$, which are approximations of exact values $q(t,\bar{a}),\bar{\mu},\bar{\lambda},\bar{A}_n,\bar{B}_n,\bar{C}_0$. Therefore, we can calculate the approximation $\Phi_{\min}^{(appr)} = \Phi\left(q(t,\tilde{a}),\tilde{\mu},\tilde{\lambda},\tilde{A}_n,\tilde{B}_n,\tilde{C}_0\right)$ for $\Phi_{\min}$. We set the constant of the estimation algorithm, $C > 1$, close to unity (for example, $C = 1.01$) and consider the extremum problem: find the quantities $a^* \in M_q$, $(\mu^*,\lambda^*,A_n^*,B_n^*,C_n^*) \in M_p$ such that

$$\frac{\|q(t,a^*)-q(t,\tilde{a})\|}{\|q(t,\tilde{a})\|} = \max\left\{\frac{\|q(t,a)-q(t,\tilde{a})\|}{\|q(t,\tilde{a})\|} : a \in M_q,\ \Phi\left(q(t,a),\mu,\lambda,A_n,B_n,C_n\right) \le C\Phi_{\min}^{(appr)}\right\}$$

In this optimization problem, the first constraint means minimization over all admissible sources, and the second restriction corresponds to minimization over all sources and parameters that give residuals close to $\Phi_{\min}^{(appr)}$. Thus, the value $\varepsilon_q = \frac{\|q(t,a^*)-q(t,\tilde{a})\|}{\|q(t,\tilde{a})\|}$ represents the greatest relative deviation of the approximate solution to inverse problem (3) from all admissible source functions for which the values of the residual is close to optimal. The degree of this proximity is determined by the number C. We take the value ε_q as a posteriori error estimate of found approximate solution $q(t,\tilde{a})$. To calculate ε_q, we need to solve the last extremum problem using known optimization methods. Similarly, one can find a posteriori error estimate for an volume velocity pulse $v(t,a) = \int_0^t q(\tau,a)d\tau$:

$$\varepsilon_v = \max\left\{\frac{\|v(t,a)-v(t,\tilde{a})\|}{\|v(t,\tilde{a})\|} : a \in M_q, \Phi\left(q(t,a),\mu,\lambda,A_n,B_n,C_n\right) \le C\Phi_{\min}^{(appr)}\right\}. \tag{5}$$

In what follows, error estimates ε_v are presented, which were obtained numerically with the help of optimization modules of MATLAB software package.

6. Numerical experiments on the error estimation for volume velocity impulses

For model problems with known solutions $q(t)$ and corresponding synthetic speech signals known with relative error δ, we can solve the problem (2) and find approximate PWL-solutions $q(t,a_\delta)$ and corresponding function $v(t,a_\delta)$. Then the accuracy of $v(t,a_\delta)$ is given by the formula

$$\Delta(\delta) = \frac{\|v(t)-v(t,a_\delta)\|_{L_2[0,T_0]}}{\|v(t)\|_{L_2[0,T_0]}}.$$

Therefore, we can comppare such a calculated accuracy with its estimate $\varepsilon_v(\delta)$ found from (5). The results of this comparison are presented in Fig.3 for different δ.

Now we describe the results concerning a posteriori error estimates in voice source recovery with actual speech signals from the database CMU ARCTIC [8]. Speech signals of two speakers were analyzed: a man (BDL speaker) and a woman (SLT speaker). The total number of analyzed voice source pulses for each speaker exceeds 22,000. The plots on Fig.4A show the calculated standard distribution density p_ε of the relative errors ε_v, which were found using formula (5) for these speakers. Standard distribution density p_Φ of relative residuals $\Phi_{\min}^{(appr)}$ for these speakers are shown in Fig.4B.

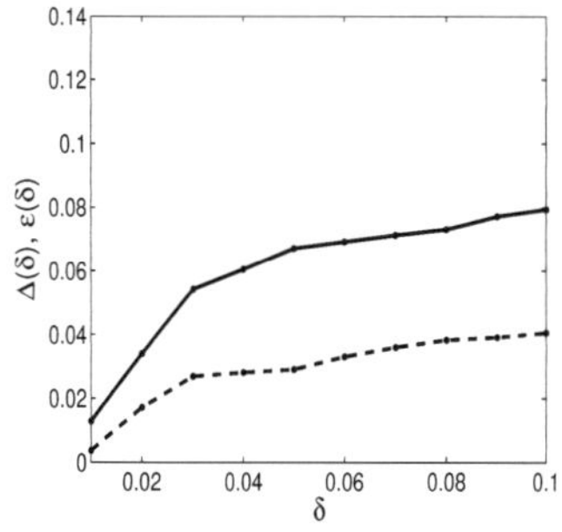

Figure 3. $\Delta(\delta)$ (- - - -) and $\varepsilon_v(\delta)$ (——).

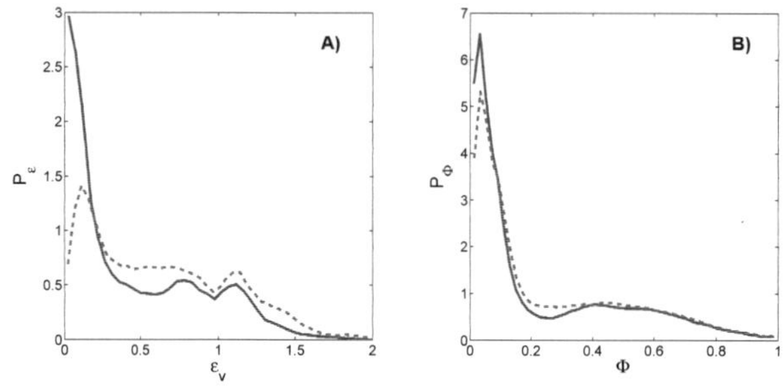

Figure 4. Standard distribution densities of: A) ε_v, and B) $\Phi_{\min}^{(appr)}$ for BDL speaker (——) and SLT speaker (- - - -).

Further numerical experiments were performed with speech signals from the database for Russian numerals. The speech was recorded under ordinary conditions in rooms with different dimensions and different natural-noise levels. Thus, the characteristics of speech signals were maximally close to the actual operating conditions for the speaker identification systems. Speech signals from 56 speakers were used in the experiments. The total number of analyzed voice source pulses is about 322,000. The corresponding results of estimating the upper bound of the errors, ε_v, and residuals, $\Phi_{\min}^{(appr)}$, for PWL-sources are shown in Fig.5A and Fig.5B correspondingly.

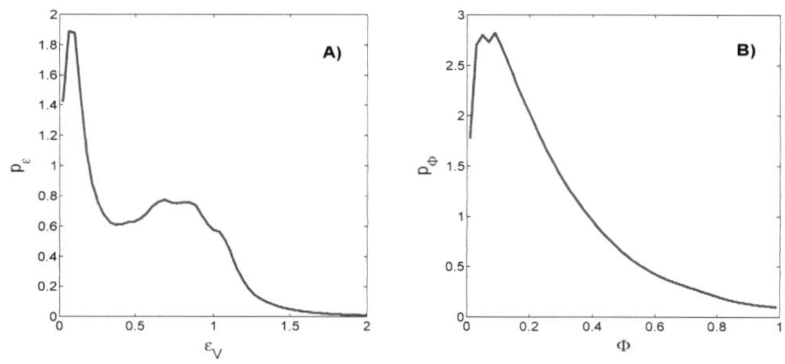

Figure 5. Standard distribution dencities of: A) ε_v, and B) $\Phi_{\min}^{(appr)}$ for Russian numerals.

7. Conclusions

1) The rather simple model (1) of speech production and the piecewise linear model of the source make it possible to reproduce the studied speech material with good accuracy. The available deviations of the measured speech signals from those calculated by the model (1) are caused both by the simplification of the description of speech processes, and by experimental errors of various kinds that arise during registration of speech and its preliminary processing (segmentation). These errors, especially those that have a systematic character, can lead to the fact that the model signal calculated on the considered pitch period by means of found PWL-source is significantly different from registered. For this period, the model (1) is not adequate to the speech data used. The above graphs of the residual distribution (Fig.4B,5B) show the proportion of such speech segments in the used speech information.

2) The most probable value of the upper error limit ε_v for found PWL voice source is estimated at approximately 7-8% for the speech material considered. The graphs in Fig.4A,5A allow us to analyze the frequency of the appearance of pulses with too large and too small error estimates, and make quantitative conclusions about the accuracy of the approximation of actual voice sources. The quantity 7-8% for source errors seems very good in comparison with rather large experimental data errors produced by actual reverberation and random errors of the order of 5-10%.
3) The values ε_v can be used to estimate the quality of voice source impulses obtained in solving the inverse problem. The use of these quantities opens up new possibilities in the task of speaker recognition. Analyzing the obtained impulses, one can discard those of them that have large error estimates ε_v. Thus, it is possible to increase the reliability of recognition and accelerate the recognition procedure.

Acknowledgments

This work was supported by the Russian Foundation for Basic Research (project no.17-01-00159a) and the Programm of Competitiveness Increase of the National Research Nuclear University MEPhI (Moscow Engineering Physics Institute); contract no. 02.a03.21.0005, 27.08.2013.

References

[1] Drugman T, Bozkurt B and Dutoit T 2012 *Comput. Speech Lang.* **26** 20
[2] Tikhonov A N, Leonov A S and Yagola A G 1998 *Non-Linear Ill-Posed Problems* vol 1,2 (London: Chapmen & Hall)
[3] Engl H W, Hanke M and Neubauer A 1996 *Regularization of Inverse Problems* (Dordrecht: Kluwer)
[4] Leonov A S 2012 *J. Inv. Ill-Posed Probl.* **20** 637
[5] Leonov A S 2012 *J. Inv. Ill-Posed Probl.* **22** 713
[6] Leonov A S and Sorokin V N 2012 *Doklady Mathematics* **85** 432
[7] Sorokin V N 2016 *Acoust. physics* **62** 244
[8] *CMU ARCTIC speech synthesis databases* http://festvox.org/cmu arctic
[9] Leonov A S and Sorokin V N 2017 *Acoust. physics* **63** 570
[10] Ivanov V K, Vasin V V and Tanana V P 2002 *Theory of Linear Ill-posed Problems and Its Applications* (Utrecht: VSP)
[11] Childers D G, Paige A and Moore A 1976 *Archives of Otolaryngology* **102** 407

IOP Conf. Series: Journal of Physics: Conf. Series **937** (2017) 012027 doi:10.1088/1742-6596/937/1/012027

On an example of a system of differential equations that are integrated in Abelian functions

M D Malykh and L A Sevastianov

Department of Applied Probability and Informatics, Peoples' Friendship University of Russia (RUDN University), Russia

E-mail: malykh_md@rudn.university

Abstract. The short review of the theory of Abelian functions and its applications in mechanics and analytical theory of differential equations is given. We think that Abelian functions are the natural generalization of commonly used functions because if the general solution of the 2nd order differential equation depends algebraically on the constants of integration, then integrating this equation does not lead out of the realm of commonly used functions complemented by the Abelian functions (Painlevé theorem). We present a relatively simple example of a dynamical system that is integrated in Abelian integrals by "pairing" two copies of a hyperelliptic curve. Unfortunately, initially simple formulas unfold into very long ones. Apparently the theory of Abelian functions hasn't been finished in the last century because without computer algebra systems it was impossible to complete the calculations to the end. All calculations presented in our report are performed in Sage.

1. Introduction

Abelian functions arose for the first time in the works of Jacobi as the natural generalization of elliptic functions [1]. Let P be a polynomial from the ring $\mathbb{Q}[x]$ with simple roots in the field $\mathbb{C}$. If the polynomial has degree 1 or 2 and variables x and u are connected by the equation

$$\int \frac{dx}{\sqrt{P(x)}} = u, \qquad (1)$$

then x can be expressed as an entire elementary function of u. If the polynomial has degree 3 or 4 then x can be expressed from equation (1) as a meromorphic function of u, this function isn't elementary, but special function called *elliptic function*. In general case the finding of the dependence $x(u)$ is called the inversion of the integral (1).

If the polynomial has degree larger then 4 then the algebraic curve

$$C: \quad y^2 = P(x), \qquad (2)$$

on a projective xy-plane is called *hyperelliptic*. We will consider integral the integral

$$u = \int \frac{dx}{\sqrt{P(x)}} = \int \frac{dx}{y}$$

as a function of point (x, y) on the curve C and and the finding of the dependence $x(u)$ is cal it the hyperelliptic integral. Jacobi wanted to solve the problem of the inversion for hyperelliptic integrals, that is to find of the dependence $x(u)$, but encountered on the following obstacle.

Theorem 1 (Jacobi) *If α and β are two roots of P then*

$$\omega = 2 \int\limits_{\xi=\alpha}^{\beta} \frac{d\xi}{\eta}$$

is a period of the hyperelliptic integral

$$\int \frac{d\xi}{\eta}.$$

Thus the number of independent periods of the hyperelliptic integral grows together with the degree of the polynomial P.

If the polynomial has degree 3 or 4 then there are exactly two complex periods ω_1 and ω_2. There are meromorphic functions with such periods, for example, Weierstrass $\wp$-function

$$\wp(u) = \frac{1}{z^2} + \sum_{n^2+m^2\neq 0} \left\{ \frac{1}{(z + m\omega_1 + n\omega_2)^2} - \frac{1}{(m\omega_1 + n\omega_2)^2} \right\}.$$

It is obvious that

$$\wp\left(\int\limits_{(a,b)}^{(x,y)} \frac{d\xi}{\sqrt{P(\xi)}} \right) \tag{3}$$

is a single-valued function defined at all points of the elliptic curve (2). There are no singularities of the integral (1) on this curve thus the expression (3) is an algebraical function of the variable x:

$$\wp(u) = r(x).$$

From this equation we can express x as algebraic function of $\wp(u)$. For this reason the inversing of the elliptic integral in elliptic functions is possible.

However if degree of P is equal to 5 then it is not more possible to express x as a meromorphic function of u because there are no single-valued functions of one variable having 4 periods. Jacobi claimed that the problem of the inversion for hyperelliptic integrals isn't solvable not only in meromorphic functions but even in any analytical expressions [1].

There are no single-valued functions of one variable having 4 periods, but there exist such functions among the functions of two variables. So Jacobi proposed the following generalization of the problem

$$\begin{cases} \int\limits_{(a_1,b_1)}^{(\xi_1,\eta_1)} \frac{d\xi}{\sqrt{P(\xi)}} + \int\limits_{(a_2,b_2)}^{(\xi_2,\eta_2)} \frac{d\xi}{\sqrt{P(\xi)}} = u, \\ \int\limits_{(a_1,b_1)}^{(\xi_1,\eta_1)} \frac{\xi d\xi}{\sqrt{P(\xi)}} + \int\limits_{(a_2,b_2)}^{(\xi_2,\eta_2)} \frac{\xi d\xi}{\sqrt{P(\xi)}} = v. \end{cases} \tag{4}$$

Weierstrass [2] proved that rational symmetric functions of the variables ξ_1, ξ_2 are really meromorphic functions of the variables u, v, and they are called Abelian functions.

Theorem 2 (Weierstrass) *The rational symmetric functions of two points (ξ_1, η_1), (ξ_2, η_2) on the hyperelliptic curve (2) are meromorphic functions of the variables u, v.*

Weierstrass in his lectures in Berlin university [2] in 1870th years shown that the theory of θ-series can be generalized on the case of several variables and that Abelian functions can be represented as ration of two θ-series. In 1880th years it was clear that this theory is natural development of the theory of elliptic functions. In lectures of Weierstrass there is only the outline of the theory of the transcendental functions arising in the theory of Abelian functions.

Applications of hyperelliptic integrals in mechanics are obvious and numerous, we can see the integral (1) even in Philosophiae Naturalis Principia Mathematica of Newton. Abelian functions appear as a trick in the works of Jacobi and in the 1860s there was the theory of Abelian functions but there were no applications in physics.

Shortly after constructing the theory of Abelian functions, Weierstrass proposed to his students to find its applications in mechanics. H. Bruns investigated at this viewpoint the three bodies problem, he proved that this problem has no integrals which are expressed in algebraic functions but even in Abelian integrals [3, 4]. L. Königsberger and S.V. Kowalewski looked for the applications of abelian functions in mechanics with Weber forces [5] and in the theory of a gyroscope [6] respectively.

We can consider the equations of a gyroscope motion as the dynamical system

$$\dot{x} = h(x, A, \dots);$$

the left side h depends of several parameters $A, \dots$. Mechanics of XVIII century found the condition on the parameters under which this system is integrated in elliptic functions. Kowalewski found the condition under which the dynamical system integrated in Abelian functions, with both arguments being polynomial functions t, that is the solution of the dynamical system has the form

$$x = \mathrm{Al}\left(f(t), g(t)\right), \tag{5}$$

where Al is an abelian function, f and g are univariable polynomials. One such case was successfully found by Kowalewski in the 1880s years and it is now known as the case of Kowalewski of a gyroscope motion [6]. Kowaliwski gyroscope really can be made thus at the end of the XIX century there is the example of real dynamical system integrating in Abelian functions. However it wasn't clear when and why these functions appear in physical problems.

In the 1890s, Painlevé given some explanation of the reasons in the scope of his theory the transcendent functions [7]. He noticed that general solutions of differential equations solvable in commonly used functions are not only meromorphic functions of the independent variable, but also algebraic functions of the constants. Painlevé managed to invert this assertion for first- and second- order equations.

Theorem 3 (Painlevé) *If the general solution y of a given second-order equation*

$$f(x, \dot{x}, \ddot{x}; t) = 0,$$

depends algebraically on the integration constants, then this solution can be expressed algebraically in transcendental functions belonging to the following categories:

(i) abelian functions of the form $\mathrm{Al}(u, v)$ and their derivatives with respect to u and v where u and v can be expressed in terms of t by the quadratures

$$u = \int h(t)dt + C_1, \quad v = \int k(t)dt + C_2;$$

(ii) the elliptic functions $\wp(u + C)$ and $\wp'(u + C)$ where u can be expressed in terms of t by the following quadrature:

$$u = \int h(t)dt,$$

IOP Conf. Series: Journal of Physics: Conf. Series **937** (2017) 012027 doi:10.1088/1742-6596/937/1/012027

(iii) the solutions of the linear or Riccati equations.

Obviously, in case (i) we have solution in the form (5), which was postulated in the begin of the investigation of Kowalewski.

There are nontrivial connections between Painlevé property, integrability of systems and property of general solutions, described in the theorem [6]. Kowalewski's gyroscope has all three properties. However inverse square system of Calogero [8] has only two last properties. In other hand the differential equation

$$y'' = 6y^2 + x$$

has only Painlevé property, but its general solution depends on the integration constants non algebraically and thus it is non classical transcendental function know as Painlevé transcendents [10]. Furthermore simplified equations for differential equations with Painlevé property can be integrated in Abelian functions [9].

The theorem 3 is also important for computer algebra. The idea of solvability in finite terms is one of those obscure notions that are commonly understood to be a historical convention and are in essence the subject of a general agreement on what this term actually means. The theorem 3 leads us to an unexpected conclusion: by fixing the algebraic properties of the general solution one can derive a class of commonly used transcendental functions. In this viewpoint Painlevé theory was given in [11].

2. Motivation

Mathematicians of 19th century considered the theory of abelian functions as the necessary completion of mathematical analysis, but after WWI works in this theory have died away. So Felix Klein wrote in 1926 [12]:

> Als ich studierte, galten die Abelschen Funktionen (...) als der unbestrittene Gipfel der Mathematik, und jeder von uns hatte den selbstverstandlichen Ehrgeiz hier selbst weiterzukommen. Und jetzt? Die junge Generation kennt die Abelschen Funktionen kaum mehr.

Now there is a number of reference books on elliptic functions [13, 14], many systems of computer algebra like Maple and Mathematica successfully apply the elliptic functions for symbolical integration. At the same time there are no reference books and manuals, no packages in computer algebra systems for the work with abelian functions. It should be noted that old textbook of Baker [15] written before the publishing of Weierstrass lectures is still in the use.

For development of this theory we need test examples. As can be seen, the Kowalewski's gyroscope gives an illustration for the Painlevé theorem, but this illustration is extremely difficult computationally. Now we want present a relatively simple example of a dynamical system that is integrated in Abelian integrals by "pairing" two copies of a hyperelliptic curve. Unfortunately, initially simple formulas unfold into very long ones. Apparently the theory of Abelian functions did not evolve in the last century because without computer algebra systems it was impossible to compute the calculations to the end.

3. Pairing two copies of a hyperelliptic curve

For a start we construct Abelian variety that is algebraic surface which can be parametrized with the help of Abelian functions. We take two constant points $(a_1, b_1), (a_2, b_2)$ and two mobile points $(\xi_1, \eta_1), (\xi_2, \eta_2)$ on the hyperelliptic curve

$$C: \quad \eta^2 = \xi(\xi-1)(\xi-2)(\xi-3)(\xi-4)$$

and define the correspondence between these points and two variables u and v by equation (4). By theorem 2 symmetric functions

$$x = \xi_1 + \xi_2, \quad y = \xi_1\xi_2, \quad z = \eta_1 + \eta_2$$

are Abelian functions of two parameters u, v.

These functions define an algebraic surface

$$S(x, y, z) = 0, \quad S \in \mathbb{Z}[x, y, z]$$

Eliminating the variables $\xi_1, \xi_2, \eta_1, \eta_2$ from system

$$x = \xi_1 + \xi_2, \quad y = \xi_1\xi_2, \quad z = \eta_1 + \eta_2, \quad \eta_1^2 = P(\xi_1), \quad \eta_2^2 = P(\xi_2),$$

we can calculate the equation of this surface:

$$\begin{aligned}
&x^{10} - 20x^9 - 10x^8y + 170x^8 + 180x^7y + 35x^6y^2 - 800x^7 - 1360x^6y - 540x^5y^2\\
&-50x^4y^3 - 2x^5z^2 + 2273x^6 + 5600x^5y + 3400x^4y^2 + 600x^3y^3 + 25x^2y^4 + 20x^4z^2\\
&+10x^3yz^2 - 3980x^5 - 13686x^4y - 11100x^3y^2 - 2790x^2y^3 - 160xy^4 - 4y^5 - 70x^3z^2\\
&-80x^2yz^2 - 10xy^2z^2 + 4180x^4 + 20380x^3y + 19649x^2y^2 + 6000xy^3 + 280y^4 + 100x^2z^2\\
&+210xyz^2 + 40y^2z^2 + z^4 - 2400x^3 - 18400x^2y - 17840xy^2 - 5092y^3 - 48xz^2\\
&-200yz^2 + 576x^2 + 9600xy + 6720y^2 - 2304y = 0
\end{aligned} \tag{6}$$

This expression is very long thus it was never written out explicitly.

4. Cross-section $z = 0$

The most noticeable structures on the surface S lies on cross-section by the plane $z = 0$. In general the cross-section of S by the plane $z = a$ is irreducible plane curve of very large genus, however this cross-section is the union of two curves: the curve

$$Q: \quad x^4 - 10x^3 - 3x^2y + 35x^2 + 20xy + y^2 - 50x - 35y + 24 = 0$$

and the parabola $Q' : 4y = x^2$. These curves on the surface S have extremely small genus.

The curve Q is elliptic curve and thus its genus is equal only 1. Points of the curve Q are singularities of the surface S. We can check this in Sage by calculation of the radical ideal

$$\sqrt{\left(S, \frac{\partial S}{\partial x}, \frac{\partial S}{\partial y}, \frac{\partial S}{\partial z}\right)} = (z, x^4 - 10x^3 - 3x^2y + 35x^2 + 20xy + y^2 - 50x - 35y + 24).$$

For interpretation of the second curve we return to variable $\xi_1, \dots$. In these variables the equation $z = 0$ means that

$$\eta_1 = -\eta_2$$

and thus

$$P(\xi_1) = P(\xi_2).$$

So this cross-section consists of two lines, one of them is

$$\xi_1 = \xi_2, \; \eta_1 = -\eta_2,$$

that is Q'. By Abel theorem two points $p, p' \in Q'$ corespondent to one value of the parameter u, e.g.

$$u_0 = \int_{(a_1,b_1)}^{(\xi_1,\eta_1)} \frac{d\xi}{\eta} + \int_{(a_2,b_2)}^{(\xi_1,-\eta_1)} \frac{d\xi}{\eta} = \int_{(a_1,b_1)}^{(\xi_1',\eta_1')} \frac{d\xi}{\eta} + \int_{(a_2,b_2)}^{(\xi_1',-\eta_1')} \frac{d\xi}{\eta}$$

and one value of the parameter v. So the correspondence between u, v and points of S is one-to-one almost everywhere, but Q' is an exceptional curve [16].

IOP Conf. Series: Journal of Physics: Conf. Series **937** (2017) 012027 doi:10.1088/1742-6596/937/1/012027

5. Generator family on S

We can imagine the surface S, indicating a pencil of curves on S. If we fix the point (ξ_2, η_2), but will move the point (ξ_1, η_1) along the curve C, then formulas

$$x = \xi_1 + \xi_2, \quad y = \xi_1\xi_2, \quad z = \eta_1 + \eta_2,$$

define a curve on S. So there is correspondence between points of C and curves on S. We can describe these curves by explicit formulas

$$\begin{cases} x\xi_2 - \xi_2^2 - y = 0, \\ \xi_2^5 - 10\xi_2^4 + 35\xi_2^3 - 50\xi_2^2 - \eta_2^2 + 24\xi_2 = 0, \\ x^5 - 10x^4 - 5x^3y + 35x^3 + 40x^2y + 5xy^2 - 50x^2 - 105xy - 20y^2 - \\ -z^2 + 2z\eta_2 - 2\eta_2^2 + 24x + 100y = 0 \end{cases}$$

This means that the curve which correspond to the point $(\xi_2, \pm\eta_2)$ lies in the plane

$$x\xi_2 - \xi_2^2 - y = 0.$$

Thus the surface S is weaved by these plane curves (as hyperboloid is weaved by right lines). This curves form an irrational pencil on the surface [16].

6. Differentials

Any rational symmetric function of two points $(\xi_1, \eta_1), (\xi_2, \eta_2)$ is rational function on S and vice versa. These functions are meromorphic with respect to the variables u, v.

On the surface S there are two linear independent holomorphic differentials, namely

$$du = \frac{d\xi_1}{\eta_1} + \frac{d\xi_2}{\eta_2} \qquad = p_{11}dx + p_{12}dy,$$
$$dv = \frac{\xi_1 d\xi_1}{\eta_1} + \frac{\xi_2 d\xi_2}{\eta_2} \qquad = p_{21}dx + p_{22}dy,$$

For calculating of p_{ij} it should be noted that

$$\begin{cases} dx = d\xi_1 + d\xi_2, \\ dy = \xi_1 d\xi_2 + \xi_2 d\xi_1, \end{cases}$$

so

$$du = \frac{1}{\xi_1 - \xi_2}\left[\frac{\xi_1 dx - dy}{\eta_1} - \frac{\xi_2 dx - dy}{\eta_2}\right]$$

and

$$dv = \frac{1}{\xi_1 - \xi_2}\left[\frac{\xi_1^2 dx - \xi_1 dy}{\eta_1} - \frac{\xi_2^2 dx - \xi_2 dy}{\eta_2}\right].$$

By elimination technique we can calculate in Sage the explicit expression for p_{ij}. For example p_{11} is ratio of

$$-x^5 + 10x^4 + 3x^3y - 35x^3 - 20x^2y - xy^2 + 50x^2 + 35xy + z^2 - 24x$$

and

$$(-x^5 + 10x^4 + 5x^3y - 35x^3 - 40x^2y - 5xy^2 + 50x^2 + 105xy + 20y^2 + z^2 - 24x - 100y)z.$$

The restriction du and dv to the general curve K on the surface S give two linear independent holomorphic differentials for the curve K, thus genus of K is more or equal to 2. This fact doesn't contradict existence of the curves Q and Q'. For example the numerator and the denominator of p_{11} are equal to 0 on Q.

7. Dynamical system

Let $\mathrm{Al}(u, v)$ be meromorphic function with 4 periods. Then

$$\mathrm{Al}\left(\int p_{11}dx + p_{12}dy, \int p_{21}dx + p_{22}dy\right)$$

is rational function on the surface S.

Thus integration of dynamical system

$$\begin{cases} p_{11}(x,y,z)\dot{x} + p_{12}(x,y,z)\dot{y} = f(t) \\ p_{21}(x,y,z)\dot{x} + p_{22}(x,y,z)\dot{y} = g(t) \end{cases}$$

with arbitrary functions $f(t)$ and $g(t)$ introduces transcendental functions of the form

$$\mathrm{Al}\left(\int f(t)dt, \int g(t)dt\right).$$

Theorem 4 *Integration of the dynamical system*

$$\begin{cases} \dot{x} = \det\begin{pmatrix} f(t) & p_{12} \\ g(t) & p_{22} \end{pmatrix} / \det\begin{pmatrix} p_{11} & p_{12} \\ p_{21} & p_{22} \end{pmatrix} \\ \dot{y} = \det\begin{pmatrix} p_{11} & f(t) \\ p_{21} & g(t) \end{pmatrix} / \det\begin{pmatrix} p_{11} & p_{12} \\ p_{21} & p_{22} \end{pmatrix} \end{cases}$$

on the surface S introduces two expressions of the form

$$\mathrm{Al}\left(\int f(t)dt, \int g(t)dt\right).$$

In mechanical problems the equation $S(x, y, z) = 0$ has the meaning of conservation law.

8. Conclusion

In last sections we have constructed the test-example of dynamical system integrating in Abelian functions and satisfying to the theorem 3 of Painleé. For use in practice we need to solve the inverse problem. We have a system of ODEs on some surface, how we can understand that this system can be integrated in Abelian functions?

Unfortunately modern CAS (lake Maple) can't find:

(i) the group of automorphisms even for curves,

(ii) holomorphic differentials on the surface, in Maple we can do this only for curves, and

(iii) families of curves on the surface with some extremal properties like generator family.

Perhaps special curves Q and Q' are a key to identification at least in practice.

Acknowledgments

The calculations in this article are performed in the computer algebra system Sage [17]. The publication was prepared with the support of the "RUDN University Program 5-100". The work was partially supported by RFBF grants No 15-07-08795, No 16-07-00556.

References

[1] Markushevich I A 2006 Introduction to the classical theory of Abelian functions. *Translation of mathematical monographs* **96** (AMS)
[2] Weierstrass K 1902 *Mathematische Werke von Karl Weierstrass* vol 4 (Berlin: Mayer&Müller)
[3] Bruns H 1887 Über die Integrale des Vielkörper-Problems *Acta. Math.* **11** 25–96
[4] Painlevé P 1975 *Œuvres de Paul Painlevé* vol 3 (Paris) pp. 666–699.
[5] Koenigsberger L 1901 *Die Principien der Mechanik* (Leipzig: Teubner)
[6] Golubev V V 1960 *Lectures on integration of the equations of motion of a rigid body about a fixed point* (Jerusalem: Israel program for scientific translations)
[7] Painlevé P 1972 *Œuvres de Paul Painlevé* vol 1 (Paris)
[8] Moser J 1978 Various aspects of integrable Hamiltonian systems. *Dynamical Systems. CIME Summer Schools* **78** (Springer)
[9] Garnier R 1918 Sur une classe de systmes differentiels abliens deduits de la thorie des quations linaires. *Rend. Circ. Mat. Palermo* **43** 155–191
[10] Umemura H 1990 Birational automorphism groups and differential equations *Nagoya Math J.* **119** 1–80
[11] Malykh M D 2015 On transcendental functions arising from integrating differential equations in finite terms *Journal of Mathematical Sciences* **209** (6) 935–952
[12] Klein F 1979 *Vorlesungen über die Entwicklung der Mathematik im 19. Jahrhundert* vol 1 (Springer)
[13] Byrd P F and Friedman M D 1971 *Handbook of elliptic integrals for engineers and scientists* (2nd ed., Springer)
[14] NIST Digital library of mathematical functions (`http://dlmf.nist.gov`)
[15] Baker H F 1996 *Abelian functions. Abel's theorem and the allied theory of theta functions* (2nd ed. Cambridge University Press, Cambridge)
[16] Enriques F 1949 Le superficie algebriche (Bologna)
[17] *SageMath, the Sage Mathematics Software System (Version 7.5.1)*, The Sage Developers, 2017, `http://www.sagemath.org`.

Axisymmetric contact between a rigid punch and a coated foundation with rough surfaces

A V Manzhirov[1,2,3,4*] **and K E Kazakov**[1,2**]

[1]Ishlinsky Institute for Problems in Mechanics of the Russian Academy of Sciences, Moscow, Russia
[2]Bauman Moscow State Technical University, Moscow, Russia
[3]National Research Nuclear University MEPhI, Moscow, Russia
[4]Moscow Technological University, Moscow, Russia

E-mail: *manzh@inbox.ru, **kazakov-ke@yandex.ru

Abstract. Axisymmetric contact problem for a coated foundation and a rigid punch is considered. It is assumed that the shape of the thin upper layer (coating) and the shape of the punch surface can be described by different rapidly changing functions. We obtain basic integral equation for this problem and additional condition. A projection method is developed for this problem. It allows one to obtain the solution of the equation with high accuracy that cannot be done by known methods.

1. Introduction

Some plane and axisymmetric contact problems for bodies with coatings were considered in [1–5]. It were some investigations of plane and axisymmetric problems for conformal contact between layered foundations and punches, contact problems for nonuniform foundations and rough punches. This paper devoted to the contact problem for foundation with homogeneous coating of variable thickness and annular punch with complex base forms for a first time.

2. Statement of the contact problem

We assume that a viscoelastic aging layer of an arbitrary thickness H with a thin coating made of another viscoelastic aging material of an thickness $h(r)$ lies on a rigid basis. We denote the moments of coating and lower layer production by τ_1 and τ_2, respectively. We also assume that the coating rigidity is less than the rigidity of the lower layer or they are of the same order of magnitude. There is smooth contact or perfect contact between layers and between the lower layer and the rigid base.

At time τ_0, the axial force $P(t)$ starts to indent a rigid annular punch of inner radius a and outer radius b into the surface of such a foundation. The function $g(r)$ describes distance between contact surfaces in nondeformable state and called backlash function. If the function $f(r)$ describes the form of the punch then $g(r) = f(r) - h(r) - h_0$, where $h_0 = \min_{r\in[a,b]}[f(r) - h(r)]$. In the case of conformal contact $g(r) \equiv 0$. Contact area is constant: it is a ring of inner radius a and outer radius b. The coating is assumed to be thin compared with the width of contact area, i.e., its thickness satisfies the condition $h(r) \ll (b - a)$.

IOP Conf. Series: Journal of Physics: Conf. Series **937** (2017) 012028 doi:10.1088/1742-6596/937/1/012028

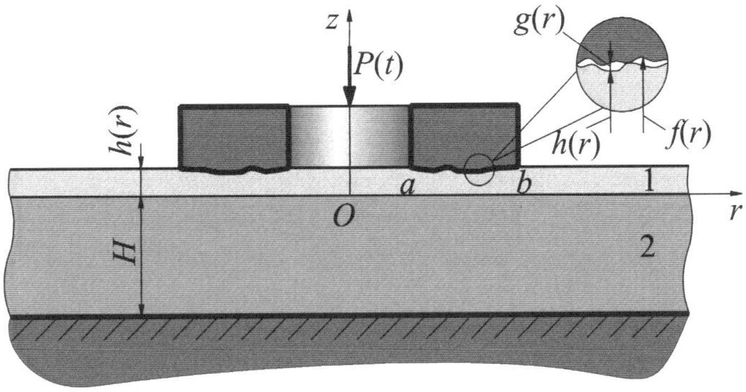

Figure 1. Axisymmetric contact interaction

As a result of this interaction, the punch is immersed in the base by the quantity $\delta(t)$. The contact interaction scheme is shown in figure 1.

The integral equations and additional condition for described problem has a form [6, 7]

$$\theta\left[\frac{q(r,t)h(r)}{E_1(t-\tau_1)} - \int_{\tau_0}^{t}\frac{q(r,\tau)h(r)}{E_1(\tau-\tau_1)}K_1(t-\tau_1,\tau-\tau_1)\,d\tau\right]$$
$$+\frac{2(1-\nu_2^2)}{H}\left[\int_a^b k_{\rm ax}\left(\frac{r}{H},\frac{\rho}{H}\right)\frac{q(\rho,t)}{E_2(t-\tau_2)}\rho\,d\rho\right.$$
$$\left.-\int_{\tau_0}^{t}K_2(t-\tau_2,\tau-\tau_2)\int_a^b k_{\rm ax}\left(\frac{r}{H},\frac{\rho}{H}\right)\frac{q(\rho,\tau)}{E_2(\tau-\tau_2)}\rho\,d\rho\,d\tau\right] = \delta(t)-g(r), \tag{1}$$

$$2\pi\int_a^b q(\rho,t)\rho\,d\rho = P(t),\quad K_k(t,\tau) = E_k(\tau)\frac{\partial}{\partial\tau}\left[\frac{1}{E_k(\tau)}+C_k(t,\tau)\right],\quad r\in[a,b],\quad t\ge\tau_0, \tag{2}$$

where $\delta(t)$ is the punch settlement; $E_k(t)$ are the Young moduli of the coating ($k=1$) and the lower layer ($k=2$) and ν_2 is Poisson's ratio of the lower layer; $\mathbf{I}$ is the identity operator; $K^{(k)}(t,\tau)$ ($k=1,\,2$) are tensile creep kernels; $C^{(k)}(t,\tau)$ ($k=1,\,2$) are the tensile creep functions; θ is a dimensionless coefficient depending on the contact conditions between coating and lower layer; in the case of a smooth coating-layer contact, we have $\theta=1-\nu_1^2$, and in the case of an perfect contact, $\theta=(1-\nu_1-2\nu_1^2)/(1-\nu_1)$, where ν_1 is Poisson's ratio of the coating; $k_{\rm ax}(r/H,\rho/H)$ is known kernel of the plane contact problem, which has the form

$$k_{\rm ax}(r,\rho)=\int_0^\infty L(u)J_0(ru)J_0(\rho u)\,du,$$

and, in the case of a smooth contact between the lower layer and the rigid base, $L(u)=(\cosh 2u-1)/(\sinh 2u+2u)$, and in the case of a perfect contact, $L(u)=(2\varkappa\sinh 2u-4u)/(2\varkappa\cosh 2u+4u^2+1+\varkappa^2)$ ($\varkappa=3-4\nu_2$) (see, for example, [8]).

If we make a change of variables according to the formulas

$$
\begin{aligned}
&(r^*)^2 = \frac{r^2-a^2}{b^2-a^2}, \quad (\rho^*)^2 = \frac{\rho^2-a^2}{b^2-a^2}, \quad t^* = \frac{t}{\tau_0}, \quad \tau_k^* = \frac{\tau_k}{\tau_0}, \quad \lambda = \frac{H}{b-a}, \quad \eta = \frac{a}{b-a},\\
&\zeta^2 = \frac{b+a}{b-a}, \quad \delta^*(t^*) = \frac{\delta(t)}{b-a}, \quad g^*(r^*) = \frac{g(r)}{b-a}, \quad m^*(r^*) = \frac{\theta}{1-\nu_2^2}\frac{h(r)}{2(b-a)},\\
&c^*(t^*) = \frac{E_2(t-\tau_2)}{E_1(t-\tau_1)}, \quad q^*(r^*,t^*) = \frac{2q(r,t)(1-\nu_2^2)}{E_2(t-\tau_2)}, \quad P^*(t^*) = \frac{P(t)(1-\nu_2^2)}{\pi E_2(t-\tau_2)(b^2-a^2)},\\
&\mathbf{F}^* f(r^*) = \int_0^1 k_{\text{ax}}^*(r^*,\rho^*) f(\rho^*)\rho^*\,d\rho^*, \quad \mathbf{V}_k^* f(t^*) = \int_1^{t^*} K_k(t^*,\tau^*) f(\tau^*)\,d\tau^*,\\
&k_{\text{ax}}^*(r^*,\rho^*) = \frac{\zeta^2}{\lambda} k_{\text{ax}}\left(\frac{r}{H},\frac{\rho}{H}\right), \quad K^{1*}(t^*,\tau^*) = \frac{E_1(t-\tau_1)E_2(\tau-\tau_2)}{E_1(\tau-\tau_1)E_2(t-\tau_2)} K_1(t-\tau_1,\tau-\tau_1)\tau_0,\\
&K^{2*}(t^*,\tau^*) = K_2(t-\tau_2,\tau-\tau_2)\tau_0,
\end{aligned} \tag{3}
$$

then we obtain the following mixed integral equation with the additional condition

$$
\begin{aligned}
&c(t)m(r)(\mathbf{I}-\mathbf{V}_1)q(r,t) + (\mathbf{I}-\mathbf{V}_2)\mathbf{F}q(r,t) = \delta(t) - g(r),\\
&\int_0^1 q(\rho,t)\rho\,d\rho = P(t), \quad r \in [0,1], \quad t \geq 1.
\end{aligned} \tag{4}
$$

Hereinafter we will omit the asterisks in formulas with dimensionless variables.

It easy to see that there exist two versions of the substitution: 1) the settlement of the punch is known (i.e., the right-hand side of the integral equation is given); 2) the indenting force is known. We will construct the solution for the case with known force $P(t)$ and unknown settlement $\delta(t)$ and contact pressures $q(r,t)$. To this end we will use projection method described in [9].

3. Solution for version with known indenting force

We will find the solution in the form

$$
q(r,t) = \frac{Q(r,t)}{\sqrt{m(r)}} - \frac{g(r)}{m(r)}(\mathbf{I}-\mathbf{V}_1)^{-1}\frac{1}{c(t)}, \tag{5}
$$

where $Q(r,t)$ is new unknown function. Then the integral equation and auxiliary condition (4) can be reduced to the following integral equation with the Hilbert–Schmidt kernel $k(x,\xi)$ [10, 11]:

$$
\begin{aligned}
&c(t)(\mathbf{I}-\mathbf{V}_1)Q(r,t) + (\mathbf{I}-\mathbf{V}_2)\mathbf{A}Q(r,t) = \frac{\delta(t)}{\sqrt{m(r)}} + \frac{\tilde{c}(t)\tilde{g}(r)}{\sqrt{m(r)}},\\
&\int_0^1 \frac{Q(\rho,t)}{\sqrt{m(\rho)}}\rho\,d\rho = \tilde{P}(t), \quad r \in [0,1], \quad t \geq 1,
\end{aligned} \tag{6}
$$

where

$$
\begin{aligned}
&\mathbf{A}f(r) = \int_0^1 k(r,\rho)f(\rho)\rho\,d\rho, \quad k(r,\rho) = \frac{k_{\text{ax}}(r,\rho)}{\sqrt{m(r)m(\rho)}}, \quad \tilde{g}(r) = \int_0^1 \frac{k_{\text{ax}}(r,\rho)g(\rho)}{m(\rho)}\rho d\rho,\\
&\tilde{c}(t) = (\mathbf{I}-\mathbf{V}_2)(\mathbf{I}-\mathbf{V}_1)^{-1}\frac{1}{c(t)}, \quad \tilde{P}(t) = P(t) + (\mathbf{I}-\mathbf{V}_1)^{-1}\frac{1}{c(t)}\int_0^1 \frac{g(\rho)}{m(\rho)}\rho\,d\rho.
\end{aligned}
$$

We seek the solution $Q(r,t)$ in the class of functions continuous in time t in the Hilbert space $L_2(0,1)$. To this end, we at first construct an orthonormal system of functions $\{p_0(r), p_1(r), p_2(r), \ldots\}$ in $L_2(0,1)$ which contains $1/\sqrt{m(r)}$ and remaining basis functions can be written as the products of functions depending on r and weight function $1/\sqrt{m(r)}$. So we will orthonormalize the system of linearly independent functions $\{1/\sqrt{m(r)}, r/\sqrt{m(r)}, r^2/\sqrt{m(r)}, \ldots\}$ in $L_2(0,1)$.

The Hilbert space $L_2(0,1)$ can be represented as the direct sum of orthogonal subspaces $L_2(0,1) = L_2^{(0)}(0,1) \oplus L_2^{(1)}(0,1)$, where $L_2^{(0)}(0,1)$ is the Euclidean space with basis $\{p_0(r)\}$ and $L_2^{(1)}(0,1))$ is the Hilbert space with basis $\{p_1(r), p_2(r), \ldots\}$. The integrand and the right-hand side of (6) can also be represented as the algebraic sum of functions continuous in time t and ranging in $L_2^{(1)}(0,1)$ and $L_2^{(2)}(0,1)$, respectively, i.e.,

$$Q(r,t) = Q_0(r,t) + Q_1(r,t), \quad \delta(t) + \tilde{c}(t)\tilde{g}(r) = \Delta_0(r,t) + \Delta_1(r,t),$$

where $Q_0(r,t)$, $\Delta_0(r,t) \in L_2^{(0)}(0,1)$, $Q_1(r,t)$, $\Delta_1(r,t) \in L_2^{(1)}(0,1)$ and

$$\begin{gathered} Q_0(r,t) = z_0(t)p_0(r), \quad \Delta_0(r,t) = \left[\sqrt{\tfrac{1}{2}}\delta(t) + \tilde{g}_0\tilde{c}(t)\right]p_0(r), \quad \Delta_1(r,t) = \tilde{c}(t)\tilde{g}_1(r), \\ \tilde{g}_0 = \sum_{l=0}^{\infty} K_{0l}\int_0^1 p_l(\rho)g(\rho)\rho\,d\rho, \quad \tilde{g}_1(r) = \tilde{g}(r) - \tilde{g}_0 p_0(r), \\ K_{ml} = \int_0^1\int_0^1 k(r,\rho)p_m(r)p_l(\rho)r\rho\,dr\,d\rho, \quad k(r,\rho) = \sum_{m=0}^{\infty}\sum_{l=0}^{\infty} K_{ml}p_m(r)p_l(\rho). \end{gathered} \tag{7}$$

The term $Q_0(r,t) \in L_2^{(0)}(0,1)$ is determined by the additional condition (6):

$$Q_0(r,t) = z_0(t)p_0(r), \quad z_0(t) = \sqrt{2}P(t), \tag{8}$$

and the term $\Delta_1(r,t) \in L_2^{(1)}(0,1)$ is known.

After introducing the orthogonal projection operators $\mathbf{P}_0$ and $\mathbf{P}_1$, mapping the space $L_2(0,1)$ onto $L_2^{(0)}(0,1)$ and $L_2^{(1)}(0,1)$, respectively, we apply the operator $\mathbf{P}_1$ to integral equation (6). As a result, we obtain the equation in $L_2^{(1)}(0,1)$ with a known right-hand side for determining $Q_1(r,t)$:

$$c(t)(\mathbf{I} - \mathbf{V}_1)Q_1(r,t) + (\mathbf{I} - \mathbf{V}_2)\mathbf{P}_1\mathbf{F}Q_1(r,t) = -(\mathbf{I} - \mathbf{V}_2)\mathbf{P}_1\mathbf{F}Q_0(r,t) + \tilde{c}(t)\tilde{g}_1(r). \tag{9}$$

It is necessary to construct its solution in the form of a series in the eigenfunctions of the operator $\mathbf{P}_1\mathbf{F}$:

$$Q_1(r,t) = \sum_{k=1}^{\infty} z_k(t)\varphi_k(r). \tag{10}$$

The system of eigenfunctions $\{\varphi_k(r)\}_{k=1,2,3,\ldots}$ of such an operator is a basis in the space $L_2^{(1)}(0,1)$. The spectral problem for the operator $\mathbf{P}_1\mathbf{F}$ can be written in the form

$$\mathbf{P}_1\mathbf{F}\varphi_k(r) = \gamma_k\varphi_k(r), \quad \varphi_k(r) = \sum_{m=1}^{\infty}\psi_{km}p_m(r), \quad k = 1,2,3,\ldots \tag{11}$$

Using formulas (7)–(10) we can obtain expressions for determining expansion functions $z_k(t)$

$(k = 1, 2, 3, \ldots)$:

$$z_k(t) = (\mathbf{I} + \mathbf{W}_k)\frac{(\mathbf{I} - \mathbf{V}_2)[g_k(\mathbf{I} - \mathbf{V}_1)^{-1}c^{-1}(t) - K_k z_0(t)]}{c(t) + \gamma_k},$$

$$K_k = \sum_{m=1}^{\infty} K_{0m}\psi_{km}, \quad g_k = \sum_{m=1}^{\infty} \psi_{km} \sum_{l=0}^{\infty} K_{ml} \int_0^1 p_l(\rho)g(\rho)\rho\, d\rho,$$

$$\mathbf{W}_k f(t) = \int_1^t R_k^*(t,\tau) f(\tau)\, d\tau, \quad k = 2,\, 3,\, \ldots,$$

where $R_k^*(t,\tau)$ $(k = 1, 2, 3, \ldots)$ are resolvents of the kernel $K_k^*(t,\tau) = [c(t)K^1(t,\tau) + \gamma_k K^2(t,\tau)]/[c(t) + \gamma_k]$.

the resulting solution has the following structure

$$q(r,t) = \frac{1}{m(r)}\left\{z_0(t)P_0(r) + \sum_{m=1}^{\infty}\left[\sum_{k=1}^{\infty}\psi_{km}z_k(t)\right]P_m(r)\right\} - \frac{g(r)}{m(r)}(\mathbf{I} - \mathbf{V}_1)^{-1}\frac{1}{c(t)},$$

where the expression in braces is smooth ($P_i(r)$ is polynomial of degree i, $p_i(r) = P_i(r)/\sqrt{m(r)}$). One can explicitly write out the weight functions $m(r)$ and $g(r)$ in the solution. Note that the coating thickness is related to $m(r)$ and backlash function is related to $g(r)$. The formulas obtained permit obtaining efficient analytic solutions for the layers with rough coatings which can be described by complicated and rapidly oscillating functions. Such a result can hardly be done by other known methods.

To find the unknown settlement, we must apply the operator $\mathbf{P}_0$ to integral equation (6). Then the formula for the settlement $\delta(t)$ has a form:

$$\delta(t) = \sqrt{2}\left\{c(t)(\mathbf{I} - \mathbf{V}_1)z_0(t) + (\mathbf{I} - \mathbf{V}_2)\left[K_{00}z_0(t) + \sum_{k=1}^{\infty} K_k z_k(t) - g_0(\mathbf{I} - \mathbf{V}_1)^{-1}\frac{1}{c(t)}\right]\right\}.$$

Conclusions

- We pose and solve axisymmetric problems of contact between viscoelastic aging foundations with rough coatings and rigid punches in the case, where the punch base surface and coating width are described by different rapidly changing functions. To this end we develop general projection method.
- The solution of the problem is obtained analytically, and, in the expressions for the contact stresses, the coating width and backlash function is distinguished explicitly, which allows one to perform computations for actual shapes of the coating surface, which are described by rapidly changing functions.
- The explicit formulas has been obtained for the punch settlement.
- Other known methods for the solution of this problem diverge with an increase in the time parameter or yield an error of up to 100% in determining contact stresses.

Acknowledgments

This work was financially supported by the Russian Science Foundation under Project No. 17-19-01257.

References

[1] Kazakov K E 2007 Contact problems for foundations with coatings *Vestnik Samar. Gos. Univ. Estestvennonauchn. Ser.* No. 4(54) 176–96

[2] Kazakov K E and Manzhirov A V 2008 Conformal contact between layered foundations and punches *Mech. Solids.* **43** (3) 512–24

[3] Manzhirov A V and Kazakov K E 2016 Contact problem for a foundation with a rough coating *Lect. Notes Engng Comp. Sci.* **2224** (1) 877–82

[4] Kazakov K E and Manzhirov A V 2017 Axisymmetric contact between an annular rough punch and a surface nonuniform foundation *Mech. Solids* **54** (4) 444–51

[5] Manzhirov A V and Kazakov K E 2017 Modeling of the contact interaction between a nonuniform foundation and rough punch *Mat. Model.* **29** (10) 95–104

[6] Manzhirov A V 1983 Axisymmetric contact problems for non-uniformly aging layered viscoelastic foundations *J. Appl. Math. Mech.* **47** (4) 558–66

[7] Manzhirov A V 1985 On a method of solving two-dimensional integral equations of axisymmetric contact problems for bodies with complex rheology *J. Appl. Math. Mech.* **49** (6) 777–82

[8] Alexandrov V M and Mkhitaryan S M 1983 *Contact Problems for Bodies with Thin Coatings and Interlayers* (Moscow: Nauka) [in Russian]

[9] Manzhirov A V 2016 A mixed integral equation of mechanics and a generalized projection method of its solution *Dokl. Phys.* **61** (10) 489–93

[10] Lyubich Yu I 1992 *Functional Analysis I: Linear Functional Analysis (Encyclopaedia of Mathematical Sciences, vol 19)* (Berlin, Heidelberg, New York: Springer Verlag)

[11] Kolmogorov A N and Fomin S V 1999 *Elements of the Theory of Functions and Functional Analysis* (New York: Dover Publications).

Dynamic design and control of a high-speed pneumatic jet actuator

S Yu Misyurin[1,2], V I Ivlev[2], G V Kreinin[2]

[1] National Nuclear Research University MEPhI, Moscow, Russia
[2] Blagonravov Institute of Machine Science, RAS, Moscow, Russia

ssmmrr@mail.ru

Abstract. Mathematical model of an actuator, consisting of a pneumatic (gas) high-speed jet engine, transfer mechanism, and a control device used for switching the ball valve is worked out. The specific attention was paid to the transition (normalization) of the dynamic model into the dimensionless form. Its dynamic simulation criteria are determined, and dynamics study of an actuator was carried out. The simple control algorithm of relay action with a velocity feedback enabling the valve plug to be turned with a smooth nonstop and continuous approach to the final position is demonstrated

1. Introduction

Besides the all others two factors complicate the analysis and optimization of the mechatronic system. These factors are the complexity of the basic mathematical model and the need to take into consideration the interaction of all subsystems. Gradually increasing of the model complexity together with its transition to the normalized form (that is to a dimensionless variables and simularity criterion, performing the function of the parameters and objectives) essentially facilitates the exploration process.

2. Actuator

For the model normalization the scales for the variables transition must be defined. The scales definition is an individual problem, especially when so called "multi-physics modeling and optimization multi-body system" [1,2] with the sub-systems of the different physical nature is executed. Such situation is typical for the majority of the mechatronic systems.

In the work noted, by reason of the complexity of the objects the optimization process considered is attached mainly to one of the subsystems, namely to the control subsystem. However the rest parts may have a significant influence upon the system behavior. That may be illustrated by the optimization of the point-to-point motion system [3-4, 8]. In such systems besides the positional accuracy a certain requirements to the motion mode of the output link is to be ensured. In [3] the general model of the system was transformed first to the lowest reduced form and normalized. That gives an opportunity to analyze all subsystems in there interaction and to find out the parameters of the actuator, transfer mechanism and control block to a first approximation. As a model complexity arises, the parameters correction is carrying out. The increase of the model complexity is carried out by taking into account actuators mechanical characteristic and stiffness, the control algorithm development, the time lag of the controller and the distributor.

It is of interest the similarity criteria set obtained as a result of the model normalization, in particular the inertial criterion. The latter as the main factor determines a motion mode of the output link, a travel time etc. In [4] the inertial criterion was employed in the optimization of the positioning system with a hydraulic drive and digital control subsystem.

The mechatronic system analyzed in this paper consists of the jet acting drive (jet engine), the composite transfer mechanism with variable transfer ratio and the control subsystem (Fig.1). Actuator with the jet pneumatic engine is intended for switching gas ball valves [5-8]. Compressed air (or gas) from the autonomous source or supply line comes through a distributing device to a hollow shaft of the engine rotor and then through a hollow arm of the rotor to nozzles located at the rotor periphery. Gas jets ejecting from the nozzles create reactive forces and driving torque on the rotor.

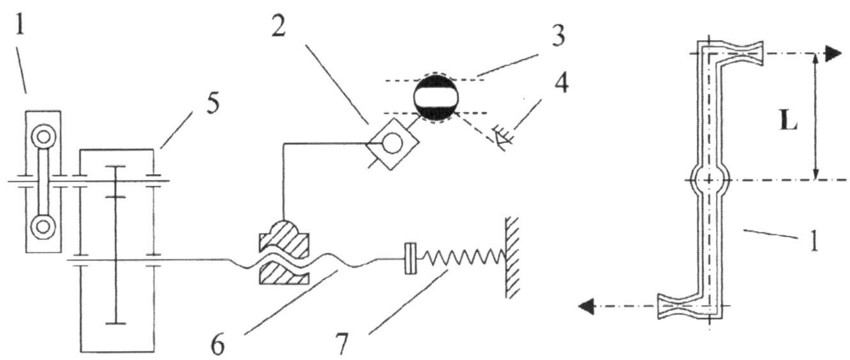

Figure 1. Ball valve drive: *1* - engine, *2* - link mechanism, *3* - ball valve, *4* - stroke limiter, *5* - reducing gear, *6* - screw mechanism, *7* - damper.

3. Conclusion

The offered normalized model of the ball valve actuator gives the opportunity to appreciate the actuator dynamics and to appoint its design parameters. It was established that the operating rate of the actuator output link is not too high but suffice for the employment of the actuator as the standby unit providing the safe pipe line shutdown in emergency situation in case of failure of the main drive.

Using the Painlev'e test one can study the analytical properties as was suggested in reference [9,10]

This work was supported by the Russian Foundation for Basic Research, in the framework of a research project 16-29-04401 "Development of methods and algorithms for the synthesis of drive complexes (robots) working in different environments, taking into account the interaction of two or more robots."

This work was supported by the MEPhI Academic Excellence Project (agreement with the Ministry of Education and Science of the Russian Federation of August 27, 2013, project no. 02.a03.21.0005).

References

[1] Samin J.C., Bruls O. and other. Multyphisics modeling and optimization multybody system. //DOI 10.1007/s 11044-007-9076-0 Multibody System Dynamics.

[2] Cotsaftis M., Keskinen E., Smooth High Precision Contact Position Control of Rotating Cylinders with Hydraulic Actuators. Proceeding of the 12th IFToMM World Congress, Besancon (France), June18-21, 2007.

[3] Kreinin G.V., Misyurin S.Yu. A Systematic Approach to Synthesis of a Drive System. ISSN 1052-6188, Journal of Machinery Manufacture and Reliability, 2011. Vol. 40, No. 6, pp. 507-511.

[4] Kreinin G.V., Misyurin S.Yu Parameters Synthesis of the Hydraulic Positioning System With The Constant Force Load // Problems of Mechanics (scientific journal of IFToMM). 2012, №3(48) c.32-37.

[5] Belousov, V.A., Nabiev, R.M., and Sayapin, V.V., *Sredstva upravleniya zaporno reguliruyushchei armaturoi truboprovodov* (Control Means for Shut off and Controlling Pipelines Equipment), St. Petersburg: Nedra, 2006, 150.
[6] Sayapin, V.V., *Privody so siruinym dvigatelem dlya zaporno reguliruyushchei apparatuiy truboprovodov* (Drives with Fluid Motor for the Shut off and Controlling Pipelines Equipment), Moscow: Tekhinform, 2003, 140.
[7] Sayapin, V.V. and Ovchinnikov, A.A., Pneumatic Emergency Drive for the Shut off and Controlling Equipment for Oil and Gas Pipelines, *Privodn. Tekhn.,* 1999, nos. 3-4, pp. 12-13.
[8] Kreinin G.V., Misyurin S.Yu. Choice of the Law for a Position Control System. ISSN 1052-6188, Journal of Machinery Manufacture and Reliability, 2012. Vol. 41, No. 4, pp. 311-336.
[9] N. A. Kudryashov; A. S. Zakharchenko. APPLIED MATHEMATICS LETTERS Volume: 32 Pages: 53-56 Published: JUN 2014.
[10] N. A. Kudryashov, M. V. Demina. APPLIED MATHEMATICS AND COMPUTATION Volume: 210 Issue: 2 Pages: 551-557 Published: APR 15 2009.

Numerical Analysis of the Elastic-Plastic Boundaries in the Thermal Stresses Theory Frameworks

E V Murashkin[1,2,3], E P Dats[4], V V Klindukhov[1]

[1] National Research Nuclear University "MEPhI" (Moscow Engineering Physics Institute), Kashirskoye shosse 31, Moscow, 115409, Russia
[2] Ishlinsky Institute for Problems in Mechanics of the Russian Academy of Sciences, Vernadsky Ave 101 Bldg 1, Moscow, 119526, Russia
[3]Bauman Moscow State Technical University, 2nd Baumanskaya Str 5/1, Moscow, 105005, Russia
4 Vladivostok State University of Economics and Service, Gogolya str. 41, Vladivostok, 690014, Russia

E-mail: evmurashkin@gmail.com

Abstract. The present study is devoted to the problem of the thermal stresses theory for a hollow thermo-elastic-plastic ball under unsteady heating. Throughout the paper the conventional elastic-plastic Prandt-Reuss model generalized on the thermal effects is used with the Tresca yield criterion linearly depending on temperature. A numerical-analytical algorithm for calculating the elastic-plastic boundaries positions and irreversible deformations is proposed. The strain-stress states during the plastic flow, repeated plastic flow and unloading are analyzed. The strain-stress state parameters like residual stresses and displacements are computed and analyzed.

1. Introduction

The prediction of residual stresses caused by uneven heating of elastoplastic material is one of the most important modern engineering problem. The main approach here is the solutions taking into account the plastic properties of materials under thermal action with the assumption of the independence of the temperature field from deformations [1-17]. Some results of the similar problems were used in the studies concerning residual stress calculations within the surface growth theory frameworks in additive manufacturing technologies (e.g., see [18, 19]) and geomechanics (e.g., see [20]). A number of problems were investigated for a stationary distributed temperature field. These solutions were constructed by comparison with isothermal elastoplastic problems solutions. New qualitative solutions can be obtained within the frameworks of the temperature stresses theory when the stresses occurs in virtue of a non-stationary thermal field gradient. The deformed state of the material can simultaneously satisfy to conditions of plastic flow and unloading in different domains. Elastoplastic boundaries separating such domains of irreversible deformation appear and disappear, moving in the direction of the thermal gradient increasing. The dependence of the yield stress on temperature leads to the appearance of high levels of residual deformations that can cause the reverse plastic flow during unloading. The study of these problems allow us to more accurately predict the distribution of residual stresses and deformations during high temperature metal forming. The processes of arising, development and disappearance of plastic flows during heating cause the necessity of elastoplastic boundaries localisation and irreversible (residual) deformations computing within unloading domains. The problems when the elastoplastic boundaries depend on the deformation process and are completely determined by the known temperature field are discussed in depth in [3-6,13,14]. They consider the problems of heating symmetric elastoplastic solids [3-6] and plates [2]. In the present paper the residual deformation within the unloading domain is derived by an envelope of the possible plastic deformations family within the plastic flow domain undepending on the residual deformations. The solution of the problem of non-stationary heating of hollow elastoplastic solids is more complicated. In this case, the boundaries position depends on the level of residual deformations and must be determined due to a system of integro-differential equations defining the conditions for the stress-strain state parameters continuity. In this paper, a numerical-analytical technique for determining the elastoplastic boundaries spatial localisation in a one-dimensional problem of heating a hollow elastoplastic ball is proposed. It allows one to calculate the positions of the elastoplastic boundaries and the distribution of plastic deformations under conditions of a linear dependence of the yield stress on temperature with high accuracy and speed. The case reverse plastic flow in the heating domain is also considered. Distributions of residual stresses and displacements forming during cooling of the material are computed.

2. Governing Equations. Boundary Value Problem Statment

Hereafter let use the conventional Prandtle–Reuss elastic-plastic model generalized on thermal effects [1, 2]. Consider thermoelastoplastic hollow ball with inner and outer radia R_0 and R_1 respectively under referential temperature $T = T_0$ and under thermal expansion conditions

$$\sigma_{rr}(R_0) = 0, \quad \sigma_{rr}(R_1) = 0, \quad T_{,r}(R_0) = 0, \quad T(R_1) = T_0(1 - e^{-xt}). \tag{1}$$

Total deformations can formulae depending on the radial displacement

$$e_{rr} = u_{r,r}, \quad e_{\phi\phi} = e_{\theta\theta} = \frac{u_r}{r}. \tag{2}$$

Then according to eqs. (2) we obtain the relationship between components of the total strain tensor

$$e_{rr} = u_{r,r} = \left(\frac{ru_r}{r}\right)_{,r} = (re_{\phi\phi})_{,r}\,. \tag{3}$$

The equilibrium equation in the spherical symmetry case is read by

$$\sigma_{rr,r} + \frac{2\left(\sigma_{rr} - \sigma_{\phi\phi}\right)}{r} = 0, \tag{4}$$

and the components of stress tensor are coupled by formulae

$$\sigma_{\phi\phi} = \sigma_{\theta\theta} = \frac{1}{2r}\left(r^2\sigma_{rr}\right)_{,r}\,. \tag{5}$$

The constitutive strain–stress equations in the spherical symmetry case are formulated by

$$e_{rr} = \alpha\left(T - T_0\right) + \frac{\left(\lambda + \mu\right)\sigma_{rr} - \lambda\sigma_{\phi\phi}}{\mu\left(3\lambda + 2\mu\right)}, \quad e_{\phi\phi} = \alpha\left(T - T_0\right) + \frac{\left(\lambda + 2\mu\right)\sigma_{\phi\phi} - \lambda\sigma_{rr}}{2\mu\left(3\lambda + 2\mu\right)}. \tag{6}$$

Throughout the present study we chose the Tresca yield criterion (maximum tangential stress one)

$$f = \max\left\{\left|\sigma_1 - \sigma_2\right|, \left|\sigma_2 - \sigma_3\right|, \left|\sigma_3 - \sigma_1\right|\right\} - 2k(T) = 0, \tag{7}$$

and the the yield stress k linearly dependent on temperature [2]

$$k = k_0\left(1 - \frac{T - T_0}{T_m - T_0}\right), \tag{8}$$

wherein k_0 is the yield stress under referential temperature T_0, T_m denotes the melting temperature of material.

3. Plastic Flow and Unloading Problems Solutions

The stress state corresponds to the edge of the Tresca's prism (7) under certain outer temperature. Then the size of plastic flow domain are given by inequalities $a(t) < r < R_1$, where $a(t)$ is the elastic-plastic boundary. Plastic deformations in this domain are furnished by

$$p_{rr} = -2p_{\phi\phi} = 2\alpha T - \frac{8\alpha k}{\nu} - \frac{6\alpha}{r^3}\int\limits_a^r T\rho^2 d\rho - \frac{2D(t)}{r^3},$$
$$D = \frac{3\alpha a^3}{(a^3 - R_0^3)}\left(\int\limits_{R_0}^{a} T\rho^2 d\rho + \frac{4R_0^3}{\nu}\int\limits_a^{R_1}\frac{k}{\rho}d\rho\right). \quad (9)$$

The slowdown of plastic deformations increasing (9) is coupled with the levelling of the temperature gradient. At some point on the surface $r = R_1$ the following condition $\dot{p}_{rr} = 0$ will fulfilled meaning the termination of the yield criterion. Thus the unloading domain $b(t) < r < R_1$ is enlarged, where $b(t)$ denotes the unloading boundary. Then for the time dependent function $D(t)$ (9) the following equation can be obtained

$$D = J\left(\frac{2\alpha}{R_0^3}\int\limits_{R_0}^{a} T\rho^2 d\rho + \frac{2\alpha}{R_1^3}\int\limits_b^{R_1} T\rho^2 d\rho - \frac{2\alpha\left(R_1^3 - b^3\right)}{R_1^3 b^3}\int\limits_a^b T\rho^2 d\rho + \frac{2\alpha}{\nu}\int\limits_a^b \frac{k}{\rho}d\rho - \int\limits_b^{R1}\frac{p'_{rr}}{\rho}d\rho\right), \quad (10)$$

where $J(a, b) = \dfrac{3a^3 b^3 R_1^3 R_0^3}{2\left(a^3 b^3 R_1^3 - R_0^3\left(a^3 b^3 + R_1^3\left(b^3 - a^3\right)\right)\right)}$.

The following nonlinear equations system can be solved to obtain the elastic-plastic boundaries positions and the function of irreversible deformation p'_{rr}

$$\begin{cases} p_{rr}(a, t) = 0, \\ p'_{rr}(b) = p_{rr}(b, t), \\ \dot{p}_{rr}(b, t) = 0. \end{cases} \quad (11)$$

Then one can approximate the integrals of deformation function p'_{rr}. Assume that the function of irreversible deformation $p'_{rr}(r)$ is smooth inside unloading domain. Then for a small time interval $\Delta t = t_1 - t_u$ the unloading domain is enlarged according to $\Delta b = b_1 - R_1$. The integrand $\dfrac{p'_{rr}(r)}{r}$ is similar to straight line in the interval Δb. This straight line passes through 2 points with coordinates $\left(b_1, \dfrac{p'_{rr}(b_1)}{b_1}\right)$ and $\left(R_1, \dfrac{p'_{rr}(R_1, t_u)}{R_1}\right)$. Integral $\int\limits_{b_1}^{R1}\dfrac{p'_{rr}}{\rho}d\rho$ is the case can be transformed by

$$\int\limits_{b_1}^{R_1}\frac{p'_{rr}}{\rho}d\rho = I(b_1) = \frac{1}{2}(b_1 - R_1)\left(\frac{p'_{rr}(b_1)}{b_1} + \frac{p_{rr}(R_1, t_u)}{R_1}\right).$$

Substituting this expression into equations (11), and using the method of successive approximations, we find the values $p'_{rr}(b_1)$, $b(\Delta t + t_u) = b_1$,$a(\Delta t + t_u) = a_1$. At the i–th time step $i\Delta t + t_u$ integral can be rewritten as

$$\int\limits_{b_i}^{R_1}\frac{p'_{rr}}{\rho}d\rho = I(b_i) = \frac{1}{2}(b_i - b_{i-1})\left(\frac{p'_{rr}(b_i)}{b_i} + \frac{p'_{rr}(b_{i-1})}{b_{i-1}}\right) + I(b_{i-1}). \quad (12)$$

Consistently calculating the values a_i, b_i, $p'_{rr}(b_i)$, at each time step, we find the elastoplastic boundaries positions and the irreversible deformation inside the unloading area. The condition $a(t_k) = B(t_k)$ is fulfilled during the calculation at some time meaning the ending of the irreversible deformation process.

Assume that at the time $t = t_r > t_u$ on the inner surface the yield criterion (7) is valid. Then for a time $t > t_r$ within vicinity of the inner syrface ($R_1 < r < c(t)$) the plastic flow domain is developed, where $c(t)$ is the elastoplastic boundary. Thus, there are both plastic flow and thermoelastic deformation (unloading) domains. The strain-stress state parameters within domain ($c(t) < r < R_2$) are derived by equations

$$\sigma_{rr} = 4\int_c^r \frac{k(\rho,t)}{\rho}d\rho + G(t), \quad \sigma_{\varphi\varphi} = 4\int_c^r \frac{k(\rho,t)}{\rho}d\rho + 2k(r,t) + G(t),$$
$$u_r = \frac{4r}{(3\lambda+2\mu)}\int_c^r \frac{k(\rho,t)}{\rho}d\rho + \frac{3}{r^2}\int_{R_1}^r \Delta(\rho,t)\rho^2 d\rho + \frac{G(t)r}{(3\lambda+2\mu)} + \frac{H(t)}{r^2}, \tag{13}$$
$$p_{rr} = 2\Delta(r,t) - \frac{6}{r^3}\int_{R_1}^r \Delta(\rho,t)\rho^2 d\rho + \frac{2k(r,t)}{\omega} - \frac{2H(t)}{r^3}.$$

For time dependent functions taking into account the new dependencies (13) we have following relations

$$m = \frac{3a^3b^3c^3R_1^3}{R_1^3(b^3c^3 - a^3(c^3-b^3)) - a^3b^3c^3},$$

$$W(t) = m\Big(\frac{1}{c^3}\int_{R_1}^c \Delta(\rho,t)\rho^2 d\rho + \frac{1}{a^3}\int_{R_1}^a \Delta(\rho,t)\rho^2 d\rho -$$

$$-\frac{1}{b^3}\int_{R_1}^b \Delta(\rho,t)\rho^2 d\rho + \frac{1}{\omega}\int_a^b \frac{k(\rho,t)}{\rho}d\rho - \frac{1}{\omega}\int_c^{R_2} \frac{k(\rho,t)}{\rho}d\rho\Big),$$

$$D(t) = \frac{m}{2}\int_b^{R_2} \frac{p'_{rr}(\rho)}{\rho}d\rho - W(t),$$

$$A = \frac{4\omega}{3c^3}D(t) + 4\int_{R_1}^c \frac{k(\rho,t)}{\rho}d\rho + \frac{4\omega}{3c^3}\int_2^{R_1} \frac{p'_{rr}(\rho)}{\rho}d\rho c, \tag{14}$$
$$B = -\frac{4}{3}\omega D(t), \quad F = -\frac{4}{3}\omega D(t), \quad H = D(t),$$

$$E = -2\omega\int_b^{R_2} \frac{p'_{rr}(\rho)}{\rho}d\rho + \frac{4\omega}{3R_2^3}\int_{R_1}^{R_2} \Delta(\rho,t)\rho^2 d\rho + \frac{4\omega D(t)}{3R_2^3},$$

$$C = 4\int_{R_1}^c \frac{k(\rho,t)}{\rho}d\rho - \frac{4\omega}{a^3}\int_{R_1}^a \Delta(\rho,t)\rho^2 d\rho -$$

$$-\frac{4\omega(c^3-a^3)D(t)}{3a^3c^3} + \frac{4\omega}{c^3}\int_{R_1}^c \Delta(\rho,t)\rho^2 d\rho.$$

The localisations of elastoplastic boundaries $a(t)$, $b(t)$, $c(t)$ and irreversible deformation $p'_{rr}(r)$ can be obtained by numerically solution of the system of equations (11) and adding equation which determines the vanishing of plastic deformation

$$p_{rr}(c,t) = 0. \tag{15}$$

Initial approximations for the system (11), (15) are given by the values $c_0 = R_2$, $a_0 = a(t_r)$, $b_0 = b(t_r)$, $p'_{rr}(b_0) = p'_{rr}(b(t_r))$.

Fig. 1 shows the stresses at the moment of the disappearance of the first plastic flow and the development of a repeated plastic flow inside the outer surface vicinity

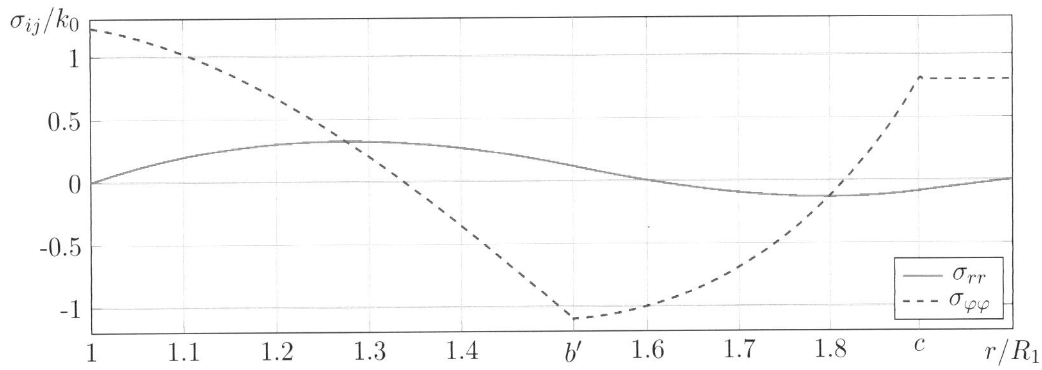

Figure 1. Field of stresses in a continuous sphere in case of occurrence of repeated plastic flow on the external surface, $R_2/R_1 = 2$, b' is the final boundary position of the unloading domain

The strain-stress state within the domain of the repeated plastic flow ($c(t) < r < R_2$) transforms into a state of neutral loading under gradual equalization of the temperature field. The repeated plastic flow boundary occupies the limiting position $c(t) = c'$. The subsequent unloading of the material occurs as a result of an increasing of the yield strength under uniform cooling. Stresses of the fully heated ball under the given temperature ($\Delta_{,r}(r,t) = 0$) are the residual ones (Fig. 2).

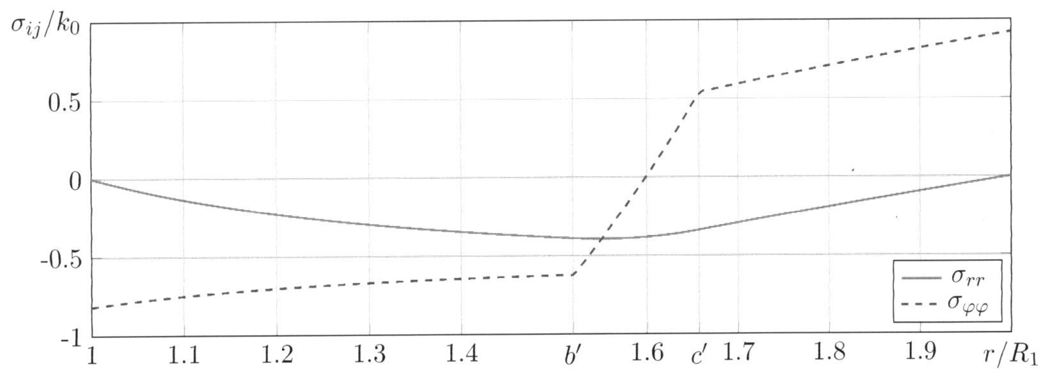

Figure 2. Residual stresses at zero temperature gradient (maximum temperature / referential temperature)

Fig. 3 shows us the radial distribution of the displacement after a uniform cooling of the ball to the referential temperature ($\Delta(r,t) = 0$)

From Fig. 3 it follows that in the case of the development of irreversible deformation processes in the material of a hollow sphere, its size after cooling to the initial temperature will be smaller than the referential size.

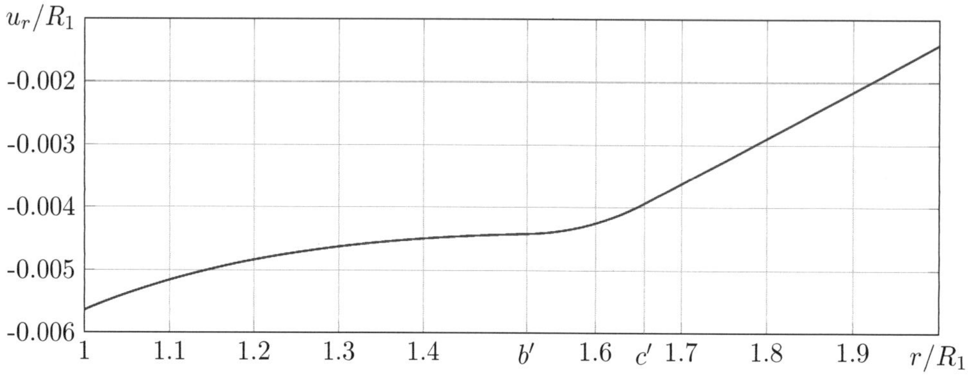

Figure 3. Residual radial displacement at referential temperature

Acknowledgment
The research was financially supported by RSF (project No. 17-19-01257).

Conclusion
- The boundary value problem within frameworks of the the conventional Prandtle–Reuss elastic-plastic model generalized on thermal effects have been solved.
- The hollow elastic-plastic ball under central symmetry unsteady heating have been considered. The numerical-analytical algorithm for calculating the elastic-plastic boundaries positions and irreversible deformations has been proposed.
- It has been shown that the strain-stress state within the domain of the repeated plastic flow has been transformed into a state of neutral loading during the temperature field equalization.
- It has been shown that the accumulated irreversible deformation in a hollow sphere lead to decreasing its size after cooling to the referential temperature.

References
[1] Boley B and Weiner J 1960 *Theory of Thermal Stresses* (New York: Wiley)
[2] Burenin A, Dats E and Murashkin E 2014 *Mech. of Solids* **49** 218
[3] Dats E, Mokrin S and Murashkin E 2015 *Lect. Not. Eng. and Comp. Sc.* **2218** 1043
[4] Dats E, Mokrin S and Murashkin E 2016 *Key Eng. Mat.* **685** 37
[5] Dats E and Murashkin E 2016 *Lect. Not. Eng. and Comp. Sc.* **2223** 69
[6] Dats E, Murashkin E and Stadnik N 2017 *Proc. Eng.* **173** 891
[7] Gamer U 1967 *Arch. Appl. Mech.* **36(3)** 174
[8] Gamer U 1988 *J. Therm. Str.* **13** 159
[9] Guven U and Altay O 2000 *Int. J. Mech. Sc.* **42** 831
[10] Odeno H 1969 *Acta Polytech. Scand. Phys. Incl. Nucleon. Series* **66** 243
[11] Odeno H 1969 *J. Mech. Eng. Sc.* **2** 284
[12] Orcan Y 1991 *Acta Mech.* **90** 61
[13] Orcan Y 1995 *Int. J. Mech. Sc.* **33** 1689
[14] Parkus H 1954 *Proc. Sec. U.S. Nat. Congr. Appl. Mech.* **32** 307
[15] Parkus H 1959 *Arch. Appl. Mech.* **28(1)** 251
[16] Perzyna P and Sawczuk A 1973 *Nucl. Eng. and Design* **24** 1-55
[17] Raniecki B 1966 *Rozpr. Inz.* **14(3)** 479
[18] Manzhirov A V 2013 *Key Eng. Mat.* **535** 89
[19] Manzhirov A V and Parshin D A 2015 *Mech. Solids* **50(6)** 657
[20] Manzhirov A V and Parshin D A 2016 *Lec. Not. Eng. and Comp. Sc.* **2224(1)** 1131

IOP Conf. Series: Journal of Physics: Conf. Series **937** (2017) 012031 doi:10.1088/1742-6596/937/1/012031

Analytical Solution of Cylindrical Wave Problem in the Frameworks of Micropolar Elasticity

E V Murashkin[1,2,3], **Y N Radayev**[2]

[1]National Research Nuclear University "MEPhI" (Moscow Engineering Physics Institute), Kashirskoye shosse 31, Moscow, 115409, Russia
[2]Ishlinsky Institute for Problems in Mechanics of the Russian Academy of Sciences, Vernadsky Ave 101 Bldg 1, Moscow, 119526, Russia
[3]Bauman Moscow State Technical University, 2nd Baumanskaya Str 5/1, Moscow, 105005, Russia

E-mail: `evmurashkin@gmail.com`, `radayev@ipmnet.ru`

Abstract. Propagation problem for coupled harmonic waves of translational displacements and microrotations along the axis of a long cylindrical waveguide is discussed at present study. Microrotations modeling is carried out within the linear micropolar elasticity frameworks. The coupled system of vector differential equations of micropolar elasticity is presented. The translational displacements and microrotations in the coupled wave are decomposed into potential and vortex parts. The coupled differential equations are then reduced to uncoupled ones. The wave equations solutions for the translational and microrotational waves potentials are obtained for a high-frequency waves in the cylindrical domain.

1. Introduction

Modelling of abnormal mechanical and multi-physical behaviour of solids (metamaterials) is an actual problem of modern Continuum Mechanics [1–4]. The metamaterials exhibit physical properties not usually found in nature. For example, there are materials with negative Poissons ratio (auxetic materials), negative thermal expansion, negative electric permittivity and the magnetic permeability. These paradoxical physical phenomena cannot be described in the frameworks of conventional Continuum Mechanics. In this case, microstructure continuum theories based on the necessity of additional (extra) freedom degrees are required [5, 6]. The problems of propagating surfaces of discontinuities of displacements and microrotations are previously discussed in number of studies [1, 2, 6–13]. In some studies [10–13] the wave problems in circular micropolar waveguides are considered. The aim of the present paper is to investigate the translational and microrotation waves propagating along an infinite cylindrical waveguide. Microrotation waves in an elastic solid are modelled within the linear micropolar elasticity frameworks by microrotations introducing in the governing equations of the Continuum Mechanics. Such a modelling can be regularly carried out in terms of the classical field theory [2], starting from the action integral and the least action variational principle.

2. Reminder of the Governing Equations

Throughout the study we employ notations used in [7] while replacing $\eta \to \alpha$. For micropolar elastic continuum, equations of motion are furnished in direct tensor representation for the case

IOP Conf. Series: Journal of Physics: Conf. Series **937** (2017) 012031 doi:10.1088/1742-6596/937/1/012031

of the absence of mass forces and mass moments

$$\begin{cases} \boldsymbol{\nabla} \cdot \boldsymbol{\sigma} = \rho \ddot{\mathbf{u}}, \\ \boldsymbol{\nabla} \cdot \boldsymbol{m} + \boldsymbol{\epsilon} \cdot \cdot \boldsymbol{\sigma} = \mathfrak{J} \ddot{\boldsymbol{\phi}}. \end{cases} \tag{1}$$

where $\boldsymbol{\nabla}$ is the three-dimensional Hamiltonian differential operator (the nabla symbol), ρ is the mass density, $\mathfrak{J}$ is a scalar dynamic characteristic of continuum (the rotational microinertia), dot over a symbol denotes partial differentiation with respect to time at fixed spatial coordinates, $\mathbf{u}$ is the translational displacement vector and $\boldsymbol{\phi}$ rotation vector, $\boldsymbol{\epsilon}$ is the three-dimensional Levi–Civita symbol (permutation symbol, antisymmetric symbol, or alternating symbol).

The constitute equations for the force stress tensor $\boldsymbol{\sigma}$ and the moment stress tensor $\boldsymbol{\mu}$ can be presented as follows

$$\boldsymbol{\sigma} = (\mu + \eta)\boldsymbol{\gamma} + (\mu - \eta)\boldsymbol{\gamma}^{\mathrm{T}} + \lambda \mathbf{I}\,\mathrm{tr}\boldsymbol{\gamma}, \qquad \boldsymbol{\mu} = (\gamma + \varepsilon)\,\boldsymbol{\kappa} + (\gamma - \varepsilon)\,\boldsymbol{\kappa}^{\mathrm{T}} + \beta \mathbf{I}\,\mathrm{tr}\,\boldsymbol{\kappa}. \tag{2}$$

The equations for the strain tensor $\boldsymbol{\gamma}$ and the bending-torsion tensor $\boldsymbol{\kappa}$ are read by

$$\boldsymbol{\kappa} = \boldsymbol{\nabla} \otimes \boldsymbol{\phi}, \qquad \boldsymbol{\gamma} = \boldsymbol{\nabla} \otimes \mathbf{u} - \boldsymbol{\phi} \cdot \boldsymbol{\epsilon}, \tag{3}$$

wherein $\mathbf{I}$ denotes the three dimensional unit tensor.

The coupled vector differential equations determining the coupled wavefields in a micropolar elastic continuum [7] can be derived after substituting constitutive equations (2) and (3) into equations of motion (1)

$$\begin{cases} (\lambda + 2\mu)\boldsymbol{\nabla}\boldsymbol{\nabla} \cdot \mathbf{u} - (\mu + \eta)\boldsymbol{\nabla} \times (\boldsymbol{\nabla} \times \mathbf{u}) + 2\eta \boldsymbol{\nabla} \times \boldsymbol{\phi} = \rho \ddot{\mathbf{u}}, \\ (\beta + 2\gamma)\boldsymbol{\nabla}\boldsymbol{\nabla} \cdot \boldsymbol{\phi} - (\gamma + \varepsilon)\boldsymbol{\nabla} \times (\boldsymbol{\nabla} \times \boldsymbol{\phi}) + 2\eta \boldsymbol{\nabla} \times \mathbf{u} - 4\eta \boldsymbol{\phi} = \Im \ddot{\boldsymbol{\phi}}, \end{cases} \tag{4}$$

where λ, μ, γ, β, ε, η denote the constitutive constants of micropolar elastic continuum; $\boldsymbol{\nabla}$ is the three dimensional Hamilton operator (Hamilton nabla); $\mathbf{u}$ is the translational displacement vector; $\boldsymbol{\phi}$ is the microrotation vector; ρ is the mass density; $\Im$ denotes the microelement inertia; dot denotes partial differentiation with respect to time.

3. Wave Problem Solution

The system (4) is coupled and can be uncoupled as follows. First, introduce the dynamic potentials of translational displacements and microrotations

$$\mathbf{u} = \boldsymbol{\nabla}\Phi + \boldsymbol{\nabla} \times \boldsymbol{\Psi}, \quad \boldsymbol{\phi} = \boldsymbol{\nabla}\Sigma + \boldsymbol{\nabla} \times \mathbf{H}, \tag{5}$$

where Φ and Σ are the scalar potentials, and $\boldsymbol{\Psi}$ and $\mathbf{H}$ are the vector potentials of translational displacements and microrotations respectively.

Second, after substituting (5) in (4) it is seen that the system (4) is satisfied if the scalar and the vector potentials fulfill the calibration conditions

$$\boldsymbol{\nabla} \cdot \boldsymbol{\Psi} = \mathbf{0}, \quad \boldsymbol{\nabla} \cdot \mathbf{H} = \mathbf{0}, \tag{6}$$

and are solutions of coupled equations

$$\begin{aligned} &\Delta\Phi - \frac{1}{c_{\parallel}^2}\ddot{\Phi} = 0, && \Delta\Sigma - \frac{1}{{}_{\mu}c_{\parallel}^2}\ddot{\Sigma} - \frac{\Omega^2}{{}_{\mu}c_{\parallel}^2}\Sigma = 0, \\ &\Delta\boldsymbol{\Psi} - \frac{1}{{}''c_{\perp}^2}\ddot{\boldsymbol{\Psi}} + 2d_{\perp}^2\boldsymbol{\nabla} \times \mathbf{H} = \mathbf{0}, && \Delta\mathbf{H} - \frac{1}{{}_{\mu}c_{\perp}^2}\ddot{\mathbf{H}} - \frac{\Omega^2}{{}_{\mu}c_{\perp}^2}\mathbf{H} + \frac{\Omega^2}{2{}_{\mu}c_{\perp}^2}\boldsymbol{\nabla} \times \boldsymbol{\Psi} = \mathbf{0}, \end{aligned} \tag{7}$$

where the following notations are used

$$\begin{aligned} &\Omega^2 = 4\eta\Im^{-1}, \quad \rho c_\parallel^2 = \lambda + 2\mu, \quad \Im_\mu c_\parallel^2 = \beta + 2\gamma, \\ &\Im_\mu c_\perp^2 = \mu + \eta, \quad \rho'' c_\perp^2 = \mu + \eta, \quad \rho' c_\perp^2 = \eta, \quad ''c_\perp^2 d_\perp^2 = 'c_\perp^2. \end{aligned} \tag{8}$$

Third, in present study only coupled high-frequency waves of thranslational displacements and microrotaions are considered ($\omega > \Omega$). In this case the following differential equations for the potentials can be obtained

$$\begin{aligned} &(\Delta + \alpha_\parallel^2)\Phi = 0, && (\Delta + \beta_\parallel^2)\Sigma = 0, \\ &(\Delta + \alpha_\perp^2)\mathbf{\Psi} + 2d_\perp^2 \nabla \times \mathbf{H} = \mathbf{0}, && (\Delta + \beta_\perp^2)\mathbf{H} + \frac{\Omega^2}{2_\mu c_\perp^2} \nabla \times \mathbf{\Psi} = \mathbf{0}. \end{aligned} \tag{9}$$

wherein the following notations have been introduced

$$\alpha_\parallel^2 = \omega^2 \big/ c_\parallel^2, \quad \beta_\parallel^2 = (\omega^2 - \Omega^2) \big/ {}_\mu c_\parallel^2, \quad \alpha_\perp^2 = \omega^2 / '' c_\perp^2, \quad \beta_\perp^2 = (\omega^2 - \Omega^2) / {}_\mu c_\perp^2. \tag{10}$$

The equations (9) for vortex potentials $\mathbf{\Psi}$ and $\mathbf{H}$ are still coupled. Uncoupled equations for vector potentials can be derived by the increasing differentiation order.

At last, for potentials $\mathbf{\Psi}$ and $\mathbf{H}$ the separate equations can be furnished by

$$\begin{aligned} &(\Delta + K_1^2)(\Delta + K_2^2)\mathbf{\Psi} = \mathbf{0}, \quad (\Delta + K_1^2)(\Delta + K_2^2)\mathbf{H} = \mathbf{0}, \quad \sigma_\perp^2 = d_\perp^2 \Omega^2 / {}_\mu c_\perp^2. \\ &K_{1,2}^2 = -\Delta_{1,2}\,, \quad 2\Delta_{1,2} = -(\alpha_\perp^2 + \beta_\perp^2 + \sigma_\perp^2) \pm \sqrt{(\alpha_\perp^2 - \beta_\perp^2 + \sigma_\perp^2)^2 + 4\beta_\perp^2 \sigma_\perp^2}, \end{aligned} \tag{11}$$

The coupled wavefields in a cylindrical domain are found in the cylindrical coordinates r, φ, z by separation of variables (see in details [8]). Thus, for the scalar potentials one can obtain

$$\Phi = C_1 I_n(p_1 r) \left\{ \begin{array}{c} \cos n\varphi \\ -\sin n\varphi \end{array} \right\} e^{\pm ikz}, \quad \sum = -C_2 I_n(p_2 r) \left\{ \begin{array}{c} \sin n\varphi \\ \cos n\varphi \end{array} \right\} e^{\pm ikz}, \tag{12}$$

wherein k denotes the wave number; C_1 and C_2 are arbitrary constants; $I_n(\cdot)$ is the Bessel function of the first kind of an imaginary argument; $p_1^2 = k^2 - \alpha_\parallel^2$, $p_2^2 = k^2 - \beta_\parallel^2$.

The vortex potentials of translational displacements and microrotations are given by the following formulas

$$\left\{ \begin{aligned} &\Psi_r = [C_3' \mathrm{I}_{n-1}(q_1 r) + C_4' \mathrm{I}_{n+1}(q_1 r) + C_3'' \mathrm{I}_{n-1}(q_2 r) + C_4'' \mathrm{I}_{n+1}(q_2 r)] \left\{ \begin{array}{c} \sin n\varphi \\ \cos n\varphi \end{array} \right\} e^{\pm ikz}, \\ &\Psi_\varphi = [C_3' \mathrm{I}_{n-1}(q_1 r) - C_4' \mathrm{I}_{n+1}(q_1 r) + C_3'' \mathrm{I}_{n-1}(q_2 r) - C_4'' \mathrm{I}_{n+1}(q_2 r)] \left\{ \begin{array}{c} \cos n\varphi \\ -\sin n\varphi \end{array} \right\} e^{\pm ikz}, \\ &\Psi_z = [C_5' \mathrm{I}_n(q_1 r) + C_5'' \mathrm{I}_n(q_2 r)] \left\{ \begin{array}{c} \sin n\varphi \\ \cos n\varphi \end{array} \right\} e^{\pm ikz}; \end{aligned} \right. \tag{13}$$

$$\left\{ \begin{aligned} &H_r = [L_3' \mathrm{I}_{n-1}(q_1 r) + L_4' \mathrm{I}_{n+1}(q_1 r) + L_3'' \mathrm{I}_{n-1}(q_2 r) + L_4'' \mathrm{I}_{n+1}(q_2 r)] \left\{ \begin{array}{c} -\cos n\varphi \\ \sin n\varphi \end{array} \right\} e^{\pm ikz}, \\ &H_\varphi = [L_3' \mathrm{I}_{n-1}(q_1 r) - L_4' \mathrm{I}_{n+1}(q_1 r) + L_3'' \mathrm{I}_{n-1}(q_2 r) - L_4'' \mathrm{I}_{n+1}(q_2 r)] \left\{ \begin{array}{c} \sin n\varphi \\ \cos n\varphi \end{array} \right\} e^{\pm ikz}, \\ &H_z = [L_5' \mathrm{I}_n(q_1 r) + L_5'' \mathrm{I}_n(q_2 r)] \left\{ \begin{array}{c} -\cos n\varphi \\ \sin n\varphi \end{array} \right\} e^{\pm ikz}, \end{aligned} \right. \tag{14}$$

wherein $C_3' - C_5'$, $C_3'' - C_5''$, $L_3' - L_5'$ and $L_3'' - L_5''$ are arbitrary constants and $q_1^2 = k^2 - K_1^2$, $q_2^2 = k^2 - K_2^2$.

Conclusion

- The constitutive and governing equations of the linear micropolar elasticity have been furnished.
- Propagation problem for coupled harmonic waves of translational displacements and microrotations along the axis of a long circular waveguide has been solved.
- The translational displacements and microrotations in the coupled wave have been decomposed into potential and vortex parts and the coupled differential equations of motion has been reduced to uncoupled ones.
- The wave problem solution has been obtained for a high-frequency waves in the cylindrical domain.

References

[1] Kovalev V, Murashkin E and Radayev Y 2017 *JPCS 788(1)* 012043
[2] Murashkin E V and Stadnik N E 2017 *MATEC Web of Conf.* **95** 12001
[3] Green A E and Naghdi P M 1992 *J. Therm. Stress.* **15** 253
[4] Zadpoor A A 2016 *Mat. Hor.* **3(5)** 371
[5] Cosserat E et F 1909 *Theorie des corps deformables* (Paris: Librairie Scientifique A. Hermann et Fils)
[6] Toupin R A 1964 *Arc. Rat. Mech. and Anal.* **17(2)** 85
[7] Nowacki W 1986 *Theory of Asymmetric Elasticity* (Oxford: Pergamon Press)
[8] Kovalev V A and Radayev Y N 2010 *Wave Problems of the Field Theory and Thermomechanics* (Saratov: Saratov University Press)
[9] Murashkin E V and Radayev Y N 2015 *Mat. Phys. and Mech.* **23** 10
[10] Smith A C 1967 *Int. J Eng. Sc.* **5(10)** 741
[11] Rao K M 1988 *Int. J Eng. Sc.* **26(2)** 135
[12] Kumar R and Deswal S 2002 *J. sound and Vibr.* **250(4)** 711
[13] Charalambopoulos A, Gergidis L N and Kartalos G 2012 *Composites Part B: Eng.* **43(6)** 2613

IOP Conf. Series: Journal of Physics: Conf. Series **937** (2017) 012032 doi:10.1088/1742-6596/937/1/012032

A Numerical Simulation of Adiabatic Shear Bands Formation in Hollow Cylinder

R V Muratov, N A Kudryashov, P N Ryabov

National Research Nuclear University MEPhI (Moscow Engineering Physics Institute), Kashirskoe shosse 31, Moscow, Russia

E-mail: nakudr@gmail.com, pnryabov@mephi.ru

Abstract. The process of torsion deformation of a hollow cylinder is considered. The mathematical model describing this process is formulated. Numerical algorithm which allows one to simulate the fully localized plastic flow is proposed. The influence of the geometrical parameters of the problem of plastic flow localization is studied. The limiting case when the problem of torsion deformation of a hollow cylinder transforms to the plane problem of deformation of infinite slab is considered. An influence of cylinder size to the formation of adiabatic shear bands is considered. It is shown that declining of inner size of cylinder results to a significant change of solution and displacement of localization area to the inner surface of a hollow cylinder.

1. Introduction

A formation of adiabatic shear bands (ASB) or localization of plastic deformations is the one of important phenomena of materials science. Adiabatic shear bands is a narrow region where high temperature and deformations are exceeded in a short period of time. The localizations of deformation are located usually in places with geometry defects or thermal heterogeneity [1] – [5]. The reason of the ASB formation is a loss of stability of plastic flow occurs due to the thermal softening effect. In conditions close to adiabatic the work of plastic deformation converted into heat which causes thermal softening of material and leads to adiabatic shear bands formation. This phenomenon is often investigated in the physical experiments with torsion deformation of a hollow cylinder. But for a numerical simulation many authors consider a limiting case: a shear deformation of an infinite slab. In this work we investigate the original problem and compare it with the limiting case.

2. Mathematical formulation of problem

We consider the process of torsion deformation of hollow cylinder which inner radius is equal to r_0 and outer radius equal to $r_0 + H$. The inner surface is fixed and the outer surface of the cylinder is rotated at a constant angular velocity.

Mathematical formulation of cylindrical (a) and plane (b) problems includes four governing equations: the Cauchy momentum equation (1), the Hooke's law (2), the equation of state or plasticity flow law (3) and the law of conservation of energy (4). So we obtain

$$a) \quad \rho r \omega_t = \frac{1}{r^2}\frac{\partial}{\partial r}\left(r^2 \tau\right), \qquad b) \quad \rho v_t = \tau_y. \tag{1}$$

IOP Conf. Series: Journal of Physics: Conf. Series **937** (2017) 012032 doi:10.1088/1742-6596/937/1/012032

$$a)\quad \tau_t - \mu r\omega_r = -\mu\dot{\gamma}, \qquad b)\quad \tau_t - \mu v_y = -\mu\dot{\gamma}, \tag{2}$$

$$\dot{\gamma} = \Phi^{-1}(\gamma, \tau, T), \qquad \gamma = \int_0^t \dot{\gamma}(s)ds, \tag{3}$$

$$C\rho T_t = k\Delta T + \tau\dot{\gamma}. \tag{4}$$

where y, r are coordinates, t is time, $\tau, \gamma, v, \omega, T$ are stress, strain, velocity, angular velocity and temperature relatively. The cylindrical problem approaches to the plane problem at $r_0 \to \infty$ and change of variables: $r = r_0 + y$, $v = r\omega$.

Boundary conditions are similar for both tasks and have the form:

$$a)\quad \omega|_{r=r_0} = 0, \quad \omega|_{r=r_0+H} = \omega_0, \quad T_r|_{r=r_0} = T_r|_{r=r_0+H} = 0, \tag{5}$$

$$b)\quad v|_{y=0} = 0, \quad v|_{y=H} = v_0, \quad T_y|_{y=0} = T_y|_{y=H} = 0. \tag{6}$$

And initial conditions:

$$a)\quad \omega(r,0) = \dot{\gamma}_0 \ln\frac{r}{r_0}, \qquad T(r,0) = T_0(r), \tag{7}$$

$$b)\quad v(y,0) = \dot{\gamma}_0 y, \qquad T(y,0) = T_0(y), \tag{8}$$

$$\gamma^p|_{t=0} = 0, \quad \tau|_{t=0} = 0, \quad \dot{\gamma}^p|_{t=0} = \dot{\gamma}_0. \tag{9}$$

The boundary value problem is solved numerically. The algorithm consist of two stages. The first stage devoted to solution of mechanical part of the problem. For this purpose we use the Courant-Isaacson-Rees scheme [6] with the Newton iterations to system (1), (2), (3). On the second stage, we solve thermal part of the problem (4).

3. Results of numerical simulation

For numerical simulations, we use high-tensile steel HY-100. The Thermophysical parameters of the HY-100 steel can be found in table 1.

μ, GPa	ρ, kg/m^3	k, W/m·K	C, J/kg·K	$\dot{\varepsilon}_y$, s^{-1}	a, K^{-1}	τ_0, MPa	m
80	7860	49.2	473	10^{-4}	$6.43 \cdot 10^{-4}$	600	0.0251

Table 1. Thermophysical parameters of HY-100 steel.

Our experiment is based on the plane problem which was investigated in the works [2], [3]. For steel we use the Litonski plasticity flow law, so $\Phi(\gamma, \dot{\gamma}, T)$ from equation (3) has the form:

$$\tau = \Phi(\gamma, \dot{\gamma}, T) = \tau_0 e^{-aT}\left(1 + \frac{\dot{\gamma}^n}{\dot{\varepsilon}_y}\right)^m. \tag{10}$$

We test block which has thickness $H = 6$ mm, strain rate $\dot{\gamma}_0 = 750$ s^{-1}. At an initial moment temperature has gaussian-shaped distribution with a peak $T_{max} = 16.2$ °C at center of a layer. Below we use dimensionless time $\gamma_{nom} = \dot{\gamma}_0 \cdot t$.

The first experiment is conducted at $r_0 = 100$ m, the ratio $y/r_0 \sim 10^{-6}$, so this is almost plane problem, the parameters $H, \dot{\gamma}_0, T_{max,0}$ correspond to the parameters in the works [2], [3]. In this case the initial perturbation results in the formation of single ASB in the center of the sample. Stages of ASB formation are shown at figures 1 and 2. Time dependence of the

IOP Conf. Series: Journal of Physics: Conf. Series **937** (2017) 012032 doi:10.1088/1742-6596/937/1/012032

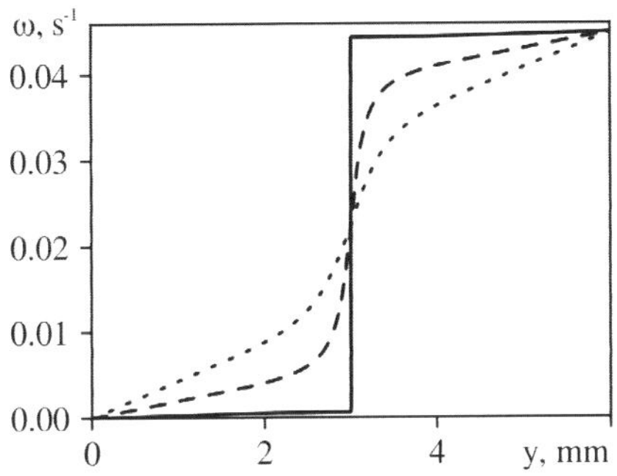

Figure 1. Angular velocity at γ_{nom} = 0.2, 0.25, 0.3 (solid, dashed, dotted lines relatively). $r_0 = 100$ m.

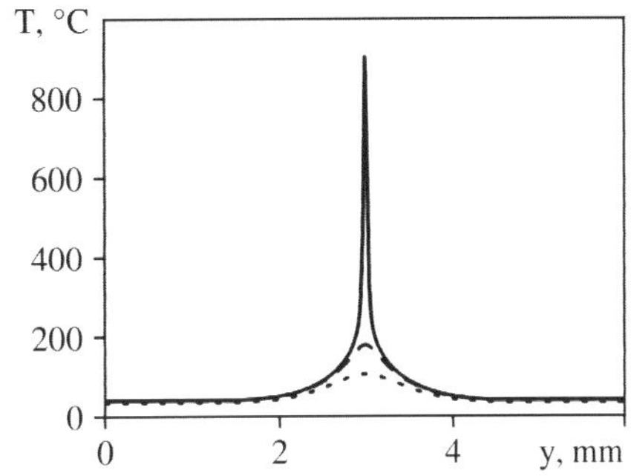

Figure 2. Temperature at γ_{nom} = 0.2, 0.25, 0.3 (solid, dashed, dotted lines relatively). $r_0 = 100$ m.

maximum temperature is presented at figure 8. We can see rapid increase of the temperature at the center of slab beginning about $\gamma_{nom} = 0.27 - 0.38$. These results are completely consistent with the results of other authors.

For plane problem graphs of temperature, stress and strain are symmetric with respect to $y = 0.5H$, but symmetry is breaking with decreasing of inner radius of the cylinder. The second experiment is conducted at $r_0 = 1$ m. Results are shown at figures 3 and 4. On the figure 4 we can see that distribution of temperature has two local maximum and new maximum is located at the inner surface. Thus we get the temperature at the inner surface is more than at the outer.

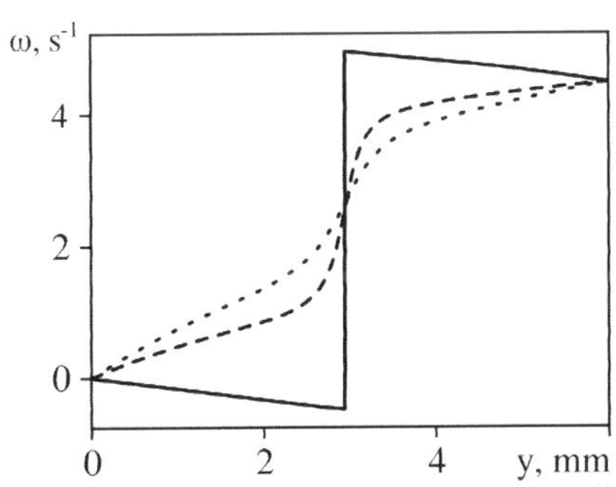

Figure 3. Angular velocity at γ_{nom} = 0.2, 0.25, 0.3 (solid, dashed, dotted lines relatively). $r_0 = 1$ m.

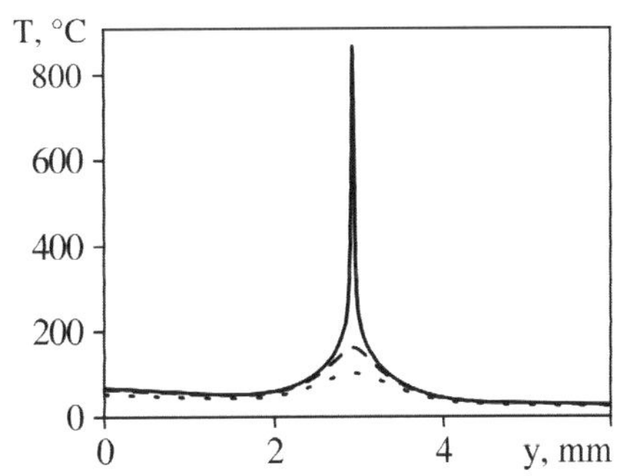

Figure 4. Temperature at γ_{nom} = 0.2, 0.25, 0.3 (solid, dashed, dotted lines relatively). $r_0 = 1$ m.

This effect is enhanced for less radius of the hole. Let's decrease inner radius r_0 to 0.5 m. In this case the temperature maximum almost immediately arise at the boundary. And at infinite time we come to stationary distribution where the whole cylinder without one point (inner surface) is moving at constant angular velocity, which equals the velocity of the outer surface, and the one point is fixed. It is demonstrated at figures 5 and 6.

However, at r_0 we also can get ASB at the center of the layer. It happens if we set a larger peak of temperature at initial moment (160 °C, for example). Let's consider a problem with stationary initial distribution of temperature. In this case we get a solution with one fixed point regardless of the inner radius. Results of experiment for $r_0 = 0.5$ m are shown at figures 9 and

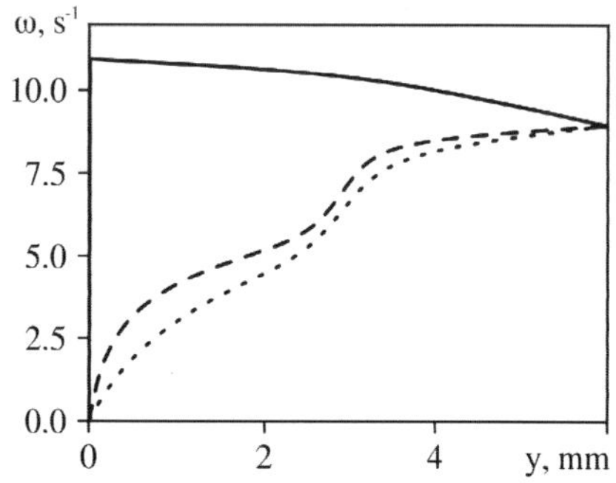

Figure 5. Angular velocity at γ_{nom} = 0.2, 0.25, 0.3 (solid, dashed, dotted lines relatively) and $r_0 = 0.5$ m.

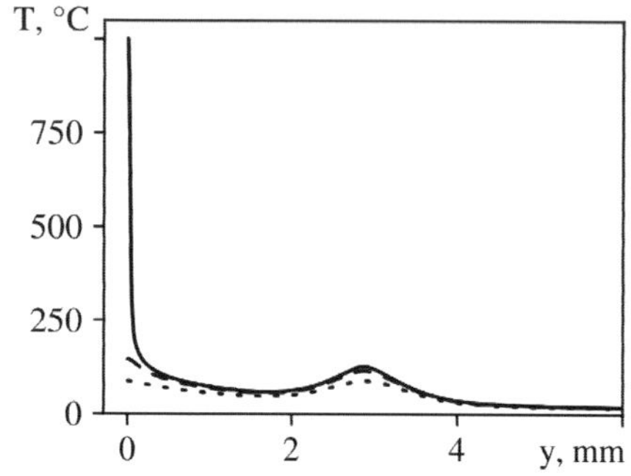

Figure 6. Temperature at γ_{nom} = 0.2, 0.25, 0.3 (solid, dashed, dotted lines relatively) and $r_0 = 0.5$ m.

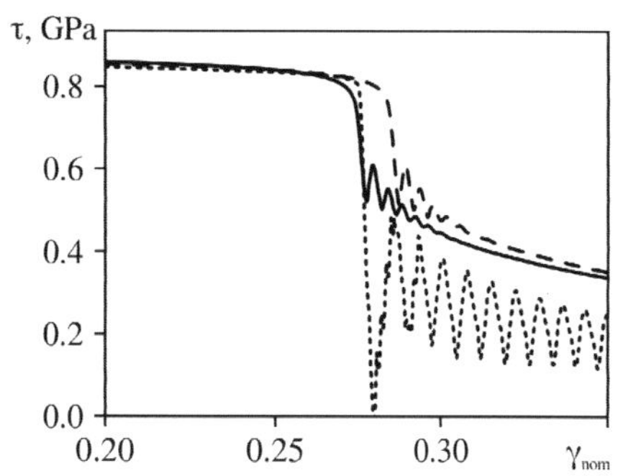

Figure 7. Time dependence of the minimum stress for r_0 = 100, 1, 0.5 m (solid, dashed, dotted lines relatively).

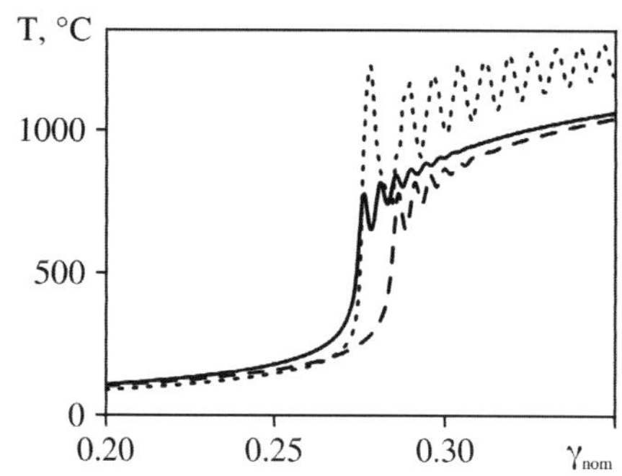

Figure 8. Time dependence of the maximum temperature for r_0 = 100, 1, 0.5 m (solid, dashed, dotted lines relatively).

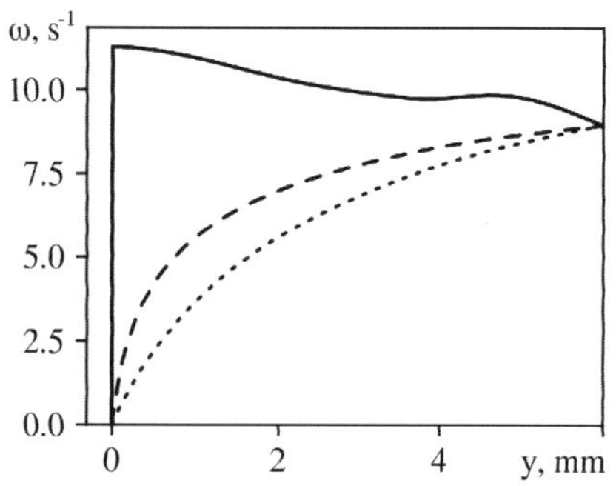

Figure 9. Angular velocity at γ_{nom} = 0.2, 0.25, 0.3 (solid, dashed, dotted lines relatively) and $r_0 = 0.5$ m.

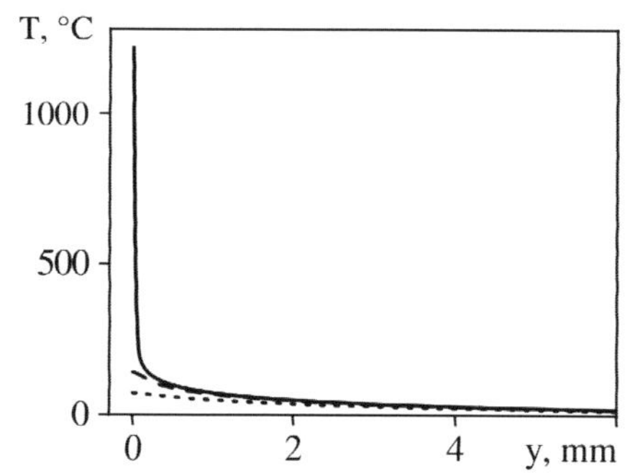

Figure 10. Temperature at γ_{nom} = 0.2, 0.25, 0.3 (solid, dashed, dotted lines relatively) and $r_0 = 0.5$ m.

10. The plane task has completely different, simple solution: the slab uniformly heats up due to internal friction. But in cylindrical case, stress at inner surface is higher, so internal friction is also higher. It leads to rapid increasing of temperature at the boundary.

It is obvious that we cannot get such results in a physical experiment. Because any material has micro defects which could become centers for localization of plastic strain, it was demonstrated in work [4]. If we has impossible continuous material, then the inner surface of sample most likely will melt, and we will not be able to fix it anymore. So in this case cylinder will become to rotate at constant angular velocity. However, our model does not take into account phase transactions.

4. Acknowledgments.

This work was supported by the Grant for support of young Russian scientists.

References

[1] Marchand A and Duffy J 1988 *J. Mech. Phys. Sol.* **36** 251
[2] Walter J W 1992 *Int. J. Plast.* **8** 657-693
[3] Zhou F, Wright T W and Ramesh K T 2006 *J. Mech. Phys. Sol.* **54** 904-926
[4] Kudryashov N A, Ryabov P N and Zakharchenko A S 2015. *J. Mech. Phys. Sol.* **76** 180-192
[5] Kudryashov N A, Ryabov P N and Muratov R V 2017. *J. Phys.: Conf. Series* **788** 012031
[6] Courant R, Isaacson E and Rees M 1952 *Comm. Pure Appl. Math.* **5** 243-255.

IOP Conf. Series: Journal of Physics: Conf. Series **937** (2017) 012033 doi:10.1088/1742-6596/937/1/012033

Reliability of the past surface temperature reconstruction methods

O V Nagornov and S A Tyuflin

National Research Nuclear University MEPhI (Moscow Engineering Physics Institute), 31 Kashirskoe shosse, 115409 Moscow, Russia

E-mail: nagornov@yandex.ru

Abstract. The analysis of the SVD method for the past surface temperature reconstructions based on the measured borehole temperatures is presented. This method has got wide application in geophysics for interpretation of the wide-world temperature profile data set. We show examples where reconstructions do not reflect the true behavior of the past surface temperatures.

1. Introduction

The proxy climate indicators are used to get information on the past climate. The most reliable data for the past surface temperature reconstructions are the borehole temperatures connected directly with the surface temperature changes while the other data result from empirical correlations and calibrations. The past surface temperature changes penetrate in the Earth depth and disturb the steady-state temperature field. Thus, analysis of deviation of the borehole temperature profile from the steady-state thermal regime allows us to reconstruct the resultant ground surface temperature. Numerous results on the past surface temperature histories based on the ground borehole temperature measurements in different places of the Earth were obtained in the last dozen years. The objective of this paper is to study the reliability of these reconstructions. We show that major past climatic changes derived by the inverse problem solutions can be lost.

2. Mathematical model and method

The underground temperature distribution is mainly determined by two types of processes. The first is the surface temperature changes and the second is the heat flux from the Earth that is subjected to the long-time geological processes [1, 2, 3]. The borehole temperature-depth profile $\chi(z)$ can be expressed by the long-term surface temperature T_0, the geothermal gradient Γ_0 and the temperature perturbation $T(z)$ caused by the ground surface temperature variations [4]:

$$\chi(z) = T_0 + \Gamma_0 z + T(z). \tag{1}$$

The climatic ground surface signal influences on the Earth's temperature field and can be described by the one-dimensional thermal diffusivity equation with initial and boundary conditions [5]:

$$\begin{cases} T_t = a^2 T_{zz}, & 0 < t < t_f, 0 < z < \infty, \\ T(0,t) = \mu(t), & 0 \leq t < t_f, \\ T(z,t_f) = 0, & 0 \leq z < \infty, \end{cases} \quad (2)$$

where a^2 is the thermal diffusivity, and $\mu(t)$ is the ground surface temperature (GST) anomaly from the steady-state temperature T_0 being at the time t_f.

If the past variations of ground surface temperature $\mu(t)$ are modeled as a series of K step changes in temperature, the solution of this problem can be found out analytically:

$$T(z) = \sum_{k=1}^{K} M_k[\mathrm{erfc}(\frac{z}{2\sqrt{a^2 t_k}}) - \mathrm{erfc}(\frac{z}{2\sqrt{a^2 t_{k-1}}})] \quad (3)$$

where M_k is the ground surface temperature at the time interval (t_k, t_{k-1}), and erfc is the complementary error function.

The inverse problem for the surface temperature reconstruction involves equations (2) and the re-determination condition $T(z,0) = \theta(z)$, where $\theta(z)$ is the measured temperature-depth profile. Then the inverse problem solution can be reduced the solution of the linear system of equations:

$$\mathbf{\Theta_j} = \mathbf{A_{jk} M_k}, \quad (4)$$

where $\mathbf{\Theta_j}$ is the temperature perturbation at depth z_j, $\mathbf{M_k}$ is the average temperature on the time interval (t_k, t_{k-1}), and $\mathbf{A_{jk}}$ is a matrix:

$$\mathbf{A_{jk}} = \mathrm{erfc}(\frac{z_j}{2\sqrt{a^2 t_k}}) - \mathrm{erfc}(\frac{z_j}{2\sqrt{a^2 t_{k-1}}}). \quad (5)$$

This system is over-determined and can be solved by the singular value decomposition (SVD) [2].

$$\mathbf{A} = \mathbf{U \Lambda V^T}, \quad (6)$$

where $\mathbf{\Lambda}$ is a $(N \times M)$ diagonal matrix whose elements are the non-zero singular values λ_r, $r = 1,...R$ (R is the rank of $\mathbf{A}$) and zero for the null singular values, $\mathbf{U}$ is an $(N \times N)$ orthonormal matrix of eigenvectors spanning data space and $\mathbf{V}$ is an $(M \times M)$orthogonal matrix of eigenvectors spanning model space. Therefore, the solution of the equation (4) can be written in the following form:

$$\mathbf{M} = \mathbf{V \Lambda^{-1} U^T \Theta}, \quad (7)$$

Matrix Λ^{-1} is a $(M \times N)$ diagonal matrix whose elements are $1/\lambda_r$, $r = 1,...R$. Thus, any error in the data will be multiplied by $1/\lambda_r$ and be amplified for the very small singular values. In order to reduce the impact of noise it is necessary to eliminate the singular values which are smaller than cutoff value.

3. Simulations

We consider influence of the GST on the temperature-depth distribution, and derive what information can be retrieved from this profile. We solve the test problem with parameters similar ones obtained from the borehole measurements. The depth of the borehole is $1000m$, step of temperature measurements is $1m$ with accuracy of 0.01 $°C$, the thermal diffusivity is

$a^2 = 10^{-6}\ m^2/s$. For simplicity we assume that the long-term surface temperature T_0 and the geothermal gradient Γ_0 are exact measured. The GST is given by the step function (Fig. 1 (a)):

$$\mu(t) = \begin{cases} 0, & 0 < t < 200, \\ 0.8, & 200 < t < 400, \\ 0, & 400 < t < 1000. \end{cases} \tag{8}$$

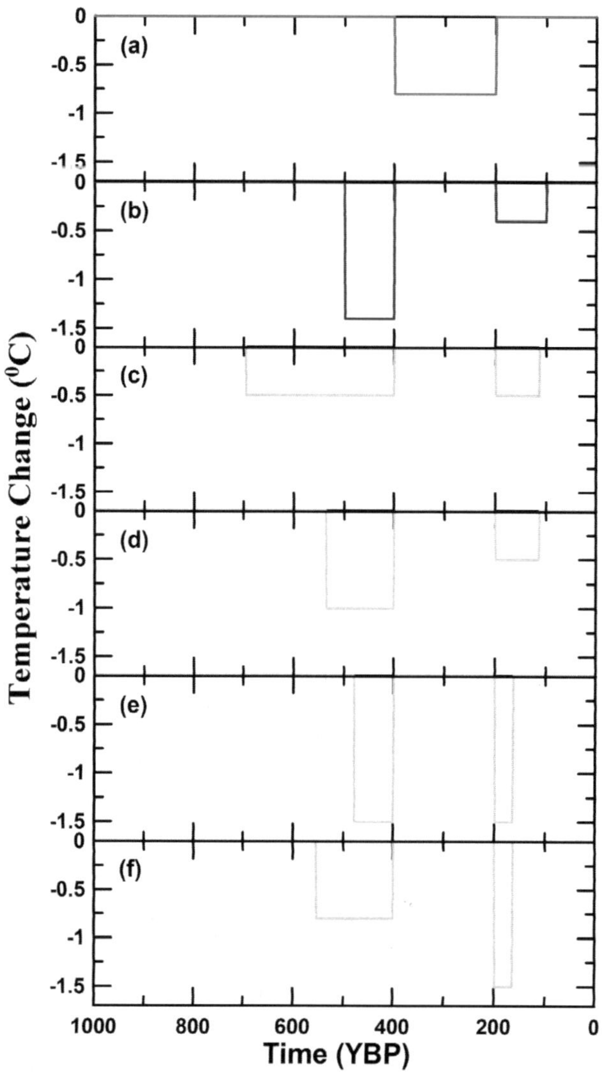

Figure 1. The test GST (Eq. 8) (a) and five chosen GST (b-f) resulting in the same temperature-depth disturbance in range of 0.01 $°C$ error.

This function is close to the Little Ice Age that is the last most significant climatic event at the Earth. The temperature-depth disturbances in the present time $t = 0$ are shown in Fig. 2. We also point out some other GST result in the similar temperature-depth profiles with accuracy 0.01 $°C$. There are infinite set of such GST. The number of these variants increases with increase of the temperature measurement steps and accuracy. Consider examples of GST in the form of two rectangle signals:

$$\mu(t) = \begin{cases} 0, & 0 < t < t_1, \\ \mu_1, & t_1 < t < 200, \\ 0, & 200 < t < 400, \\ \mu_2, & 400 < t < t_2, \\ 0, & t_2 < t < 1000. \end{cases} \tag{9}$$

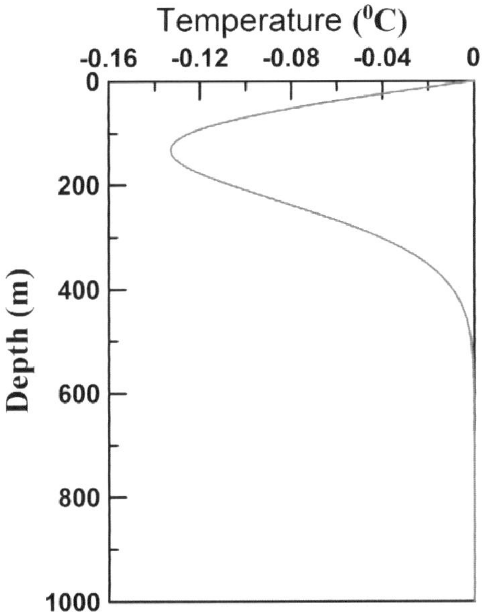

Figure 2. The temperature-depth disturbances for test GST (Eq. 8).

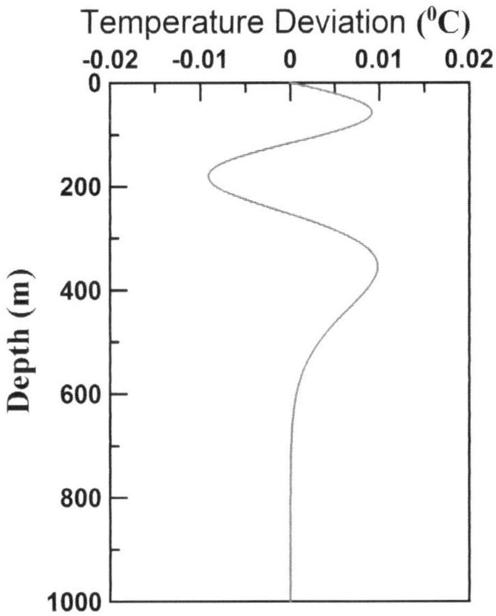

Figure 3. Deviation of the temperature-depth profiles corresponding to GST (Fig. 1(a) and (b)).

One can see that the chosen (9) and test (8) temperatures differ significantly each other. They have different intervals where each signal is zero (Fig. 1(b-f)). With measurement accuracy it is not possible to determine which GST results in the measured temperature-depth profile: the temperature (8) with climatic signal at $200 < t < 400$ years or the temperature (9) with two climatic signals at $t_1 < t < 200$ and $400 < t < t_2$ years.

In all previous papers based on the step GST, the reconstructed temperatures were interpreted as the average temperatures at the one time interval [1, 2, 3, 4]. These reconstructions are not corrected (Fig. 1). This situation can not be fixed due to increase time intervals corresponding to the father past time. It is not possible uniquely to determine average temperature at each interval (Fig. 1(b)) at the existing accuracy Fig. 3. This interval length is used for the past surface reconstructions based on data from the Global Database of Borehole Temperatures and Climate Reconstructions.

The other test problem is based on the GST shown in Fig. 4 (black) and used in several papers [4, 6]. This function was based on typical results obtained from analyses of the geothermal data in Eastern Canada. It was checked to estimate quality of the reconstructions by thye SVD method. The borehole depth is equal to $600m$ and the thermal diffusivity is $a^2 = 10^{-6}\ m^2/s$ in this examples. The temperature anomaly in borehole due to GST is shown in Fig. 5. We also consider four other GST accordingly to Fig. 4. These GST result in the same borehole-depth profile in range of the temperature measurement accuracy 0.01 °C. Deviations of the four temperature-depth prtofiles from the test temperature (Fig. 5) are shown in Fig. 6.

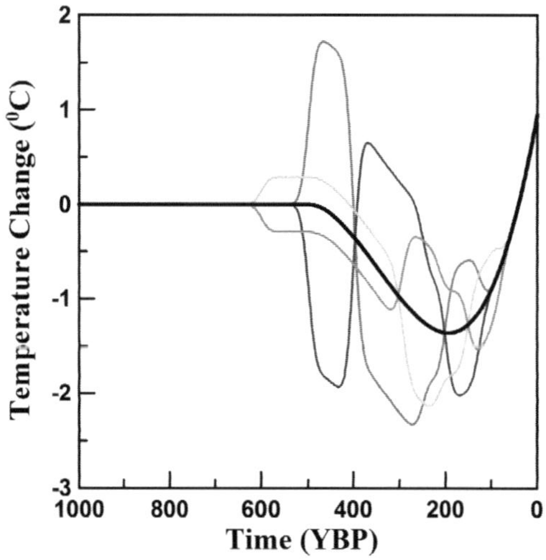

Figure 4. The test GST [4] (black) and four GST resulting in the same temperature-depth disturbance in range of 0.01 °C error.

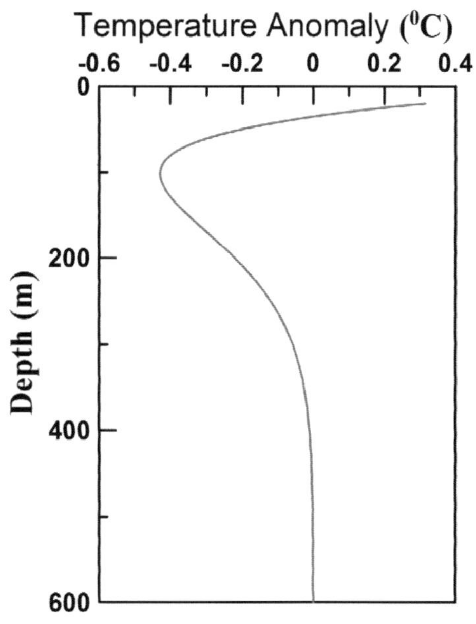

Figure 5. The temperature-depth disturbances for the test GST (Fig. 4, black).

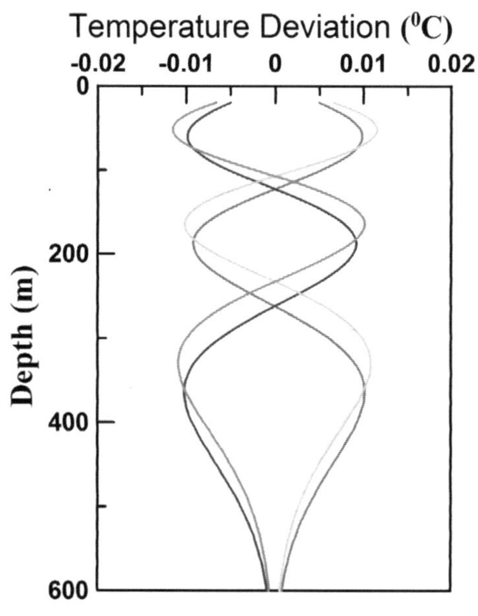

Figure 6. Deviation of the temperature-depth profiles corresponding to GST (Fig. 4).

The solution of the inverse problem for these five temperature anomaly by the piece-wise functions. Each function consists of twenty steps of fifty years duration. In the SVD method

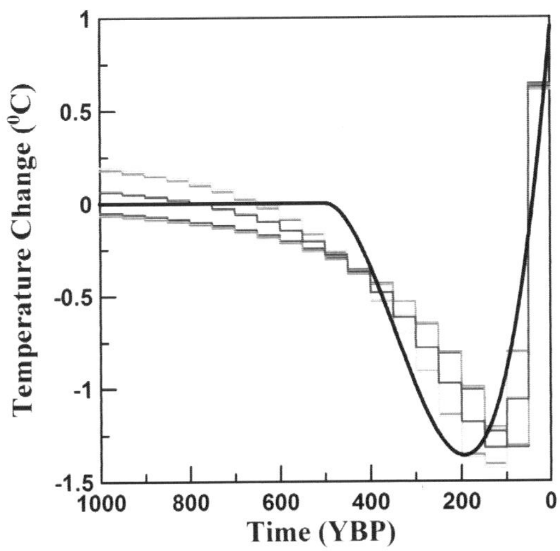

Figure 7. The reconstructed GST based on the SVD method for temperature-depth profiles (Fig. 4).

an eigenvalue cutoff set at 0.025 similar to [4]. The reconstructed temperatures are shown in Fig. 7. Their comparison with GST (Fig. 4) allows us to conclude that important features of the climatic signals at the surface are significantly filtrated (Fig. 7). All reconstructed temperatures are smoothed so that reconstructed temperature at 50 years intervals is not often average surface temperature at the at same interval. It is most noticeable in the range of 200-400 YBP.

4. Conclusions

The Earth thickness filtrates significantly high-frequency components of the surface temperature signals. As a result the surface temperature history information is significantly lost in the measured temperature-depth borehole profiles. We derived examples of the inverse problem for the GST reconstructions where climatic signals can not be found out by the SVD method. Therefore, the numerous paleotemperature reconstructions have to be reconsidered. The correction of the GST reconstructions can be done if there are additional data such as the tree ring width data, isotope data and others [7].

References

[1] Shen P Y and Beck A E 1991 Least squares inversion of borehole temperature measurements in functional space *Journal of Geophysical Research* **96** 19965–19979

[2] Mareschal J C and Beltrami H 1992 Evidence for Recent Warming from Perturbed Geothermal Gradients: Examples from eastern Canada *Climate Dynamics* **6** 135–143

[3] Pollack H N and Huang S 2000 Climate Reconstruction from Subsurface Temperatures *Annual Review of Earth and Planetary Sciences* **28:1** 339–365

[4] Beltrami H Matharoo G S and Smerdon J E 2015 Ground surface temperature and continental heat gain: uncertainties from underground *Environmental Research Letters* **10** 014009

[5] Carslaw H S and Jaeger J C 1959 *Conduction of Heat in Solids, 2nd ed.* (Oxford Univ. Press, New York)

[6] Beltrami H Smerdon J E Matharoo G S and Nickerson N 2011 Impact of maximum borehole depths on inverted temperature histories in borehole paleoclimatology *Clim. Past* **7** 745–756

[7] Nagornov O V Nikitaev V G Pronichev A N Tyuflin S A and Bukharova T I 2016 Analysis of past surface temperature reconstructions based on the tree-ring chronologies and borehole temperature measurements *AIP Conference Proceedings* **1738** 230006

Multicriteria adaptation principle on example of groups of mobile robots

A P Nelyubin[1], S Yu Misyurin[2]

[1] Mechanical Engineering Research Institute RAS, 4 Malyi Kharitonievski pereulok, Moscow 101990, Russia
[2] National Research Nuclear University MEPhI, 31 Kashirskoe shosse, Moscow 115409, Russia

nelubin@gmail.com

Abstract. The article presents a multicriteria approach to the adaptation of groups of search, explore or research robots to unknown and volatile environment conditions. The basis of this approach is the application of multicriteria analysis both at the design stage of a group of mobile robots and at the stage of its adaptation in real-time conditions. It is proposed to maintain a variety of robots by properties and by optimality criteria in order to take into account the preferred mode of operation.

1. Introduction

In [1] we proposed a multi-criteria approach to the design of complex robotic systems capable of adapting to dynamically changing operating conditions. In this article, the general principles of multicriteria adaptation are illustrated by examples of groups of mobile robots. Such groups of robots can be used for the purpose of carrying out search, explore or research work in hard-to-reach or completely inaccessible areas. For example, in search and rescue operations in emergency situations (earthquakes, destruction of structures, fires, floods). Another example is the exploration of natural resources, the discovery of oil, gas, minerals. Mobile robots can explore the relief of the seabed, craters of volcanoes, the surfaces of other planets, high layers of the atmosphere. Such technologies of the future as "smart dust" [2], self-organizing robot swarm [3], or surgery with microscopic robots can also be regarded. Other examples of mobile micro-robots applications are military robots, secure systems, smart home, telecare systems [4, 5].

The operating conditions of the robotic system can vary greatly. The environment is characterized by different permeability, chemical composition, temperature, radiation background. The passability of robots also depends on the relief of the surfaces, the structure of the cavities. In addition to the fact that these conditions are usually hard and aggressive, they can still be heterogeneous and vary dynamically in a wide range. Moreover, in a number of problems, the possible values of environmental parameters are not known in advance. This imposes high demands on the reliability of the robots being developed.

In addition to reliability, there are other criteria for the quality and efficiency of both individual robots and the robotic system as a whole. These criteria depend on the purpose of the system and the tasks it performs. And in different periods of time the system and individual robots can solve different tasks. For example, an investigation of the whole region, moving from place to place, transporting goods, specialized work. Performance criteria can include speed, passability, positioning accuracy, accuracy of measurements and transmitted information, and other specialized indicators.

2. Multicriteria adaptation principles

Available literature on the designing adaptive systems using the word "multicriteria" simply lists a number of criteria by which the system can be optimized. Then optimization is performed either by individual criteria independently, or the main criterion is selected, and the remaining criteria are used as constraints.

The approach we propose is to use multicriteria analysis tools both at the design stage of the system and at the stage of its adaptation to the changing environmental conditions in the real-time mode of operation.

One of the main principles of designing adaptable systems is the possibility of reconfiguring the system during its operation mode [6-7]. In the case when the designed system consists of a set of mobile robots, new opportunities for such a reconfiguration appear. In particular, one can change:

- configuration of individual robots,
- composition of a group of robots,
- common centralized control system for a group of robots (if any),
- mechanism of interaction of robots with each other.

The main principle of multicriteria adaptation is taking into account changes in preferences between optimality criteria depending on changing conditions of the system functioning [1]. And the system reconfiguration should be carried out not only in accordance with the changed parameters of the environment and the task, but also in accordance with the changed preferences.

The problem of designing a group of mobile robots provides many opportunities for implementing the main principle of multicriteria adaptation. For example, we can:

- maintain a variety of robots in terms of parameters and properties,
- divide groups of robots into subgroups by specialization,
- vary the composition of a group of robots.

These advantages can allow automatic and decentralized multicriteria adaptation of a group of robots. For these purposes, bio-inspired approaches [1], such as genetic algorithms [8] and particle swarms [9], can be used.

Thus, maintaining the diversity and heterogeneous of robots induces competition mechanisms among individual robots in order to optimize the configuration of robots. In addition to competition, collaboration mechanisms can also be used to perform common tasks by a group of robots [14-15]. For example, by executing a leader-follower scenario [1, 16].

It is important to emphasize that maintaining the diversity of robots allows the use of ideas of multicriteria adaptation at the hardware level. In particular, it is concerns the idea of creating a Pareto-optimal coalition of robots at the stage of multicriteria design of the system [1].

3. Sensor networks

One of the promising areas for the development of robotics technology is the creation of sensor networks, consisting of mini-sensors exchanging information via wireless peer to peer communication [10-12]. Areas of application are security and personalization systems, smart home, telecare and others [4, 5, 10]. They can also be used for search, explore or research.

The optimality criteria of such systems are the following: density of the signal coverage; data accuracy; fault tolerance; reliability of the system, the system life span; low cost and power consumption, small size, lightweight.

In [11] the main criterion is distinguished as the life span of the entire system operation without recharging. In this case, the remaining criteria act as constraints. In particular, the complete coverage of the network must be ensured. For this, special algorithms are used to calculate this coverage, which do not require knowledge of the sensors coordinates, but use wireless signals for exchanging information between the sensors. Adaptation in [11] is regarded just as self-tuning of the sensors location in order to provide the required level of coverage. In this case, the initial location of the nodes and the target is undefined and may change.

Within the example of sensor networks, it is possible to demonstrate the ideas of the proposed multicriteria adaptation approach. In real problems, such sensor systems must operate in different modes, in which preferences for the optimality criteria of the system can be very different. Here are some examples:

1) Security systems spend most of the time in the monitoring and waiting mode. At this time, economy and reliability are more important at a given level of coverage by a signal. But at certain points in time, the system goes into the mode of threat recognition or personalization. And then the criteria for the data quality and accuracy are brought to the forefront.

2) Search and rescue systems operate part of the time in target (victim) detection mode. In this case, the size of the coverage area by the signal, the speed and efficiency of the space investigation of the robot group are important. But when localizing the target, accuracy of data transmission and fault-tolerance become more important.

Ideas in the construction of sensor networks can be used in the design of a group of mobile robots that perform other functions than sensing. In such a system, a mini-sensor can be either an individual robot or a part of a more complex mobile robot.

4. Distributed multifunctional systems

Many works are devoted to the design of adaptive modular robots [7, 13-14]. These are complex systems, consisting of separate modules, physically connected to each other. Modules can be heterogeneous and perform various functions. Such a system can also be considered a group of robots. Among the applications of modular robots are complex manipulators, mobile robots, multifunctional systems.

As a mechanism for self-adapting the modular robot, the feedback mechanism on the sensor-actuator pair is often considered. Several approaches have been proposed in this direction based on the interaction of neighbor modules or robots, in particular multi-agent algorithms called distributed consensus [14-15]. However, such firmware adaptation mechanisms do not take into account the multicriteria nature of the problem.

In [16-17] a more complex approach to designing the architecture of a distributed system of heterogeneous robots is proposed. In addition to self-adaptation mechanisms, the concept of self-management of the system is considered. Special attention is paid to the development of software for such systems, including complex algorithms for finding solutions, as well as machine learning [17].

Such a complex approach allows us to use the ideas of multicriteria adaptation of robot groups proposed in [1], as it opens up opportunities for integrating robot software with multicriteria optimization and analysis systems. For this purpose we develop Visual Analytics Tool system [18].

5. Conclusion

The proposed ideas of multicriteria adaptation of mobile robotic systems can enhance flexibility and robustness of such systems in dynamically changing conditions. This allows to transfer part of multicriteria optimization routine from design phase to the real-time phase of system operation. So we can change configurations in accordance with changing conditions and multicriteria preferences, or even optimize some parameters on-line.

The study was financially supported by the RFBR project No. 16-29-04401.

References

[1] Misyurin S Yu, Nelyubin A P, Ivlev V I 2017 Multicriteria adaptation of robotic groups to dynamically changing conditions. *IOP Conf. Series: J. of Physics* vol 788(1), 012027.

[2] Warneke B, Last M, Liebowitz B and Pister K S J 2001 Smart Dust: Communicating with a Cubic-Millimeter. *Computer* vol 34 pp 44–51

[3] Rubenstein M, Cornejo A and Nagpal R 2014 Programmable self-assembly in a thousand-robot

swarm. *Science* vol 345 issue 6198 pp 795–799

[4] Essa I A 2000 Ubiquitous sensing for smart and aware environments. *IEEE Personal Communication* vol 7(5) pp 47–49

[5] Noury N, et al. 2000 Porcheron Monitoring behavior in home using a smart fall sensor. *IEEE-EMBS Special Topic Conf. on Microtechnologies in Medicine and Biology* pp 607–610

[6] Christoforou E G, Muller A, Phocas M C, Matheou M and Arnos S 2015 Design and Control Concept for Reconfigurable Architecture. *J. of Mechanical Design* vol 137, 042302

[7] Yim M, Shen W, Salemi B, Rus D, Moll M, Lipson H, Klavins E and Chirikjian G S 2007 Modular Self-Reconfigurable Robot Systems. *IEEE Robotics & Automation Magazine* pp 43–52

[8] Mitchell M 1996 An Introduction to Genetic Algorithms (Cambridge, MA: MIT Press)

[9] Poli R, Kennedy J and Blackwell T 2007 Particle Swarm Optimization: An Overview. *Swarm Intelligence* vol 1(1) pp 33–57

[10] Rabaey J M, Ammer M J, da Silva J L, Patel D, and Roundy S 2000 PicoRadio supports ad hoc ultra-low power wireless networking. *IEEE Computer Magazine* vol 33(7) pp 42–48

[11] Zhang H, Nixon P, Dobson S 2008 Multi Criteria Adaptation Mechanisms in Homological Sensor Networks. *11th IEEE Singapore Int. Conf. on Communication Systems ICCS 2008* pp 937-942

[12] Tian D, and Geogranas N D 2002 A coverage-preserving node scheduling scheme for large wireless sensor networks. *ACM International Workshop on Wireless Sensor Networks and Applications WSNA-2002* pp 32–41

[13] Yu C, Nagpal R 2010 A Self-Adaptive Framework for Modular Robots in Dynamic Environments: Theory and Applications. *Int. J. of Robotics Research* vol 30(8) pp 1015–1036

[14] Yu C, Nagpal R 2009 Self-adapting modular robotics: A generalized distributed consensus framework. *In: IEEE Int. Conf. on Robotics and Automation (ICRA)* pp 1881–1888

[15] Olfati-Saber R, Fax J, and Murray R 2007 Consensus and cooperation in networked multi-agent systems. *In: Proc. of IEEE* vol 95(1) pp 215–233

[16] Edwards G, Garcia J, Tajalli H, Popescu D, Medvidovic N, Sukhatme G, Petrus B 2009 Architecture-driven self-adaptation and self-management in robotics systems. *In: ICSE Workshop on Software Engineering for Adaptive and Self-Managing Systems (SEAMS)* pp 142–151

[17] Hartmann J, Stechele W, Maehle E 2013 Self-adaptation for Mobile Robot Algorithms Using Organic Computing Principles. *In: Kubatova H., Hochberger C., Danek M., Sick B. (eds) Architecture of Computing Systems – ARCS 2013. Lecture Notes in Computer Science* vol 7767 Springer, Berlin, Heidelberg

[18] Nelyubin A P, Galkin T P, Galaev A A, Popov D D, Misyurin S Yu, Pilyugin V V 2017 Usage of Visualization in the Solution of Multicriteria Choice Problems. *Scientific Visualization* vol 9(5)

Modelling the influence of noise of the image sensor for blood cells recognition in computer microscopy

V G Nikitaev, O V Nagornov, A N Pronichev, E V Polyakov and V V Dmitrieva

National Research Nuclear University MEPhI (Moscow Engineering Physics Institute), 31 Kashirskoe shosse, 115409 Moscow, Russia

vgnikitayev@mephi.ru

Abstract. The first stage of diagnostics of blood cancer is the analysis of blood smears. The application of decision-making support systems would reduce the subjectivity of the diagnostic process and avoid errors, resulting in often irreversible changes in the patient's condition. In this regard, the solution of this problem requires the use of modern technology. One of the tools of the program classification of blood cells are texture features, and the task of finding informative among them is promising. The paper investigates the effect of noise of the image sensor to informative texture features with application of methods of mathematical modelling.

1. Introduction

One of the relevant problems of medicine is blood cancer. The first step in the diagnosis of these diseases is the analysis of blood smears. The effectiveness of existing automated hematology analyzers is still not high enough, so their work must be supplemented by microscopic analysis of blood smears and aspirates of bone marrow [1]. Usually this work is done by a skilled doctor of clinical laboratory diagnostics. Such a study is time consuming, requires a large exertion, and the result is subjective, depends on the skill of the doctor and sometimes incorrect [2].

The use of modern computer technology can help in solving this problem, and would serve as a useful tool [3-5]. The application of decision-making support systems would reduce the subjectivity of the diagnostic process, and as a result avoid errors, resulting in often irreversible changes in the patient's condition [6].

Software for classification of pathological cells used texture features. Note that the input images for analysis system are distorted by noise [7]. In this regard, it is necessary to assess the degree of influence of noise on texture features and how it affects on the results of classification of cell types[8].

In order to study the influence of this factor in different ranges of values texture features we proposed to use the method of mathematical modeling. A mathematical model of image formation of the nuclei of the blood cells was offered for reflecting the different nature of the structure of chromatin to various cell types [9]. It was a result of analysis of real images of different types of blood cells [10].

2. Materials and methods

The following model of image of the blood cell nucleus was proposed. Image field is covered with a conventional grid, a step which is set as parameter, circles are in the nodes with random brightness, the center of which is deviated from the node to a random value. The radius of the circles, their brightness and distance along the axes x and y are normally distributed, their mean value and standard deviation are specified as model parameters. Areas of intersection are colored by weighted sum of brightness of the circles [11].

As noise models was used values of a random variable, distributed according to the normal law. The model is described by the equation:
$g(x, y) = f(x, y) + \eta(x, y)$, where $f(x, y)$ is original image, $g(x, y)$ is the noisy image, $\eta(x, y)$ is additive noise and independent of the signal with Gaussian probability density function. The value of the mathematical expectation of the noise was equal to zero, and the standard deviation value is specified as a model parameter in the present study.

It is possible to create images with texture typical for the structure of the chromatin of the nuclei of various types of blood cells by using the proposed model. This achieves compliance as visually perceived images of the model and the real blood cells and measured on these images quantitative texture features.

Texture features "energy", "inertia moment", "maximum probability", "local homogeneity", "entropy" were used. They are computed on base of normalized matrix of spatial adjacency (Haralic matrix).

The image model generation was produced for sampling images with random values of model parameters. That provided the variation of the values of texture characteristics in each of the cell types. Noise with varying degrees of intensity was added to images on the next stage of the modeling. The intensity of the noise was set to the standard deviation for the normal distribution with zero expectation. The noise standard deviation was from the series of values: 0,2,4,6,8,10,12,14,16,18 gradations of brightness.

Examples of real and model images for blast and lymphoid cells are shown on figure 1.

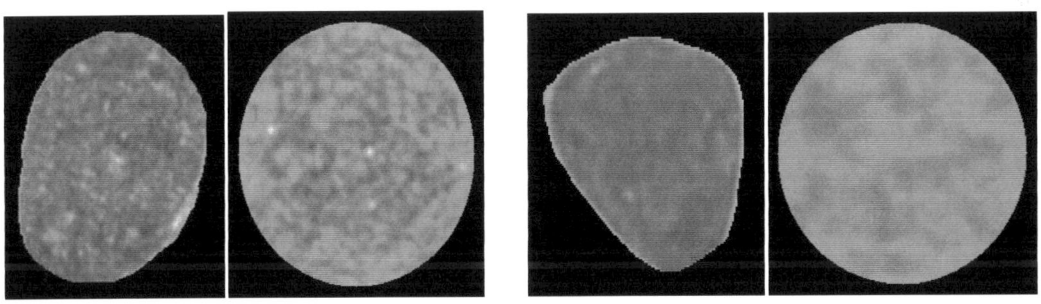

Figure 1. Examples of real and model images for blast and lymphoid blood cells: – first и third picture are the nucleus of blast and lymphoid blood cells, the second and fourth picture are model images.

3. Results and discussion

The sample contained 100 images for each noise level and the cell type. The results are shown in graphs of the dependencies of the values of the texture features from the noise standard deviation. The chart shows the results for the two types of cells – blasts and lymphoid blood cells. Each class for the noise level is represented by three points – the average value of the texture attribute and the range limits, which are specified by double value of the standard deviation of the feature for the sample at some noise level.

The analysis of the experimental data revealed that the best separation in the classification of cell types makes use of the texture feature "energy". Graphs of the texture feature "energy" depending on the noise level for the two types of blast and lymphoid cells are shown on figure 2.

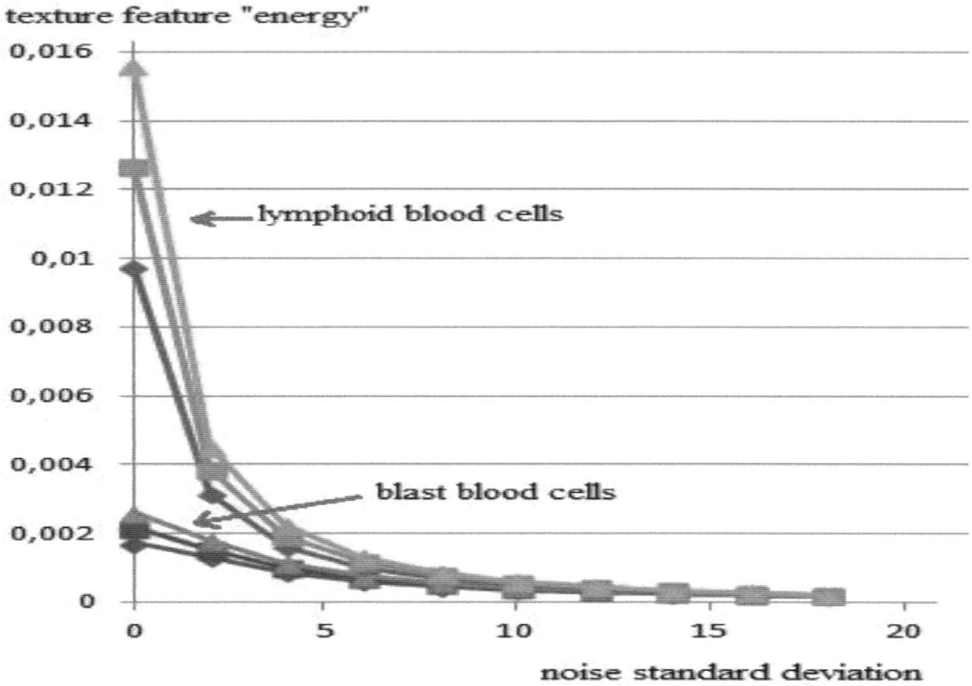

Figure 2. Texture feature "energy" depending on the noise level

It can be seen from the graph in figure 2, that the value of the texture feature "energy" decreases with increasing intensity of the noise simultaneously for both cell types, while the differences between them are also reduced. And when the level of noise standard deviation is above 10 brightness levels the differences between two cells type are fully absent.

4. Conclusion

On the results of the model experiment we can draw the following conclusions:

- if we have noise rise, texture characteristics of chromatin in images of the nuclei of the blood cells become less distinguishable for different types of cells, and the value of texture features begins to describe not the structure of chromatin, but it describes noise texture in the image of the nucleus of the blood cells;
- the blood cell images obtained in the systems with various kind of image sensors with different noise levels in the image is not correct to compare the value of texture features (images of the same cell photographed in different systems will have different texture features);
- the parameters of the classifier based on the measurement of texture features should be configured for a specific system, because these parameters depend on the noise level in the image, and when you will replace the image sensor in the computer microscopy system, you should check the change of the noise level in the image, as the change in intensity requires a change of the parameters of the classification of blood cells, and characteristics of classification accuracy of the blood cells will be changed too.

Acknowledgement

The reported study was funded by RFBR according to the research project № 17-07-01496.

IOP Conf. Series: Journal of Physics: Conf. Series **937** (2017) 012035 doi:10.1088/1742-6596/937/1/012035

References

[1] Nikitaev V G, Nagornov O V, Pronichev A N, Polyakov E V and Dmitrieva V V 2015 *WSEAS transactions on biology and biomedicine* **12** 16
[2] Pronichev A N, Polyakov E V, Tupitsyn N N, Frenkel M A and Mozhenkova A V 2017 *J. Phys.: Conf. Ser.* **784** 012003
[3] Nikitaev V G, Nagornov O V, Pronichev A N , Polyakov E V, Sel'chuk V Y, Chistov K S, Blyndar V N , Dmitrieva V V, Zaytsev S M and Gordeev V V 2015 *Measurement techniques* **57** (10) 1203
[4] Zakharenko Y V, Nikitaev V G, Polyakov E V and Seldyukov S O 2017 *J. Phys.: Conf. Ser.* **798** 012127
[5] Blindar V N, Nikitaev V G, Polyakov E V and Matveeva I I 2017 *J. Phys.: Conf. Ser.* **798** 012129
[6] Nikitaev V G, Pronichev A N, Polyakov E V, Dmitrieva V V, Tupitsyn N N, Frenkel M A and Mozhenkova A V 2017 *J. Phys.: Conf. Ser.* **784** 012042
[7] Polyakov E V, Nikitaev V G and Seldyukov S O 2017 *J. Phys.: Conf. Ser.* **798** 012130
[8] Polyakov E V and Nikitaev V G 2017 *J. Phys.: Conf. Ser.* **798** 012128
[9] Mozhenkova A V, Tupitsin N N, Frenkel M A, Falaleeva N A, Nikitaev V G and Polyakov E V 2017 *J. Phys.: Conf. Ser.* **798** 012126
[10] Nikitaev V G, Nagornov O V, Pronichev A N, Polyakov E V, Sel'chuk V Y, Chistov K S, Blyndar V N, Dmitrieva V V and Gordeev V V 2014 *Measurement techniques* **57** (5) 560
[11] Nikitaev V G, Nagornov O V, Pronichev A N, Dmitrieva V V and Polyakov E V 2017 *J. Phys.: Conf. Ser.* **788** 012056

Matching of energetic, mechanic and control characteristics of positioning actuator

N Y Nosova[1,2], S Yu Misyurin[1,2], G V Kreinin[2]

[1] National Nuclear Research University MEPhI, Moscow, Russia
[2] Blagonravov Institute of Machine Science, RAS, Moscow, Russia

ssmmrr@mail.ru

Abstract. The problem of preliminary choice of parameters of the automated drive power channel is discussed. The drive of the mechatronic complex divides into two main units – power and control. The first determines the energy capabilities and, as a rule, the overall dimensions of the complex. The sufficient capacity of the power unit is a necessary condition for successful solution of control tasks without excessive complication of the control system structure. Preliminary selection of parameters is carried out based on the condition of providing the necessary drive power. The proposed approach is based on: a research of a sufficiently developed but not excessive dynamic model of the power block with the help of a conditional test control system; a transition to a normalized model with the formation of similarity criteria; constructing the synthesis procedure.

1. Introduction

Positioning actuator consists of two main parts, i.e. power and control units (Fig.1). Its dynamics and specified tracking performance depend on coordination of characteristics both units. The complexity of the control system depends directly on energetics and mechanics of the power unit. In one's turn the later determines the energetic activity, overall dimensions of an actuator as a whole device and their capability in generating of the modes desired.

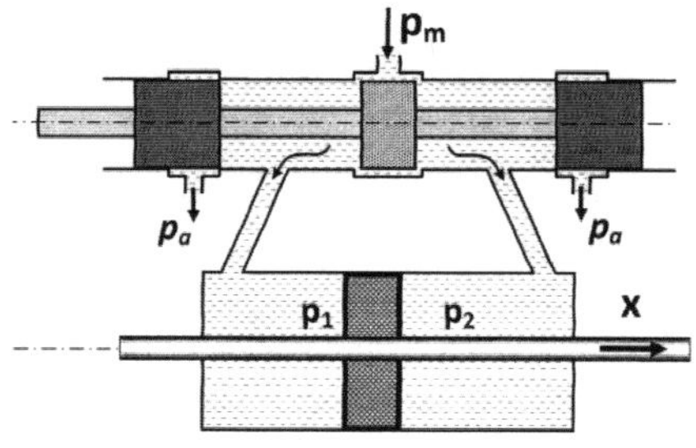

Figure 1. Positioning actuator.

There are a lot of publications, where design procedure of the actuator is reduced to only control system design without optimization of the power unit. Hence it follows the great variety of relatively complicate control systems with the neurocontrollers, self-tuning fuzzy PID controllers etc., intended for devices of the same type [1 -3]. But, on the other hand, there is very difficult to design both units taken as a whole because of the considerable complexity of actuator dynamics.

Some authors proposed a compromised methodology based on the separate but the interconnected designing procedure of the power and control units [4-7]. The reference mode of the output link motion, typical for the actuator designed, is taken as a basis of the methodology considered. As a reference mode for positioning actuator some kind point-to point movement of the output link can be selected. The vibration mode can be used as a reference mode for the actuator operated in tracking mode, etc. In such cases the structure and parameters of the power unit may be determined on account for the degree of two modes (reference and actual mode) correlation on the base of an analysis of the actuator dynamic normalized model simulation in the specified mode. Information obtained is used further for the revealing of the basic relations between the similarity criteria of the system for the tradeoff solution within specified constraints.

2. The choice of parameters.

The parameters selection on account for two mentioned modes correlation, means that the actuator performance is tuned namely on the reference mode given. It gives the opportunity to use the reverence mode signal as a signal controlling the output link movement on the main part of the travel up to the end position. In some point near the end of the stroke, the reference mode signal is substituted automatically for any type specified positioning signal.

As stated before, the design procedure is based on the simulation results, in general case, of nonlinear dynamic model of the actuator. The model of the positioning actuator considered here, is presented in normalized dimensionless form. As a general requirements for point-to point law of movement can be, e.g., specified travel time t_s, near to zero velocity at the end of the process, gradually acceleration mode over the first half of the stroke and the gradually deceleration mode over the second half of the stroke. The law of movement describing as $\hat{\nu} = 0,5(1-\cos\hat{\omega}\hat{\tau})$ meets the requirements mentioned; $\hat{\omega} = \pi / \hat{\tau}_s$ is normalized, conditional frequency of the process, $\hat{\tau}_s$ is normalized travel time. The designing procedure is illustrated further by designing of the point-to-point positioning actuator with electric motor, linear mechanical characteristic, inertial and power loads.

The dynamic model $\ddot{\nu} = \chi - \dot{\nu}/\Lambda - \lambda_L\dot{\nu} + \chi_L$ of the power unit is expressed in the output coordinates, where $\chi = \gamma - \dot{\nu}/\Lambda$ represents the reduced motor moving force. The terms χ_L and $\lambda_L\dot{\nu}$ represent constant load and fluid friction load, where $\lambda_L = 0,2\Lambda$. The index value is in consequence of an fluid friction force makes 1/10 of the actuator starting force when $\dot{\nu} = .5\Lambda$. The variable γ characterizes the active part of the reduced motor moving force realized as a response to the input control signal $\bar{\gamma}$ with time delay $\dot{\gamma} = (1/\tau_e)(\bar{\gamma} - \gamma)$, where τ_e is time constant of the control system.

Using the Painlev'e test one can study the analytical properties as was suggested in reference [8-10]

The main purpose of the current study is to present the methodology of an application of the reference signal as a control signal in point-to-point positioning actuator. As a reference mode, mentioned above the law $\hat{\nu} = 0,5(1-\cos\hat{\omega}\hat{\tau})$ was selected. On the main part of the stroke (the region of active reference mode) the control signal $\bar{\gamma} = \hat{\kappa}_1(\hat{\nu} - \nu) + \hat{\kappa}_2(\hat{\dot{\nu}} - \nu)$ is activated. Near the end point the reference signal is substituted for special positioning signal, for example, for $\bar{\gamma} = \kappa_1(\nu_e - \nu) - \kappa_2\dot{\nu}$, where $\hat{\nu}_e$ is the end coordinate of positioning object. The dynamic model simulation showed, that the distance between the reference and actual modes depends on two main factors - travel time τ_s (the main factor) and moving force χ (that is a loop gain $\hat{\kappa}_1$); decreasing of τ_s leads to decreasing of the distance between curves mentioned and, as will be shown further, decreasing of the motor power needed. The

IOP Conf. Series: Journal of Physics: Conf. Series **937** (2017) 012036 doi:10.1088/1742-6596/937/1/012036

process performance depends, as well, on the similarity criteria of the system: normalized no-load velocity of the output link Λ and load factors χ_L and λ_L.

Figure 3 illustrates a correlation between the reference and actual modes. It can be seen that curves are disposed fairly close each other since the magnitude of τ_s is close to its lower limit. Some diagrams showing total control process performance are represented in Figure 2. Here both parts of the control mode (reference and final control) are shown as integrated process for diverse χ_L. Figure 4 gives a notion of the system performance on condition that the load is changed while the output link is moving. In that case, the load is changed sharply from zero to .4 and back to zero in the middle of the stroke. It can be seen that this action has prompted only a rather small violation in velocity curve while the displacement curve has no signs of the disturbance. This way can be used to study any other environmental effects on a system performance.

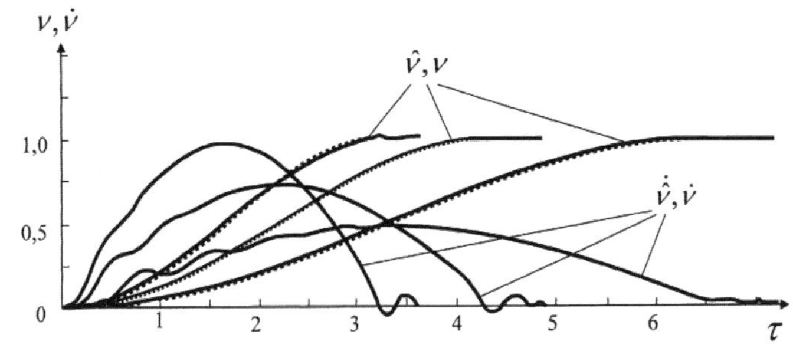

Figure 2.

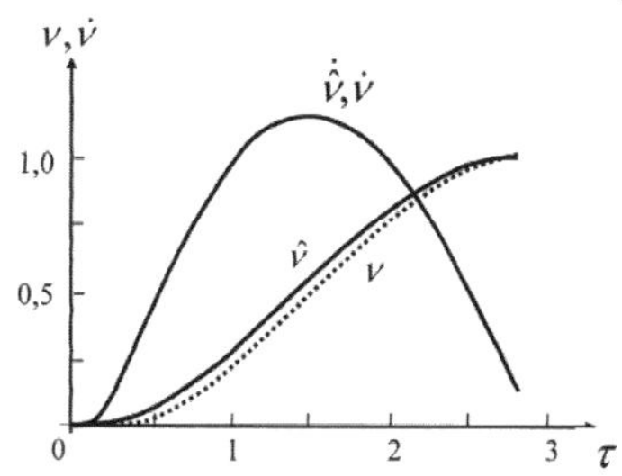

Figure 3.

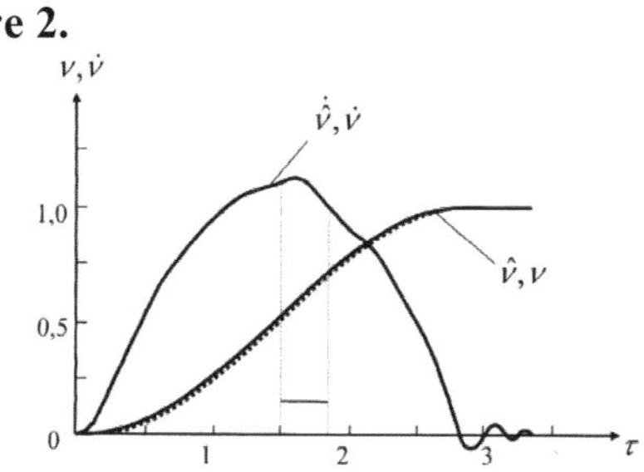

Figure 4.

In conclusion the procedure of parameters estimation based on the simulation results is to be considered. The required power of the motor W is determined in general case as [6 – 7]

$$W = P_x \dot{x}_h = \frac{m_y s_y^2}{t_s^3}\left(\Lambda \tau_s^3 + \frac{x_h^2 t_s^2}{s^2} (\alpha_{xy} \tau_s / \Lambda) \right)$$

where m_x, m_y are input and output inertial load, $s_y = y_e - y_i$, t_s is a stroke time desired, $\dot{x}_h$ is an no-load velocity of an motor, $\alpha_{xy} = m_x / m_y$, P_x is a starting moving force of a motor. If $\alpha_{xy} \approx 0$, then $W_{\min}$ is to be determined using a relationship $\Lambda \tau_s^3 = (\Lambda \tau_s^3)_{\min} \approx 23 + 400 \chi_L^2$, ($\chi_L = 0 - 0{,}4$). Mechanism gear ratio is defined according to $i = \dot{x}_h t_s / 3 s_y$. The more the value of $\dot{x}_h$, the more noticeable is an effect of α_{xy}. The $W \approx W_{\min}$ is defined in this case using the relation given above, assuming $\tau_s = 3 + 10\chi_L^2$ и $\Lambda = 1$. Mechanism gear ratio is defined using relationship $i = \dot{x}_h t_s / s_y \tau_s$, where $\tau_s = 3 + 10\chi_L^2$. The deigning procedure is considered in details in [6-7].

Methodology of the matching performance characteristics of the actuator is based on the assumption that the totality its performance modes can be represented by some only mode of general type. So the point-to-point positioning mode can be represented by any mode having desired time of the motion and monotonic profile of the transient processes on the symmetrical acceleration and deceleration parts, In continuous position (or other variable) tracking systems. The moving target following mode can be considered as some general mode as well.

3. Conclusion

In this paper is considered the point-to-point positioning actuator characterized by the general mode $y = 0{,}5(1 - \cos\Omega t)$, where $\Omega = \pi / t_s$ is the conditional frequency of the process, t_s is the stroke time, y is the displacement coordinate of the output link. The proposed mode proved to be very suitable since it contains only parameter Ω and do not change its state when the initial and end positions are varied. It gives the possibility to employ the selected mode in cases when the initial conditions are varied. For reasons mentioned this mode was assigned as a basic mode and is called the reference mode.

The same reference mode was employed for the power unit parameters estimation of the point-to-point actuator with hydraulic motor, but not used for the fulfillment of the control functions [6,7]. In this paper the usage of the proposed reverence mode as control mode acting up to coming near the end position is considered. The graphs representing the positioning process performance shows the fairly accurate reproduction of the motion desired on account that normalized stroke time is over its lower limit. The reference mode reproduction is supplemented by relatively simple control mode, acting on the end part of the stroke. But it is separate problem to be solved.

It is to be noted that in the parameters estimation the dynamic model (linear or nonlinear) of the system was used intensively. The transition to the dimensionless form facilitates essentially the designing procedure and gives an additional chance for the matching of the dynamic and control performance of the actuator.

This work was supported by the Russian Foundation for Basic Research, project №15-08-05493-a.

This work was supported by the MEPhI Academic Excellence Project (agreement with the Ministry of Education and Science of the Russian Federation of August 27, 2013, project no. 02.a03.21.0005).

References

[1] K.Nikzad, J. Chahoussi, L. Paul. Actuators dynamics and delay compensation using neurocontroller. Journ. of Engineering Mechanics Vol.122, N10, 1996.

[2] Zulfatman and M.F. Rahmat. Application 0f self-tuning fuzzy PID controller on industrial hydraulic actuator. International Journ. on smart sensing and intelligent systems, Vol.2, N2, June, 2009.

[3] Z. Wang, J. Shao, J. Lin, G, Han. Research on controller design and simulation of electro-hydraulic servo-system. Proc. of the 2009 JEEE International Conference on Mechatronics and Automation. August 9-12, Changchun, Chine.

[4] B.I. Petrov, B.I. Polkovnikov, L.B. Pobinovich. Dynamics of servo drives. Moscow, 1982, 496p.

[5] G.V.Kreinin and S.Yu.Misyurin. Dynamics and Synthesis of the parameters of a Positioning Drive. ISSN 1052-6188, Journal of Machinery Manufacture and Reliability, 2009, Vol. 38, No. 6, pp.523-530

[6] S. Yu. Misyurin and G. V. Kreinin. Power Optimization Criteria of a Mechanical Unit of an Automated Actuator. ISSN 1028_3358, Doklady Physics, 2015, Vol. 60, No. 1, pp. 15–18.

[7] S. Yu. Misyurin and G. V. Kreinin. Dynamics and design of a power unit with a hydraulic piston actuator. ISSN 1028_3358, Doklady Physics, 2016, Vol. 61, No. 7, pp. 354–359.

[8] N. A. Kudryashov. JOURNAL OF PHYSICS A-MATHEMATICAL AND GENERAL Volume:

35 Issue: 21 Pages: 4617-4632 Article Number: PII S0305-4470(02)33994-5 Published: MAY 31 2002.
[9] N. A. Kudryashov; A. S. Zakharchenko. APPLIED MATHEMATICS LETTERS Volume: 32 Pages: 53-56 Published: JUN 2014
[10] N. A. Kudryashov, M. V. Demina. APPLIED MATHEMATICS AND COMPUTATION Volume: 210 Issue: 2 Pages: 551-557 Published: APR 15 2009

IOP Conf. Series: Journal of Physics: Conf. Series **937** (2017) 012037 doi:10.1088/1742-6596/937/1/012037

Inverse problem for the equation with n-times integrated semigroup

D G Orlovsky

National Research Nuclear University MEPhI (Moscow Engineering Physics Institute), Moscow

E-mail: odg@bk.ru

Abstract. We consider an abstract differential equation of the first order with unbounded linear operator $u'(t) = Au(t) + f(t)$ in a Banach space X. For this equation the Cauchy problem is studied with initial data $u(0) = x$ at $t = 0$. We assume that the operator A generates an n-times integrated semigroup $V(t)$. The inverse problem to determine the nonhomogeneous member is considered under the assumption that this term has the following representation $f(t) = p$, where p is an unknown element of the space X. This problem belongs to the class of inverse problems. Inverse problems for abstract differential equations was discussed initially with this inhomogeneous structure member but for equations with an operator generating a C_0-semigroup. An additional condition $u(T) = y$ is specified for the determination of the unknown p, where y is a given element of the space X. Thus we get the two-point problem, which had not been considered previously for integrated semigroups. The question of existence and uniqueness of the classical solution of the inverse problem is studied. Sufficient conditions of correct solvability of the inverse problem are obtained. Explicit formula to determine the unknown element in the differential equation is given.

Let us consider the abstract Cauchy problem for differential equations in Banach space X

$$\begin{cases} u'(t) = Au(t) + p, & 0 \leqslant t \leqslant T, \\ u(0) = x, \\ u(T) = y. \end{cases} \tag{1}$$

We will assume that A is a linear closed operator with a dense domain and this operator generates an n-times integrated non-degenerate semigroup $V(t)$. The elements x, y, $p \in X$. It should be noted that the term "integrated" is conditional and he is, generally speaking, not responding the integration of strongly continuous semigroups in Banach spaces. The concept of integrated semigroups was first introduced in [1] and [2]. Unlike an ordinary semigroup, for which $V(t)V(s) = V(t+s)$, integrated semigroup satisfies another identity

$$V(t)V(s) = \frac{1}{(n-1)!}\left(\int\limits_t^{t+s}(t+s-\tau)^{n-1}V(\tau)d\tau - \int\limits_0^s(t+s-\tau)^{n-1}V(\tau)d\tau\right).$$

This fact complicates the theory of such operator functions, n-times integrated semigroup is any operator function $V(t)$ that satisfies a specified the above identity and strongly continuous in t.

Note here that instead of the equality $V(0) = I$, which are valid for ordinary semigroups, we now have a different ratio $V(0) = 0$. The generator A of the semigroup is defined also more complicated way: $y = Ax$ if and only if

$$V(t)x = \int_0^t V(\tau)y d\tau + \frac{t^n}{n!}x \quad \text{for every } t \geqslant 0.$$

N-fold integral of any C_0-semigroup $S(t)$ is the simplest example of n-times integrated semigroups

$$V(t) = \int_0^t dt_1 \int_0^{t_1} dt_2 \ldots \int_0^{t_{n-1}} S(t_n)dt_n.$$

Integrated semigroups have many properties similar to ordinary semigroups. For example, $V(t)$ is bounded in the operator norm on any finite interval, $V(t)$ and $V(s)$ commute for all $t, s \geqslant 0$, integrated semigroup commutes with its generator in its domain. However, some properties of ordinary semigroups are not transferred to the integrated semigroups. For example, there are integrated semigroups which are not exponentially bounded (however, if the generator of the semigroup has nonempty resolvent set, then this integrated semigroup has a finite exponential type), there are integrated semigroups with not densely defined generators. In this article, we assume nondegeneracy integrated semigroups: equality $V(t)x = 0$ for all $t \geqslant 0$ imply $x = 0$ (not all integrated semigroups are nondegenerate).

We give first the basic information about the direct problem

$$u'(t) = Au(t) + f(t), \quad u(0) = x, \quad t \geqslant 0. \tag{2}$$

Integrated semigroups are a useful tool for the study abstract of the Cauchy problem (2), well-posed provided that $x \in D(A^{n+1})$ (for example, for the case when A – elliptic differential operator in $L_p(R^n)$).

In this article we will only consider classical solutions of the Cauchy problem. This means that the function $u(t) \in C^1([0;T];X) \cap C([0;T];D(A))$ and satisfies (2). We will use the method of solution of the abstract Cauchy problem (2), in which it is reduced to an integral equation

$$y(t) = A\int_0^t y(s)ds + \frac{t^n}{n!}x + \int_0^t \frac{(t-s)^n}{n!} f(s)ds. \quad (3).$$

In the work [2] established the following connection between solutions of problems (2) and (3). If the operator A is the generator of the n-times integrated semigroup, the function $f \in C^{n+1}([0,T];X)$, $x \in D(A)$, $x_1 = Ax + f(0) \in D(A), \ldots, x_n = Ax_{n-1} + f(0) \in D(A)$, then the function

$$y(t) = V(t)x + \int_0^t V(t-s)f(s)ds$$

$n+1$ times continuously differentiable and is a solution of the integral equation (3), moreover, the problem (2) has a unique solution $u(t)$, with

$$u(t) = y^{(n)}(t).$$

IOP Conf. Series: Journal of Physics: Conf. Series **937** (2017) 012037 doi:10.1088/1742-6596/937/1/012037

We note that the above statement means that the integrated semigroup can be obtained as an evolutionary family (solving operator) the integral equation (3).

We also need the following formula for differentiation: if $x \in D(A)$, then

$$\frac{d}{dt}(V(t)x) = V(t)Ax + \frac{t^{n-1}}{(n-1)!}x,$$

(this formula contains for example, in [3]).

Feature of problem (1) is the fact that an element p included in the right part of the differential equation is unknown. Therefore, in this problem it in addition to the solution of the equation $u(t)$ it is necessary to find also the unknown element. Such problems have received the name of inverse problems (see for example [4]). The condition $u(0) = x$ is usual Cauchy data. The equality $u(T) = y$ provides additional information (the so-called overspecified data). This additional information allows to determine the unknown element of p. Due to the fact that the solution $u(t)$ is defined in two points $t = 0$ and $t = T$ such inverse problems have received the name two-point problems. These two-point problems was actually the first inverse problems for abstract differential equations. Previously, such problems were considered only for equations with the operator A, generating a C_0-semigroups (see [5] and [6]).

Recall that the Cauchy problem (1) we discussed in class

$$u \in C^1([0,T],X) \cap C([0,T],D(A)).$$

This problem for the case of strongly continuous semigroups ($n = 0$) has initiated the consideration of many reverse task. Read more about this is mentioned in [4]. Here we turn to the study of this problem for the case of integrated semigroups. In the work [5] a similar problem was considered for the case $n = 1$, but with a different nonhomogeneous term of $f(t) = tp$ instead of $f(t) = p$. For the new model we get the result on the solvability of the inverse problem for arbitrary natural n.

The first two equations in the inverse problem (1) constitute a direct problem (2) with a constant nonhomogeneous term of $f(t) = p$. Obviously, this nonhomogeneous term satisfies the condition of solvability of the direct problem $f \in C^{n+1}([0,T];X)$ for any n.

Before proceeding to the inverse problem, we consider some properties of the direct problem

$$\begin{cases} u'(t) = Au(t) + f(t), & 0 \leqslant t \leqslant T, \\ u(0) = x. \end{cases} \tag{4}$$

Lemma. Let $f \in C^{n+1}([0,T];X)$ and there is a sequence

$$x_0, x_1, ..., x_n$$

for which

$$x_0 = x,\ x_{k+1} = Ax_k + f^k(0),\ (k = 0, 1, ..., n-1).$$

Then the solution of problem (4) is given by the formula

$$u(t) = V(t)x_n + \sum_{k=0}^{n-1} \frac{t^k}{k!} x_k + \int_0^t V(t-s) f^{(n)}(s) ds. \tag{5}$$

Proof. According to [2] the solution of problem (4) is closely related to the solution of integral equations

$$y(t) = A \int_0^t y(s)ds + \frac{t^n}{n!} x + \int_0^t \frac{(t-s)^n}{n!} f(s)ds. \tag{6}$$

The solution of equation (6) is a function

$$y(t) = V(t)x + \int_0^t V(t-s)f(s)ds, \tag{7}$$

and solution of the problem (4) is given by the formula

$$u(t) = y^{(n)}(t).$$

Differentiating equation (7) n times and using formula

$$\frac{d}{dt}V(t)x = V(t)Ax + \frac{t^{n-1}}{(n-1)!}x$$

we get the equality (5). The Lemma is proved

Let us consider the inverse problem (1) and formulate the main result.

Theorem. Let $x \in D(A^{n+1})$, $y \in D(A^n)$ and the operator

$$V(T)A^{n-1} + \sum_{k=1}^{n-1} \frac{T^k}{k!} A^{k-1}$$

is invertible. Then the solution of the inverse problem exists and is unique.

Proof. Assuming $f(t) = p$ we find a sequence

$$x_0 = x, \ x_k = A^k x + A^{k-1} p \quad (k = 1, \dots, n).$$

Consequently, the solution of the direct problem

$$u(t) = V(t)A^n x + V(t)A^{n-1}p + \sum_{k=0}^{n-1} \frac{t^k}{k!} A^k x + \sum_{k=1}^{n-1} \frac{t^k}{k!} A^{k-1} p. \tag{8}$$

Putting in the equality (8) $t = T$, we obtain the equation for the unknown element of p

$$B(T)p = g,$$

where

$$B(t) = V(t)A^{n-1} + \sum_{k=1}^{n-1} \frac{t^k}{k!} A^{k-1},$$

$$g = y - V(T)A^n x - \sum_{k=0}^{n-1} \frac{T^k}{k!} A^k x$$

By the condition of the theorem the operator $B(T)$ is invertible, so the solution of the inverse problem exists and is unique and, moreover, this solution is given by the explicit formulas

$$\begin{cases} p = B(T)^{-1}g, \\ u(t) = V(t)A^n x + B(t)B(T)^{-1}g. \end{cases}$$

The theorem is proved.

We note a special case of a once integrated semigroup. The solution of the inverse problem for this case is given by the formula:

$$\begin{cases} p = V(T)^{-1}(y - V(T)Ax - x), \\ u(t) = V(t)Ax + x + V(t)V(T)^{-1}(y - V(T)Ax - x). \end{cases}$$

It is important to compare the result with the classical result for strongly continuous semigroups (see [5]). We restrict here to the case of invertible operator A.

Integrated semigroup $V(t)$ is expressed through a strongly continuous semigroup $U(t)$ representing solving operator of the problem (2), according to the formula

$$V(t) = A^{-1}(U(t) - I).$$

therefore, the condition of invertibility of the operator $V(T)$ is equivalent to the invertibility of the operator $U(t) - I$. It is the previously known result (see [5]). We must also note that stronger results were obtained for strongly continuous semigroups. These results apply separately to the uniqueness of the solution and separately to the existence of the solution of the inverse problem (1).

So, for example, for uniqueness it is sufficient to points

$$\lambda_k = 2\pi ik/T, \quad k \in Z, \quad k \neq 0$$

do not lie in the point spectrum of the operator A (the invertibility of the operator A is not actually required). Additional conditions required for the existence of a solution (see [5]). The complete results has been obtained for equations with analytic semigroups. For problems with integrated semigroups this issue has to be studied.

References

[1] Arendt W 1987 Resolvent positive operators *Proc. London Math. Soc.* B **54** 321-349

[2] Arendt W 1987 Vector-valued Laplace trabsforms and Cauchy problem *Israel J.Math.* B **59** 327-352

[3] Arendt W, Batty J K, Hieber M, Neubrander F 2001 Vector-valued Laplace transforms and Cauchy problems *Monogr. Math.* vol 96 (Basel: Birkhauser)

[4] Prilepko A I, Orlovsky D G, Vasin I A 1999 *Methods for solving inverse problems in mathematical physics* (New York, Basel: Marcel Dekker)

[5] Rundell W 1980 Determination of an unknown nonhomogeneous term in a linear partial differential equation from overspecified boundary data *Appl. Anal.* B **10** 231-242

[6] Eidelman Y S 1993 Two-point boundary value problem for the differential equation with a parameter *Dokl. AN USSR* B **4** 15-18

Quasistatic problems for piecewise-continuously growing solids with integral force conditions on surfaces expanding due to additional material influx

D A Parshin[1,2] **and A V Manzhirov**[1,2,3,4]

[1] Laboratory of Modelling in Mechanics of Solids, Institute for Problems in Mechanics of the Russian Academy of Sciences, Prospekt Vernadskogo 101-1, Moscow 119526, Russia
[2] Department of Applied Mathematics, Bauman Moscow State Technical University, 2-ya Baumanskaya ulitza 5, Moscow 105005, Russia
[3] Department of Higher Mathematics, National Research Nuclear University MEPhI, Kashirskoye Shosse 31, Moscow 115409, Russia
[4] Department of Higher Mathematics, Moscow Technological University (MIREA), Prospekt Vernadskogo 78, Moscow 119454, Russia

E-mail: parshin@ipmnet.ru, manzh@inbox.ru

Abstract. The piecewise continuous processes of additive forming of solids are studied. The being formed solids exhibit properties of deformation heredity and aging. The approaches of linear mechanics of growing solids in the framework of the theory of viscoelasticity of the homogeneously aging isotropic media are applied. Nonclassical boundary-value problems for describing the mentioned processes with the integral satisfaction of force conditions on some expanding due to the influx of additional material parts of the formed solid surface are investigated. A proposition about the commutativity of the time-derived integral operator of viscoelasticity with a limit depending on the solid point with the integration over an arbitrary, expanding due to the growth, surface inside or on the boundary of the growing solid is given. This proposition provides a way to construct the solution of the corresponding growing solids mechanics problem on the basis of Saint-Venant principle. The solution will retrace the evolution of the stress-strain state of the solid under consideration during and after the process of its additive formation. An example of applying the announced technic to modelling the processes of additive forming solids of conical shape under simultaneous action of end loads that are statically equivalent to an axial time-varying force is demonstrated.

1. Introduction

The additive formation of solids is realized in a wide variety of natural and technological processes. Many of these processes can be considered as continuous growing processes, such that during the formation of a solid an infinitely thin layer of additional material joins to its surface each infinitely small period of time. In the course of additive processes different factors influence on solids being formed and cause their deformation. The development of stress-strain state of such solids is impossible to describe within the framework of classical concepts of continuum mechanics in principle. This is due to the lack of any configuration of the continuously growing solid which could be associated with introduction of the strain measures. An adequate description of mechanical behavior of solids deforming in processes of their continuous growing can be given on the basis of approaches and methods of mechanics of growing solids being actively developed nowadays [1–11].

In the proposed study we consider the situation when the solid being additively formed exhibits the properties of deformation heredity (viscoelasticity) and aging (weakening the deformation properties over time regardless stresses existing in the solid), and therefore, during pauses in the growing process as well as after the final cessation of growth the solid continues to change its stress-strain state. This situation is quite difficult to simulate as rheological manifestations in the deformation response of the material continuously interact with mechanical reactions of the solid on the developing in time process of adding new material elements to it.

2. The material properties and its description

We consider the homogeneously aging isotropic linearly viscoelastic material described by the equation of state [12]

$$\mathbf{T}(\mathbf{r},t) = \mathcal{H}^{-1}_{\tau_0(\mathbf{r})}\big[2\mathbf{E}(\mathbf{r},t) + \chi\mathbf{1}\,\mathrm{tr}\,\mathbf{E}(\mathbf{r},t)\big] \tag{1}$$

were $\tau_0(\mathbf{r})$ is the time of occurrence of stresses at the point of the solid with the radius-vector $\mathbf{r}$; $\mathbf{T}$ and $\mathbf{E}$ are the stress and small strain tensors at the point $\mathbf{r}$ at the time instant t, $\mathbf{1}$ is the unit tensor of the second rank; $\chi = 2\nu/(1-2\nu)$, and $\nu = \mathrm{const}$ is the Poisson's ratio. The linear time-operator $\mathcal{H}_s^{-1} = G(t)\big(\mathcal{I} + \mathcal{N}_s\big)$ is inverse to the linear operator $\mathcal{H}_s = \big(\mathcal{I} - \mathcal{L}_s\big)G(t)^{-1}$ with the real parameter $s \geqslant 0$. Here $G(t)$ is the elastic shear modulus, $\mathcal{I}$ is the identity operator,

$$\mathcal{L}_s f(t) = \int_s^t f(\tau)K(t,\tau)\,d\tau, \quad \mathcal{N}_s f(t) = \int_s^t f(\tau)R(t,\tau)\,d\tau,$$

$$K(t,\tau) = G(\tau)\,\frac{\partial\Delta(t,\tau)}{\partial\tau}, \qquad \Delta(t,\tau) = \frac{1}{G(\tau)} + \omega(t,\tau).$$

$K(t,\tau)$ and $R(t,\tau)$ are the kernels of creep and relaxation, $\Delta(t,\tau)$ and $\omega(t,\tau)$ are the specific strain function and the creep measure for pure shear ($t \geqslant \tau \geqslant 0$). It is accepted by definition that $\omega(\tau,\tau) \equiv 0$.

We denote $g^\circ(\mathbf{r},t) = \mathcal{H}_{\tau_0(\mathbf{r})}\,g(\mathbf{r},t)$ for arbitrary function $g(\mathbf{r},t)$ of solid point $\mathbf{r}$ and time t, and we denote $f^\circ(t) = \mathcal{H}_{t_0} f(t)$ for arbitrary function of time $f(t)$ which is not associated with specific points of considered solid.

Note that the defining relations written out above were developed especially for the description of processes of concrete deformation. However, they are also well suited to describe the mechanical behaviour of some rocks, as well as polymers, soils, ice.

3. Commutativity of the time-derived viscoelasticity operator with the integration over an expanding surface

The state equation (1) is used in the present work to describe the mechanical behaviour of a solid which is built up with additional material being loaded simultaneously with its attaching to the solid. Obviously, in this case, the function $\tau_0(\mathbf{r})$ in (1) will be determined in the following way. In the originally existing (before accreting) part of the solid it will be identically equal to the time moment t_0 of loading of this part. In the additional part of the solid, formed during accreting, it will coincide with the distribution $\tau_*(\mathbf{r})$ of moments of attaching particles $\mathbf{r}$ of additional material to the solid.

We investigate mechanical problems for growing solids in quasistatic statement and in the approximation of small strains and displacements. The latter let us consider the time-variable space domain occupied with the growing solid to be known at any time instant and prescribed by a specific simulated growing process. So the part of the growing solid surface to which the additional material inflows during this process moves in the space in a known manner. We will name this part of the solid surface the *growth surface*.

We suppose also that the problem in question contains integral force conditions on some expanding due to the influx of additional material parts of the growing solid surface. We can prove the following proposition:

Let Ω_0 and Ω_{A} be two arbitrary limited surfaces inside or on the boundary of a solid, subordinated to the state equation (1) *and formed in a process of piecewise-continuous accretion in N stages $t \in [t_{2k-1}, t_{2k})$ $(k = 1, \dots, N)$ of continuous growth with arbitrary long pauses between them. The surface Ω_0 lies entirely within the original (existing before accreting) part of the solid considered. The surface Ω_{A} lies entirely in the additional part of the solid and is obtained by motion in space of a curve $\Gamma(t)$, $t \in [t_1, +\infty)$, which belongs to the current growth surface of the solid at every moment t of its continuous accreting and is fixed in the pauses between the stages of continuous accreting, i.e. outside the time intervals $[t_{2k-1}, t_{2k})$. Let $g(\mathbf{r}, t)$ be an arbitrary function defined in the points $\mathbf{r}$ of surfaces Ω_0 and Ω_{A} for $t \geqslant \tau_0(\mathbf{r})$. Assume that $g\big(\mathbf{r}, \tau_*(\mathbf{r})\big) = 0$. Then, when $t > t_1$ the formula*

$$\frac{d}{dt}\left[\int_{\Omega(t)} g(\mathbf{r}, t)\, dS\right]^{\circ} = \int_{\Omega(t)} \frac{\partial g^{\circ}(\mathbf{r}, t)}{\partial t}\, dS$$

will be fair, where the expanding in time (disconnected in general) surface $\Omega(t)$ combines the surface Ω_0 and that part of the surface Ω_{A}, which has already been formed by the time $t \geqslant t_0$: $\Omega(t) = \Omega_0$ if $t \in [t_0, t_1]$, and $\Omega(t) = \Omega_0 \cup \big\{\Gamma(\tau)\,\big|\, t_1 \leqslant \tau \leqslant t\big\}$ if $t \in (t_1, +\infty)$.

The formulated proposition gives a way to solve the corresponding problem for the growing solid on the basis of Saint-Venant principle as we show below.

We omit the proof of the proposition due to the limited volume of this paper. Note that the surfaces Ω_0 and Ω_{A} considered in the proposition may have arbitrary curvature. Meanwhile their boundaries may not have common points. In the special case it is possible that $\Omega_0 = \varnothing$. Forming a surface Ω_{A} curves $\Gamma(t)$ can be both closed and unclosed. In particular, the surface Ω_{A} may "circle" original part of the solid or form a "tube" enveloping only the material of the additional part of the having been formed solid.

4. An example of solving a problem with integral force conditions on the expanding surface of a viscoelastic growing solid

Applying the above announced proposition to solving the growing solids mechanics problems with integral force conditions is demonstrated in the present work on the example of modelling additive processes of formation of the relatively long in the axial direction conical solids. It is assumed that in the process of formation of the solid its end surfaces are acted with loading statically equivalent to the axial tension–compression force which can change over time. Forming the solid under consideration is carried out by means of its thickening in the radial direction due to the influx of additional material to the conic side surface. This process is piecewise continuous, i.e. consists of arbitrary number of stages of continuous accreating alternating with arbitrary long pauses during which the influx of the material does not take place. With the help of the given proposition the solution of the boundary-value problem for the considered mechanical model is constructed in the presented work in a closed analytical form.

4.1. Description of the problem

Let there be a conical solid of rotation which length l significantly exceeds its transverse dimensions. It is made from isotropic homogeneous aging linearly viscoelastic material subordinated to the constitutive equation (1). Take the moment of this material nucleation be the start of timing t.

IOP Conf. Series: Journal of Physics: Conf. Series **937** (2017) 012038 doi:10.1088/1742-6596/937/1/012038

At the moment $t = t_0$ a load is applied to the ends of the existing solid. We believe that at every moment of time $t \geqslant t_0$ it is statically equivalent to axial forces acting in the central points of the ends and varying with time following the law $P(t)$. We will consider positive the magnitude of tensile end force.

Some time after the application loading at the time $t = t_1$ we start the process of gradual axisymmetric thickening of the considered conical solid by adding the additional material to its lateral initially free from stresses surface. Thickening occurs in such a way that in each time moment the accreted body maintains the shape of a right circular truncated cone of length l. This process is piecewise continuous in time, i.e. it consists of N consecutive phases of continuous accreting $t \in [t_{2k-1}, t_{2k})$ $(k = 1, \ldots, N)$, separated by pauses of arbitrary duration. At the stages of continuous accreting an infinitely thin layer of material attaches to the solid each infinitely small period of time. The added material is supposed identical to the original one. In pauses the influx of additional material to the solid does not take place and its lateral surface is free from stresses. In the process of piecewise continuous accreting and after its completion time-varying central axial forces $P(t)$ continue to act to the end surfaces of the cone.

We will investigate the evolution of stress-strain state of the considered conical solid under specified conditions of loading before the start, during and upon the completion of the described process of accreting.

Changing the geometry of the considered conical solid due to its piecewise-continuous accreting is completely defined obviously by defining laws of increasing the radii of its ends in time. Denote them by $a(t)$ and $b(t)$, $t \geqslant t_0$. These functions are continuous, non-decreasing and constant outside intervals $[t_{2k-1}, t_{2k})$.

Let the reference plane of a cylindrical polar coordinate system coincides with that end of the cone which radius was denoted by $a(t)$. Place the beginning of coordinates O in the center of this end and extend coordinate axis Oz perpendicular to it inside the cone. Denote the polar radius and the angle as ρ and φ. If $\{\mathbf{e}_\rho, \mathbf{e}_\varphi, \mathbf{k}\}$ is normalized local basis of the introduced cylindrical coordinate system (ρ, φ, z), then the radius-vector of an arbitrary point of the solid can be represented in the form $\mathbf{r} = \mathbf{e}_\rho(\varphi)\,\rho + \mathbf{k}\,z$.

Moving due to the influx of additional material (accreting) the lateral surface of the cone under consideration is described by the equation $\rho = \Lambda(z, t)$, where $\Lambda(z, t) = a(t) \cdot (1 - z/l) + b(t) \cdot z/l$. The trace of its passing in the space forms an additional part of the considered solid. At time moments $t \in [t_{2k-1}, t_{2k})$ $(k = 1, \ldots, N)$ the lateral surface represents the actual growing surface of the cone, i.e. it is the level surface t of the function $\tau_*(\mathbf{r})$. Unit vectors of the external (directed from the axis of the cone) normal line to this surface form a vector field $\mathbf{n}(\mathbf{r}) = \mathbf{e}_\rho(\varphi) \cos\alpha\big(\tau_*(\mathbf{r})\big) - \mathbf{k} \sin\alpha\big(\tau_*(\mathbf{r})\big)$, in the additional part of the solid, where $\alpha(t) = \arctan\{[b(t) - a(t)]/l\}$ is the current polarstar angle of the growing cone.

4.2. Boundary problem on the stage before accreting

Before the start of accretion the stress-strain state of the considered conical solid can be determined on the basis of the theory of viscoelasticity of homogeneously aging isotropic solids [13] and the principle of Saint-Venant from the solution of the following classical mechanical boundary value problem with integral force condition on its end surface, $t_0 \leqslant t \leqslant t_1$:

$$\nabla \cdot \mathbf{T} = \mathbf{0}, \quad 0 \leqslant \rho < \Lambda(z, t_0); \qquad \mathcal{H}_{t_0}\mathbf{T} = 2\mathbf{E} + \chi \mathbf{1}\,\mathrm{tr}\,\mathbf{E}, \quad \mathbf{E} = (\nabla\mathbf{u}^{\mathrm{T}} + \nabla\mathbf{u})/2;$$
$$\mathbf{n} \cdot \mathbf{T} = \mathbf{0}, \quad \rho = \Lambda(z, t_0); \qquad \int_{\{z=l\}} \left\| \begin{matrix} \mathbf{k} \cdot \mathbf{T} \\ \mathbf{e}_\rho \rho \times (\mathbf{k} \cdot \mathbf{T}) \end{matrix} \right\| dS = \left\| \begin{matrix} \mathbf{k}\,P(t) \\ \mathbf{0} \end{matrix} \right\|. \tag{2}$$

Here $\mathbf{u}(\mathbf{r}, t)$ is the vector field of displacements. To exclude displacement components not causing deformation of the solid we are to impose the conditions of fixing the neighborhood of the center point of one of its end surfaces: $\mathbf{u} = \mathbf{0}$ and $\nabla \times \mathbf{u} = \mathbf{0}$ for $\rho = 0$, $z = 0$. We require these conditions to be satisfied after the start of the process of the considered solid accretion as well.

The boundary value problem (2) can be reformulated for values $\mathbf{u}$, E, T°, $t_0 \leqslant t \leqslant t_1$:

$$\begin{aligned} &\nabla \cdot \mathsf{T}^\circ = \mathbf{0}, \quad 0 \leqslant \rho < \Lambda(z, t_0); \quad \mathsf{T}^\circ = 2\mathsf{E} + \chi \mathbf{1} \operatorname{tr} \mathsf{E}, \quad \mathsf{E} = (\nabla \mathbf{u}^{\mathrm{T}} + \nabla \mathbf{u})/2; \\ &\mathbf{n} \cdot \mathsf{T}^\circ = \mathbf{0}, \quad \rho = \Lambda(z, t_0); \quad \int_{\{z=l\}} \begin{Vmatrix} \mathbf{k} \cdot \mathsf{T}^\circ \\ \mathbf{e}_\rho \rho \times (\mathbf{k} \cdot \mathsf{T}^\circ) \end{Vmatrix} dS = \begin{Vmatrix} \mathbf{k} P^\circ(t) \\ \mathbf{0} \end{Vmatrix}. \end{aligned} \tag{3}$$

In the boundary value problem (3) time t is not a significant variable but acts only as a parameter. We can see that the problem (3) turned out to be mathematically equivalent to the similar classical mechanical problem of the equilibrium of a linearly elastic truncated circular cone of permanent composition with free lateral surface $\rho = \Lambda(z, t_0)$, $z \in [0, l]$, being under the action of axial forces centrally applied to its ends. The radii of the ends of the cone and the value of forces acting on it depend on a real parameter t. This formal coincidence is gained by substituting in the problem (3) the values P° to the value of tensile force related to the shear modulus, and the tensor T° to the stress tensor related to the shear modulus. We can construct the closed analytical solution of the described classical problem if we combine known specific solutions of the theory of elasticity [14]. We will not trace out the constructed analytical solution in this succinct paper.

4.3. Boundary problem on the stage of piecewise-continuous accretion

Due to the objective lack of natural (unstressed) configuration in the growing solid the kinematic description of the process of its deformation that is traditional in the mechanics of deformable solids is not suitable for this solid. However, it is clear that the particles of the new material after the attaching to the surface of growth continue to move as a part of continuous, even though growing, solid. This means that in the region of space occupied by the whole growing solid at this time, the enough smooth velocity field $\mathbf{v}(\mathbf{r}, t)$ of the motion of its particles is uniquely determined. Therefore, the problem of such a body deformation can be put in terms of velocity. In this case in the formulation of the defining relations of the material a tensor of velocities of deformation

$$\mathsf{D}(\mathbf{r}, t) = (\nabla \mathbf{v}^{\mathrm{T}} + \nabla \mathbf{v})/2 \tag{4}$$

may play a part of the deforming process characteristics. The adopted equation of state (1) can be rewritten by using this tensor in the form [15]:

$$\mathsf{S} = 2\mathsf{D} + \chi \mathbf{1} \operatorname{tr} \mathsf{D}, \tag{5}$$

where we have introduced the so-called tensor of velocities of operator stresses $\mathsf{S}(\mathbf{r}, t) = \partial \mathsf{T}^\circ / \partial t$.

The approach requires knowing the whole story of changing the state of additional material elements up to their inclusion in the composition of the solid considered. In the studied in the present work process of accreting the additional material is supposed to be initially free of stresses (see Subsection 4.1). In other words, we believe that the additional material begins to deform directly in the time of its attaching to the formed body, and the attaching layers of additional material to the surface of the body does not cause the appearance nonzero stresses in the formed solid near the surface of its growth:

$$\mathsf{T} = \mathbf{0}, \quad \rho = \Lambda(z, t), \quad t \in [t_{2k-1}, t_{2k}) \ (k = 1, \ldots, N). \tag{6}$$

Note that condition (6) provides the equality to zero of the stress vector $\mathbf{n} \cdot \mathsf{T}$ at the current growth surface, i.e. unload of this surface.

The condition of instantaneous local equilibrium in the growing body has obviously the same form as in the classical solid of permanent composition. In the considered case of mass forces absence this condition is expressed by the standard equation $\nabla \cdot \mathsf{T} = \mathbf{0}$. It is possible to show [15] that for the simulated growth process (in the absence of load on the future and the actual surface of solid growth during the whole process of its deformation) this equation generates similar differential equation for the tensor $\mathbf{S}$:

$$\nabla \cdot \mathbf{S} = \mathbf{0}. \tag{7}$$

Equation (7) is fair at every moment of time $t > t_1$ in the region of space occupied by the whole growing body at this moment. It should be emphasized that this equation is not a trivial consequence of the standard equilibrium equation, as in the case of growing the body the integral operator $\mathcal{H}_{\tau_0(\mathbf{r})}$ and the operator of divergence $(\nabla \cdot\,)$ do not commute in general because of the principal dependence of time τ_0 of the occurrence of stresses in the growing solid from the point of this solid $\mathbf{r}$.

One can also show, following [15] that from the specific boundary condition (6) on the moving surface of growth $\rho = \Lambda(z,t)$ the condition on the components of the tensor $\mathbf{S}$ implies for every kth step of continuous accreting which is similar in appearance to the standard boundary condition for the stresses:

$$\mathbf{n} \cdot \mathbf{S} = \mathbf{0}, \quad \rho = \Lambda(z,t), \quad t \in [t_{2k-1}, t_{2k}). \tag{8}$$

In the pauses between stages of continuous growth and after the completion of growing the non-traditional condition (6) on the lateral surface of the cone should be replaced by the classical condition of equality to zero of the stress vector on this surface: $\mathbf{n} \cdot \mathbf{T} = \mathbf{0}$. Acting on this condition with the operator $\mathcal{H}_{\tau_0(\mathbf{r})}$ and differentiating the result by time t, we see that the boundary condition (8) saves force even out of time intervals $[t_{2k-1}, t_{2k})$. However, it has a completely different mechanical nature in this case.

The above-formulated relations (7), (5), (4), (8) for the quantities $\mathbf{v}$, $\mathbf{D}$, $\mathbf{S}$ are to be complemented with the same integral force conditions as the boundary problem (2) contains — but here for the cone end surface $\{z = l\}$ expanding over time t due to the additional material influx. After that we get the boundary problem that describes the process of deforming the considered conical solid on all the temporary beam $t > t_1$ after the beginning of its accreting. The main difficulty by solving this problem is that the integral conditions are formulated for the stress tensor $\mathbf{T}$, not for the tensor of velocities of operator stresses $\mathbf{S}$.

However we can overcome the mentioned difficulty by using the proposition given in Section 3. As a surface $\Omega(t)$ in the being solved problem of accreting a conical solid it is necessary to consider the flat surface constituting one end side of the growing cone $z = l$ for $t \geqslant t_0$. The surface Ω_{A} in this case is annular, and its forming curves $\Gamma(t)$ are concentric circles $\rho = b(t)$. The surface Ω_0 is a circle $0 \leqslant \rho \leqslant b_0$. Then according to the proposition we have

$$\frac{\partial}{\partial t}\left[\int_{\{z=l\}} \begin{Vmatrix} \mathbf{k}\cdot\mathbf{T} \\ \mathbf{e}_\rho \rho \times (\mathbf{k}\cdot\mathbf{T}) \end{Vmatrix} dS\right]^{\circ} = \int_{\{z=l\}} \begin{Vmatrix} \mathbf{k}\cdot\mathbf{S} \\ \mathbf{e}_\rho \rho \times (\mathbf{k}\cdot\mathbf{S}) \end{Vmatrix} dS, \quad t > t_1.$$

As a result we can supply the following boundary value problem on the stage of piecewise-continuous accretion of the considered conical solid, $t > t_1$:

$$\begin{gathered} \nabla \cdot \mathbf{S} = \mathbf{0}, \quad 0 \leqslant \rho < \Lambda(z,t); \quad \mathbf{S} = 2\mathbf{D} + \chi \mathbf{1}\,\mathrm{tr}\,\mathbf{D}, \quad \mathbf{D} = (\nabla \mathbf{v}^{\mathrm{T}} + \nabla \mathbf{v})/2; \\ \mathbf{n} \cdot \mathbf{S} = \mathbf{0}, \quad \rho = \Lambda(z,t); \quad \int_{\{z=l\}} \begin{Vmatrix} \mathbf{k}\cdot\mathbf{S} \\ \mathbf{e}_\rho \rho \times (\mathbf{k}\cdot\mathbf{S}) \end{Vmatrix} dS = \begin{Vmatrix} \mathbf{k}\,\partial P^{\circ}(t)/\partial t \\ \mathbf{0} \end{Vmatrix}. \end{gathered} \tag{9}$$

When solving the problem (9) we are to provide rigid fixing the neibourhood of the coordinates origin O throughout the whole process of deformation of considered growing solid, i.e. to meet the following conditions for the vector field of velocities $\mathbf{v}(\mathbf{r},t)$ in this neighbourhood: $\mathbf{v} = \mathbf{0}$ and $\nabla \times \mathbf{v} = \mathbf{0}$ for $\rho = 0$, $z = 0$.

We can see that the boundary problem (9) contains the time variable t only as a real parameter and also is mathematically identical to the problem (3): the analogues of the desired quantities S, D, $\mathbf{v}$, and of the known function $\Lambda(z,t)$ and $\partial P^\circ(t)/\partial t$ in the problem (9) are the desired quantities T°, E, $\mathbf{u}$, and the known function $\Lambda(z,t_0)$ and $P^\circ(t)$ in the problem (3). Therefore, we can obtain the closed-form analytical solution of the problem (9) (see Subsection 4.2).

4.4. Determination of the stress evolution in the accreted solid

After we have solved the boundary problem (9) we know the evolution of the velocity vector $\mathbf{v}$ and the tensor S of velocities of operator stresses at each point $\mathbf{r}$ of the considered piecewise continuously accreted aging viscoelastic conical solid on the time beam

$$t > \tau_1(\mathbf{r}) = \begin{cases} t_1, & 0 \leqslant \rho < \Lambda(z,t_0), \\ \tau_*(\mathbf{r}), & \Lambda(z,t_0) \leqslant \rho < \Lambda(z,t_{2N}). \end{cases}$$

This beam covers the entire history of deformation of the neighborhood of a given point in the composition of the formed solid after the beginning of the process of the solid accretion.

After solution of the problem (3) we know the evolution of the displacement vector $\mathbf{u}$ and the operator stress tensor T° at the points of the original part of the considered solid on the time segment $t \in [t_0, t_1]$, i.e. before the beginning of the accretion process.

So the evolution of the operator stress tensor T° at any point $\mathbf{r}$ of the solid for all $t \geqslant \tau_1(\mathbf{r})$ can be recovered by using the integration procedure:

$$\mathsf{T}^\circ(\mathbf{r},t) = \mathsf{T}^\circ\big(\mathbf{r},\tau_1(\mathbf{r})\big) + \int_{\tau_1(\mathbf{r})}^{t} \mathsf{S}(\mathbf{r},\tau)\,d\tau.$$

We have $\mathsf{T}^\circ\big(\mathbf{r},\tau_1(\mathbf{r})\big) = \mathbf{0}$ in the additional part of the solid in consequence of the condition (6).

When we know the complete evolution of the tensor T° at a point $\mathbf{r}$ of the considered solid, i.e., the values of this tensor since the moment $t = \tau_0(\mathbf{r})$ of occurrence of stresses at a given point, we can find the complete evolution of the stresses tensor T at this point by using the inverse to $\mathcal{H}_{\tau_0(\mathbf{r})}$ transformation $\mathcal{H}^{-1}_{\tau_0(\mathbf{r})}$:

$$\frac{\mathsf{T}(\mathbf{r},t)}{G(t)} = \mathsf{T}^\circ(\mathbf{r},t) + \int_{\tau_0(\mathbf{r})}^{t} \mathsf{T}^\circ(\mathbf{r},\tau)\,R(t,\tau)\,d\tau.$$

If we use a particular approximation for the creep kernel $K(t,\tau)$ the expression for the respective relaxation kernel $R(t,\tau)$ may not be known in the closed form or be too bulky. Then the procedure of reconstructing the evolution of the tensor T by numerical treatment of the Volterra equation

$$\frac{\mathsf{T}(\mathbf{r},t)}{G(t)} - \int_{\tau_0(\mathbf{r})}^{t} \frac{\mathsf{T}(\mathbf{r},\tau)}{G(\tau)}\,K(t,\tau)\,d\tau = \mathsf{T}^\circ(\mathbf{r},t),$$

for example, by the method of quadratures [16], will be less expensive from a computational point of view and may be even more precise.

Acknowledgments

This work was supported by the Russian Foundation for Basic Research (under grants Nos. 18-01-00920-a, 18-01-00770-a, and 17-51-45054-IND_a), and by the Department of Energetics, Mechanical Engineering, Mechanics and Control Processes of the Russian Academy of Sciences (Program No. 12 OE).

References

[1] Manzhirov A V 2013 Mechanics of growing solids and phase transitions *Key Engineering Materials* **535–536** 89–93

[2] Manzhirov A V and Lychev S A 2014 Mathematical modeling of additive manufacturing technologies *Lecture Notes in Engineering and Computer Science: Proc. World Congr. on Engineering 2014, WCE 2014, 2–4 July 2014 London* pp 1404–1409

[3] Manzhirov A V and Parshin D A 2006 Accretion of a viscoelastic ball in a centrally symmetric force field *Mech. Solids* **41** (1) 51–64

[4] Manzhirov A V and Parshin D A 2006 Modeling the accretion of cylindrical bodies on a rotating mandrel with centrifugal forces taken into account *Mech. Solids* **41** (6) 121–134

[5] Manzhirov A V and Parshin D A 2015 Arch structure erection by an additive manufacturing technology under the action of the gravity force *Mech. Solids* **50** (5) 559–570

[6] Manzhirov A V and Parshin D A 2015 Influence of the erection regime on the stress state of a viscoelastic arched structure erected by an additive technology under the force of gravity *Mech. Solids* **50** (6) 657–675

[7] Manzhirov A V and Parshin D A 2016 Accretion of spherical viscoelastic objects under self-gravity *Lecture Notes in Engineering and Computer Science* **2224** (1) 1131–1135

[8] Manzhirov A V and Parshin D A 2016 Application of prestressed structural elements in the erection of heavy viscoelastic arched structures with the use of an additive technology *Mech. Solids* **51** (6) 692–700

[9] Parshin D A and Hakobyan V N 2017 Additive manufacturing of a cylindrical arch of viscoelastic aging material under gravity action at various modes of the process *Procedia IUTAM* **23** 66–77.

[10] Parshin D A 2017 The technological stresses in a vaulted structure built up on a falsework *Mechanics for Materials and Technologies* ed H Altenbach, R V Goldstein and E Murashkin *Advanced Structured Materials* **46** (Springer) pp 377–386

[11] Parshin D A 2017 Analytic solution of the problem of additive formation of an inhomogeneous elastic spherical body in an arbitrary nonstationary central field of forces *Mech. Solids* **52** (5) doi: 10.3103/S0025654417050089

[12] Arutyunyan N Kh 1966 *Some Problems in the Theory of Creep* (Oxford: Pergamon Press)

[13] Arutyunyan N Kh and Manzhirov A V 1999 *Contact Problems of the Theory of Creep* (Erevan: Izdat. Inst. Mekhaniki NAN RA) [in Russian]

[14] Lurie A I 1955 *Spatial Problems of the Theory of Elasticity* (Moscow: Gos. Izdat. Tekhniko-Teoretich. Lit.) [in Russian]

[15] Manzhirov A V 1995 The general non-inertial initial-boundaryvalue problem for a viscoelastic ageing solid with piecewise-continuous accretion *J. Appl. Math. Mech.* **59** (5) 805-816

[16] Polyanin A D and Manzhirov A V 2008 *Handbook of Integral Equations* 2nd ed (Chapman & Hall/CRC Press)

Application of the bridged crack model for evaluation of materials repairing and self-healing

M Perelmuter
Institute for Problems in Mechanics of RAS
prospect Vernadskogo 101-1, Moscow, 119526 Russia

E-mail: perelm@ipmnet.ru

Abstract. The bridged crack model is used for analysis of repairing and self-healing of cracked structures. Material repairing is treated as insertions of external ligaments into cracks or placement of the reinforcing patches over cracks. Bonds destruction and regeneration at the crack bridged zone is evaluated by the thermo-fluctuation kinetic theory. The healing time is dependent on the chemical reaction rate of the healing agent, the crack size and the external loads. The decreasing of the stress intensity factors is used as the measure of the repairing and healing effects. The mathematical background of the problem solution is based on the methods of the singular integral-differential equations. The model can be used for the evaluation of composite materials durability.

1. Introduction

Actions of the external loads, aggressive environments and physical fields lead to degradation of the physical-mechanical properties of materials and to loss of structures integrity. Maintenance and recovery of serviceability of materials and structures is the fundamentally scientific and applied problem. Models of cracks formation and evolution, based on the crack surfaces interaction accounting (a zone of crack surfaces interaction is named usually as 'process zone' or 'cohesive-bridged zones'), allows us to investigate processes of destruction in heterogeneous mediums (materials) at various scale levels. Models of a crack with the developed bridged zone can be used also in the analysis of the processes connected with repairing of damaged structures and self-healing of materials.

At presence of the macroscopic cracks-defects (flaws) which can be detected visually or by means of special sensors, the repairing of serviceability is possible, in particular, with use of special fillers or repairing cover plates [1]. Accumulation of microscopic defects inside materials and also inside inaccessible parts of structures can lead to catastrophic destruction. In such cases can be effectively used self-healed materials [2-3]. Self-healing capability of materials is especially important when the human intervention is complicated, for example, in space, or in extreme physical or chemical conditions. This process of self-healing is very important and for nanocomposites which have excellent functional properties, but it can be degraded under extreme service conditions. Self-healing of cracks in materials can occur under special external affects (mainly, the heating which is making active chemical or physical processes in a material [4]).

IOP Conf. Series: Journal of Physics: Conf. Series **937** (2017) 012039 doi:10.1088/1742-6596/937/1/012039

The other approach is based on the development of new composite materials containing special components (inclusions) which provide self-recovery of cracked material. For example, in the polymeric material made on the basis of epoxy matrix, capsules with "a healing liquid" (dicyclopentadiene), and also microcapsules of catalyzing substance are introducing. In an initial state the catalyst is not in contact with dicyclopentadiene . If there is a crack in a material that grows, it destroys walls of capsules and captures catalyst microcapsules. Dicyclopentadiene outflows and due to capillary effects fills a crack and comes into contact with the catalyst. The polymerisation reaction of dicyclopentadiene leads to liquid solidification and bonds healing. "Healed" material is capable to stand the loads similar to new material (the restoration factor of mechanical properties is about 75%-90% [5]).

An alternative approach to self-healing is based on inserting of hollow fibers system into a material (by analogy to vascular system of living organisms), containing a special reagent [6-7]; level of material restoration in the latter case attain up to 100%. The technique of inserting nanotubes into a polymeric composite material which becomes electroconductive, shows high efficiency. On changing nanotubes conductance system it is possible to determine a place of defect formation and to provide its self-healing due to local heating at an electric current getting through [8]. Nanotubes which are introduced into a polymeric matrix can also contain "healing liquid". In this case they are simultaneously supporting fibers, forming of nanocomposite and provide material healing at crack formation [9].

Numerical and analytical estimations of various factors influence on materials repairing and self-healing processes are necessary for increasing these procedures efficiency. We note, also, that in many cases the experimental investigations can be extremely long and labor-consuming. It is possible to mark out three basic stages of process material repairing and self-healing: 1) formation and growth of cracks in a heterogeneous material under the external loading; 2) insertion into cracks or pores of the repairing media or self-healing agent; 3) strengthening and restoration of ligaments between cracks or pores surfaces, leading to a material or structure restoration.

Experimental results have been obtained on each of these stages, but physical-mechanical models and numerical techniques are only started to develop. Known numerical techniques are based, mostly, on the finite element method [10-12].

The model of a crack with bonds in the bridged zone allows one to determine the crack resistance and the adhesive strength of joints between different materials on the basis of micromechanical properties of the bonds [13-15]. The development of this model with consideration of time-variation in the physical and mechanical properties in the crack bridged zone allows one to estimate the long-term strength and variation over time of materials resistance characteristics [16].

In the present paper the bridged crack model is used for modeling of cracks repairing and materials healing. For cracks repairing we consider quasistatic problem for a bridged crack with increased ligaments stiffness. Healing of material due to bonds restoration is considered on the basis of Zhurkov's kinetic model of thermal fluctuation fracture [17-18] together with the crack bridged zone model.

The estimation of bonds restoration inside a crack is based on the following assumptions:

- at the initial time instant the polymerization process of healing agent inside of a crack is starting;
- polymerization process rate is defined by kinetic theory;
- restored bonds between crack surfaces are considered as a part of the crack bridged zone;
- variation of bonds density over time in that zone is considered on the basis of the thermal fluctuation mechanism;
- bonds rigidity is proportional to their density at every point of the crack bridged zone.

2. Crack bridged model

The plane elasticity problem on a crack at the interface of two dissimilar joint half-planes is considered. The crack surface interaction exists in the bridged zones, $\ell - d \le |x| < \ell$ (Fig.1). As a simple mathematical model of the crack surfaces interaction we will assume that the linearly elastic bonds act through out the crack bridged zones.

Denote by $\sigma(x)$ the stresses arising in the bonds

$$\sigma(x) = \sigma_{yy}(x) - i\sigma_{xy}(x), \qquad i^2 = -1, \tag{1}$$

where $\sigma_{yy}(x)$ and $\sigma_{xy}(x)$ are the normal and shear components of bonds stress. The crack opening, $u(x)$ at $\ell - d \le |x| < \ell$ is determined as follows [13]

$$u(x) = u_y(x) - iu_x(x) = \frac{H}{E_b}(\phi_1(x)\sigma_{yy}(x) - i\phi_2(x)\sigma_{xy}(x)), \tag{2}$$

where $u_{y,x}(x)$ are the projections of the crack opening on the coordinate axes (Fig.2), H is a linear scale related to the thickness of the intermediate layer adjacent to the interface, E_b is the effective elasticity modulus of the bond and $\varphi_{1,2}$ are dimensionless functions of the coordinate x.

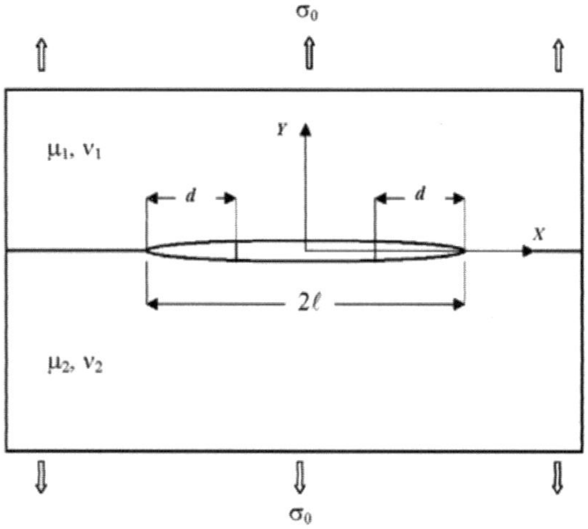

Figure1. Plate with bridged interfacial crack of length 2ℓ under tension loading with two bridged zone of length d.

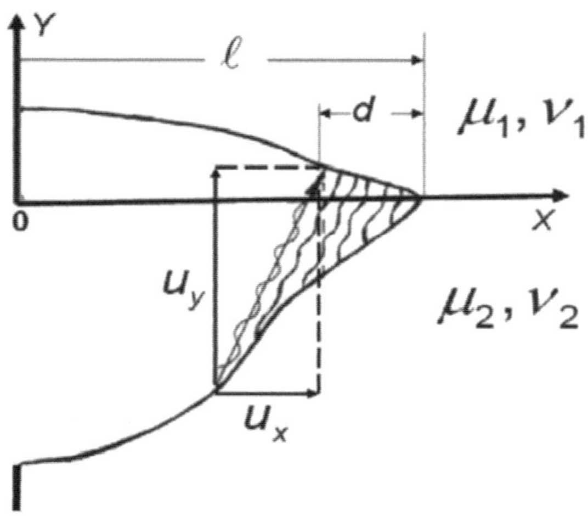

Figure2. Normal and shear crack opening at the bridged zone edge.

By incorporating linearity of the problem one can represent the crack opening $u(x)$ as follows

$$u(x) = u_o(x) - u_b(x), \tag{3}$$

where $u_0(x)$, $u_b(x)$ are the crack opening caused by the external loads and bond stresses $\sigma(x)$, respectively.

By incorporating formulae (2)-(3) one can obtain a system of integral-differential equations relative to bonds stress $\sigma(x)$.

Introduce the new variable, $s = x / \ell$, and differentiate relation (3) to obtain

$$\frac{H}{\ell}\frac{\partial}{\partial s}\left[\phi_1(s)\sigma_{yy}(s) - \phi_2(s)\sigma_{xy}(s)\right] + u_b'(s)E_b = u_o'(s)E_b, \tag{4}$$

where the right side of this relation is the given function of the coordinate.

Taking into account the results given in [13, 19].one can obtain the system of two singular integral-differential equations (SIDE) relative to bonds stresses $\sigma_{yy}(x)$ and $\sigma_{xy}(x)$ in the following form [13]

$$T_{ij}(s,\sigma)\frac{dq_j(s)}{ds}+W_{ij}(s,\sigma)q_j(s)+\varepsilon\int_{1-d/\ell}^{1}G_{ij}(s,t)q_j(t)dt=Z_i(s),\ i,j=1,2 \tag{5}$$

where $q_j(s)$ are unknown function depending on bond stresses as follow

$$\sigma_{yy}(s)-i\sigma_{xy}(s)=(q_y(s)-iq_x(s))\sqrt{1-s}\left(\frac{1-s}{1+s}\right)^{-i\beta}, \tag{6}$$

The explicit relations for the equations (5) kernels T_{ij}, W_{ij}, G_{ij}, Z_i are presented in [13], parameters ε and β in (5)-(6) depend on the materials and bonds properties

For numerical solution of the equations system we use a collocation scheme with piecewise quadratic approximation unknown of the bond stresses. See details in [13].

Having the distribution of the bond stresses $\sigma_{yy}(x)$ and $\sigma_{xy}(x)$ over the crack bridged zone one can calculate the stress intensity factors (SIF) K_I, K_{II} following to [20]

$$K_{\rm I}+iK_{\rm II}=\lim_{\delta\to 0}\sqrt{2\pi\delta}\left(\sigma_{yy}(\delta)+i\sigma_{xy}(\delta)\right)\delta^{-i\beta}, \tag{7}$$

where $\sigma_{yy}(\delta)$ and $\sigma_{xy}(\delta)$ are the stresses ahead the crack tip caused by the external loads and by the bonds stresses, δ represents the small distance to the crack tip.

On the other hand, the SIF can be written as follows

$$K_{\rm I}+iK_{\rm II}=(K_{\rm I}^{ext}+K_{\rm I}^{\rm int})+i(K_{\rm II}^{ext}+K_{\rm II}^{\rm int}), \tag{8}$$

where $K_{I,II}^{ext}$ and $K_{I,II}^{\rm int}$ are the SIF caused by the external loads and the bond stresses.

By incorporating the formula for the stress distribution ahead the interface crack tip under arbitrary loads [19] and using (7)-(8) we obtain for the external tension load σ_0 (K_B is SIF modulus)

$$K_{\rm I}+iK_{\rm II}=\frac{\sqrt{\pi\ell}}{(2\ell)^{i\beta}}\left[\sigma_0(1+2i\beta)-\frac{2\cosh(\pi\beta)}{\pi}\int_{1-d/\ell}^{1}\frac{(q_y(t)+itq_x(t))}{\sqrt{1-t^2}}dt\right],\quad K_B=\sqrt{K_I^2+K_{II}^2} \tag{9}$$

3. Kinetics of bonds at the crack bridged zone

The restoration of bonds inside of cracks can be considered on the basis of the Zhurkov's fluctuation model [17, 18]. According to the fluctuation theory of fracture the lifetime of bonds (τ_B) at the point x of the crack bridged zone under the external loading is the exponential function

$$\tau_B=\tau_o\exp\left(\frac{U_o-A(x)}{kT}\right) \tag{10}$$

where τ_0 is the characteristic time (10^{-13}-10^{-12} s), k is the Boltzmann constant, T is absolute temperature, U_o is the energy of bonds destruction, $A(x)$ is the work of the deformation per one intermolecular bond at the point x of the crack bridged zone.

Bonds lifetime at the crack bridged zone is the function of the bond position. It's assumed that the time increasing of the bonds surface density $n(x,t)$ is governed by the equation

$$n(x,t) = N_0 \exp\left(+\frac{t}{\tau_B(x)} \right), \tag{11}$$

where $\tau_B(x)$ is the lifetime of the bond and N_o is a small initial bond density inside of a crack. The increasing of bonds surface density over time can be modelled by the changing of bonds properties. Let's denote the bond stiffness by k_s. Then the effective stiffness of bonds per unit of an area in the crack bridging zone, k, is determined as follows

$$k(x,t) = k_s n(x,t) = k_0 \exp\left(+\frac{t}{\tau_B(x)} \right), \tag{12}$$

where $k_0 = k_s N_0$ is the initial effective stiffness of bonds in the crack bridged zone. Since the effective bonds compliance $c(x,t)$ is the value reciprocal to its stiffness one can write that

$$c(x,t) = c_0 \exp\left(-\frac{t}{\tau_B(x)} \right). \tag{13}$$

This relation enables us to model the kinetics of bonds restoration in the crack bridged zone by means of the bonds compliance variation over time. The system of SIDE for the solving of static crack bridging problems [13] was extended for the time-steps scheme [16]. On each time step the bond compliances depend on the density of bonds according to relation (13) and the system of SIDE is solved numerically. If the initial crack reaches sufficient size in the polymer composite with the healing microcapsules and/or with shape-memory alloy wires then the self-healing process is started. On this stage of bonds restoration the problem for the crack bridged zone is solved. The healing time is dependent on the chemical reaction rate of the healing agent, wires properties, crack size and the external loads. The decreasing of the stress intensity factors is used as the measure of the healing effect. The model can be used for the evaluation of healed composite materials durability. The results of computation modelling of interfacial crack repairing and bonds healing are presented.

4. Computational results

The above proposed approach was used for several problems analysis. A crack on the interface between different materials under the external tension σ_0 was considered. It was assumed that at the initial time instant (when crack surfaces are free of constraints) some healing process is activated inside a crack and bridges between the crack surfaces are built.

The numerical calculations were performed for plane strain conditions and the following elastic constants of the joint materials and bonds (*Cu*-epoxy polymer): $E_1 = 25GPa$, $E_2 = 135GPa$; $E_b = E_2$, $\nu_1 = \nu_2 = 0.35$. The purpose of calculations is the dependence analysis of the repairing and self-healing processes efficiency (the measure of efficiency is the level of SIF at the crack tip) on the bridged zone length (the crack filling with bonds) and on the bonds stiffness.

In Fig. 3 for different values of the relative bonds stiffness $\kappa = \ell/H$ the dependencies of the SIF module (it can be obtained from eq. (9)) versus the relative bridged zone length are shown. For bonds with relative stiffness more than 10 the repairing efficiency reaches the saturation if the crack has filled with bonds more than for the half of it's length.

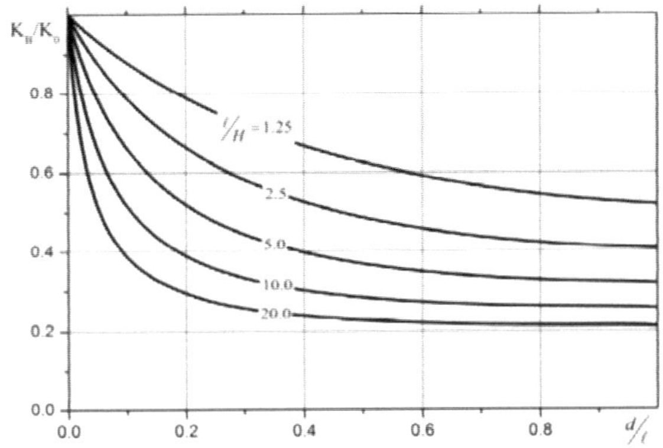

Figure 3. SIF module versus relative bridged zone length, $K_0 = \sigma_0\sqrt{\pi\ell}$.

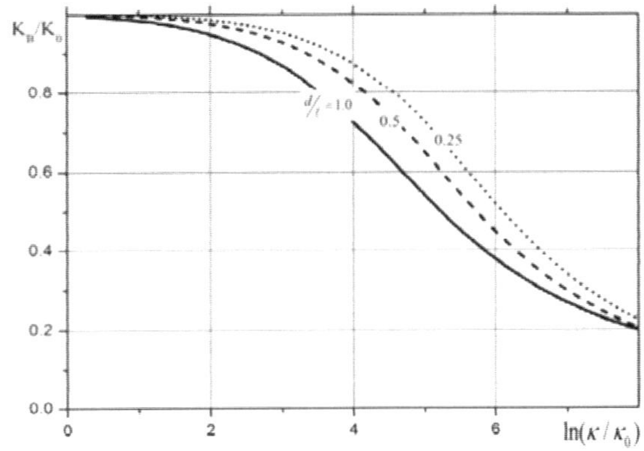

Figure 4. SIF module versus relative bond stiffness, $K_0 = \sigma_0\sqrt{\pi\ell}$.

The evolution of the healing process as the dependence of the SIF module versus the relative bond stiffness (in logarithmic scale) is shown in Fig. 4, where the scale factor $\kappa_0 = 1$ corresponds to $\ell = H$. Saturation of the repairing effect is observed for bonds with rather big stiffness.

The next result shows the healing process for the whole crack over time. Computation of the SIF dependence on the healing time were performed for the following problem parameters: the activation energy of the healing process - $U_M = 100\ kJ/mole$, $\tau_0 = 10^{-12} s$ (see relation (10)), the initial bonds density is $n_0 = 10^{17} m^{-2}$, the size of the healed crack is $2\ell = 10^{-5} m$. The SIF dependence on the healing time (in logarithmic scale) is shown in Fig. 5. During the healing time the initial bonds stiffness increases by 20 times. Therefore, the SIF module decreases due to bonds restoration.

The results of computations on the basis of the crack bridged model allow us to estimate the SIF module variation due to the repairing of cracks and bonds restoration over time. The results based on the kinetic approach to bonds restoration might be helpful for durability analysis of adhesion joints. Since the computational parameters of the kinetic model strongly depend on the initial data (which is caused by the exponential dependence of the durability in formula (10)), the comparative analysis of different healing agents and processes under appropriate loading conditions is of greatest practical interest in this method.

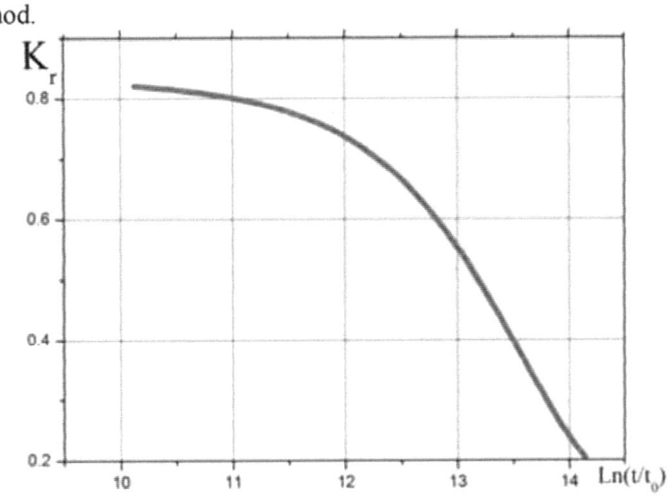

Figure 5. Relative SIF modulus due to bonds restoration, $K_r = (K_0 - K_b)/K_0$, $t_0 = 1s$.

Acknowledgements
Financial support for this research was partly provided by RFBR, Project No. 17-08-01312.

References

[1] Wang C H 2000 Fatigue crack closure analysis of bridged cracks representing composite repairs, *Fract. Engng. Mater. Struct* **23**, 477-488

[2] White S R, Sottos N R, Geubelle P H, Moore J S, Kessler M R, Sriram S R, Brown E N and Viswanathan S, 2001 Autonomic healing of polymer composites, *Nature* **409**, 794–797

[3] Syrett J A, Becer C R and Haddleton D M 2010 Self-healing and self-mendable polymers, *Polymer Chemistry* **1**, 978-987

[4] Volynsky A L and Bakeev N F 2009 Healing of an interphase surface in polymeric systems, *High-molecular compounds, Series A* **51**, 1783-1816

[5] Blaiszik B J, Sottos N R, White S R 2008 Nanocapsules for self-healing materials, *Composites Science and Technology* **68**, 978-986

[6] Toohey, K S, Sottos N R, Lewis J A, Moore J S and White S R 2007 Self-healing materials with microvascular networks, *Nature*, 6, 581-585

[7] Trask R S, Williams G J and Bond I P 2007 Bioinspired Self-Healing of Advanced Composite Structures Using Hollow Glass Fibres, *J. Royal Society, Interface* **4**, 363-371

[8] Zhang W, Sakalkar V. and Koratkara N. 2007 In situ health monitoring and repair in composites using carbon nanotube additives, *Applied physics letters* **91**, 133100-133102

[9] Lanzara G, Yoon Y, Liu H, Peng S and Lee W-I 2009 Carbon nanotube reservoirs for self-healing materials, *Nanotechnology* **20**, 335704-7

[10] Maiti S and Geubelle P. H. 2006 Cohesive modelling of fatigue crack retardation in polymers - Crack closure effect, Eng. Fract. Mech. **73**, 22-41

[11] Ural A, Krishnan V R and Papoulia K D 2009 A cohesive zone model for fatigue crack growth allowing for crack retardation *Int. J. of Solids and Structures* **46**, 2453-2462

[12] Ozaki S, Osada T and Nakao W 2016 Finite element analysis of the damage and healing behavior of self-healing ceramic materials, *Int. J. of Solids and Structures* **100–101**, 307-318

[13] Goldstein R V and Perelmuter M N 1999 Modeling of bonding at the interface crack, *Int. J. Fracture* **99** 53-79

[14] Perelmuter M N 2007 A criterion for the growth of cracks with bonds in the end zone *J. Appl. Math. Mech.* **71** (1) 137-153

[15] Goldstein R V and Perelmuter M N 2009 Modeling of crack resistance of composite materials (in Russian), *Computational continuum mechanics* **2**, 22-39

[16] Goldstein R V and Perelmuter M N 2012 Kinetics of crack formation and growth on the material interface, *Mech. Solids* **47**, 400–414

[17] Zhurkov S N 1965 Kinetic concept of the strength of solids *Int. J. Fracture Mech.* **1,** 311-323

[18] Regel V G, Slutsker A I and Tomashevskii E E 1974 *Kinetic Nature of Strength of Solids* (Moscow: Nauka), in Russian

[19] Slepijn L I 1981 *Mechanics of Cracks* (Leningrad: Sydostroenie), in Russian

[20] Rice J R 1988 Elastic fracture mechanics concepts for interfacial cracks *Trans. ASME J. Appl. Mech.* **55** 98-103

IOP Conf. Series: Journal of Physics: Conf. Series **937** (2017) 012040 doi:10.1088/1742-6596/937/1/012040

On Representation of Riesz–space–valued Functions by Fourier Series on Multiplicative Systems

S V Petrov[1] **and V M Prostokishin**[1,2]

[1] National Research Nuclear University MEPhI (Moscow Engineering Physics Institute), 31 Kashirskoe shosse, 115409 Moscow, Russia

[2] Gazprom VNIIGAZ LLC, box 130, 115583 Moscow, Russia

E-mail: VMProstokishin@mephi.ru

Abstract. The result that was established in [1] for Walsh series with coefficients belonging to a Riesz–space is generalized for multiplicative systems. Conditions for recovering the coefficients of a convergent series on the multiplicative systems from its sum by Fourier formulae are considered. For these formulae the Henstock–Kurzweil integral for Riesz–space–valued functions defined with respect to the special basis $\mathcal{B}_{\mathbb{P}}$ will be used.

In this work the result that was established in [1] for Walsh series [2] with coefficients belonging to a Riesz–space [3] is generalized for multiplicative systems. These systems are very popular for solving applied tasks nowadays [4],[5]. In particular, we consider conditions, for which such series are some generalized Fourier series, where in order to write down Fourier's formulae we use the Henstock–Kurzweil type integral [1, 6]. This integral is defined with respect to the basis $\mathcal{B}_{\mathbb{P}}$ generated by intervals $\{\delta_r^{(k)}\}$ of the special form [2]:

$$\delta_r^{(k)} = \left[\frac{r}{m_k}, \frac{r+1}{m_k}\right), \quad k = 0, 1, 2, \ldots, \quad 0 \leqslant r \leqslant m_k - 1.$$

We recall the definition of the Henstock–Kurzweil type integral with respect to a basis for Riesz–space–valued functions following [1].

Let $[a, b]$ be a fixed interval of $\mathbb{R}$ and $\mathcal{I}$ be the family of all subintervals of $[a, b]$. A basis on $[a, b]$ is any subset $\mathcal{B}$ of $\mathcal{I} \times [a, b]$ such that $(I, x) \in \mathcal{B}$ implies $x \in I$.

An interval I for a given basis $\mathcal{B}$ is called a $\mathcal{B}$–interval if $(I, x) \in \mathcal{B}$, for some $x \in I$. Furthermore, assume that $[a, b]$ is a $\mathcal{B}$–interval. For a set E such that $\emptyset \neq E \subset [a, b]$, put

$$\mathcal{B}[E] = \{(I, x) \in \mathcal{B} : x \in E\}.$$

For $\emptyset \neq E \subset [a, b]$ by Δ_E denote the directed set of all positive real valued functions defined on E and endowed with the natural ordering: for given two functions δ_1 and δ_2, $\delta_1 \leqslant \delta_2$ iff $\delta_1(x) \leqslant \delta_2(x)$ for all $x \in E$.

A function $\delta \in \Delta_E$ is often said to be a gage on E.

For a given gage $\delta \in \Delta$, put

$$\mathcal{B}_\delta = \{(I, x) \in \mathcal{B} : I \subset (x - \delta(x), x + \delta(x))\}.$$

Let $\emptyset \neq E \subset [a,b]$. A finite subset Π of $\mathcal{B}[E]$ is called a $\mathcal{B}$–decomposition on E if for every distinct elements (I', x') and (I'', x'') of Π, the corresponding intervals I' and I'' are nonoverlapping, that is, their interiors are pairwise disjoint. If else $\coprod_{(I,x)\in\Pi} I = [a,b]$, for $\Pi \subset \mathcal{B}$, then we say that Π is a $\mathcal{B}$–partition of $[a,b]$.

Note that $\mathcal{B}_\delta$ is also a basis on $[a,b]$, so it can define the set $\mathcal{B}_\delta[E]$. Thus a $\mathcal{B}_\delta$–partition can be well defined.

Let $\mathcal{B}$ be a fixed basis on $[a,b]$. A function $f\colon [a,b] \to R$ is Henstock–Kurzweil integrable on a $\mathcal{B}$–interval $E \subset [a,b]$ with respect to $\mathcal{B}$ if there exists an element $Y \in R$ such that

$$\inf_{\delta\in\Delta_{[a,b]}} \left(\sup \left\{ \left| \sum_{(I,x)\in\Pi} f(x)|I| - Y \right| : \Pi \text{ is a } \mathcal{B}_\delta\text{–partition of } E \right\} \right) = 0.$$

This element Y is called $H_{\mathcal{B}}$–integral and is denoted by $(H_{\mathcal{B}}) \int_E f = Y$.

Now we recall the definition of the multiplicative systems using the $\mathbb{P}$–digit expansion for the real numbers; by definition, put

$$m_0 = 1, \quad m_j = \prod_{s=1}^{j} p_s,$$

where p_s is a member of a sequence of the set of natural numbers

$$\mathbb{P} = \{p_1, p_2, \ldots, p_j, \ldots\}, \quad p_j \geqslant 2,\ j \geqslant 1,$$

and any integer $n \geqslant 0$ has the form

$$n = \sum_{i=1}^{k} \alpha_j m_{j-1}, \quad 0 \leqslant \alpha_j \leqslant p_j,\ j = 1, 2, ..., k.$$

Take the point $x \in [0,1)$ and consider the series $x = \sum\limits_{j=1}^{\infty} \frac{x_j}{m_j}$, where $0 \leqslant x_j \leqslant p_{j-1},\ j \geqslant 1$. Here for $\mathbb{P}$–digit rational x we use only the finite expansion. The point x is called the $\mathbb{P}$–digit rational point if there exists a finite representation $x = \sum\limits_{j=1}^{K} \frac{x_j}{m_j}$, $K \in \mathbb{N}$. Using the introduced notations, by definition, put

$$\chi_n(x) = \exp\left(2\pi i \sum_{j=1}^{k} \frac{\alpha_j x_j}{p_j}\right).$$

Proposition 1. *Under the condition of $n < m_k$, the functions $\chi_n(x)$ are constant on the intervals*

$$\delta_r^{(k)} = \left[\frac{r}{m_k}, \frac{r+1}{m_k}\right), \quad k = 0, 1, 2, \ldots, \quad 0 \leqslant r \leqslant m_k - 1.$$

For our purpose, with pointwise (o)–converges (order converges) we use (u)–converges on a set (see [1]), according to the definition:

Let Λ be any nonempty set, R be Dedekind complete Riesz space and $D = N^\Lambda$. The sequence of R–valued functions $(S_n\colon \Lambda \to R)_n$ (u)*–converges* to the function $S\colon \Lambda \to R$ (with respect to order convergence) if there exists an (o)–net $(a_\nu)_{\nu\in D}$ such that for any $\nu \in D$ the following condition holds:

$$\sup\{|S_n(x) - S(x)| : x \in \Lambda,\ n \leqslant \nu(x)\} \leqslant a_\nu.$$

Proposition 2. *It is evident that* (u) *- convergence of a sequence on a set implies* (o) *- convergence of this sequence at each point of the set.*

Let $S_n = \sum_{j=0}^{n-1} a_j\chi_j$ be the partial sum of a series in the multiplicative system $\{\chi_j\}_j$, where coefficients a_j belong to a Riesz–space R.

Hereafter in this paper, by $\mathcal{P}$ we denote the basis $\mathcal{B}_{\mathbb{P}}$ consisting of all pairs (I, x), where $x \in I = \left(\frac{i}{m_k}, \frac{i+1}{m_k}\right)$ and $k \in N \cup \{0\}$, and $i = 0, 1, \ldots, m_{k-1}$.

The Fourier series of a $H_{\mathcal{P}}$–integrable function f is a series $S_n = \sum_{j=0}^{n-1} \hat{f}(n)\chi_j(x)$, where $\hat{f}(n) = \int_0^1 f(t)\overline{\chi_n(t)}\,dt$.

Since the system $\{\chi_n\}_n$ is an orthonormal system (see [2]), it follows that $\hat{S}_{m_k}(n) = \int_0^1 S_{m_k}(t)\overline{\chi_n(t)}\,dt = \int_0^1 \sum_{j=0}^{m_k-1} a_j\chi_j(t)\overline{\chi_n(t)}\,dt = \sum_{j=0}^{m_k-1} a_j \int_0^1 \chi_j(t)\overline{\chi_n(t)}\,dt = \sum_{j=0}^{m_k-1} a_j \int_0^1 \chi_{j\ominus n}(t)\,dt = a_n$. Thus we have

Proposition 3. *For a number* $n, n < m_k$, *the coefficients* a_j *of a series* $S_n(x) = \sum_{j=0}^{n-1} a_j\chi_j(x)$ *on the system* $\{\chi_n\}_n$ *are Fourier coefficients of the partial sums* $S_{m_k}(x)$.

Proof. Denoting by the $\overline{\chi_{ni}}$ value of the function $\overline{\chi_n}$ on $\delta_i^{(k)}$, we have

$$a_n = \int_0^1 S_{m_k}(t)\overline{\chi_n(t)}\,dt = \sum_{j=0}^{m_k-1} \int\limits_{\delta_j^{(k)}} S_{m_k}(t)\overline{\chi_n(t)}\,dt = \sum_{j=0}^{m_k-1} \overline{\chi_{nj}} \int\limits_{\delta_j^{(k)}} S_{m_k}(t)\,dt =$$

$$\sum_{j=0}^{m_k-1} \overline{\chi_{nj}}\psi(\delta_j^{(k)}) = \sum_{j=0}^{m_k-1} \overline{\chi_{nj}}(H_{\mathcal{P}}) \int\limits_{\delta_j^{(k)}} f(t)\,dt = \sum_{j=0}^{m_k-1} (H_{\mathcal{P}}) \int\limits_{\delta_j^{(k)}} f(t)\overline{\chi_{nj}}\,dt =$$

$$\sum_{j=0}^{m_k-1} (H_{\mathcal{P}}) \int\limits_{\delta_j^{(k)}} f(t)\overline{\chi_n(t)}\,dt = \sum_{j=0}^{m_k-1} (H_{\mathcal{P}}) \int\limits_{\delta_j^{(k)}} f(t)\overline{\chi_n(t)}\,dt = (H_{\mathcal{P}}) \int\limits_{\delta_0^1} f(t)\overline{\chi_n(t)}\,dt$$

Since for any n there exists k such that $n < m_k$, it follows that the first equality in the chain of the formulae above is true and the proposition is proved. □

Our main result is as follows.

Theorem 1. *If* R *is a regular Riesz–space and the series*

$$\sum_{j}^{\infty} a_j\chi_j, \quad a_j \in R \tag{*}$$

(u) *- converges to a function* f *on a set* $[0,1] \setminus E$, *where* E *is a countable subset of the interval* $[0,1]$, *then* f *is* $H_{\mathcal{B}}$ *- integrable on* $[0,1]$, *and the series (*) is the Fourier series of* f *in the sense of the* $H_{\mathcal{B}}$ *- integral.*

For proving this theorem the function $\psi(\delta_i^{(k)}) = \int_{\delta_i^{(k)}} S_{m_k}$ of the intervals $\delta_i^{(k)}$ is used, where S_{m_k} is the partial sum of the series (*). This function is additive with respect to the measure algebra, generated by the intervals $\{\delta_r^{(k)}\}$. This follows from the equality

$$\psi(\delta_r^{(k)}) = \psi(\delta_{rp_{k+1}}^{(k+1)}) + \psi(\delta_{rp_{k+1}+1}^{(k+1)}) + \cdots + \psi(\delta_{rp_{k+1}+p_{k+1}-1}^{(k+1)}) = \sum_{s=rp_{k+1}}^{(r+1)p_{k+1}-1} \psi(\delta_s^{(k+1)}), \tag{**}$$

where $\delta_r^{(k)} = \bigcup_{s=rp_{k+1}}^{(r+1)p_{k+1}-1} \delta_s^{(k+1)}$.

Theorem 1 follows as a result of applying the statement of Theorem 2 (see [1]) to the function ψ. First, we recall some definitions.

The function τ is called (o)–continuous at a point $x_0 \in [a,b]$ with respect to the basis $\mathcal{B}$ if

$$\inf_\delta [\sup\{|\tau(I)| : (I, x_0) \in \mathcal{B}_\delta[\{x_0\}]\}] = 0.$$

The function τ is (o)–continuous on E with respect to the basis $\mathcal{B}$ if it is (o)–continuous at every point $x_0 \in E$.

The function τ is called (u)–differentiable on E with respect to the basis $\mathcal{B}$ if there exists a function $g\colon E \to R$ such that

$$\inf_\delta \left[\sup\left\{\left|\frac{|\tau(I)|}{|I|} - g(x)\right| : (I,x) \in \mathcal{B}_\delta[E]\right\}\right] = 0.$$

This function g is called the (u)–derivative with respect to the basis $\mathcal{B}$ of τ on E.

Theorem 2. *Let R be a regular Riesz–space, $\mathcal{B}$ a fixed basis, $f\colon [a,b] \to R$ and let τ be the R - value $\mathcal{B}$ - interval function, such that for some countable set $Q \in [a,b]$ the function f is the (u) - derivative of τ with respect to the basis $\mathcal{B}$ on the set $[a,b] \setminus Q$ and τ is (o) - continuous with respect to the basis $\mathcal{B}$ on Q. Then f is $H_{\mathcal{B}}$ - integrable over $[a,b]$, and*

$$(H_{\mathcal{B}}) \int_a^b f = \tau([a,b]).$$

From analysis of the proof of Theorem 4.2 from the work [1], it follows that to prove this theorem, in our case it is necessary and sufficient to have $\tau([0,1]) = \sum_{i=1}^q \tau(I_i)$ for any $\mathcal{B}_\delta$-partition Π on $[0,1]$, where $\Pi = \{(I_i, x_i),\ i = 1, \ldots, q\}$. According our notations we here have $\tau = \psi$, $\mathcal{B} = \mathcal{P}$.

We claim that $\tau([0,1]) = \sum_{i=1}^q \tau(I_i)$ for any partition Π on $[0,1]$. Indeed, this follows from the fact that a $\mathcal{P}_\delta$–partition Π on $[0,1]$ consists of intervals $\delta_r^{(k)}$ and from the equality (**), where the function ψ is replaced by the function τ.

Thus, Theorem 2 can be proved similarly to Theorem 4.2 from [1] and Theorem 1 can be proved similarly to Theorem 5.4, using Proposition 3.

It is important to note that the elements of a Riesz–space can multiply by real numbers only, to preserve the axioms of an order.

In our case, the functions $\chi_n(x)$ can be complex–valued. Thus, we must give some definitions. We define the product of two numbers $z \in \mathbb{C}$ and $r \in R$ as

$$r \cdot z = r(a + ib) = ra + i \cdot rb,$$

where r is a element of the Riesz–space and $z = a + ib$, $a, b \in \mathbb{R}$. Here, addition is applied separately to each component as it is usually done for complex numbers.

By definition, the numbers $r_1 \cdot z_1$ and $r_1 \cdot z_1$ are equal if $Re\, r_1 \cdot z_1 = Re\, r_2 \cdot z_2$ and $Im\, r_1 \cdot z_1 = Im\, r_2 \cdot z_2$, where $z_1, z_2 \in \mathbb{C}$ and $r_1, r_2 \in R$.

If $f(x) = f_1(x) + if_2(x)$, where $f_1(x), f_2(x)$ are Riesz–space–valued functions, $x \in \mathbb{R}$, then $f_1(x) = Ref(x)$, $f_2(x) = Imf(x)$ and by definition, put

$$\int f(x)\,dx = \int f_1(x)\,dx + \int f_2(x)\,dx.$$

Hypothesis. *If a series $\sum_j^\infty a_j\chi_j$, where $a_j \in R$, (u) - converge to a Riesz–space–valued function (not complex–valued), then $S_{m_k}(x) \in R$, $k = 0, 1, \ldots$.*

Acknowledgements. The authors are grateful to Professor V. A. Skvortsov for constant attention to this work and to Professor A. I. Rubinstein for useful discussions.

IOP Conf. Series: Journal of Physics: Conf. Series **937** (2017) 012040 doi:10.1088/1742-6596/937/1/012040

References

[1] Boccuto A, Skvortsov V A 2003/04 Henstock–Kurzweil type integration of Riesz–space–valued functions and applications to Walsh series *Real Anal. Exch.* **29** N 1 pp 419–38
[2] Golubov B, Efimov A and Skvortsov V 1991 *Walsh Series and Transforms: Theory and Applications. Mathematics and its Applications*, Book 64 (Springer) p 368
[3] Vulih B Z 1967 *An Introduction to Functional Analysis* [in Russian] (Moscow: Nauka) p 415
[4] Ship F, Wade W R and Simon P 1990 *An Introduction to Dyadic Analysis.* (CRC Press) 1990 p 528
[5] Kenny D A, Kashy D A and Cook W L 2006 *Dyadic Data Analysis. Methodology in the Social Sciences* (The Guilford Press) p 458
[6] Lukashenko T P, Skvortsov V A and Solodov A P 2010 *Generalized Integrals* [in Russian] (Moscow: LIBRICOM) p 280

Nonlinear reaction-diffusion equations with delay: some theorems, test problems, exact and numerical solutions

A D Polyanin[1,2,3], **V G Sorokin**[1,3]

[1] Institute for Problems in Mechanics, Russian Academy of Sciences, Moscow, Russia
[2] National Research Nuclear University MEPhI, Moscow, Russia
[3] Bauman Moscow State Technical University, Moscow, Russia

E-mail: polyanin@ipmnet.ru, vsesor@gmail.com

Abstract. The paper deals with nonlinear reaction-diffusion equations with one or several delays. We formulate theorems that allow constructing exact solutions for some classes of these equations, which depend on several arbitrary functions. Examples of application of these theorems for obtaining new exact solutions in elementary functions are provided. We state basic principles of construction, selection, and use of test problems for nonlinear partial differential equations with delay. Some test problems which can be suitable for estimating accuracy of approximate analytical and numerical methods of solving reaction-diffusion equations with delay are presented. Some examples of numerical solutions of nonlinear test problems with delay are considered.

Keywords: reaction-diffusion equations with delay, partial differential equations with delay, exact solutions, nonlinear test problems, numerical solutions

1. Introduction

Differential equations with delay arise in biology, biophysics, biochemistry, chemistry, medicine, control, climate model theory, fluid mechanics, heat conduction and other areas [1–5]. It is noteworthy that these equations occur in the mathematical theory of artificial neural networks, whose results are used for signal and image processing and in image recognition problems [6, 7].

The paper deals with nonlinear reaction-diffusion equations with one or several delays, which allow modelling systems and processes when the rate of change of the unknown depends not only on the current state of the system but also on the values of the unknown at certain times in the past.

2. Theorems of generation of exact solutions for equations with one delay

Consider a class of nonlinear reaction-diffusion equations with delay

$$u_t = a\Delta u + f(u - kw) + ug(u - kw) + wh(u - kw), \qquad w = u(\mathbf{x}, t - \tau), \tag{1}$$

where $f(z)$, $g(z)$, and $h(z)$ are arbitrary functions and $\tau > 0$, $a > 0$, and k are free parameters.

Theorem 1. With $k > 0$, Eq. (1) admits a solution

$$u = \theta(\mathbf{x}) + e^{c_1 t}\xi(\mathbf{x}, t), \qquad c_1 = (\ln k)/\tau,$$

IOP Conf. Series: Journal of Physics: Conf. Series **937** (2017) 012041 doi:10.1088/1742-6596/937/1/012041

where the function $\theta = \theta(\mathbf{x})$ satisfies the independent stationary equation

$$a\Delta\theta + f(\zeta) + \big(g(\zeta) + h(\zeta)\big)\theta = 0, \qquad \zeta = (1-k)\theta; \tag{2}$$

and the function $\xi = \xi(\mathbf{x}, t)$ is a τ-periodic solution of the equation

$$\xi_t = a\Delta\xi + \Big(g(\zeta) + \frac{1}{k}h(\zeta) - c_1\Big)\xi, \qquad \xi(\mathbf{x}, t) = \xi(\mathbf{x}, t+\tau), \tag{3}$$

which is linear with respect to ξ.

With $k = 1$, which corresponds to $\zeta = 0$, Eq. (2) becomes linear and inhomogeneous; shifting θ by a constant, we obtain the Helmholtz equation, whose exact solutions can be found in [8].

With $k \neq 1$, the simplest solution of Eq. (2) is a constant $\theta_0 = \zeta_0/(1-k)$, where ζ_0 is a root of the algebraic (transcendental) equation $(1-k)f(\zeta_0) + \zeta_0\big(g(\zeta_0) + h(\zeta_0)\big) = 0$.

In the one-dimensional case, the general solution of Eq. (2) can be expressed in terms of integrals. For a number of functions $f(\zeta)$, $g(\zeta)$, and $h(\zeta)$, exact solutions of Eq. (2) are presented in [9] (the one-dimensional case) and [10] (the two- and three-dimensional cases).

Eq. (3) admits stationary solutions $\xi_0 = \xi_0(\mathbf{x})$, as well as τ-periodic solutions of the form

$$\xi_n = \varphi_n(\mathbf{x})\cos(\beta_n t) + \psi_n(\mathbf{x})\sin(\beta_n t), \qquad \beta_n = \frac{2\pi n}{\tau}, \qquad n = 1,\ 2,\ \ldots$$

With $g(\zeta) = h(\zeta) \equiv 0$, Eq. (3) becomes a linear equation with constant coefficients and does not depend on θ. Formulas of τ-periodic solutions of one-dimensional equations of type (3) are presented in [11, 12].

Theorem 2. With $k < 0$, Eq. (1) admits a solution

$$u = \theta(\mathbf{x}) + e^{c_2 t}\eta(\mathbf{x}, t), \qquad c_2 = (\ln|k|)/\tau,$$

where the function $\theta = \theta(\mathbf{x})$ satisfies Eq. (2); and the function $\eta = \eta(\mathbf{x}, t)$ is a τ-aperiodic solution of the equation

$$\eta_t = a\Delta\eta + \Big(g(\zeta) + \frac{1}{k}h(\zeta) - c_2\Big)\eta, \qquad \zeta = (1-k)\theta, \qquad \eta(\mathbf{x}, t) = -\eta(\mathbf{x}, t+\tau), \tag{4}$$

which is linear with respect to η.

Exact τ-aperiodic solutions of Eq. (4) are sought in the form

$$\eta_n = \varphi_n(\mathbf{x})\cos(\beta_n t) + \psi_n(\mathbf{x})\sin(\beta_n t), \qquad \beta_n = \frac{\pi(2n-1)}{\tau}, \qquad n = 1,\ 2,\ \ldots$$

With $g(\zeta) = h(\zeta) \equiv 0$, Eq. (4) becomes a linear equation with constant coefficients and does not depend on θ. Formulas of τ-aperiodic solutions of one-dimensional equations of type (4) are presented in [11, 12].

Theorem 3. With $k = 1$, Theorem 1 is valid. In addition, Eq. (1) admits a solution

$$u = t\varphi(\mathbf{x}) + \psi(\mathbf{x}), \tag{5}$$

where the functions $\varphi = \varphi(\mathbf{x})$ and $\psi = \psi(\mathbf{x})$ satisfies the system of stationary equations

$$\begin{aligned} &a\Delta\varphi + \varphi\big(g(\tau\varphi) + h(\tau\varphi)\big) = 0, \\ &a\Delta\psi + \psi\big(g(\tau\varphi) + h(\tau\varphi)\big) = \varphi - f(\tau\varphi) + \tau\varphi h(\tau\varphi). \end{aligned} \tag{6}$$

In the one-dimensional case, the general solution of the first equation in (6) can be expressed in terms of integrals for arbitrary $g(z)$ and $h(z)$. For the two-dimensional and three-dimensional cases, some exact solutions are presented in [10].

Remark 1. Exact solutions for one-dimensional PDEs of type (1) and other reaction-diffusion equations including those of hyperbolic type can be found in [11–15].

3. Exact solutions for a class of equations with several delays

Consider a class of more complex nonlinear reaction-diffusion equations with several delays

$$u_t = a\Delta u_{xx} + bw_0 + f(u - w_1, \dots, u - w_n), \quad w_j = u(\mathbf{x}, t - \tau_j), \tag{7}$$

where f is an arbitrary function; $\tau_j \geq 0$, $j = 0, 1, \dots, n$; $a > 0$.

Exact solution 1. Eq. (7) admits an additive separable solution

$$u = \varphi(\mathbf{x}) + \psi(t),$$

where the function $\varphi(\mathbf{x})$ satisfies the Helmholtz equation

$$a\Delta\varphi(\mathbf{x}) + b\varphi(\mathbf{x}) = 0; \tag{8}$$

and the function $\psi(t)$ is a solution of an ordinary differential equation with several delays

$$\psi'(t) = b\psi(t - \tau_0) + f\big(\psi(t) - \psi(t - \tau_1), \dots, \psi(t) - \psi(t - \tau_n)\big).$$

Exact solutions of the Helmholtz equation can be found in [8].

Remark 2. Exact solutions for some ODEs are presented in [15]. Some numerical methods for ODEs and related issues are studied in [16–18].

Exact solution 2. Eq. (7) admits a solution of the form (5), where the function $\varphi = \varphi(\mathbf{x})$ satisfies Eq. (8); and $\psi = \psi(\mathbf{x})$ is a solution of the inhomogeneous Helmholtz equation

$$a\Delta\psi + b\psi = \varphi - f(\tau_1\varphi, \dots, \tau_n\varphi).$$

4. Principles of construction and use of test problems

The following principles are useful in the process of construction, selection and use of test problems, designed for estimating accuracy of approximate analytical and numerical methods of solving nonlinear partial differential equations with delay.

1. Test problems are constructed on the basis of exact solutions of a considered or wider class of nonlinear partial differential equations with delay.
2. Test problems and their solutions should contain a number of free parameters, which can be varied within wide limits.
3. It is better to choose simple test problems that have solutions in elementary functions.
4. One should use several test problems to estimate accuracy of numerical methods.
5. It is helpful to consider test problems with larger gradients of the unknowns in the initial data or boundary conditions (for example, rapidly oscillating initial data).
6. It is useful to consider rapidly growing solutions for sufficiently large times.
7. It is desirable to test numerical methods near the values of the parameters that determine the region of instability (ill-posed problems).

Well-chosen test problems make it possible to compare and improve efficient numerical methods and to filter out unusable ones. The theorems described in Section 2 allow obtaining exact solutions, which afterwards are used to construct corresponding test problems for reaction-diffusion equations with delay.

5. Test problems and their exact solutions

Consider a one-dimensional equation of type (1) with a nonlinear kinetic function

$$u_t = au_{xx} + bu\big(1 - s(u - kw)\big), \quad w = u(x, t - \tau). \tag{9}$$

IOP Conf. Series: Journal of Physics: Conf. Series **937** (2017) 012041 doi:10.1088/1742-6596/937/1/012041

Eq. (9) contains five parameters: $a > 0$, b, s, k, and $\tau > 0$. In a degenerate case, with $k = 0$ or $\tau = 0$, Eq. (9) is referred to as the Fisher's equation.

Test problem 1. Eq. (9) with the following initial and boundary conditions:

$$\begin{aligned}
&u(x,t) = e^{ct}\big(\cos(\pi x/2) + 2\sin(\pi x/2)\big), \quad c = (\ln k)/\tau, \quad 0 \le x \le 1, \quad -\tau \le t \le 0;\\
&u(0,t) = e^{ct}, \quad t \ge 0;\\
&u(1,t) = 2e^{ct}, \quad t \ge 0.
\end{aligned} \tag{10}$$

With $b = (\ln k)/\tau + a\pi^2/4$ and $k > 0$, test problem (9), (10) has the exact solution

$$u(x,t) = e^{ct}(\cos(\pi x/2) + 2\sin(\pi x/2)), \quad c = (\ln k)/\tau, \quad 0 \le x \le 1, \quad t > 0. \tag{11}$$

Test problem 2. Eq. (9) with the following initial and boundary conditions:

$$\begin{aligned}
&u(x,t) = \cosh^{-2}(x) + e^{ct}\cosh^3(x), \quad c = (\ln k)/\tau, \quad 0 \le x \le 1, \quad -\tau \le t \le 0;\\
&u(0,t) = 1 + e^{ct}, \quad t \ge 0;\\
&u(1,t) = \cosh^{-2}(1) + e^{ct}\cosh^3(1), \quad t \ge 0.
\end{aligned} \tag{12}$$

With $b = -4a$, $k = e^{5a\tau}$, and $s = \frac{3}{2(1-k)}$, test problem (9), (12) has the exact solution

$$u(x,t) = \cosh^{-2}(x) + e^{ct}\cosh^3(x), \quad c = (\ln k)/\tau, \quad 0 \le x \le 1, \quad t > 0.$$

6. Numerical solutions of test problem (10), (11) based on the method of lines

The idea of the method of lines is to reduce a partial differential equation to a system of ordinary differential equations by approximation of the spatial derivatives by finite differences [19, 20].

Let us define spatial mesh points as $x_m = mh$, where $m = 0, 1, \ldots, M$ and $h = 1/M$ is a step size. Then we proceed with discretization of the initial-boundary value problem and get a system of $M - 1$ regular ordinary differential equations and two boundary conditions. The resulting problem with the initial conditions can be solved by well-known numerical methods, such as the implicit Runge-Kutta methods.

The resulting discretized system for test problem (9), (10) has the form

$$\begin{aligned}
&(u_m)'_t = a\delta_{xx}u_m + bu_m\big(1 - s(u_m - kw_m)\big), \quad m = 1, \ldots, M-1, \quad 0 \le t \le T;\\
&u_0(t) = e^{ct}, \quad 0 \le t \le T;\\
&u_M(t) = 2e^{ct}, \quad 0 \le t \le T;\\
&u_m(t) = e^{ct}(\cos(\pi x_m/2) + 2\sin(\pi x_m/2)), \quad m = 0, 1, \ldots, M, \quad -\tau \le t \le 0,
\end{aligned} \tag{13}$$

where δ_{xx} is a finite difference operator, $\delta_{xx}u_m = (u_{m+1} - 2u_m + u_{m-1})/h^2$; T is a time interval of calculations. The discretized system for test problem (9), (12) is obtained in the same way.

The numerical experiment has been carried out in the computer algebra system Mathematica. The system has been solved by the predefined function `NDSolve`, which is a general numerical differential equation solver of Mathematica. The option `Method` of the function `NDSolve` allows choosing the implicit Runge-Kutta method with the Lobatto IIIC coefficients and the second order of local accuracy. This helps to solve stiff systems that naturally occur after discretization of a partial differential equation with delay. Information about initialization of differential equations with delay, implementation of the method of lines, as well as description of the function `NDSolve` can be found on the webpages listed in the footnote[1].

[1] `http://reference.wolfram.com/language/tutorial/NDSolveDelayDifferentialEquations.html`
`http://reference.wolfram.com/language/tutorial/NDSolveMethodOfLines.html`

The calculations have been carried out from $t = 0$ to $t = 50\tau$ with the following values of parameters: $a = 1$, $k = 0.5$, $s = 0.2$.

The numerical solutions of system (13) for $M = 50$ and delay times $\tau = 0.05$ and $\tau = 0.5$ together with exact solutions (11) of test problem (9), (10) are shown in logarithmic scale on Figure 1. The difference between the exact and numerical solutions is visible only when both of them are practically zero.

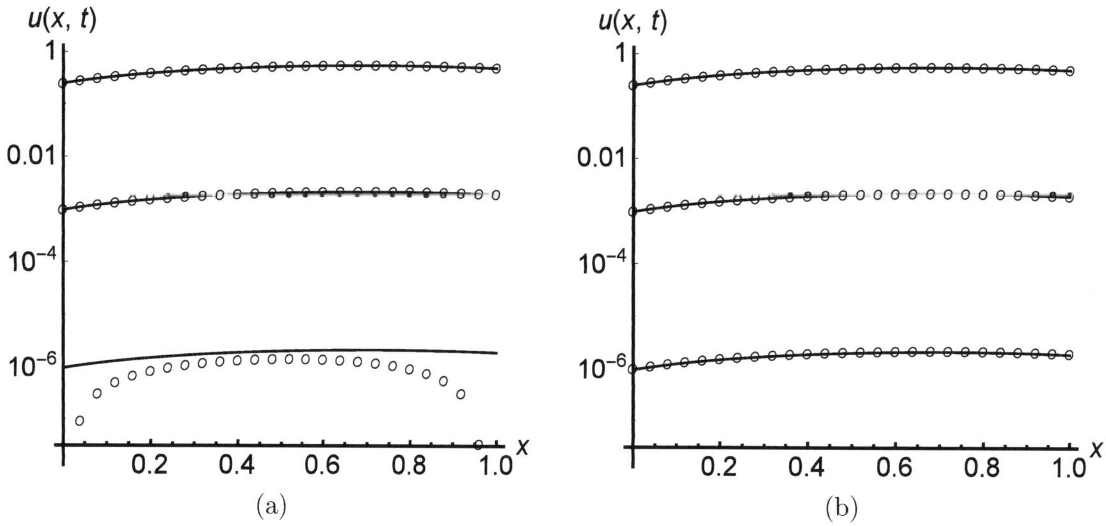

Figure 1. Graphs in logarithmic scale of the exact solution (solid line) and numerical solution with $M = 50$ (circles) of problem (9), (10) for $a = 1$, $k = 0.5$, $s = 0.2$ and a) $\tau = 0.05$, b) $\tau = 0.5$. The upper, middle and lower lines correspond to $t_1 = 2\tau$, $t_2 = 10\tau$, and $t_3 = 20\tau$ respectively.

The absolute errors of the numerical solutions of problem (9), (10) for $M = 10$, $M = 50$, and $M = 100$ and delay times $\tau = 0.05$, $\tau = 0.1$, and $\tau = 0.5$ are presented in Table 1. The percent errors for this case are not relevant due to the solution approaches zero thus the program deals with very small numbers.

Table 1. Absolute errors of the numerical solutions of problem (9), (10).

τ	$M = 10$	$M = 50$	$M = 100$
0.05	$2.8 \cdot 10^{-4}$	$1.3 \cdot 10^{-5}$	$4.5 \cdot 10^{-6}$
0.1	$4.6 \cdot 10^{-4}$	$2.0 \cdot 10^{-5}$	$6.1 \cdot 10^{-6}$
0.5	$9.9 \cdot 10^{-4}$	$4.1 \cdot 10^{-5}$	$1.2 \cdot 10^{-5}$

7. Numerical solutions of the problem with homogeneous boundary conditions and stationary initial data based on the method of lines

Consider Eq. (9) with the following initial and boundary conditions:

$$\begin{aligned} &u(0,t) = u(1,t) = 0, \quad t > 0; \\ &u(x,t) = 10\sin(\pi x), \quad 0 \le x \le 1, \quad -\tau \le t \le 0. \end{aligned} \tag{14}$$

The calculations have been carried out from $t = 0$ to $t = 10$ for two qualitatively different cases: $\tau = 0$ and $\tau = 0.5$. The numerical solutions of problem (9), (14) with $M = 50$ are shown on Figure 2. The parameters have the following values: $a = 1$, $b = 10$, $k = 0.5$, $s = 0.2$. In case without delay, the solution approaches zero faster than in case with delay. The maximum absolute difference between the two solutions is 3.1 and it is achieved at $t \approx 0.51$ and $x = 0.5$.

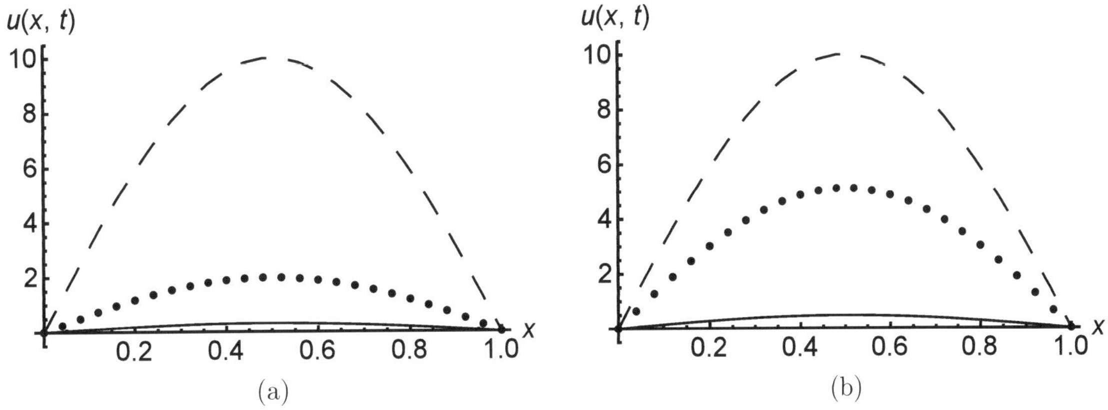

Figure 2. Graphs of the numerical solutions of problem (9), (14) for $a = 1$, $b = 10$, $k = 0.5$, $s = 0.2$ with a) $\tau = 0$ and b) $\tau = 0.5$. The dashed, dotted and solid lines correspond to $t_1 = 0$, $t_2 = 0.5$, and $t_3 = 5$ respectively.

Acknowledgments

The work is partically supported by the Russian Foundation for Basic Research (project No. 16-08-01252).

References

[1] Wu J 1996 *Theory and Applications of Partial Functional Differential Equations* (New York: Springer)
[2] Huang J and Zou X 2002 *J. Math. Anal. Appl.* **271** 455-66
[3] Wang X and Li Z 2007 *Nonlinear Anal. Theory Methods Appl.* **67** 2699-711
[4] Polyanin A D and Zhurov A I 2013 *Int. J. Non-Linear Mech.* **54** 115-26
[5] Polyanin A D and Zhurov A I 2013 *Int. J. Non-Linear Mech.* **57** 116-22
[6] Wang L and Gao Y 2006 *Phys. Lett.* A **350** 342-48
[7] Lu J G 2008 *Chaos Solitons Fractals* **35** 116-25
[8] Polyanin A D and Nazaikinskii V E 2016 *Handbook of Linear Partial Differential Equations for Engineers and Scientists, Second Edition* (Boca Raton: CRC Press)
[9] Polyanin A D and Zaitsev V F 2003 *Handbook of Exact Solutions for Ordinary Differential Equations, 2nd Edition* (Boca Raton: Chapman & Hall/CRC Press)
[10] Polyanin A D and Zaitsev V F 2012 *Handbook of Nonlinear Partial Differential Equations* (Boca Raton: CRC Press)
[11] Polyanin A D and Zhurov A I 2014 *Commun. Nonlinear. Sci. Numer. Simul.* **19** 417-30
[12] Polyanin A D and Zhurov A I 2014 *Int. J. Non-Linear Mech.* **62** 33-40
[13] Polyanin A D and Zhurov A I 2014 *Commun. Nonlinear. Sci. Numer. Simul.* **19** 409-16
[14] Polyanin A D, Sorokin V G and Vyazmin A V 2015 *Theor. Found. Chem. Eng.* **49** 622-35
[15] Polyanin A D and Sorokin V G 2015 *Appl. Math. Lett.* **46** 38-43
[16] Bellen A and Zennaro M 2003 *Numerical Methods for Delay Differential Equations* (Oxford: Oxford university press)
[17] Kuang J and Cong Y 2005 *Stability of Numerical Methods for Delay Differential Equations* (Beijing: Science Press)
[18] Shampine L F and Thompson S 2009 *Delay Differential Equations: Recent Advances and New Directions* (New York: Springer) chapter 9 pp 245-71
[19] Van Der Houwen P J, Sommeijer B P and Baker C T H 1986 *IMA J. Num. Anal.* **6** 1-23
[20] Rihan F A 2010 *Numer. Methods Partial Differ. Equ.* **26** 1556-71

The method of non-local transformations: Applications to blow-up problems

A D Polyanin[1,2,3], I K Shingareva[4]

[1] Institute for Problems in Mechanics, RAS, 101 Vernadsky Avenue, bldg 1, 119526 Moscow, Russia
[3] National Research Nuclear University MEPhI, 31 Kashirskoe Shosse, 115409 Moscow, Russia
[2] Bauman Moscow State Technical University, 5 Second Baumanskaya Street, 105005 Moscow, Russia
[4] University of Sonora, Blvd. Luis Encinas y Rosales S/N, Hermosillo C.P. 83000, Sonora, México

E-mail: polyanin@ipmnet.ru, inna@mat.uson.mx

Abstract. The method for numerical integration of Cauchy problems for ODEs with blow-up solutions is described. It is based on introducing a new non-local variable that reduces a single nth-order ODE to a system of first-order coupled ODEs. This method leads to problems whose solutions are presented in parametric form and do not have blowing-up singular points; therefore the standard fixed-step numerical methods can be applied. The efficiency of the proposed method is illustrated with two test problems. It is shown that the first Painlevé equation with suitable initial conditions have non-monotonic blow-up solutions.

Keywords: nonlinear differential equations, non-local transformations, blow-up problems, numerical solutions, Painlevé equations

1. Introduction

We will consider Cauchy problems for ODEs, whose solutions tend to infinity at some finite value of the independent variable $t = t_*$, where t_* does not appear explicitly in the differential equation under consideration and it is not known in advance. Similar solutions exist on a bounded interval and are called *blow-up solutions*. This raises the important question for practice: how to determine the position of a singular point t_* and the solution in its neighborhood by numerical methods.

In the general case, monotonic blow-up solutions, having a power singularity, can be represented in a neighborhood of the singular point t_* in the form

$$x \simeq A(t_* - t)^{-\beta}, \qquad \beta > 0, \tag{1}$$

where it is assumed that $t_0 \le t < t_*$. For such solutions we have $\lim\limits_{t\to t_*} |x| = \infty$ and $\lim\limits_{t\to t_*} |x'_t| = \infty$. Differentiating (1), we obtain the derivatives near the singular point

$$x'_t \simeq A\beta(t_* - t)^{-\beta-1}, \qquad x''_{tt} \simeq A\beta(\beta+1)(t_* - t)^{-\beta-2}. \tag{2}$$

The formulas (1)–(2) remain valid for non-monotonic blow-up solutions if there is a neighborhood on the left of the singular point ($t_1 \le t < t_*$, where $t_0 \le t_1$), in which the solution is monotonic.

The direct application of the standard fixed-step numerical methods to blow-up problems leads to certain difficulties because the range of variation of the independent variable is unknown in advance [1]. One of the basic ideas of numerical integration of blow-up problems consists in the application of an appropriate transformation at the initial stage, which leads to the equivalent problem whose solutions have no singularities at a priori unknown point. In particular, for such problems with monotonic solutions, the hodograph transformation can be used [2]. Some special methods of numerical integration of blow-up problems are described, for example, in [3–7]. This paper generalizes the results of [9–11], in which to numerical integration of these problems we used the method of non-local transformations.

2. Blow-up Cauchy problems for second-order ODEs

2.1. Solution method based on the introduction of a non-local variable

The Cauchy problem for a second-order differential equation has the form

$$x''_{tt} = f(t, x, x'_t) \quad (t > t_0); \tag{3}$$

$$x(t_0) = x_0, \qquad x'_t(t_0) = x_1. \tag{4}$$

We assume that the problem (3)–(4) has a non-monotonic blow-up solution.

We introduce a new non-local variable ξ by means of the first-order ODE and the initial condition:

$$\xi'_t = g(t, x, x'_t, \xi), \qquad \xi(t_0) = 0. \tag{5}$$

Here $g = g(t, x, x'_t, \xi)$ is the *regularizing function* that can be varied.

We represent the second-order equation (3) in the form of an equivalent system of two equations of the first order

$$x'_t = y, \qquad y'_t = f(t, x, y). \tag{6}$$

By using (5), we pass in (6) and (4) from t to a new independent variable ξ. As a result, the Cauchy problem (3)–(4) is transformed to the following problem for the system of three equations:

$$\begin{gathered} t'_\xi = \frac{1}{g(t,x,y,\xi)}, \qquad x'_\xi = \frac{y}{g(t,x,y,\xi)}, \qquad y'_\xi = \frac{f(t,x,y)}{g(t,x,y,\xi)} \qquad (\xi > 0); \\ t(0) = t_0, \qquad x(0) = x_0, \qquad y(0) = x_1. \end{gathered} \tag{7}$$

Choosing an appropriate function $g = g(t, x, y, \xi)$ in (7), we can obtain the Cauchy problem whose solution will not have blow-up singularities and which can be integrated by applying the standard fixed-step numerical methods [13–15].

In the particular case, when $g = g(t, x, x'_t)$, the function ξ by integrating the equation (5), can be expressed in terms of the solution of the Cauchy problem (3)–(4) in the form of an integral

$$\xi = \int_{t_0}^{t} g(t, x, y)\, dt, \qquad x = x(t), \qquad y = y(t). \tag{8}$$

For numerical solution of blow-up problems it is reasonable to apply the regularizing functions of the form

$$g = G(|y|, |f|), \tag{9}$$

where $f = f(t, x, y)$ is the right-hand side of the equation (3) and $y = x'_t$. We impose the following conditions on the function $G = G(u, v)$:

$$G > 0; \quad G_u \geq 0, \quad G_v \geq 0, \quad G \to \infty \text{ as } u + v \to \infty, \tag{10}$$

where $u \geq 0$, $v \geq 0$. In this case the non-local variable (8) will be monotonically increasing and unbounded with increasing t.

Further, in the main, we will consider a five-parameter class of the regularizing functions of the form

$$g = (1 + k_1|y|^{p_1} + k_2|f|^{p_2})^q,$$

where $k_1 \geq 0$, $k_2 \geq 0$ $(k_1 + k_2 \neq 0)$, $p_1 > 0$, $p_2 > 0$, and $q > 0$. The particular case $g = \sqrt{1 + y^2 + f^2}$ $(k_1 = k_2 = 1,\ p_1 = p_2 = 2,\ q = 1/2)$ corresponds to the arc-length transformation [3].

Remark 1. Non-local transformations of a special form were used in [16, 17] for finding exact solutions and linearizing some ordinary differential equations of the second order.

2.2. Conditions that the regularizing functions must satisfy. Examples of regularizing functions
Let $g = g(t, x, x'_t)$. Then the formula (8) is valid. Since the transformed problem (7) should not have blow-up singularities, it is necessary that its solution $x = x(t)$, $y = y(t)$ satisfies the following condition:

$$\lim_{t \to t_*} I = \infty, \qquad I = \int_{t_0}^{t} g(t, x, y)\, dt. \tag{11}$$

We have

$$I = I_1 + I_2, \quad I_1 = \int_{t_0}^{t_1} g(t, x, y)\, dt, \quad I_2 = \int_{t_1}^{t} g(t, x, y)\, dt, \tag{12}$$

where $t_1 > 0$ is a point sufficiently close to the blow-up point t_* $(t_1 < t_*)$. The convergence or divergence of the integral I is determined by the convergence or divergence of the integral I_2.

1°. *Regularizing functions of the first type.* First, we consider regularizing functions of the form $g = g(|y|) > 0$, which in addition to the normalization condition $g(0) = 1$ satisfy the asymptotic condition of power growth for large $|y|$:

$$g(|y|) \to C|y|^\alpha \quad \text{as} \quad |y| \to \infty \quad \text{with} \quad \alpha > 0 \quad (C > 0). \tag{13}$$

We use the asymptotics (2) and (13) to analyze the convergence or divergence of the integral I_2 in (12). We have

$$I_2 \simeq |A\beta|^\alpha C \int_{t_1}^{t} (t_* - \tau)^{-\alpha(\beta+1)}\, d\tau. \tag{14}$$

Hence it follows that the integral I_2 diverges if $\alpha(\beta + 1) \geq 1$ that is equivalent to the condition

$$\alpha \geq \frac{1}{\beta + 1}. \tag{15}$$

For the most common singularity of the solution that has a first-order pole, which corresponds to the value $\beta = 1$, we should choose $\alpha \geq \frac{1}{2}$. Since $\beta > 0$, then $\alpha = 1$ is suitable for any blow-up solution of the power (and logarithmic) type.

The asymptotics as $|y| \to \infty$ of regularizing functions of the form

$$g = (1 + k|y|^p)^q \qquad (k > 0,\ p > 0,\ q > 0) \tag{16}$$

is determined by the value $\alpha = pq$ in (13). For the function (16) with $p = 2$ and $q = 1/2$ we have $\alpha = 1$ and the inequality (15) holds for any positive β. For the function (16) with $p = 1$ and $q = 1/2$ we have $\alpha = 1/2$. In this case, the inequality (15) holds for a first-order pole, which is determined by the value $\beta = 1$, and also for all $\beta \geq 1$ (that is, for integer poles of any order).

2°. *Regularizing functions of the second type.* We now consider regularizing functions of the form $g = g(|z|) > 0$, where $z = x''_{tt} = f(t, x, y)$, which in addition to the normalization condition $g(0) = 1$ satisfy the asymptotic condition of power growth for large $|z|$:

$$g(|z|) \to C|z|^\alpha \quad \text{as} \quad |z| \to \infty \quad \text{with} \quad \alpha > 0 \quad (C > 0). \tag{17}$$

We use the asymptotics (2) and (17) to analyze the convergence or divergence of the integral I_2 in (12). We have

$$I_2 \simeq |A\beta(\beta+1)|^\alpha C \int_{t_1}^{t} (t_* - \tau)^{-\alpha(\beta+2)}\, d\tau. \tag{18}$$

It follows that the integral I_2 diverges if $\alpha(\beta + 2) \geq 1$, or

$$\alpha \geq \frac{1}{\beta + 2}. \tag{19}$$

For the most common singularity of the solution that has a first-order pole, which corresponds to the value $\beta = 1$, we should choose $\alpha \geq 1/3$. Since $\beta > 0$, then $\alpha \geq 1/2$ is suitable for any blow-up solution of the power (and logarithmic) type.

The asymptotics as $|z| \to \infty$ of regularizing functions of the form

$$g = (1 + k|z|^p)^q \qquad (k > 0,\ p > 0,\ q > 0) \tag{20}$$

is determined by the value $\alpha = pq$ in (17). For the function (20) with $p = 1$ and $q = 1/2$ or $p = 1/2$ and $q = 1$ we have $\alpha = 1/2$ and the inequality (19) holds for any positive β. For the function (20) with $p = 1$ and $q = 1/3$ or $p = 1/3$ and $q = 1$ we have $\alpha = 1/3$. In this case, the inequality (19) holds for a first-order pole, which is determined by the value $\beta = 1$, and also for all $\beta \geq 1$ (that is, for integer poles of any order).

Remark 2. A comparison of the exact solutions of a number of test problems and the corresponding numerical solutions obtained by the method of non-local transformations shows that the most efficient regularizing functions are those that have the least admissible value of α, the exponent in the asymptotics (13) and (17), and are determined by the equality sign in (15) and (19).

3°. *Regularizing functions of the mixed type.* In a similar way, we can define the domain of divergence of regularizing functions of the mixed type $g = g(|y|, |z|)$, starting from the asymptotics of this function as $|y| + |z| \to \infty$. In particular, the function $g = (1 + |y| + |z|)^{1/3}$ can be used if a solution has a singularity in the form of a pole of any integer order ($\beta = 1, 2, \ldots$).

Example 1. Consider a Cauchy problem for the first Painlevé equation [12]

$$x''_{tt} = 6x^2 + t \quad (t > 0); \qquad x(0) = a, \quad x'_t(0) = b. \tag{21}$$

For $a > 0$ and $b < 0$, the problem (21) has non-monotonic blow-up solutions.

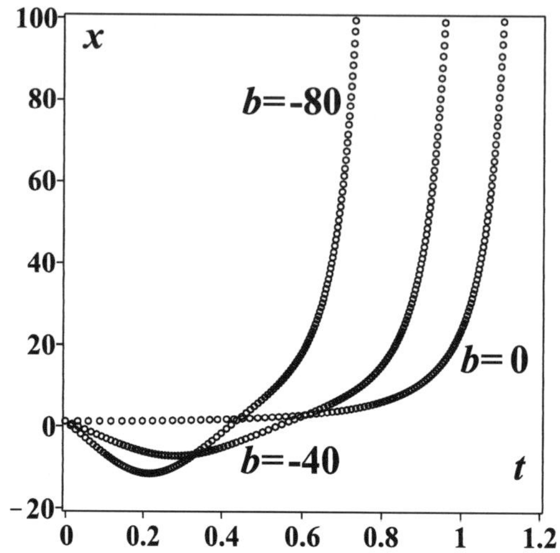

Figure 1. Numerical solutions of the problem (21) for the first Painlevé equation with $a = 1$, regularizing function $g = (1 + |y| + |f|)^{1/3}$, and $h = 0.01$

Numerical solution of the problem (21) for the first Painlevé equation with $a = 1$ and the three values of the parameter b ($b = 0$, $b = -40$, $b = -80$) which is obtained by integrating the transformed system (7) with $f = 6x^2 + t$ and $g = (1 + |y| + |f|)^{1/3}$ by the Runge–Kutta method of the fourth-order approximation for the fixed stepsize $h = 0.01$ are shown by circles in Fig. 1. For $b = -40$ and $b = -80$, the solutions of the first Painlevé equation have a non-monotonic character and exist in the finite regions $0 \le t < t_* = 1.0577704$ and $0 \le t < t_* = 0.8323860$, respectively. Reducing the stepsize by one-half, the maximum module of difference between the numerical solutions (with $h = 0.01$ and $h = 0.005$) for problem (21) for $x \le 100$ with $b = -40$ is equal to 0.0000033 and for this problem for $x \le 100$ with $b = -80$ is equal to 0.0000044. It should be noted that $\min x = x(t_m) = -7.3590292$ with $t_m = 0.2829017$ for $b = -40$ and $\min x = x(t_m) = -11.6922543$ with $t_m = 0.2170987$ for $b = -80$.

Remark 3. It can be shown that the Cauchy problem for the second Painlevé equation

$$x''_{tt} = 2x^3 + tx + c \quad (t > 0); \qquad x(0) = a, \quad x'_t(0) = b,$$

for $a > 0$ and $b < 0$, can have non-monotonic blow-up solutions.

Remark 4. Several blow-up test problems and their non-monotonic exact solutions are described in [11].

3. Blow-up Cauchy problems for higher-order ODEs

3.1. General description of the method of non-local transformations

We consider the Cauchy problem for the nth-order differential equation of the general form

$$\begin{gathered} x_t^{(n)} = f(t, x, x'_t, \dots, x_t^{(n-1)}) \quad (t > t_0); \\ x(t_0) = x_0, \quad x'_t(t_0) = x_0^{(1)}, \quad \dots, \quad x_t^{(n-1)}(t_0) = x_0^{(n-1)}, \end{gathered} \tag{22}$$

where $x_t^{(k)} = d^k x/dt^k$ ($k = 3, \dots, n$).

As before, we introduce a new non-local independent variable ξ by means of the first-order differential equation (with respect to ξ) and the initial condition:

$$\xi'_t = g(t, x, x'_t, \dots, x_t^{(n-1)}, \xi), \qquad \xi(t_0) = 0. \tag{23}$$

Here $g = g(t, x, x'_t, \dots, x_t^{(n-1)}, \xi)$ is a regularizing function that can be varied.

Now we represent the Cauchy problem for one nth-order equation (22) in the form of an equivalent problem for a system consisting of n equations of the first order (of special form):

$$\begin{gathered}(x_1)'_t = x_2, \quad (x_2)'_t = x_3, \quad \dots, \quad (x_{n-1})'_t = x_n, \quad (x_n)'_t = f(t, x_1, x_2, \dots, x_n); \\ x_1(t_0) = x_0, \quad x_2(t_0) = x_0^{(1)}, \quad \dots, \quad x_n(t_0) = x_0^{(n-1)},\end{gathered} \tag{24}$$

where $x_1 \equiv x$. Then, by using (23), we pass from t to a new independent variable ξ in (24). As a result, the Cauchy problem for one equation of the nth order (22) is transformed to the following problem for a system consisting of $(n+1)$ st equation of the first order:

$$\begin{gathered}t' = \frac{1}{g}, \quad x'_1 = \frac{x_2}{g}, \quad x'_2 = \frac{x_3}{g}, \quad \dots, \quad x'_{n-1} = \frac{x_n}{g}, \quad x'_n = \frac{f}{g}; \\ t(0) = t_0, \quad x_1(0) = x_0, \quad x_2(0) = x_0^{(1)}, \quad \dots, \quad x_n(0) = x_0^{(n-1)},\end{gathered} \tag{25}$$

where $f = f(t, x_1, x_2, \dots, x_n)$, $g = g(t, x_1, x_2, \dots, x_n, \xi)$, and the prime denotes the derivative with respect to ξ.

In the particular case, when $g = g(t, x, x'_t, \dots, x_t^{(n-1)})$, the function ξ by integrating the equation (23), can be expressed in terms of the solution of the Cauchy problem (23) in the form of an integral

$$\xi = \int_{t_0}^{t} g(t, x_1, x_2, \dots, x_n)\, dt, \qquad x_k = x_k(t) \equiv x_t^{(k)}. \tag{26}$$

3.2. Examples of regularizing functions for nth-order equations

From (1), we obtain an approximate formula for the derivative of an arbitrary order in a neighborhood of the singular point t_*:

$$x_t^{(n)} \simeq A_n(t_* - t)^{-\beta-n}, \qquad A_n = A\beta(\beta+1)\cdots(\beta+n-1). \tag{27}$$

We consider regularizing functions of the form $g = g(|f|) > 0$, where f is the right-hand side of the equation (22). Suppose, in addition to the normalization condition $g(0) = 1$, that the function g satisfies the asymptotic condition of power growth for large $|f|$:

$$g(|f|) \to C|f|^{\alpha} \quad \text{as} \quad |f| \to \infty \quad \text{with} \quad \alpha > 0 \quad (C > 0). \tag{28}$$

We use the asymptotics (27) and (28) to analyze the convergence or divergence of the integral I_2 in (12). Arguing as in Section 2.2, we can show that the integral I_2 diverges if the following condition is satisfied:

$$\alpha \geq \frac{1}{\beta + n}. \tag{29}$$

Since $\beta > 0$, then $\alpha = 1/n$ is suitable for any blow-up power-type solution. For the most common singularity of the solution that has a first-order pole, which corresponds to the value $\beta = 1$, we can choose $\alpha = 1/(n+1)$ (this value of α is also suitable for a pole of any integer order).

The asymptotics as $|f| \to \infty$ of regularizing functions of the form

$$g = (1 + k|f|^p)^q \qquad (k > 0, \ p > 0, \ q > 0) \tag{30}$$

is determined by the value $\alpha = pq$ in (28). For the function (30) with $p = 1$ and $q = 1/n$ or $p = 2$ and $q = 1/(2n)$ we have $\alpha = 1/n$ and the inequality (29) holds for any positive β. For the function (30) with $p = 1$ and $q = 1/(n+1)$ or $p = 1/(n+1)$ and $q = 1$ we have $\alpha = 1/(n+1)$. In this case the inequality (29) holds for a first-order pole, which is determined by the value $\beta = 1$, and also for all $\beta \geq 1$ (that is, for integer poles of any order).

Below, in more detail, we describe the procedure for applying the method for third-order ODEs.

3.3. Equations of the third order. A test problems and its numerical integration

We consider the Cauchy problem for the third-order differential equation

$$x'''_{ttt} = f(t, x, x'_t, x''_{tt}) \quad (t > 0); \qquad x(0) = x_0, \ \ x'_t(0) = x_1, \ \ x''_{tt}(0) = x_2, \tag{31}$$

which is equivalent to the problem for a system consisting of the three first-order equations:

$$\begin{gathered} x'_t = y, \quad y'_t = z, \quad z'_t = f(t, x, y, z) \qquad (t > 0); \\ x(0) = x_0, \quad y(0) = x_1, \quad z(0) = x_2. \end{gathered} \tag{32}$$

Introducing a non-local variable ξ by means of the differential equation

$$\xi'_t = g(t, x, x'_t, x''_{tt}, \xi), \qquad \xi(0) = 0,$$

we transform the system (32) to the following form:

$$\begin{gathered} t'_\xi = \frac{1}{g}, \quad x'_\xi = \frac{y}{g}, \quad y'_\xi = \frac{z}{g}, \quad z'_\xi = \frac{f}{g} \qquad (\xi > 0); \\ t(0) = 0, \quad x(0) = x_0, \quad y(0) = x_1, \quad z(0) = x_2, \end{gathered} \tag{33}$$

where $f = f(t, x, y, z)$ and $g = g(t, x, y, z, \xi)$.

Let us consider various possibilities for choosing the function g in the system (33).

1°. We can take $g = y/x$. In this case, the system (33) is simplified, since the second equation is directly integrated, and taking into account the second initial condition, we obtain $x = x_0 e^\xi$.

2°. We can take $g = z/y$. In this case, the system (33) is simplified, since the third equation is directly integrated, and we obtain $y = x_1 e^\xi$. Taking into account the relations (32), we also have $\xi = \ln(x'_t/x_1)$.

3°. Also, we can take $g = f/z$. In this case, the system (33) is also simplified, since the fourth equation is directly integrated, and we obtain $z = x_2 e^\xi$.

4°. We can take $g = \big(1 + k_1|y|^{p_1} + k_2|z|^{p_2} + k_3|f|^{p_3}\big)^q$ for $k_m \geq 0$, $p_m > 0$, and $q > 0$. The case $k_m = 1$, $p_m = 2$, and $q = 1/2$ corresponds to the method of the arc-length transformation [3].

The transformations corresponding to the first three Items, 1°, 2°, and 3°, will be called the special *exp-type transformations*, they lead to the solutions, in which the variable t tends exponentially rapidly to a blow-up point t_*. These transformations can not be applied if for non-monotonic solutions (since the function g function can vanish in such cases). For non-monotonic solutions one can use the last function from Item 4° (for example, with $p_m = 1$, $q = 1/2$ or $p_m = 1$, $q = 1/3$).

Example 2. We consider in more detail the test Cauchy problem of the form

$$x'''_{ttt} = 6x^4 \qquad (t > 0); \quad x(0) = x'_t(0) = 1, \quad x''_{tt}(0) = 2. \tag{34}$$

The exact solution of this problem is determined by the formula $x = 1/(1 - t)$. It has a power-type singularity (a first-order pole) at the point $t_* = 1$ and does not exist for $t > t_*$.

In Table 1, a comparison of the efficiency of the numerical integration methods, based on various nonlocal transformations of the form (5) is presented by using the example of the test blow-up problem for the third-order ODE (34). The comparison is based on the number of grid points needed to make calculations with the same maximum error, % = 0.01.

Transformation	Function g	Max. interval $\xi_{\max}$	Stepsize h	Grid points N
Arc-length, Item 4°	$g=(1+y^2+z^2+f^2)^{1/2}$	1996500.000	0.3630	5500000
Hodograph	$g=y$	98.987	0.0790	1253
Special exp-type, Item 3°	$g=f/z$	13.800	0.1150	120
Nonlocal, Item 4°	$g=(1+\|f\|)^{1/4}$	7.222	0.0675	107
Special exp-type, Item 1°	$g=y/x$	4.590	0.0450	102
Nonlocal, Item 4°	$g=(1+\|y\|+\|z\|)^{1/3}$	6.030	0.0603	100
Special exp-type, Item 2°	$g=z/y$	9.212	0.0940	98

Table 1. Various types of analytical transformations applied for numerical integration of the problem (34) with a given accuracy (percent error is 0.01 for $1 \le x \le 100$) and their basic parameters (maximum interval, stepsize, grid points number).

It can be seen that the arc-length transformation is the least effective, since the use of this transformation is associated with a large number of grid points. In particular, when using the last four transformations, you need 52401–56122 times less of a number of grid points. The hodograph transformation has an intermediate (moderate) efficiency.

4. Brief conclusions

The method for numerical integration of blow-up problems for ODEs $x_t^{(n)} = f(t, x, x'_t, \ldots, x_t^{(n-1)})$, based on the introduction of a new non-local independent variable ξ, which is related to the original variables t and x by the equation $\xi'_t = g(t, x, x'_t, \ldots, x_t^{(n-1)}, \xi)$, is described. With a suitable choice of the regularizing function g, the proposed method leads to problems whose solutions are presented in parametric form and do not have blowing-up singular points; therefore the transformed problems allow the application of the standard fixed-step numerical methods. The numerical integration of test problems has shown high efficiency of methods based on non-local transformations of a special type (and the practical inapplicability of the arc-length transformation for numerical solving blow-up problems with ODEs of high order). It is shown that the first Painlevé equation with suitable initial conditions have non-monotonic blow-up solutions.

It is important to note that the method of non-local transformations can be useful for numerical integration of other problems with large solution gradients (for example, in problems with solutions of boundary-layer type).

References

[1] Stuart M, Floater M S 1990 *European J. Applied Math.* **1** 47–71.
[2] Acosta G, Durán G, Rossi J D 2002 *Computing* **68** 343–373.

[3] Hirota C, Ozawa K 2006 *J. Comput. Applied Math.* **193** 614–637.
[4] R. Meyer-Spasche 1998 *J. Comput. Appl. Mathematics* **97** 137–152.
[5] Budd C J, Leimkuhler B, Piggott M D 2001 *Appl. Numer. Math.* **39** 261–288.
[6] Baris J, Baris P, Ruchlewicz B *J. Math. Sciences* 2008 **149** 1369–1375.
[7] Cho C-H 2011 *Appl. Math. Letters* **24** 49–54.
[8] Dlamini P G, Khumalo M 2012 *Math. Problems Eng.* **2012** 162034
[9] Polyanin A D, Shingareva I K 2017 *Int. J. Non-Linear Mechanics* **95** 178–184.
[10] Polyanin A D, Shingareva I K 2017 *Vestnik NIYaU MIFI* **6** 405–424 (in Russian).
[11] Polyanin A D, Shingareva I K 2018 *Appl. Math. Letters* **76** 123–129.
[12] Polyanin A D, Zaitsev V F 2017 *Handbook of Ordinary Differential Equations: Exact Solutions, Methods, and Problems* (Boca Raton–London: CRC Press/Chapman & Hall)
[13] Butcher J C 1987 *The Numerical Analysis of Ordinary Differential Equations: Runge-Kutta and General Linear Methods* (New York: Wiley-Interscience)
[14] Lambert J D 1991 *Numerical Methods for Ordinary Differential Systems* (New York: Wiley) 1991.
[15] Shingareva I K, Lizárraga-Celaya C 2009 *Maple and Mathematica. A Problem Solving Approach for Mathematics, 2nd Edition* (Wien – New York: Springer)
[16] Muriel C, Romero J L 2010 *J. Physics A: Math. Theor.* **43** 434025
[17] Kudryashov N A, Sinelshchikov D I 2016 *Appl. Math. Letters* **57** 114–120.

Further progress in control of localized nonlinear waves

A V Porubov[1,2,3], **I D Antonov**[1,3], **A L Fradkov**[1,2]
[1]Institute of Problems in Mechanical Engineering, Bolshoy 61, V.O., Saint Petersburg 199178, Russia
[2]St. Petersburg State University, 7-9, Universitetskaya nab., V.O., Saint Petersburg 199034, Russia
[3]St. Petersburg State Polytechnical University, Polytechnicheskaya st., 29, Saint Petersburg 195251, Russia

E-mail: alexey.porubov@gmail.com

Abstract. Previously found mechanical system is studied which consists of the nonlinearly elastic layer subjected to external loading. The last is assumed to be a distributive control which causes in particular, localization of longitudinal nonlinear strain waves. Various types of localized nonlinear waves are achieved due to a suitable choice of the external loading. Wave profiles arising at variations around this optimal choice of the control are studied.

1. Introduction

The localized wave solutions to nonlinear equations contain restrictions concerning their shapes, velocities and initial phases. Only particular solutions of nonitegrable nonlinear wave equations may be obtained by direct methods [1,2]. More general localized wave solutions can be obtained only numerically, however, their stable propagation depends on many factors including choice of initial and boundary conditions.

Recently, a method of control of nonlinear waves has been developed [4–6]. The algorithm of distributed feedback control allows us to achieve the desired profile of localized wave and its stable propagation of permanent shape and velocity. In particular, it was found [4] that the most efficient control of the solution of the sine- Gordon equation is achieved when the control is organized as follows,

$$U_{tt} - U_{xx} + \sin U + u(x,t) = 0 \tag{1}$$

where the control function is

$$u(x,t) = \gamma \left(\alpha_1 (U - U^*) - \alpha_2 (U_t - U_t^*)\right),$$

where $U(x,t)^*$ is the target function or a desired wave solution.

A natural question arises whether such control terms relate to reality. For this purpose, a mechanical system has been described in [6] whose model equation is Eq. (1). This system is a horizontal nonlinearly elastic layer laying on a visco-elastic foundation and subjected to an external loading from the upper lateral surface. Then propagation of shear strain waves along the layer is described by Eq. (1) where the control function is described by external loading.

However, propagation of longitudinal waves in the layer is described by another model equation with a control. In this paper, the new algorithm of control following from the mechanical problem is examined. A tendency to some localized waves of permanent shape and velocity is revealed, and suitable loading is found. Variations in the wave shape due to the variations in the loading around its suitable form are studied.

2. Statement of the problem

The model nonlinear equation for longitudinal strain waves in a layer lying on the visco-elastic foundation and subjected to an external loading is [6]:

$$\rho U_{tt} - \frac{4\mu(\lambda+\mu)}{(\lambda+2\mu)^2} U_{xx} + \beta \sin\left(\frac{2}{h}U\right) + v(x,t) \;=\; 0, \tag{2}$$

where $v(x,t)$ is the control function,

$$v(x,t) \;=\; \frac{\lambda}{\lambda+2\mu}\, k_d\, \left(\alpha_1 (U-U^*)_x + \alpha_2\, (U_t - U_t^*)_x\right).$$

At the same time, the target function U^* is connected with the tangentional external stress,

$$f_\tau \;=\; \frac{\lambda}{\lambda+2\mu}\, k_d\, (\alpha_1\, U_x^* + \alpha_2\, U_{xt}^*),$$

Other notations are: λ and μ are the Láme coefficients, $k_d > 0$ is a friction coefficient at the upper lateral surface, α_1,α_2 are the coefficients of the visco-elastic foundation model.

Equation (2) can rewritten in a canonical form (1) by introducing scales for the variables,

$$x \;=\; \sqrt{\frac{2\mu h(\lambda+\mu)}{\beta(\lambda+2\mu)^2}}\;\tilde{x},\; t = \sqrt{\frac{\rho h}{2\beta}}\;\tilde{t}, U = \;\frac{h}{2}\,\tilde{U}$$

Let us introduce

$$\tilde{\gamma} \;=\; \frac{\lambda k_d h}{2\beta(\lambda+2\mu)},\; \tilde{\alpha_1} \;=\; \sqrt{\frac{\beta(\lambda+2\mu)^2}{2\mu h(\lambda+\mu)}}\;\alpha_1,\; \tilde{\alpha_2} \;=\; \frac{\beta(\lambda+2\mu)}{h}\sqrt{\frac{1}{\rho\mu(\lambda+\mu)}}\;\alpha_2.$$

Then omitting tildes for simplicity, one obtains the control function of the form

$$u(x,t) = \gamma\left(\alpha_1 (U-U^*)_x + \alpha_2\, (U_t - U_t^*)_x\right). \tag{3}$$

We will consider the control problem of achievement of the desired wave profile from initially undisturbed state which differs from our previous studies where transformation of some nonzero initial disturbance was studied.

The first choice of the target function is

$$U^*(x,t) = A\, \mathrm{sech}^2(k(x - Wt - x_1)) - \frac{1}{2} A\, \mathrm{sech}^2(k(x - Wt - x_2)), \tag{4}$$

where W is a constant phase velocity, A and k are the profile parameters. The profile of this wave consists of a hump and a hole parts that correspond to the tensile and compression parts respectively, see the dashed profile in the first sketch in Fig.1.

Another choice of the target function corresponds to the breather-shaped wave:

$$U^* = 4\, \mathrm{sech}(x - a\, Vt - x_0) \sin(b\, x - Vt - x_1) \tag{5}$$

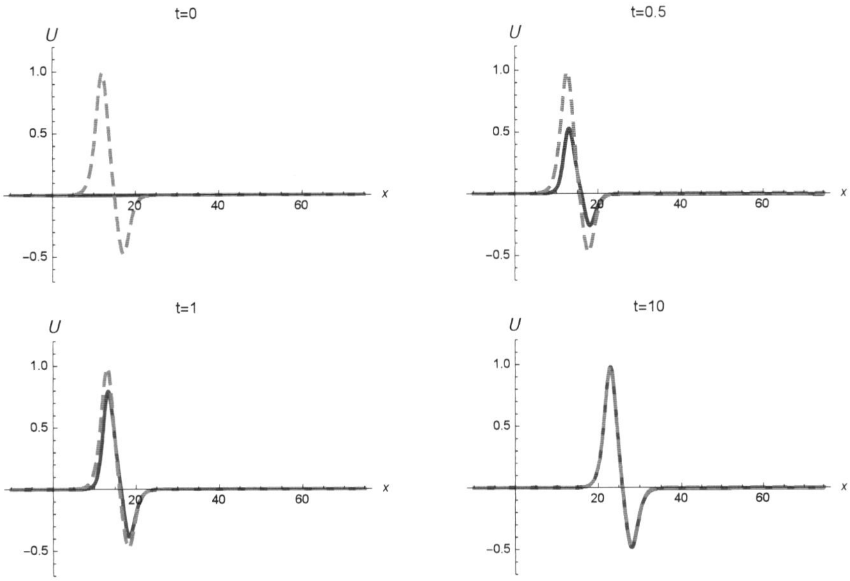

Figure 1. Generation of the localized wave with compression and tensile parts at $\gamma = 5$. Shown by dashed line is the desired traveling wave (4).

Both (4) and(5) do not correspond to analytical solutions of the sine-Gordon equation. The waves with such profiles do not exist without control.

In both cases zero initial conditions are assumed,

$$U(x,0) = 0, \quad U_t(x,0) = 0.$$

and zero boundary conditions are

$$U \to 0 \quad \text{at} \quad x \to -\infty$$

$$U \to 0 \quad \text{at} \quad x \to \infty.$$

3. Achievement of the desired wave profile

Numerical simulations of Eq.(1) with the control function defined by Eq. (3) were preformed using the tools of Wolfram Mathematica with the NDSolve procedure.

The parameters of the target function (4) are chosen equal to $W = 1.1, A = 1, k = 0.5, x_1 = 12, x_2 = 17$. The control parameters depend on the features of the layer material and foundation. We formally assume fixed values for $\alpha_1 = 50$, $\alpha_2 = 40$ and vary the value of γ. One can otherwise vary the values of foundation parameters keeping the fixed value of γ.

Shown in Fig. 1 is the generation of the wave (4) due to the control at $\gamma = 5$. One can see that initially undisturbed state gradually fills in the desired wave shape shown by dashed line. At time $t = 10$ the waves coincide and further calculations demonstrate stable propagation of the wave shown in the last sketch in Fig. 1 with the velocity W.

The similar study is performed for the target function (5). The target function parameters are assumed to be equal to $a = 0.5, b = 5, x_0 = 0, x_1 = 0, V = 0.95$, while the control parameters α_i remain of the same value. The simulations at $\gamma = 1.5$ are shown in Fig. 2. One can see that again the profile of the nonlinear wave is generated from undisturbed initial state by the control, the wave gradually achieves the desired wave shape of the breather and further propagates keeping its shape and velocity according the target function (5).

IOP Conf. Series: Journal of Physics: Conf. Series **937** (2017) 012043 doi:10.1088/1742-6596/937/1/012043

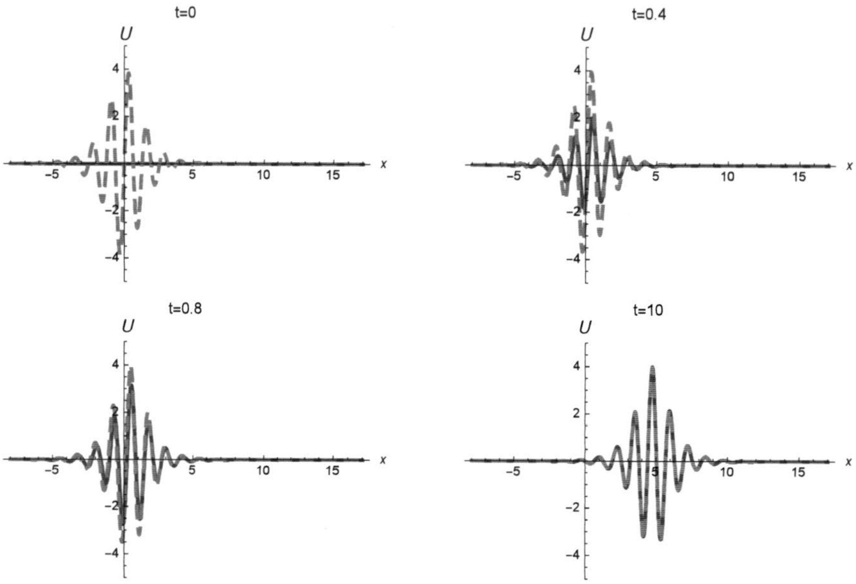

Figure 2. Generation of the breather by the control at $\gamma = 1.5$. Shown by dashed line is the desired traveling wave (5).

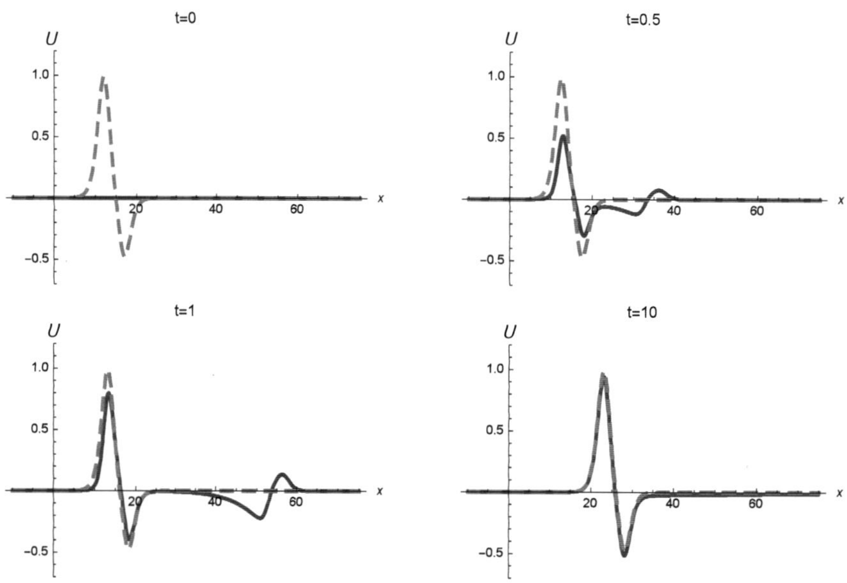

Figure 3. Generation of compression and tensile wave with $\gamma = 1$. Shown by dashed line is the desired traveling wave (4).

4. Variations in the desired wave

We achieve the desired localized waves propagation at special values of the control parameters. One can note that variation in these values result in temporal arising of additional parts in the wave profile. Thus, for γ_1 the generation of the wave (4) is accompanied by an additional variations in the wave profile as shown in Fig. 3. One can see that this defect visible at $t = 0.5$ and $t = 1$ disappears at $t = 10$ and further propagation of the wave is similar to the case $\gamma = 5$.

IOP Conf. Series: Journal of Physics: Conf. Series **937** (2017) 012043 doi:10.1088/1742-6596/937/1/012043

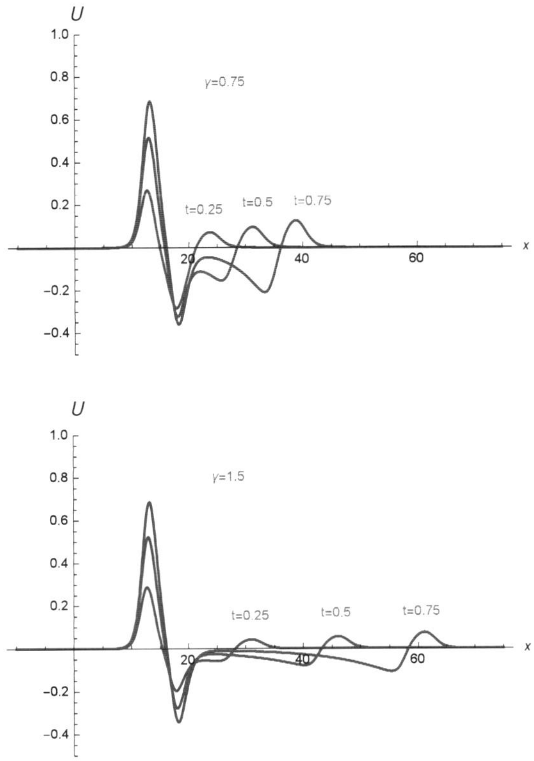

Figure 4. Deviation from the desired profile (4) at different values of γ.

The dynamics of the defect evolution for different values of γ is shown in Fig. 4 for $\gamma = 0.75$ and $\gamma = 1.5$. In both cases one can see that the defect is developing for some time and goes away from the desired wave but then it gradually disappears as shown in the last sketch in Fig. 3.

5. Conclusion

The new algorithm of distributed feedback control arising from the real mechanical system, is examined. It is shown that it works efficiently similar to our previous algorithms. variations in the desired shape of the wave may appear for some time but they decay as time further goes on. Future work concerns the study of a similar mechanical system but for the materials having di-atomic crystalline structure. In this case one anticipates the control of the coupled nonlinear wave equations [5].

Acknowledgments

The work was performed in IPME RAS, supported by the Russian Science Foundation (grant 14-29-00142).

References

[1] Kudryashov N A 2012 *Commun.Nonlinear Sci. Numer.Simulat.* **17** 2248.
[2] Kudryashov N A, Loguinova N B 2008 *Appl. Math. Comp.* **205** 396.
[3] Porubov A, Andrievsky 2017 *Philos. Trans. of The Royal Soc. A* **375** 20160212
[4] Porubov A V, Antonov I D, Fradkov A L, Andrievsky B R 2016 *Intern. Journ. of Non-Linear Mech.* **86** 174.
[5] Porubov A V, Antonov I D 2017 *Journ. of Phys. Conf. Ser.* **788** 012029.
[6] Porubov A V, et al. 2017 *Mech. Re. Commun.*, http://dx.doi.org/10.1016/j.mechrescom.2017.07.014

On the Analog of the Theorem Bary–Stechkin of the Conjugate Functions for the Dyadic Group

A I Rubinshtein[1]

National Research Nuclear University MEPhI (Moscow Engineering Physics Institute), 31 Kashirskoe shosse, 115409 Moscow, Russia

E-mail: rubinshtein_aleksandr@mail.ru

Abstract. We consider the problem of the modulus of continuity for the analogy conjugate function in the case of functions defined on the dyadic group. It was shown that in this case there is no analogy for Bary–Stechkin theorem.

Let $G = \{x = (x_1, x_2, \dots), x_k{=}0; 1\}$ be the Abelian group with operation defined by

$$x \dot{+} y = (x_1, x_2, \dots) \dot{+} (y_1, y_2, \dots) = z = (z_1, z_2, \dots),$$

where $z_k = x_k + y_k (\mod 2)$, while $f(x) : G \to \mathbb{R}$.

A topology of G defined of the system of subgroups

$$U_{k-1} = \Big\{x = (\underbrace{0, \dots, 0}_{k-1}, x_k, \dots)\Big\}, \quad k = 1, 2, \dots$$

is a basic system of neighborhoods of $x = 0$.

It is established in [1, 5] that the operator

$$(Kf)(x) = -\lim_{m\to\infty} \int_{G\setminus U_m} (f(x \dot{+} t) - f(x))K(t)d\mu(t),$$

where for $t \neq 0$

$$K(t) = 2^k, \quad t \in U_{k-1} \setminus U_k, \qquad k = 1, 2, \dots$$

and $\mu(t)$ — normalized Haar measure on G.

It is obvious that operator $(Kf)(x)$ appears a certain analogue of the conjugate operator $f \to \tilde{f}$, $f \in L(T)$. The criterion of the modulus of continuity $\omega(\delta, f)_{C(T)}$ that $\omega(\delta, \tilde{f}) \sim \omega(\delta, f)$ is established in [2]. (It is established I.I. Privalov that $\tilde{f} \in \mathrm{Lip}\,\alpha$, if $f \in \mathrm{Lip}\,\alpha$, $0 < \alpha < 1$). It is established the criterion of modulus of continuity

$$\omega(\delta, f)_{L_p(G)} = \left\{\omega_n^{(p)}(f) = \sup_{h\in U_n} \left(\int_G |f(x \dot{+} t) - f(t)|^p d\mu(t)\right)^{1/p}, \ 1 \le p \le \infty\right\}$$

[1] This work is partially supported by the program of increasing of competitiveness of the National Nuclear Research University MEPhI, project No. 02.a03.21.0005 from 27.08.2013

in [3], and for $L_2(T)$ by O.V. Besov and S.B. Stechkin.

It is established in [4], that if

$$f(x) \sim c_0 + \sum_{n\geq 0} \sum_{k=2^n}^{2^{n+1}-1} c_k w_k(x), \quad c_k = c_k(f) = \int_G f(t) w_n(t)\, d\mu(t),$$

when $\{w_n(x)\}$ is system Walsh–Paly (system of Pontryagin characters G), than

$$\sum_{k\geq 2^n} |c_k(f)|^2 \leq \frac{1}{2}\left(\omega_n^2(f)\right). \tag{1}$$

And it is established in [1] that

$$(Kf)(x) \sim \sum_{n\geq 0} (n+2) \sum_{k=2^n}^{2^{n+1}-1} c_k(f) w_k(x).$$

It is clear that there is not exist analogy of Bary–Stechkin theorem for $(Kf)(x)$. But exists a connection between results about $\omega(Kf)_{L_2(G)}$ and $\omega(f)_{L_2(G)}$.

Let

$$g(x) = A_0 + \sum_{n\geq 0} A_n \sum_{k=2^n}^{2^{n+1}-1} w_k(x).$$

It is established in [6] (Lemma 3) that

$$\omega_n^{(p)}(g) = \begin{cases} \sup\limits_{k\geq n} \left(2 \sum\limits_{s\geq k+1} \left| \sum\limits_{k+1\leq \nu\leq s} 2^\nu (A_{\nu-1} - A_\nu) \right|^p \cdot 2^{-s-1} \right)^{1/p} & \text{if } 1 \leq p < \infty; \\ \sup\limits_{l\geq k\geq n} \left| \sum\limits_{k+1\leq \nu\leq l} 2^\nu (A_{\nu-1} - A_\nu) \right| & \text{if } p = \infty. \end{cases}$$

Let $p = \infty$ and $A_\nu \searrow 0$. Than

$$\omega_n^{(\infty)}(g) = \sum_{\nu\geq n+1} 2^\nu (A_{\nu-1} - A_\nu)$$

and

$$\omega_n^{(\infty)}(g) - \omega_{n+1}^{(\infty)}(g) = 2^{n+1}(A_{\nu-1} - A_\nu).$$

It is clear that than

$$A_n = \sum_{k\geq n} 2^{-n-1} \left(\omega_n^{(\infty)}(g) - \omega_{n+1}^{(\infty)}(g)\right).$$

And (see [5])

$$\omega_n^{(\infty)}(Kg) = \sum_{\nu\geq n+1} 2^\nu (A_{\nu-1} \cdot \log(\nu-1) - A_\nu \cdot \log \nu).$$

It is shown that not exist analogy of Bary–Stechkin theorem for $(Kf)(x)$ in $L_\infty(G)$.

Let $p = 1$ and $A_\nu \searrow 0$. Than (see $\omega_n^{(p)}(g)$ for p=1)

$$\omega_n^{(1)}(g) = A_n \qquad \left(= o\left(1/\log n \right) \right)$$

IOP Conf. Series: Journal of Physics: Conf. Series **937** (2017) 012044 doi:10.1088/1742-6596/937/1/012044

and

$$\omega_n^{(1)}(Kg) = A_n \cdot \log n.$$

Again is not exist analogy of Bary–Stechkin theorem on conjugate functions on the dyadic group (for $L_1(G)$).

It is obvious by the formula for $\omega_n^{(p)}(g)$ (and $\omega_n^{(p)}(Kg)$) that is not exist analogy of Bary–Stechkin theorem and for $L_p(G)$, $1 < p < \infty$.

Abstract of this result see in [7].

References

[1] A I Rubinstein About functions on the dyadic group and Walsh series, Analysis Mathematica, 49(2015), 1–2, 73–81.

[2] N K Bary, S B Stechkin Ambsten approximations and differential properties of conjugate functions, Transactions of Moscow Mathematical Society, 5(1956), 483–521 [in Russian].

[3] A I Rubinstein Moduli of continuity of functions, defined on a zero-dimensional group, Mat. Zametki, 23(1978), 3, 379–388 [in Russian], English Transl. in Mathematical Notes, 23(1978).

[4] N Y Vilenkin,A I Rubinstein A I One theorem of S.B. Stechkin in absolute convergence and series of systems characters of zero-dimensional groups, Izvestia Vusov, Mathematica, 9(1975), 3–9 [in Russian].

[5] A I Rubinstein On a certain integral operator acting on functions defined on the dyadic group, Eurasian Mathematical Journal, 7, 1(2016), 68–73.

[6] M F Timan,A I Rubinstein On the inclusion of the classis functions definite on the zero-dimensional groups, Izvestia Vusov, Mathematica, 8(1980), 66–76 [in Russian].

[7] A I Rubinstein On the Analog of the Bary–Stechkin Theorem of the conjugate Functions for the Dyadic Group, Sixth international conference "The problems of Mathematical Physics and Mathematical modeling", Book of abstracts, 2017, Moscow, Russia, 210–211.

IOP Conf. Series: Journal of Physics: Conf. Series **937** (2017) 012045 doi:10.1088/1742-6596/937/1/012045

About normal distribution on SO(3) group in texture analysis

T I Savyolova[1], S V Filatov[1]

[1] National Research Nuclear University MEPhI, Kashirskoe highway, 31, 115409, Moscow, Russia

E-mail: *feelshadow5@gmail.com*

Abstract. This article studies and compares different normal distributions (NDs) on SO(3) group, which are used in texture analysis. Those NDs are: Fisher normal distribution (FND), Bunge normal distribution (BND), central normal distribution (CND) and wrapped normal distribution (WND). All of the previously mentioned NDs are central functions on SO(3) group. CND is a subcase for normal CLT-motivated distributions on SO(3) (CLT here is Parthasarathy's central limit theorem). WND is motivated by CLT in R^3 and mapped to SO(3) group. A Monte Carlo method for modeling normally distributed values was studied for both CND and WND. All of the NDs mentioned above are used for modeling different components of crystallites orientation distribution function in texture analysis.

1. Introduction

There are a number of different approaches for calculating crystallites orientation distribution function (CODF) of polycrystalline materials by obtaining the sample of pole figures (PF) using X-ray or neutron methods or by using EBSD (Electron Backscattering Diffraction) methods [1-4]. One of those methods is a components method or standard function approximation method [3, 4]. Different distributions are used as central functions, but the most attractive ones are normal distributions on SO(3) group [4-7]. Articles [8, 11, 12] introduce a new kind of ND on SO(3) group. Article [8] compares (numerically and by properties) known NDs on SO(3), which are central functions and are able to be presented as characters of representations of SO(3) group. Articles [9-10] also review and compare known NDs on SO(3) group.

This article conducts the comparison between that kind of NDs on SO(3). Normal distributions are computed for different values of concentration parameter, asymptotic behaviour is analysed for small values of concentration parameter, the presence of properties characteristic for normal distributions on SO(3) group is studied. As well as the possibility of applying these NDs in texture analysis is considered. That includes computational complexity of the algorithm and the correctness of application of Monte Carlo method for modeling the EBSD experiment for studied NDs.

2. Central normal distributions on SO(3) group

Central functions on SO(3) can be expressed as a series with terms as characters of representations of SO(3) group

$$\chi_l(t) = \frac{\sin[t(l+\frac{1}{2})]}{\sin(\frac{t}{2})}, \qquad l = 0, 1, \dots, \qquad t \in (-\pi, \pi]$$

that form and orthogonal function system with respect to the invariant measure [11].

IOP Conf. Series: Journal of Physics: Conf. Series **937** (2017) 012045 doi:10.1088/1742-6596/937/1/012045

$$dq = \frac{1}{\pi}\sin^2(\frac{t}{2})\mathrm{dt}$$

3. Normal distributions

The article studies following distributions:

3.1. Fisher normal distribution (FND)

$$\frac{\mathrm{e}^{\mathrm{k_F^2}\ \cos(t)}}{I_0(k_F^2) - I_1(k_F^2)} \frac{1}{\pi} sin^2\left(\frac{t}{2}\right) \quad (1)$$

$I_0(z),\ I_1(z)$ - modified Bessel functions.

3.2. Bunge normal distribution (BND)

$$C(k_B)\ exp(-\frac{k_B^2 t^2}{2})\ \frac{1}{\pi} sin^2\left(\frac{t}{2}\right) \quad (2)$$

$C(k_B)$ – normalization constant.

3.3. Central normal distribution (CND)

$$\sum_{l=0}^{\infty}(2l+1)\exp\left(\frac{-l(l-1)}{2k_G^2}\right)\frac{\sin((l+\frac{1}{2})t)}{\sin(\frac{t}{2})}\ \frac{1}{\pi} sin^2\left(\frac{t}{2}\right) \quad (3)$$

3.4. Wrapped normal distribution (WND)

$$\sum_{l=-\infty}^{+\infty}\frac{k^3}{\sqrt{2\pi}}(2l\pi - t)^2 \exp\left(\frac{-k^2(2l\pi - t)^2}{2}\right) \quad (4)$$

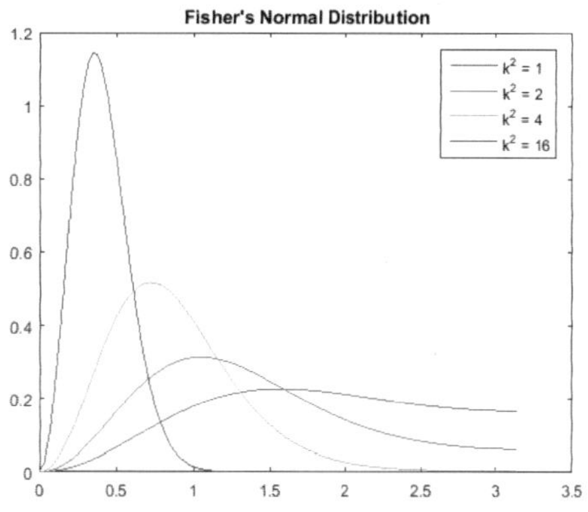

Figure 1. Fisher normal distribution

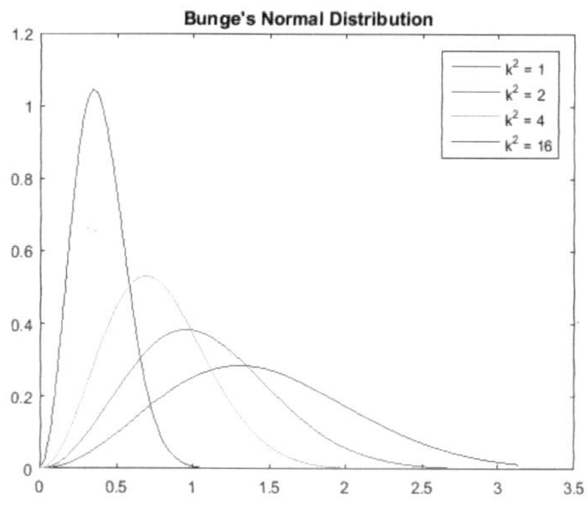

Figure 2. Bunge normal distribution

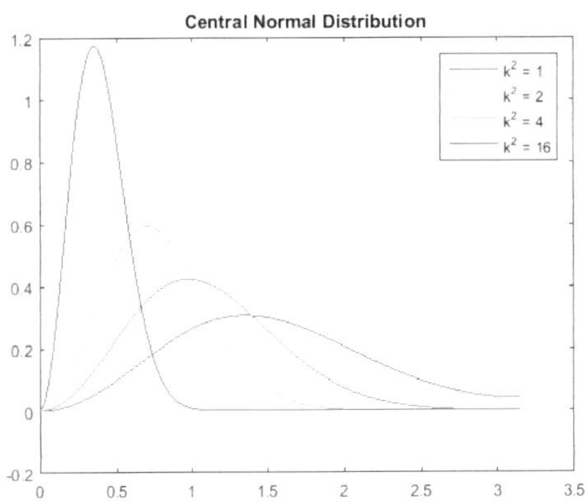

Figure 3. Central normal distribution

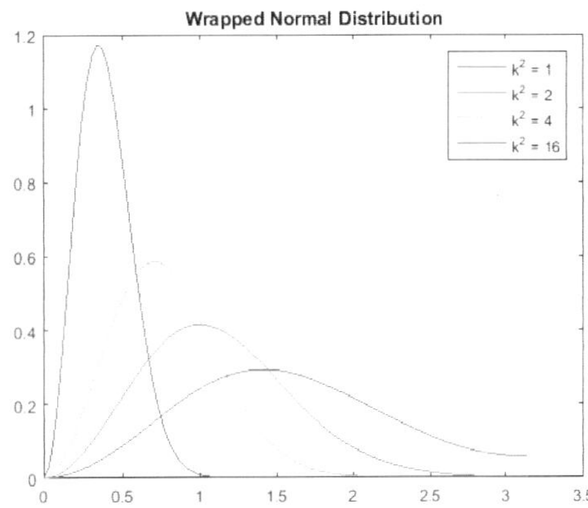

Figure 4. Bunge normal distribution

Distributions (1)-(3) were used as model functions in texture analysis before [1-7]. WND (4) is a recently introduced ND, obtained as a result of modeling a Markov chain using Monte Carlo method on SO(3) group [8].

In ([8], p. 4) distribution (4) is explained as distributional limit for compositions of large numbers of independent, small random rotations and it is pointed out that WND statistically corresponds to Maxwell-Boltzmann distribution.

WND is motivated by CLT in R^3 and converted to SO(3) group through exponential mapping.

Let's point out main properties of distributions (1)-(4):

- FND and BND are able to give the best likelihood estimation (maximum likelihood estimation method, that is used in statistics to define unknown parameters of a model by the sample) and able to maximize entropy [2].
- CND is an infinitely divisible distribution [6]. It has a CLT motivation (CLT here is Parthasarathy's central limit theorem on SO(3) group)and is able to be obtained as a result of Brownian motion.
- WND is obtained as a result of modeling a Markov chain using Monte Carlo method on SO(3) group.

Figures 1-4 show plots of distributions (1)-(4) for $k = 2^m, m = 0,1,2,3$.

4. Asymptotic analysis of different NDs on SO(3) group

Let's take WND (4). With $l = 0$ expression (4) transforms into

$$f(t)dt = \frac{k^3}{\sqrt{2\pi}} t^2 e^{\frac{-k^2t^2}{2}} dt, \qquad |t| < \pi \tag{5}$$

With the $k^2t^2 = u$ variable substitution in (5) it turns into

$$f_1(t)dt = \frac{1}{2\sqrt{2\pi}} u^{\frac{1}{2}} e^{\frac{-u}{2}} du \tag{6}$$

Distribution (6) coincides with χ_3^2 distribution (with 3 degrees of freedom). Using the table of χ^2 values ([14], p.572), it was found out that with the value of concentration parameter k and confidence interval corresponding to expressions (7)

$$|t| \leq \frac{\sqrt{7.8}}{k} \leq \pi \qquad k \geq \frac{\sqrt{7.8}}{\pi} \tag{7}$$

only one term (5) of the series (4) approximates WND with the p-value (confidence value) of $\beta = 0.95$.

With that said, it should be mentioned, that previously introduced WND plot (figure 4) displays wrapped normal distribution calculated using expression (5) for $k^2 = 2,4,16$. And uses expression (4) for $k^2 = 1$ considering only the terms that correspond to $l = -1,0,1$.

To create a CND plot (figure 3) only first 100 terms of the series (3) were used to calculate the distribution.

Article [6] states that expression (8) approximates CND (where $\varepsilon^2 = \frac{1}{2k^2}$)

$$f(t)dt = \frac{\sqrt{\pi}}{\varepsilon^3}\, erfc\left(\frac{\varepsilon}{2}\right) \exp\left(\frac{\varepsilon^2}{4}\right) \exp\left(\frac{t^2}{4\varepsilon^2}\right) \frac{t}{2} \frac{1}{\pi} \sin\left(\frac{t}{2}\right) dt, \qquad |t| \leq \pi \tag{8}$$

With $\frac{t^2}{2\varepsilon^2} = u$ substitution, and as $t \to 0$, (8) transforms into

$$f(u)du = \frac{1}{2}\, e^{\frac{1}{2k^2}}\, erfc\left(\frac{1}{2\sqrt{2}k}\right) f_{\chi_3^2}(u)du$$

Using the same variable substitution $k^2t^2 = u$ and taking into account $t \to 0$, it can be shown that FND and BND are approximated by χ_3^2distribution with three degrees of freedom.

- FND

$$f(u)du = \frac{e^{k^2}\sqrt{2\pi}}{8\pi k^3 (I_0(k^2) - I_1(k^2))} f_{\chi_3^2}(u)du$$

- BND

$$f(u)du = C(k)\frac{\sqrt{2\pi}}{8\pi k^3} f_{\chi_3^2}(u)du$$

Figures 5a and 5b show plots for separate terms ($l = 0, \pm 1, \ldots$) of the series (4) with $k = 2,\ k = \frac{3}{4}$ and $-\infty < t < +\infty$.

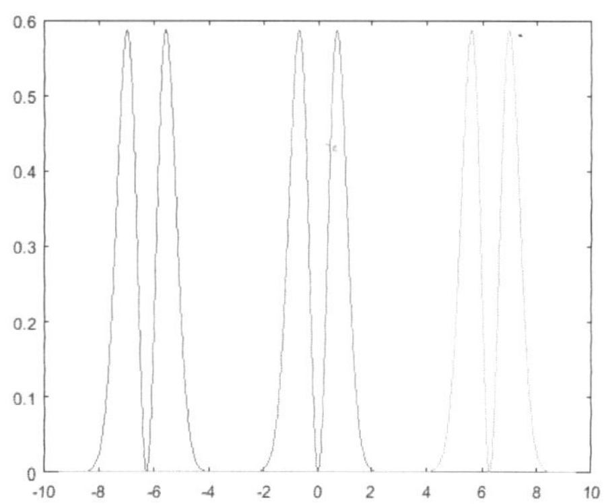

a

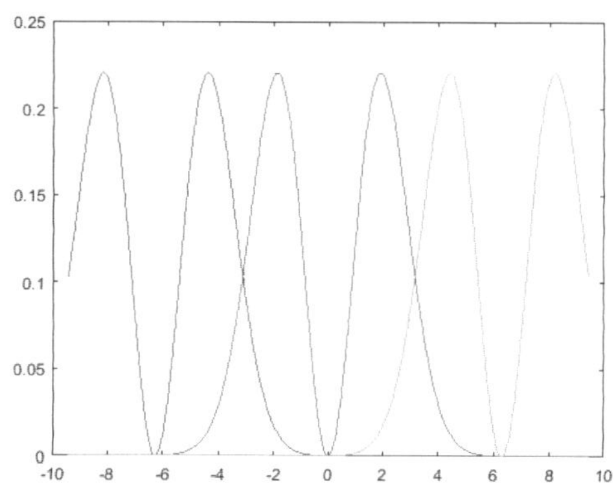

b

Figure 5. WND separate terms *a*) k=2 *b*) k=0.75

Figure 5a shows, that for $k = 2$ each plot of each term appears to be almost isolated from the others with the period of 2π. In that case the conditions (7) are met and it is enough to use only one term of series (4), where $l = 0$, to calculate WND for $|t| \leq \pi$. On the other hand, figure 5b shows that for $k = \frac{3}{4}$ plots seem to intersect with each other, therefore the influence of the "neighbor" terms has to be accounted for.

In that case, it is sufficient to use two terms of series (4) (where $l = 0, 1$) to calculate WND. That is because on $[0; \pi]$ the term corresponding to $l = 0$ gives a 0.5 of the "weight" of the whole function, whereas the term corresponding to $l = 1$ gives the remaining part of "weight" (see the table of χ^2 values with 3 degrees of freedom: for $k = \frac{3}{4}$ the confidence interval $|t| \leq (\frac{\pi}{2})^2 \approx 2{,}46$, and the possibility $p = \frac{1}{2}$) ([14], p.572). It becomes obvious, that for $k \geq 2$ it is possible to use only one term corresponding to $l = 0$ in order to calculate WND, however, for $k \leq \frac{1}{2}$ two or more terms have to be used for the same purpose.

5. Monte Carlo method

Monte Carlo methods for calculating ND on SO(3) group could be used to model orientations of separate crystallites of polycrystalline materials [4]. These distributions are convenient to use in EBSD (Electron Backscattering Diffraction) experiment in order to study the texture of a material's microstructure as well as its macrostructure [4, 8].

Article [4] introduces Monte Carlo method for calculating ND's on SO(3) based on Parthasarathy's CLT. Article [8] describes Monte Carlo method for calculating WND based on CLT in $R^3$and mapping to SO(3) group.

Let's study the examples of Monte Carlo method application for calculating CND [4] and compare modeling results with WND. The calculations were carried out for $\varepsilon = 1, \frac{1}{4}$ $(\varepsilon^2 = \frac{1}{2k^2})$, with sample size $N = 10^4$, number of convolutions n = 20 and number of bins in a histogram nBins = 30.

The following results were obtained after applying the Monte Carlo method and calculating Pearson's test statistic:

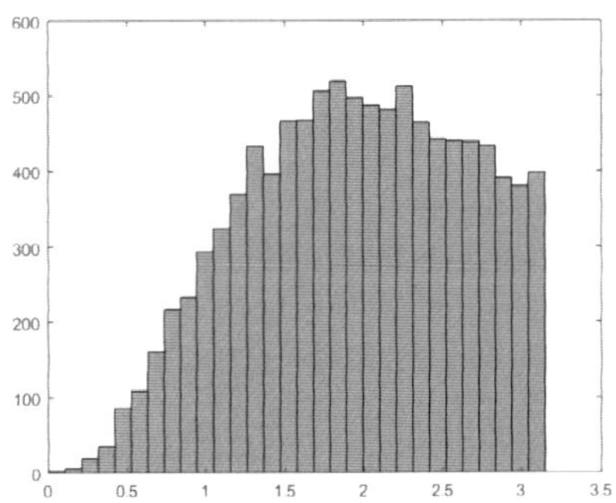

Figure 6. CND $\varepsilon = 1$.

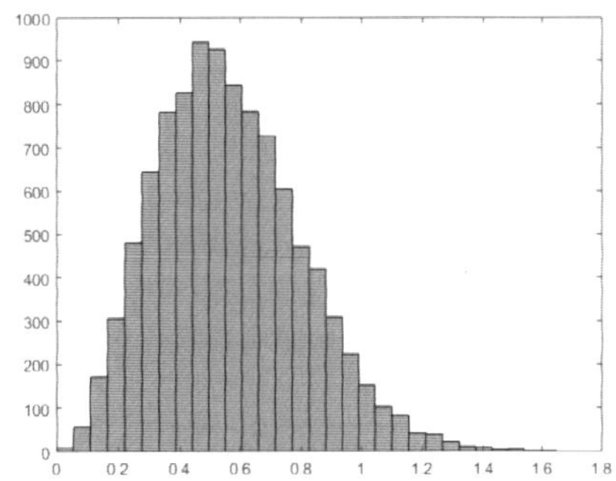

Figure 7. CND $\varepsilon = \frac{1}{4}$.

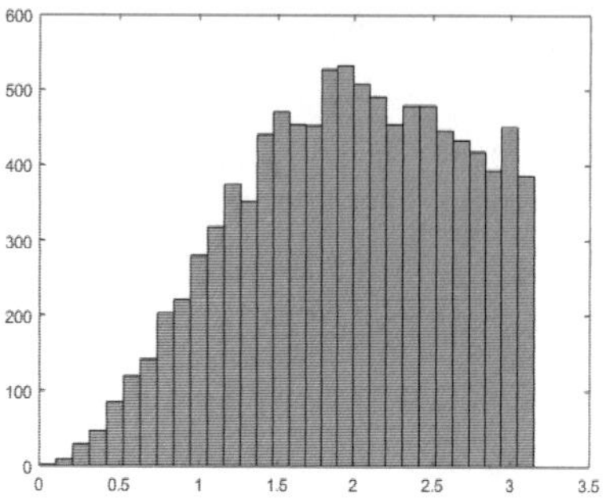

Figure 8. WND $\varepsilon = 1$.

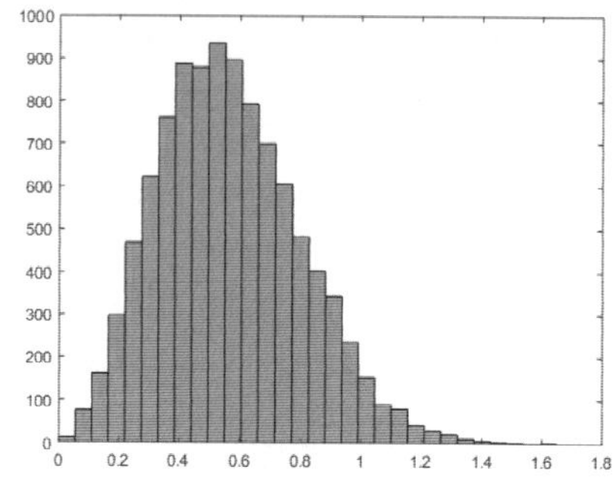

Figure 9. WND $\varepsilon = \frac{1}{4}$.

In the case depicted by figure 6 Pearson's test statistic W = 25.55 does not exceed the critical value of χ^2= 42.56. Therefore the hypothesis (that there is no difference between theoretical distribution and experimental distribution of the sample) is accepted.

In the case depicted by figure 7 Pearson's test statistic W = 21.77 does not exceed the critical value of χ^2= 42.56. Therefore the hypothesis (that there is no difference between theoretical distribution and experimental distribution of the sample) is accepted.

In the case depicted by figure 8 Pearson's test statistic W = 180.21 exceeds the critical value of χ^2= 42.56. Therefore the hypothesis (that there is no difference between theoretical distribution and experimental distribution of the sample) is rejected.

In the case depicted by figure 9 Pearson's test statistic W = 26.78 does not exceed the critical value of χ^2= 42.56. Therefore the hypothesis (that there is no difference between theoretical distribution and experimental distribution of the sample) is accepted.

To sum up, for CND the hypothesis is accepted for all of the considered values of ε ($\varepsilon = 1, \frac{1}{4}$), however for WND, the hypothesis is accepted only in the case of $\varepsilon = \frac{1}{4}$, while for $\varepsilon = 1$ it is rejected according to Pearson's chi-squared test.

Hence it can be concluded that CND and WND approximate each other for small values of ε and differ otherwise (or approximate each other for large values of concentration parameter k and differ otherwise).

6. Texture analysis application

Normal distributions on SO(3) are widely used in texture analysis to solve problems of CODF restoration by its PF [1-7]. CODF is presented as a sum of ND's with certain weights. Articles [1] use ND (2), [3, 6] – ND (1). Article [8] suggests using WND (4) for same purposes.

PF for CND (3) is calculated as follows ([4], p.106):

$$P_{\vec{h}}(\vec{y}, q_0, \varepsilon) = \sum_{l=0}^{\infty}(2l+1)\exp(-l(l+1)\varepsilon^2)\, P_l(\vec{h}, q_0\vec{y}) \tag{9}$$

where $\vec{h}$ - desired crystallographic direction, $q_0 \in SO(3)$ - center of ND, $\vec{y} \in S^2$ – point on two-dimensional sphere, ε – concentration parameter ($k \leq \frac{1}{\varepsilon}$), $P_l(z), l = 0,1, \dots$ - Legendre polynomials.

With $\varepsilon \rightarrow 0+$ PF (9) is approximated by expression (8)

$$P(cos\theta)sin\theta d\theta \approx \frac{1}{\varepsilon^2}\exp\left(\frac{\theta^2}{4\varepsilon^2}\right)\theta d\theta \tag{10}$$

In the light of the fact that with $t \rightarrow 0+$ ND (3) and ND (4) approximate each other, expression (10) can be used to closely calculate PF for large values of k (small values of ε).

Presenting WND (4) as a series of terms, where terms themselves are characters of representations of SO(3) group, and using ([15], p. 509), values of coefficients are calculated as

$$C_l = \left(1 + \frac{l^2}{k^2}\right)\exp\left(-\frac{l^2}{2k^2}\right) - \left(1 + \frac{(l+1)^2}{k^2}\right)\exp\left(-\frac{(l+1)^2}{2k^2}\right)$$

Thus WND (4) has a significant difference from CND (3) in presenting it as a series with terms as characters of representations of SO(3) group.

Article [16] gives an example of CODF approximation for texture description using just one component of CND (3). The texture was obtained as a result of deformation of monocrystalline alloy Ni-Ti. Two pole figures were used to obtain the parameters of CND: center coordinates $q_0 = (56.92^o; 75.89^o; 65.78^o)$ (Euler rotation angles), component weight $M_0 = 1.51$, concentration parameter $\varepsilon = 0.025$ ($\varepsilon^2 = (2k_G^2)^{-1}$). Figure 10 depicts the kind of PF that corresponds to expression (10) without measure $sin\theta d\theta$ and only its even part, because it is well known that CODF reconstructs ambiguously ([4], p. 96).

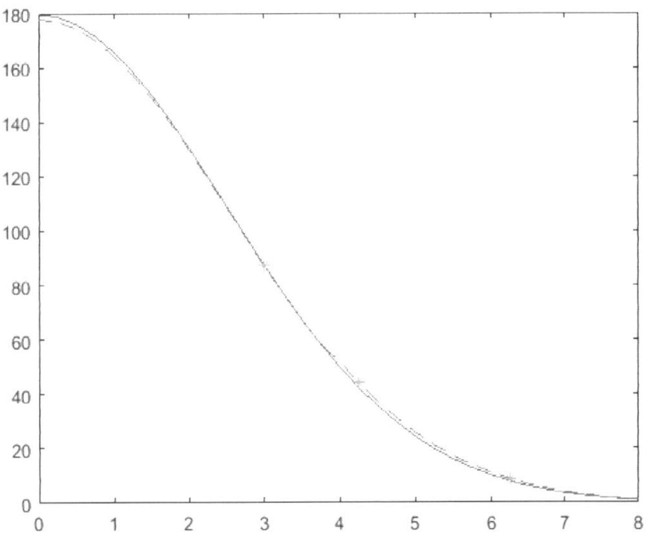

Figure 10. Pole figure

Figure 10 has R in degrees plotted on the x-axis. Stars on figure 10 correspond with experimentally measured PF {110} values obtained by X-ray method, solid line represents calculated results.

Each of the NDs (1)-(4) can be equally used as a model function for CODF approximation due to their close similarity for the large values of concentration parameter k.

Article [8] presents confidence intervals for concentration parameter k for different values of p-value (or confidence value), in order to consider them when EBSD measurements are analyzed. EBSD measurements produce a set of crystallite orientations in polycrystal. That set allows to identify different structure and texture characteristics of a material. Very rarely a material has a texture that could be characterized by CODF similar to the one mentioned before in Ni-Ti example. Mostly texture is characterized by a number of components, including the axial ones [1, 3, 4, and 6]. Which is why it seems doubtful that confidence intervals presented in [8] would find its practical use.

7. Conclusion

This article considers normal distributions on SO(3) group that are able to be represented as central functions [13]. Distributions (1)-(3) are broadly used in texture analysis for modeling crystallite orientation distribution function ([1-7]). Articles [8, 11 and 12] introduce wrapped normal distribution

(4). WND (4) is compared with distributions (1)-(3), and properties, connected with the notion "normal" (the one in probability theory and mathematical statistics on SO(3) group) are analyzed.

WND (4) is characterized by four parameters. Three of them define the position of the "center" and the fourth is the concentration parameter. WND (4) is motivated by CLT that was obtained by consecutive rotations of two-dimensional sphere S^2 on a small angle uniformly distributed on $(-\pi; \pi]$.

This article studies how many terms of series (4), depending on the value of parameter k, are needed to obtain a credible result when calculating WND. It appears that for $k \geq \frac{\sqrt{7.8}}{\pi}$ it is sufficient to use only one term of the series (4) (when $l = 0$) to calculate an accurate result with p-value (or confidence value) 0.95.

Monte Carlo method was studied for both CND and WND. The results show that according to Pearson's chi-squared test for WND (4) the hypothesis is accepted for small values of ε, for large values it is rejected. (Hypothesis read as follows: there is no difference between theoretical distribution and experimental distribution of the sample.) For CND hypothesis is accepted for all the values of ε. Which prompts a conclusion, that CND and WND approximate each other for small values of ε. Not so otherwise.

CND (3), unlike WND (4), is only a subcase of a normal distribution on SO(3) group that is characterized by 9 parameters: the position of the "center" and positive semi-definite 3×3 matrix (analogue of a covariance matrix) [5].

Let's point out that corresponding definitions of ND and CLT are provided for a bigger class of groups – SO(n) groups ($n \geq 2$). And for that class of groups Monte Carlo method has been developed in order to solve different problems in texture analysis connected with studying the results of EBSD experiment [4, 6].

The article also shows that FND (1) and BND (2) could also be approximated by $\chi_3^2$for small values of parameter t.

In conclusion, it is important to mention that all of the studied distributions (1)-(4) can be used to solve different problems in texture analysis.

8. References

[1] Bunge H J 1996 *Textures and Microstructures* **25** 71-108
[2] Schaeben H 1996 *Textures and Microstructures* **25** 171-181
[3] Helming K, Schmidt D and Ullemeyer K 1996 *Textures and Microstructures* **25** 211-222.
[4] Savyolova T I, Ivanova T M and Sypchenko M V 2012 *MEPhI* 268
[5] Savyolova T I 1984 *Industrial laboratory* **50** 48-52
[6] Bucharova T I and Savyolova T I 1993 *Textures and Microstructures* **21** 161-176
[7] Nikolayev D I and Savyolova T I 1997 *Textures and Microstructures* **29** 201-233
[8] Y Qiu 2013 *Graduate Thesis and Dissertations* **13007** 70
[9] Vardeman S B, Bingham M A, Nordman D J, Qiu Y and Du C 2015 *http://www.public.iastate.edu/~vardeman/RotationsOverview.pdf*
[10] Qiu Y, Nordman D J and Vardeman S B 2014 *Computational Statistics and Data Analysis* **71** 520-529
[11] Bingham M A, Nordman D J and Vardeman S B 2009 *Journal of the American Statistical Association* **104** 1385-1397
[12] Bingham M A, Vardeman S B and Nordman D J 2009 *Bayesian Analysis* **4** 607-630
[13] Vilenkin N Y 1965 *Science* 588
[14] Amosova N N, Kuklin B A, Makarova S B and Maksimov Y D 2001 *Spb.: Ivan Fedorov* 588
[15] Gradshteyn I S and Ryzhik I M 1971 *Science* 1108
[16] Nikolaev D I and Savyolova T I 1989 *Metals* 165-169
[17] Filatov S V and Savyolova T I 2017 *Sixth International Conference. Problems of Mathematical Physics and Mathematical Modeling* 130

A comparison of Data Driven models of solving the task of gender identification of author in Russian language texts for cases without and with the gender deception

A Sboev[1,2], **I Moloshnikov**[1], **D Gudovskikh**[1], **and R Rybka**[1]
[1] National Research Centre "Kurchatov Institute", Moscow, Russian Federation
[2] MEPhI National Research Nuclear University, Moscow, Russian Federation

E-mail: sag111@mail.ru, ivan-rus@yandex.ru, dvgudovskikh@gmail.com, rybkarb@gmail.com

Abstract. In this work we compare several data-driven approaches to the task of author's gender identification for texts with or without gender imitation. The data corpus has been specially gathered with crowdsourcing for this task. The best models are convolutional neural network with input of morphological data (f1-measure: 88%±3) for texts without imitation, and gradient boosting model with vector of character n-grams frequencies as input data (f1-measure: 64% ± 3) for texts with gender imitation. The method to filter the crowdsourced corpus using limited reference sample of texts to increase the accuracy of result is discussed.

1. Introduction

Nowadays a great amount of information circulates in the Internet, but only part of it is truthful. However, for many practical fields, such as forensic, security, and others, to have truthful information is important. There are different formulations of the task of information deception identification. In this paper we investigate the task of gender identification of Russian text author in texts with gender deception (i. e. when the authors tried to mimic other gender) in comparison with the same task for texts without deception. While the latter case is widely represented in scientific literature, see for example [1, 2, 3, 4, 5, 6, 7, 8], sources for the former case are absent. The closest formulation is issued in [9], where the task of presence of any deception in men and women texts is investigated.

The gender identification task was a part of the annual competition PAN 2017 [1] which was devoted to text author profiling for such languages as English, Arabic, Spanish, and Portuguese. That task was to determine author gender by Twitter messages. The corpus contained 500 tweets from 100 authors. Training was performed on 60% of the samples, and the remaining 40% were used for testing according to the competition rules. Classical methods of machine learning showed the best results for the English and Spanish model based on SVM [2], for the Arabic language the classification model was constructed on the basis of logistic regression [3]. Deep neural networks showed the best result for the Portuguese language, their topologies were described in the work [4]. The authors used recurrent neural networks (RNN) for words and convolutional layers (CNN) for symbols. In the network topologies, full-connected words, maxpooling and attention mechanism were used. That approach combined different text representations to classify the author's gender. Formally, these results show that today classical methods demonstrate the best result for the problem of gender identification. However, the approaches based on neural networks demonstrated their applicability for this task even on small data sets.

Regarding the gender deception task, the closest example is the work [9], where the research was concentrated on any lie in texts from men and women. The accuracy of the best model was just about 5% above baseline, based on Support Vector Machine (SVM) and the combination of semantic features (number of words and word sequences belonging to certain semantic classes) and word forms. The standard deviation was not shown in the article. Training data was collected using Amazon Mechanical Turk [10], the size of corpus was 7168 sentences (3584 truths, 3584 lies) from 512 users.

Moreover, the PAN 2017 competition results, where the Russian language track based on RusProfiling and Gender Imitation corpora took place, are presented in [6]. Unfortunately, the test using the Gender Imitation of this track had the special formulation: the learning sample consisted of Twitter texts, but the testing texts were formed by uniting in one text both the textual part without deception and the textual part with deception from the Gender Imitation corpus. It did not allow to evaluate correctly the accuracy of identifying text author gender only on texts with gender deception.

All the datasets used in our work are described in section 2. The section 4 describes the used methods of machine learning with different sets of input features described in section 3. The description of the experiments without gender imitation (subsection 5.1), with gender imitation (5.2) and with filter algorithm (5.3) and the accuracy obtained in them is presented in the section 5. Finally, we present the discussion of the results in 6 and conclusion in 7.

2. Used corpora

We used a crowdsourcing platform to collect texts imitating other gender. In addition, various corpora with texts without imitation of gender were used, and manually collected Russian Gender Imitation corpus [5], [6], which contains 125 documents from men and 269 documents from women, as the test case. The main crowdsourcing corpus GI_cs contains 1161 men and 2043 women. The users who made markup on the crowdsourcing platform used the methodology used to create the Gender Imitation corpus: each author wrote 3 texts on a theme chosen from the following list: a description of oneself and one's potential partner for a dating site, an attempt to persuade an unfamiliar person to come, or a negative tour review. The task was that these three texts needed to be written in certain different styles: Text A in the authors natural style, Text B imitating other gender style, and Text C in a different style but without gender imitation. Subcorpus GI_cs_a_b was obtained from the main corpus GI_cs, by deleting C texts.

In addition, we used the RusPersonality corpus (RusPer) [11], the RusProfiling (RusProf) [12] corpus containing texts from various social networks Twitter (further denoted as Tw), Facebook (FB) and LiveJournal (LJ) and reviews. To increase the number of examples in these subcorpuses, the combined texts of each user were divided into examples of 15 sentences: Tw_s (5062 documents from men, 2450 documents from women), LJ_s (1632 documents from men, 1624 documents from women), FB_s (868 documents from men, 749 documents from women).

3. Features

The following sets of features were used in combination with models of Section Models:

- **TF-IDF for n-grams(TF-IDF)**: The vector of n-gram frequencies, characterizing a document d, is obtained by applying by the TF-IDF formula $\mathrm{tf}-\mathrm{idf}(t,d)=\mathrm{tf}(t,d)\times \mathrm{idf}(t)$ to each n-gram t of the collection D. Here t is a character n-gram from 3 to 8 characters long, $\mathrm{tf}(t,d)$ is the number of times t occurs in d; $\mathrm{idf}(t) = \log \frac{1+n_d}{1+\mathrm{df}(d,t)} + 1$. Here n_d is the total number of texts, and $\mathrm{df}(d,t)$ is the number of texts that contain t. We tried different values of minimal $\mathrm{df}(d,t)$ in the range from 1 to 100, the best is 1.

- **Word2vec**: This is an already existing model that was preliminarily trained on a collection of random Russian web pages crawled in December 2014, contains 9 million documents in total. Corpus size is 660,628,738 tokens. Model was trained using the Continuous Skip-Gram algorithm. Vector dimensionality was set to 500, window size 2. Lemmas occurring less than 30 times were ignored [13].
- **Sequences feature (Seq. feat)**: Texts are represented as sequences of words with a full set of morphological tags (person, gender, part of speech, etc.), all one-hot encoded.
- **Linguistic Inquiry and Word Count (LIWC)**: a psycho-social dictionary [14]. As said in [15], "It reveals the links between the personality characteristics of the author and quantitative parameters of the text (number of the words of some of the parts of speech, of some lexical-thematic groups, the frequency of punctuation marks, etc.)"

4. Models

4.1. Conventional machine learning methods

- **Support Vector Machine (SVM)**. We used the classifier based on a support vector machine with linear kernel. The following hyperparameters were used: regularization parameter $C = 1$, $L2$-norm used in the penalization and squared hinge-loss function.
- **Gradient boosting classifier(GB)** trained with the following parameters: learning rate is 0.05, the number of boosting stages to perform is 300, the minimum number of samples to split an internal node is 19, the maximum tree depth is 12.
- **Random Forest classifier (RF)**. The parameters of this model are selected using a genetic algorithm, the following are set: the Gini index, the number of counts taken in calculating the best division is 55%, the minimum number of edges on the node 8, the minimum number of examples for the division of a node 5, the number of trees to be calculated 100.

Standard models were trained using 10-fold cross-validation, dividing the set into 80% training and 20% unused data, as testing was conducted on a separate corpus.

4.2. Deep neural network

Model 1 is based on convolution and fully connected layers:

(i) **Convolutional Neural Network (CNN)**: 128 neurons, window size = 2, activation function = Relu
(ii) **Maxpooling layer**: window = 4, step = 4
(iii) **CNN**: 128 neurons, window = 2, activation function = Relu
(iv) **Maxpooling layer**: window = 4, step = 4
(v) **CNN**: 128 neurons, window = 2, activation function = Relu
(vi) **Maxpooling layer**: window = 4, step = 4
(vii) **CNN**: 128 neurons, window = 2, activation function = Relu
(viii) **GlobalMaxPooling**
(ix) **Dropout 0.5**
(x) **Fully-connected MLP layer**: 128 neurons, activation function = Relu
(xi) **Dropout 0.5**
(xii) **Output layer - MLP**: activation function = softmax

IOP Conf. Series: Journal of Physics: Conf. Series **937** (2017) 012046 doi:10.1088/1742-6596/937/1/012046

Network training was performed with early stopping (if the error on the validation set grows during 15 epochs, then stop). After training the weights from the best epoch were saved. We used the Rmsprop algorithm [16] and learning rate of 0.001.

5. Experiments

In all experiments, the models are estimated by F1-score (denoted as F1 in tables) and standard deviation (F1_std in tables), obtained from the test data. The test and training samples are balanced by the number of examples in each of the classes, so the baseline (bl in tables) is 0.5. As the test set we used a balanced set of "A" and "B" of corpus Gender Imitation. Since separate corpora are used as learning and test sets, the intersection of the authors of texts is avoided. A number of experiments with different algorithms of machine learning based on combinations of training samples are described below.

5.1. Experiments 1

The goal of these calculations is to state the level of accuracy of gender identification of text author in texts without gender deception. The model learning has been made, on base of both separate parts corpora and the extended variant of its combinations. The model testing has been made only on texts without gender deception, that is, on part 'A' of the Gender Imitation corpus.

Table 1. The results of experiment 1 on subset 'A' of GI

#	Train dataset	Model	Features	F1	F1_std	Delta(bl)
1	GI_cs_a_b	GB	TF-IDF	0.65	0.03	0.15
2	GI_cs_a	GB	TF-IDF	0.78	0.02	0.28
3	GI_cs_a	SVM	TF-IDF	0.73	0.02	0.23
4	GI_cs; RusPer; Reviews; Tw_s; LJ_s; FB_s	Model 1	Seq. feat	0.88	0.03	0.38
5	RusPer; Reviews ;Tw_s; LJ_s; FB_s;	Model 1	Seq. feat	0.84	0.02	0.34
6	GI_cs	Model 1	Seq. feat	0.87	0.03	0.37
7	Tw_s; LJ_s; FB_s;	Model 1	Seq. feat	0.82	0.02	0.32
8	Tw; LJ; FB	Model 1	Seq. feat	0.77	0.04	0.27
9	GI_cs; RusPer; Reviews	Model 1	Seq. feat	0.88	0.03	0.38
10	RusPer; Reviews	Model 1	Seq. feat	0.79	0.03	0.29
11	GI_cs_a_b	Model 1	Seq. feat	0.86	0.03	0.36

5.2. Experiment 2.

In this case we exploited the approach of the previous experiment based on learning on different sets of corpora but with testing on part 'B' of the Gender Imitation corpus, with gender deception.

5.3. Experiment 3.

Calculations based on models, analogous to the ones used in second experiment 5.2, have been made on base of texts filtered from the crowdsourcing corpus. For filtering the special model is used, learned on different text samples of the reference corpus Gender Imitation.

IOP Conf. Series: Journal of Physics: Conf. Series **937** (2017) 012046 doi:10.1088/1742-6596/937/1/012046

Table 2. The results of experiment 2 on subset 'B' of GI

#	Train dataset	Model	Features	F1	F1_std	Delta(bl)
1	GI_cs_a_b	SVM	TF-IDF	0.54	-	0.04
2	GI_cs_a_b	GB	TF-IDF	0.64	0.03	0.14
3	GI_cs	Model 1	Word2vec	0.45	-	-0.05
4	GI_cs_a_b	SVM	LIWC	0.53	0.2	0.03
5	GI_cs_a_b	DT	LIWC	0.51	0.04	0.01
6	GI_cs_a_b	GD	LIWC	0.46	0.07	-0.04
7	GI_cs_a_b	RF	LIWC	0.63	0.02	0.13

(i) Learning Model 1 (SVM) on Gender Imitation corpus Gender Imitation using TF-IDF features. The 'GI dataset' group of columns in Table 3 shows the parts of the corpus used to learn the filtering model.

(ii) Using the model learned on the previous step, the crowdsourcing corpus has been filtered, taking only correctly predicted examples.

(iii) Training the model 2 (SVM) on the filtered data of 'GI_cs dataset' of Table 3 using the TF-IDF features.

(iv) The testing was conducted on the texts from Gender Imitation. The results of testing the model 2 on the full body of Gender Imitation are presented in Table 3.

Table 3. The results of experiment 3 using the "data filtering" algorithm.

#	GI dataset	GI_cs dataset	F1 (a)	F1 (b)	F1 (c)
1	a	a,b,c	0.87	0.57	0.67
2	b	a,b,c	0.48	0.87	0.55
3	c	a,b,c	0.67	0.69	0.95
4	a,b	a,b,c	0.84	0.93	0.66
5	a,c	a,b,c	0.81	0.59	0.86
6	b,c	a,b,c	0.61	0.83	0.82
7	a,b,c	a,b,c	0.84	0.83	0.84
8	a	a,b	0.79	0.49	0.60
9	b	a,b	0.45	0.83	0.49
10	c	a,b	0.61	0.73	0.93
11	a,b	a,b	0.80	0.90	0.60
12	a,c	a,b	0.77	0.58	0.82
13	b,c	a,b	0.61	0.86	0.78
14	a,b,c	a,b	0.80	0.81	0.80
15	a	a	0.85	0.45	0.64
16	b	a	0.52	0.71	0.65
17	c	a	0.65	0.59	0.91
18	a,b	a	0.86	0.71	0.67
19	a,c	a	0.81	0.53	0.79
20	b,c	a	0.67	0.76	0.78

6. Results

Results of the first experiment demonstrate an accuracy level for the task of author gender identification on texts without gender imitation. It follows from Table 1, that the best model for this task is the convolutional neural network with input data represented as a morphological features sequence, and it gives an F1-score equal to 0.88 with standard deviation of 0.03 by cross-validation.

Results of the second experiment point out that complicated neural networks with morphological data are ineffective for texts with gender imitation, and in this case the best model is gradient boosting with TF-IDF character n-gram vectors as input data, which demonstrates the F1-score, equal to 0.64 with standard deviation of 0.03. It is 0.22 lower than in case of texts without the gender imitation. One possible reason of significant accuracy falling is the difference in the features distribution between Gender Imitation and crowdsource corpora. Section 5.3 gives partial answer on this. As could be seen, training on filtered crowdsourcing data gives a significant growth in F1-score on texts with gender imitation. Moreover, accuracy of the gender identification could reach 0.93 depending on training sample composition and subsample selection from A, B, and C.

7. Conclusion

For tasks with and without gender imitation, different models of author gender identification are effective. Further accuracy growth requires sufficiently large training samples. The most perspective way to obtain it is to use crowdsourcing, but manual filtering and checking of these texts is a labour-consuming task. The proposed approach to solve the task is a way to get a larger corpus by crowdsourcing with the same features distribution as a comparatively small reference sample and use it for training the new model of increased accuracy.

8. Acknowledgements

This research is supported by the Russian Science Foundation, project No 16-18-10050. This work has been carried out using computing resources of the federal collective usage center Complex for Simulation and Data Processing for Mega-science Facilities at NRC 'Kurchatov Institute', http://ckp.nrcki.ru/

References

[1] F Rangel, P Rosso, M Potthast, and B Stein 2017 Overview of the 5th author profiling task at pan 2017: Gender and language variety identification in twitter *Working Notes Papers of the CLEF*
[2] A Basile, G Dwyer, M Medvedeva, J Rawee, H Haagsma and Malvina Nissim 2017 N-gram: New groningen author-profiling model *arXiv preprint arXiv:1707.03764*
[3] M Martinc, IzaSkrjanec, K Zupan, and S Pollak 2017 Pan 2017: Author profiling-gender and language variety prediction *Notebook for PAN at CLEF*
[4] Y Miura, T Taniguchi, M Taniguchi, and T Ohkuma 2017 Author profiling with word+ character neural attention network *Notebook for PAN at CLEF*
[5] T Litvinova, P Seredin, O Litvinova and O Zagorovskaya 2017 Differences in type-token ratio and part-of-speech frequencies in male and female russian written texts *In Proceedings of the Workshop on Stylistic Variation* 73
[6] Markov I et al 2017 The Winning Approach to Cross-Genre Gender Identification in Russian at RUSProfiling 2017 *Training* **600**
[7] Sboev A et al 2016 Deep Learning Network Models to Categorize Texts According to Author's Gender and to Identify Text Sentiment *Computational Science and Computational Intelligence (CSCI) International Conference on IEEE* pp 1101-1106
[8] Sboev A et al 2016 Machine Learning Models of Text Categorization by Author Gender Using Topic-independent Features *Procedia Computer Science* **101** pp 135-142
[9] V P erez-Rosas and R Mihalcea 2015 Experiments in open domain deception detection*EMNLP* 11201125
[10] Amazon Mechanical Turk 2012 *Amazon mechanical turk. Retrieved August* **17**

[11] T Litvinova, O Litvinlova, O Zagorovskaya, P Seredin, A Sboev and O Romanchenko 2016 "Ruspersonality": A russian corpus for authorship profiling and deception detection *In Intelligence, Social Media and Web (ISMW FRUCT) IEEE* 7

[12] RusProfiling Lab 2017 Rusprofiling corpus of russian texts [online] *http://rusprofilinglab.ru/rusprofiling-at-pan/corpus/*

[13] A Kutuzov and E Kuzmenko 2017 Building web-interfaces for vector semantic models with the webvectors toolkit *EACL 2017* 99

[14] Yla R Tausczik and James W Pennebaker 2010 The psychological meaning of words: Liwc and computerized text analysis methods *Journal of language and social psychology* **29(1)** 54

[15] O Litvinova, P Seredin, T Litvinova and J Lyell 2017 Deception detection in russian texts. *In Proceedings of the Student Research Workshop at the 15th Conference of the European Chapter of the Association for Computational Linguistics* 52

[16] Lin Wu, Chunhua Shen and Anton van den Hengel 2016 Personnet: Person re-identification with deep convolutional neural networks *arXiv preprint arXiv:1601.07255*

Reciprocal expansion of modified Bessel function in simple fractions and obtaining general summation relationships containing its zeros

V B Sherstyukov and E V Sumin[1]

National Research Nuclear University MEPhI (Moscow Engineering Physics Institute), 31 Kashirskoe shosse, 115409 Moscow, Russia

E-mail: `shervb73@gmail.com`, `sumevbir@gmail.com`

Abstract. Modified Bessel functions of the first kind $I_\nu(z)$ (Infeld functions) where $\nu > -1$ are considered. Due to the constraint on the parameter ν, all zeros of the function $I_\nu(z)$ except $z = 0$ are simple, located on the imaginary axis by symmetric pairs and form an infinite countable set. On the basis on previous research of the authors dealing with general Bessel functions of the first kind $J_\nu(z)$, a question about reciprocal expansion $1/I_\nu(z)$ in series of simple fractions of a certain structure (Krein's series) is studied. The general formulas to calculate of special infinite sums containing degrees of Infeld function zeros are an important consequence of obtained expansion in simple fractions of the value $1/I_\nu(z)$ with any $\nu > -1$. The possibility of concrete definition of established summation relationships at different values of parameters and their connection with analogous relationships for the Bessel functions of the first kind $J_\nu(z)$ is discussed.

In the present work using the modified Bessel function as an example we consider one method for calculating some infinite sums compiled by the function zeros. The method bases on following general result about reciprocal expansion of an entire function in special series of simple fractions (Krein's series) (see, e.g. [1], [2]).

Theorem A. *Let $L(z)$ be an entire function of the exponential type with set of simple zeros $Z(L) = (z_k)_{k=1}^{\infty}$, $0 \notin Z(L)$, located in the certain strip of the complex plane $\mathbb{C}$. Let $L(z)$ has also a non-negative indicator:*

$$h_L(\theta) \equiv \overline{\lim_{r\to+\infty}} \frac{\ln|L(r\,e^{i\theta})|}{r} \geqslant 0, \qquad 0 \leqslant \theta \leqslant 2\pi,$$

and for certain $p \in \mathbb{Z}_+$ the numerical series converges

$$\sum_{k=1}^{\infty} \frac{1}{|L'(z_k)|\,|z_k|^{p+1}}.$$

[1] The work of the authors is partially supported by the program to improve the competitiveness of the National Research Nuclear University MEPhI, project No 02.a03.21.0005 from 27.08.2013

IOP Conf. Series: Journal of Physics: Conf. Series **937** (2017) 012047 doi:10.1088/1742-6596/937/1/012047

Then meromorphic function $F(z) \equiv 1/L(z)$ can be expanded in Krein's series of the order p converging absolutely and uniformly on compact subsets of the domain $\mathbb{C} \setminus Z(L)$. There is

$$F(z) \equiv \frac{1}{L(z)} = P(z) + z^p \sum_{k=1}^{\infty} \frac{1}{L'(z_k)\, z_k^p\, (z - z_k)}, \qquad z \in \mathbb{C} \setminus Z(L). \tag{1}$$

If $p = 0$ then $P(z) \equiv 0$, but if $p \in \mathbb{N}$ then $P(z)$ is a polynomial calculating by formula

$$P(z) = \sum_{m=0}^{p-1} \frac{F^{(m)}(0)}{m!}\, z^m. \tag{2}$$

It is shown in [2] that if the conditions of theorem A are satisfied then expansion (1) allows deriving the universal summation relationships:

$$\sum_{k=1}^{\infty} \frac{1}{L'(z_k)\, z_k^m} = -\frac{F^{(m-1)}(0)}{(m-1)!}, \qquad m \in \mathbb{N}, \quad m \geqslant p+1. \tag{3}$$

The derivation idea is very simple and dates back to Euler. The left and right sides of the identity (1) must be expanded in a Maclaurin's series and then you ought to compare the coefficients at equal degrees of the variable z. The first few relationships at $m = 0, \ldots, p-1$ will be agreed with formula (2). Further increasing of values m leads to formulas (3). In the previous article of the authors [3] general results concluded in expressions (1), (3), have applied to an expansion question of the reciprocal to the Bessel function of the first kind $J_\nu(z)$ of arbitrary index $\nu > -1$ in Krein's series and to the calculation of "regularized" sums of degrees of reciprocals of the function zeros. First, interest to the similar sums springed in the works of Euler, Rayleigh and Kapitsa. The current state of the problem is presented in reviews [4], [5].

This paper aims to extend the results of [3] to modified Bessel functions of the first kind (Infeld functions):

$$I_\nu(z) \equiv i^{-\nu} J_\nu(iz) = \sum_{n=0}^{\infty} \frac{1}{n!\, \Gamma(\nu+n+1)} \left(\frac{z}{2}\right)^{\nu+2n}, \qquad z \in \mathbb{C}, \quad \nu > -1. \tag{4}$$

Using only the technique of the cited paper [3], we can give an independent exposition of the results for the function $I_\nu(z)$. However, due to the simple connection (4) with the function $J_\nu(z)$ it is more convenient to apply directly the proposition established in [3, Theorem 2]. We formulate one in a form proposed in [2].

Theorem B. *Let $J_\nu(z)$ be the Bessel function of the first kind:*

$$J_\nu(z) = \sum_{n=0}^{\infty} \frac{(-1)^n}{n!\, \Gamma(\nu+n+1)} \left(\frac{z}{2}\right)^{\nu+2n}, \qquad z \in \mathbb{C}, \quad \nu > -1,$$

and $\gamma_{\nu,k}$ are its positive zeros forming the sequence

$$0 < \gamma_{\nu,1} < \gamma_{\nu,2} < \ldots < \gamma_{\nu,k} < \ldots, \qquad \lim_{k\to\infty} \gamma_{\nu,k} = +\infty. \tag{5}$$

We determine the quantities

$$a_{\nu,2m} = \left(\frac{z^\nu}{J_\nu(z)}\right)^{(2m)}(0), \qquad m \in \mathbb{Z}_+. \tag{6}$$

We define the number p by formula

$$p = \left[\frac{2\nu+1}{4}\right] + 1 \tag{7}$$

(square brackets mean the integer part). Let $P(z)$ be identically zero if $-1 < \nu < -1/2$ and one specified by the rule

$$P(z) = \sum_{m=0}^{p-1} a_{\nu,\,2m}\,\frac{z^{2m}}{(2m)!},$$

if $\nu \geqslant -1/2$. Then there is the expansion

$$\frac{1}{J_\nu(z)} = z^{-\nu}P(z) - 2z^{2p-\nu}\sum_{k=1}^{\infty}\frac{1}{J_{\nu+1}(\gamma_{\nu,\,k})\,\gamma_{\nu,\,k}^{2p-\nu-1}\,(z^2-\gamma_{\nu,\,k}^2)}, \tag{8}$$

converging absolutely and uniformly on any compact subset in $\mathbb{C}$ that doesn't contain points $z = 0$ and $z = \pm\,\gamma_{\nu,\,k}$. Herewith for any $\nu > -1$ the summation relationships are satisfied

$$\sum_{k=1}^{\infty}\frac{1}{J_{\nu+1}(\gamma_{\nu,\,k})\,\gamma_{\nu,\,k}^{2m-\nu+1}} = \frac{a_{\nu,\,2m}}{2(2m)!}, \qquad m = p,\,p+1,\,\ldots\,, \tag{9}$$

where the quantities $a_{\nu,\,2m}$ and the number p are determine in (6), (7) respectively.

According to the general Bessel functions theory [6, Ch. 15] at $\nu > -1$ all zeros of Infeld function $I_\nu(z)$ except $z = 0$ are simple, purely imaginary and form the symmetric sequence $(\pm\,i\gamma_{\nu,\,k})_{k=1}^{\infty}$ evidently connected with sequence (5). The set of non-trivial zeros of the function $I_\nu(z)$ is conveniently written in the form $(\pm\,\mu_{\nu,\,k})_{k=1}^{\infty}$ where $\mu_{\nu,\,k} = i\gamma_{\nu,\,k}$.

To apply theorem B to the modified Bessel function $I_\nu(z)$ it's necessary to take into account several technical aspects. Firstly, we fix (for definiteness, the principal) branch of z^ν. Secondly, replacing z in (8) by iz and taking into account that

$$J_\nu(iz) \equiv e^{\frac{i\nu\pi}{2}}\,I_\nu(z),$$

we obtain the preliminary presentation

$$\frac{1}{I_\nu(z)} = z^{-\nu}P(iz) + 2(-1)^p\,z^{2p-\nu}\sum_{k=1}^{\infty}\frac{1}{J_{\nu+1}(\gamma_{\nu,\,k})\,\gamma_{\nu,\,k}^{2p-\nu-1}\,(z^2+\gamma_{\nu,\,k}^2)}. \tag{10}$$

Thirdly, again using (4) in the form

$$J_{\nu+1}(z) \equiv e^{\frac{i(\nu+1)\pi}{2}}\,I_{\nu+1}(-iz),$$

and using parity of the function $I_{\nu+1}(z)\,z^{-\nu-1}$, we write

$$J_{\nu+1}(\gamma_{\nu,\,k})\,\gamma_{\nu,\,k}^{-\nu-1} = e^{\frac{i(\nu+1)\pi}{2}}\,I_{\nu+1}(-i\gamma_{\nu,\,k})\,\gamma_{\nu,\,k}^{-\nu-1} = I_{\nu+1}(-i\gamma_{\nu,\,k})\,(-i\gamma_{\nu,\,k})^{-\nu-1} =$$

$$= I_{\nu+1}(i\gamma_{\nu,\,k})\,(i\gamma_{\nu,\,k})^{-\nu-1} = I_{\nu+1}(\mu_{\nu,\,k})\,\mu_{\nu,\,k}^{-\nu-1}.$$

Hence, for any $k \in \mathbb{N}$ we have

$$J_{\nu+1}(\gamma_{\nu,\,k})\,\gamma_{\nu,\,k}^{2p-\nu-1} = (-1)^p\,I_{\nu+1}(\mu_{\nu,\,k})\,\mu_{\nu,\,k}^{2p-\nu-1}. \tag{11}$$

Substituting (11) into (10) we have

$$\frac{1}{I_\nu(z)} = z^{-\nu}P(iz) + 2z^{2p-\nu}\sum_{k=1}^{\infty}\frac{1}{I_{\nu+1}(\mu_{\nu,k})\,\mu_{\nu,k}^{2p-\nu-1}\,(z^2-\mu_{\nu,k}^2)}.$$

We rewrite expression $P(iz)$ for the polynomial $P(z)$ from theorem B in terms of modified Bessel function $I_\nu(z)$. To this aim the coefficients $a_{\nu,2m}$ of formula (6) can be represented in the form

$$a_{\nu,2m} = (-1)^m\left(\frac{z^\nu}{I_\nu(z)}\right)^{(2m)}(0) \equiv (-1)^m\, b_{\nu,2m},$$

where at $-1<\nu<-1/2$ we will have

$$Q(z) \equiv P(iz) = \sum_{m=0}^{p-1} a_{\nu,2m}\,\frac{(-1)^m z^{2m}}{(2m)!} = \sum_{m=0}^{p-1} b_{\nu,2m}\,\frac{z^{2m}}{(2m)!},$$

where p is given by formula (7). We formulate the result of the reasoning.

Theorem. *Let $I_\nu(z)$ be the modified Bessel function of the first kind:*

$$I_\nu(z) = \sum_{n=0}^{\infty}\frac{1}{n!\,\Gamma(\nu+n+1)}\left(\frac{z}{2}\right)^{\nu+2n}, \qquad z\in\mathbb{C},\quad \nu>-1,$$

and $\mu_{\nu,k}$ are its zeros located on the positive imaginary half-axis (in growth to infinity of their imaginary parts). We define the quantities

$$b_{\nu,2m} = \left(\frac{z^\nu}{I_\nu(z)}\right)^{(2m)}(0), \qquad m\in\mathbb{Z}_+, \tag{12}$$

and the number $p=[(2\nu+1)/4]+1$. Further, let the polynomial $Q(z)$ be identically zero if $-1<\nu<-1/2$ and one is given by the rule

$$Q(z) = \sum_{m=0}^{p-1} b_{\nu,2m}\,\frac{z^{2m}}{(2m)!}, \tag{13}$$

if $\nu \geqslant -1/2$. Then, we have following expansion

$$\frac{1}{I_\nu(z)} = z^{-\nu}Q(z) + 2z^{2p-\nu}\sum_{k=1}^{\infty}\frac{1}{I_{\nu+1}(\mu_{\nu,k})\,\mu_{\nu,k}^{2p-\nu-1}\,(z^2-\mu_{\nu,k}^2)}, \tag{14}$$

converging absolutely and uniformly on any compact subset in $\mathbb{C}$ not containing the points $z=0$ and $z=\pm\mu_{\nu,k}$. Herewith for any $\nu>-1$ the summation relationships are satisfied

$$\sum_{k=1}^{\infty}\frac{1}{I_{\nu+1}(\mu_{\nu,k})\,\mu_{\nu,k}^{2m-\nu+1}} = -\frac{b_{\nu,2m}}{2(2m)!}, \qquad m\in\mathbb{Z},\quad m\geqslant p. \tag{15}$$

As an illustration we consider the case $\nu = 0$ applying the last theorem to the modified Bessel function of zero index:

$$I_0(z) = J_0(iz) = \sum_{n=0}^{\infty} \frac{1}{(n!)^2} \left(\frac{z}{2}\right)^{2n}, \qquad z \in \mathbb{C}.$$

Here is

$$p = 1, \quad b_{0,\,2m} = \left(\frac{1}{I_0(z)}\right)^{(2m)}(0), \quad Q(z) = b_{0,\,0} = \frac{1}{I_0(0)} = 1$$

according to general formulas (7), (12), (13). Therefore the presentation (14) and the summation relationships (15) will be respectively written in the form

$$\frac{1}{I_0(z)} = 1 + 2z^2 \sum_{k=1}^{\infty} \frac{1}{I_1(\mu_{0,\,k})\,\mu_{0,\,k}\,(z^2 - \mu_{0,\,k}^2)},$$

$$\sum_{k=1}^{\infty} \frac{1}{I_1(\mu_{0,\,k})\,\mu_{0,\,k}^{2m+1}} = -\frac{1}{2(2m)!}\left(\frac{1}{I_0(z)}\right)^{(2m)}(0), \qquad m \in \mathbb{N}.$$

In particular, for $m = 1$ we have

$$\sum_{k=1}^{\infty} \frac{1}{I_1(\mu_{0,\,k})\,\mu_{0,\,k}^3} = \frac{1}{8}.$$

Taking into account of obvious connection

$$I_1(\mu_{0,\,k})\,\mu_{0,\,k}^3 = J_1(\gamma_{0,\,k})\,\gamma_{0,\,k}^3, \qquad k \in \mathbb{N},$$

we obtain the value of the sum

$$\sum_{k=1}^{\infty} \frac{1}{J_1(\gamma_{0,\,k})\,\gamma_{0,\,k}^3} = \frac{1}{8},$$

that agrees with the particular result from [3]. In the general situation

$$\nu > -1, \quad m \in \mathbb{N}, \quad m \geqslant [(2\nu + 1)/4] + 1$$

the sums of the form (15) can be calculated using the connection

$$\sum_{k=1}^{\infty} \frac{1}{I_{\nu+1}(\mu_{\nu,\,k})\,\mu_{\nu,\,k}^{2m-\nu+1}} = (-1)^{m+1} \sum_{k=1}^{\infty} \frac{1}{J_{\nu+1}(\gamma_{\nu,\,k})\,\gamma_{\nu,\,k}^{2m-\nu+1}},$$

which we found at proof of the last theorem and using the concrete value of the sums of the form (9) obtained in [3].

References

[1] Sherstyukov V B 2011 Expanding the reciprocal of an entire function with zeros in a strip in a Krein series *Sbornik: Mathematics.* **202** 1853–1871

[2] Sherstyukov V B 2016 *Asymptotic properties of entire functions with roots in an angle* D. Sc. Dissertation. – Moscow State University, Moscow, 224 pp. (Russian)

[3] Sherstyukov V B and Sumin E V 2015 Application of Krein's series to calculation of sums containing zeros of the Bessel functions *Comput. Math. Math. Phys.* **55** 572–579

[4] Kerimov M K 1999 The Rayleigh Function: Theory and Methods of Its Calculation *Comput. Math. Math. Phys.* **39** 1883–1925

[5] Kerimov M K 2008 Overview Some New Results Concerning the Theory and Applications of the Rayleigh Special Function *Comput. Math. Math. Phys.* **48** 1454–1507

[6] Watson G N 1944 *A Treatise on the Theory of Bessel Functions* (Cambridge: Cambridge University Press)

IOP Conf. Series: Journal of Physics: Conf. Series **937** (2017) 012048 doi:10.1088/1742-6596/937/1/012048

Phase averaging method for the modeling of the multiprobe and cutaneous cryosurgery

K E Shilnikov, N A Kudryashov, I Y Gaiur

National Research Nuclear University MEPhI, Department of Applied Mathematics, Moscow, Russia

E-mail: shilnikov.k@yandex.ru

Abstract. In this paper we consider the problem of planning and optimization of the cutaneous and multiprobe cryosurgery operations. An explicit scheme based on the finite volume approximation of phase averaged Pennes bioheat transfer model is applied. The flux relaxation method is used for the stability improvement of scheme. Skin tissue is considered as strongly inhomogeneous media. Computerized planning tool is tested on model cryotip-based and cutaneous cryosurgery problems. For the case of cutaneous cryosurgery the method of an additional freezing element mounting is studied as an approach to optimize the cellular necrosis front propagation.

1. Introduction

During the last decades the cryosurgery methods has been considerably developed and has become a competitive approach for human cancer treatment. The prediction of the cryosurgery results is one of the actual problems of the computational medicine. The cryosurgical treatment is known to be based on the tissue necrosis effect.

For the cryotip based cryosurgery necrosis effect is caused by the ice ball propagation from the tips of cryoneedles in the tissue to treat, in the case of cutaneous cryosurgery the ice ball propagates for the contact region between the cutaneous surface and liquid nitrogen being sprayed on the skin. For the cutaneous cryosurgery cancer tissue, which is located under the skin surface, often have a quite complicated shape, so the total destruction of the tumor may lead to the significant damage to close-fitting healthy tissue. On the other hand, incomplete destruction of the tumor tissue lead to the disastrous increasing of the tumor growth speed.

This paper is concerned with numerical study of the possibilities of the cutaneous cryosurgery based on the combined effect from liquid nitrogen spraying and introducing of cryotips in order to optimize the tissue necrosis front propagation to the shape of tumor tissue. In order to take into account the peculiarities of physical processes in the bio-tissues which take place under the cryogenic treatment on the cell scale the phase averaged Pennes bioheat transfer model is used to describe macroscopical evolution of the thermal regime in target region. The explicit finite volume scheme for indexed convex grids with flux relaxation is applied for the numerical computation of the problem under study.

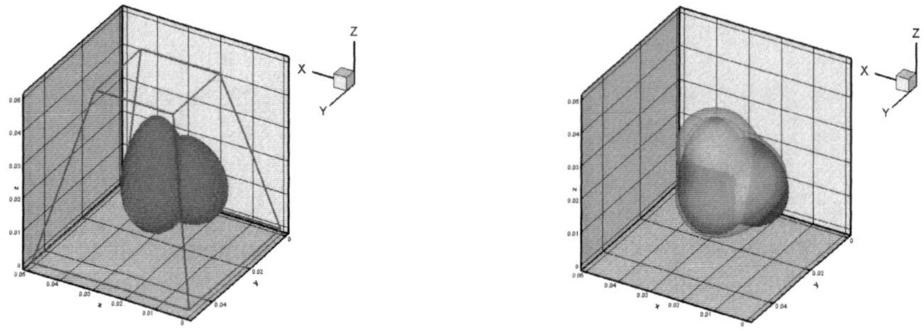

Figure 1. Left: Geometry of considered problem. Right: Localization of freezing zone and Cooper's isosurface relatively to the tumor tissue

2. Phase averaging procedure

According to the phase averaging method the cryogenic heat transfer processes in biotissue are described by 2-parametrically modified Pennes model [1, 2]:

$$\begin{aligned} &C(T;T_\alpha,T^*)\tfrac{\partial}{\partial t}T = div\left(\overline{k}(T;T_\alpha,T^*)\nabla T\right) + \rho_b\omega_b C_b(T_b - T) + q_{met}, \\ &t > 0, \mathbf{r} \in D \subset R^3, \end{aligned} \quad (1)$$

Heat conductivity coefficient is linearly interpolated on the interval (T_α, T_c).

After introducing $\delta = (T^* - T_\alpha)/(T_c - T_\alpha)$ phase averaged heat capacity of the biotissue can be approximated as follows:

$C(T;T_\alpha,\delta) = C_1(T)$ if $T \leq T_\alpha$;

$C(T;T_\alpha,\delta) = \frac{\rho L}{T_c - T_\alpha} + \delta C_1(T_\alpha) + (1-\delta)C_2$ if $T_\alpha < T < T_c$ and

$C(T;T_\alpha,\delta) = C_2$ otherwise.

The indices 1 and 2 relate to frozen and unfrozen states correspondingly.

In this way the parameters of the phase averaging are the lower temperature of the phase change T_α and T^*, which has sense of the normalized phase change temperature in terms of the Stefan problem. The upper phase change temperature is assumed to be known value $T_c = 273K$. For the cryoprobe based cryosurgery problem the optimal phase averaging parameter values are $T_\alpha = 265K, \delta = 0.6$.

In case of cutaneous cryosurgery the piecewise constant fields T_α and δ are considered: $(T_\alpha, \delta) = (260, 0.7)$ in 2mm range from cryo-active surfaces and $(T_\alpha, \delta) = (265, 0.5)$ otherwise.

3. The results of multiprobe cryosrugery modeling

An explicit scheme based on the finite volumes method is used for the numerical modeling of cryosurgery [3, 4, 5]. The flux relaxation approach [6] is involved for the stability improvement of explicit finite difference scheme. The last one leads to a decreasing of the computations time by taking a higher time steps.

The geometry of the considered tumor tissue is shown on Figure 1 (left). The resulting localization of the tissue necrosis front and Cooper's isosurface of the temperature when the mecrosis starts compared to the tissue shape are shown on the Figure 1 (right). The temperature

IOP Conf. Series: Journal of Physics: Conf. Series **937** (2017) 012048 doi:10.1088/1742-6596/937/1/012048

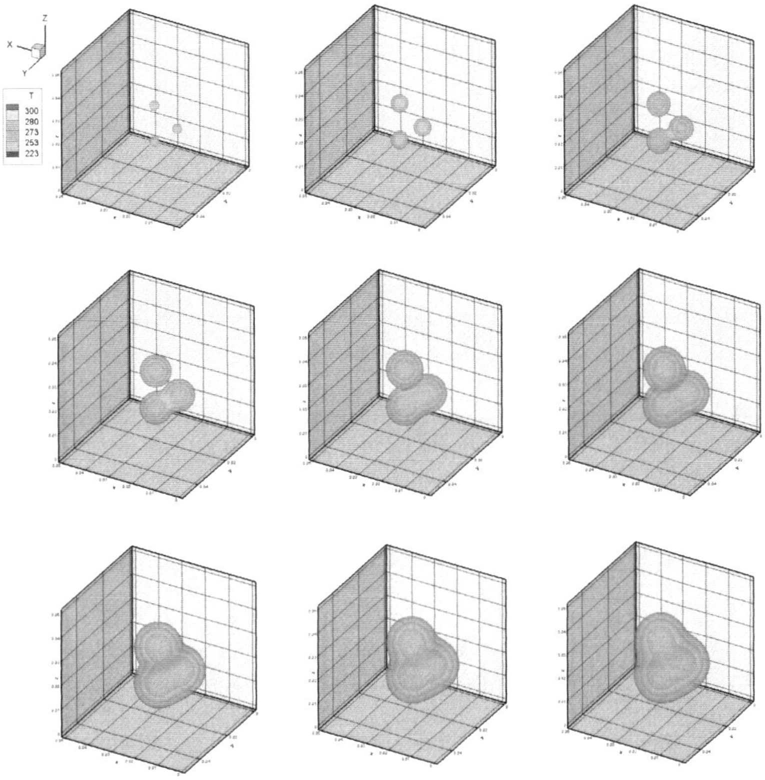

Figure 2. Critical temperature isosurfaces propagation

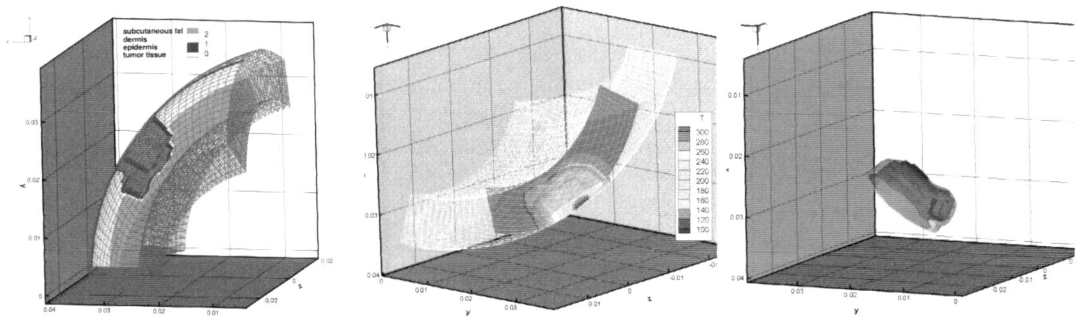

Figure 3. Cutaneous cryosurgery modeling. Left: Computation area. Center: Resulting temperature distribution. Right: Tumor/cryodamaged tissue regions.

of the working surface of 3 cryotips was taken as $\Phi(t) = \max(300 - At, 100)K$ with freezing speed $A = 80K/min$. Assuming the geometry of the tissue the sliced pyramid(5 and 2.5 cm foundations, 5 cm height) was taken as the computational area with 10^6 cells discretization.

The evolution of the critical temperature zones with 30 sec time steps is shown on the Figure 2

4. Numerical modeling of the cutaneous cryosurgery with correcting cryotip

In current work we assume the skin tissue to consist of 3 layers: epidermis (0.5mm), dermis (3.5mm), subcutaneous fat (5mm). Tumor tissue bio-physical parameters are assumed to have the same with the dermis tissue. We consider the introducing of correcting cryotip in tumor tissue for the optimization of the propagation of tissue necrosis front from the skin surface under the LN_2 spray. The values of thermophysical parameters for the skin tissue was taken from [7]. Correcting cryotip's working surface is assumed to have a sphere shape (radius is 2mm). Center slice of the computation area geometry is shown on the Figure 3(Left).

Numerical experiments shown that due to flux relaxation mechanism the stability condition may be reduced from $\tau \sim h^2$, which is peculiar to explicit schemes for parabolic equations, to $\tau \sim h^{\frac{3}{2}}$.

The temperature of the working surface of the correcting cryotip was taken as $\Phi(t) = \max(300 - At, 100)K$ with freezing speed $A = 80K/min$. Quasi-gradient method was applied to find an optimal position of the cryotip which would decrease the damage to the healthy tissue.

The resulting temperature distribution in the target region is shown be means of the center slice of the computation area on the Figure 3(Center). The resulting tissue necrosis region (yellow) and tumor tissue (blue) are shown of Figure 3(Right).

5. Conclusion

The phase averaged macroscopic model with scalar averaging fields for the cryogenic bioheat transfer is proposed as the tool for predicting the results of cryosurgery with respect to peculiarities of phase change processes in biological solutes. Introducing of the additional cryotip into the skin tissue during the cutaneous cryosurgery is considered. Numerical experiments on the model problems showed that the developed planning tool can cause a valuable increasing of effectiveness and safety of cryosurgery operations performing.

6. Acknowledgments

This work was supported by Russian Foundation For Basic Research, grant no. 16-31-00348.

References

[1] Pennes H H, 1948 Journal of Applied Physiology Vol.1 No.2 93-122
[2] Kudryashov N A, Shilnikov K E 2017 IOP Conf. Series: Journal of Physics: Conf. Series 788 012023 doi:10.1088/1742-6596/788/1/012023
[3] Kudryashov N A, Shilnikov K E 2016 AIP Conference Proceedings 1738, 230009; doi: 10.1063/1.4952018
[4] Kudryashov N A, Shilnikov K E 2015 Journal of Computational and Applied Mathematics 290 259-267
[5] Shilnikov E V 2014 Preprints of Keldysh Institute of Applied Mathematics RAS 33 URL: http://library.keldysh.ru/preprint.asp?id=2014-33 (in russian)
[6] Chetverushkin B N Shilnikov E V 2010 CD proceedings of the II International conference on Engineering Optimization (EngOpt-2010)
[7] Feng Sun, G.X. Wang, K. M. Kelly, G. Aguilar 2005 Proceedings of ASME International Mechanical Engineering Congress and Exposition (IMECE-2005)

On solvability for inverse problem of compact support source determination for the heat equation

V V Soloviev and D S Tkachenko[1]

National Research Nuclear University MEPhI (Moscow Engineering Physics Institute), 31 Kashirskoe shosse, 115409 Moscow, Russia

E-mail: soloviev.vyacheslav@gmail.com, tds-m@yandex.ru

Abstract. An inverse problem of reconstructing the source for the heat equations on a plane is considered. As an "overdetermination" (additional information about the solution of the direct problem) a trace of it's solution is given on a line segment inside of a bounded region. We give sufficient conditions for uniqueness of the solution of the task at hand, prove Fredholm alternative and sufficient conditions for existence and uniqueness of solution of the task. The studying of the problem is performed in the spaces of functions satisfying Hölder condition.

1. The problem statement

Let $T > 0$, $0 < \alpha < 1$ be a fixed number, and there is a bounded domain Ω with a smooth boundary of class $C^{2,\alpha}$ on the plane with real coordinates of points $x = (x_1, x_2)$. In the space of points (x, t) we define a cylinder $\Omega_T = \Omega \times (0, T]$ with side boundary $\Gamma_T = \partial\Omega \times [0, T]$. In the closed cylinder $\overline{\Omega}_T = \overline{\Omega} \times [0, T]$ we consider the first boundary value problem for the heat equation, that is, the problem of determining the function $u : \overline{\Omega}_T \to \mathbb{R}$ from the following conditions:

$$(Lu)(x, t) = u_t(x, t) - (\Delta u)(x, t) = f(x_1)h(x, t) + g(x, t), \qquad (x, t) \in \Omega_T, \tag{1}$$

$$u(x, t) = \mu(x, t), \qquad (x, t) \in \Gamma_T, \qquad u(x, 0) = \varphi(x), \qquad x \in \overline{\Omega}. \tag{2}$$

In equation (1), $g,\ h : \overline{\Omega}_T \to \mathbb{R}$, $\mu : \overline{\Gamma}_T \to \mathbb{R}$, $f : \mathbb{R} \to \mathbb{R}$ are given functions. The function f is assumed to have compact support, more precisely, there exist numbers $a < b$ and this function is nonzero only on the interval $[a, b]$. In addition, we assume that a number c is given such that for the interval $\omega = (a, b) \times c$ the inclusion is valid:

$$\overline{\omega} = [a, b] \times c \subset \Omega.$$

For convenience of the following constructions we define the sets:

$$P = \{(x_1, x_2) : \ a < x_1 < b \ \ x_2 \in \mathbb{R}\},$$

$$\overline{P} = \{(x_1, x_2) : \ a \leq x_1 \leq b \ \ x_2 \in \mathbb{R}\},$$

[1] The work of the authors is supported by the Program of Competitiveness Increase of the National Research Nuclear University MEPhI (Moscow Engineering Physics Institute); project No 02.a03.21.0005 from 27.08.2013

$$P_T = P \times (0,\, T], \quad \overline{P}_T = \overline{P} \times [0,\, T].$$

We define the following space of functions that satisfy the Hölder condition (for the definition of all standard Hölder functions used here, see [1], p. 16):

$$U(\Omega_T) = C(\overline{\Omega}_T) \cap C^{2+\alpha,\,1+\frac{\alpha}{2}}\left(\overline{P}_T \cap \Omega_T\right) \cap C^{2+\alpha,\,1+\frac{\alpha}{2}}\left(\Omega_T \setminus \overline{P}_T\right) \cap C^{1,\,0}_{x,\,t}\left(\Omega_T\right).$$

In the future, when formulating the inverse problem for equation (1), we will assume that the solution of the direct problem (1), (2) lies in the class of functions $U(\Omega_T)$. Since the class of functions $U(\Omega_T)$ is not generally accepted for solutions of equation (1), we present here the existence and uniqueness theorem for the solution of problem (1) - (2) in this class.

Theorem 1. *Let the inclusions are valid:* $\varphi \in C(\overline{\Omega})$, $\mu \in C(\Gamma_T)$, $h,\ g \in C^{\alpha,\,\frac{\alpha}{2}} \in C(\overline{\Omega}_T)$, $f \in C^{\alpha}[a,\, b]$, *the matching conditions* $\mu(x,\, 0) = \varphi(x)$, $x \in \partial\Omega$ *hold. Then problem (1) - (2) has a unique solution in the class of functions* $U(\Omega_T)$.

The theorem follows from the results of the monograph [1], c. 256.

Remark *An important property of the class of functions* $U(\Omega_T)$, *which will be used in the formulation of the inverse problem for equation (1), is that if we denote* $u(x_1,\, c,\, T) = \chi(x_1)$, *then for the function* $\chi(x_1)$ *the inclusion is valid* $\chi(x_1) \in C^{2,\,\alpha}[a,\, b]$.

In order to formulate the inverse problem for equation (1), we assume that in addition to the function $u(x,\, t)$, the function $f(x_1)$ is also unknown, but there is an additional information on the function $u(x,\, t)$. We assume that in addition to the condition (2), the trace of the function $u(x,\, t)$ on the interval $\overline{\omega}$ is also known (this kind of information in the formulation of the inverse problem is usually called an overdetermination). In this way, we start to study the inverse problem of determining the pair of functions $(u,\ f) \in U(\Omega_T) \times C^{\alpha}[a,\, b]$ from the conditions:

$$(Lu)(x,\, t) = f(x_1)h(x,\, t) + g(x,\, t), \qquad (x,\, t) \in \Omega_T, \tag{3}$$

$$u(x,\, t) = \mu(x,\, t), \qquad (x,\, t) \in \Gamma_T, \qquad u(x,\, 0) = \varphi(x), \qquad x \in \overline{\Omega}, \tag{4}$$

$$u(x_1,\, c,\, T) = \chi(x_1), \qquad x_1 \in [a,\, b]. \tag{5}$$

The inverse problem for the heat equation with compact support sources was considered in [2]. In that paper, the question of the uniqueness of its solution with another overdetermination in an unbounded domain was studied.

Other statements of the inverse problems of determining the sources for the heat equation, which are close to problem (3) - (5), and also the history of the study of such problems, see [3] - [6]. In this paper, for the inverse problem (3) - (5) sufficient conditions for unique solvability and the formulation of the Fredholm alternative are presented.

2. Main results

We consider an important matter for any inverse problem, that is the sufficient conditions for the uniqueness of its solution. For problem (3) - (5) the following uniqueness theorem holds.

Theorem 2. *Suppose that the conditions* $h,\ h_t,\ h_{x_2} \in C^{\alpha,\,\frac{\alpha}{2}}(\overline{\Omega}_T)$, $h(x,\, 0) = 0,\ \ x \in \partial\Omega$, $h(a,\, x_2,\, 0) = h(b,\, x_2,\, 0) = 0,\ \ (a,\, x_2),\ (b,\, x_2) \in \Omega,\ \ |h(x_1,\, 0,\, T)| \geq h_T > 0$, *where* h_T *is a positive constant, are satisfied for the function* $h(x,\, t)$ *Then there exists a number* $\nu_0 > 0$ *such that if the condition* $b - a < \nu_0$ *is satisfied, then the inverse problem (3) - (5) can not have two different solutions.*

The proof of Theorem 2 is based on the representation of the solution of problem (3) - (5) as a sum of heat potentials.

The nature of the solvability of the inverse problem (3) - (5) is closely related to the solvability of the homogeneous inverse problem corresponding to this one. We call a homogeneous inverse problem corresponding to the inverse problem (3) - (5) the similar problem (3) - (5) where $\varphi = 0$, $\mu = 0$, $g = 0$, $\chi = 0$, that is, the problem of defining a pair of functions $(u,\ f) \in U(\Omega_T) \times C^{\alpha}[a,\ b]$ from the following conditions:

$$(Lu)(x,\ t) = f(x_1)h(x,\ t), \qquad (x,\ t) \in \Omega_T, \tag{6}$$

$$u(x,\ t) = 0, \qquad (x,\ t) \in \Gamma_T, \qquad u(x,\ 0) = 0, \qquad x \in \overline{\Omega}, \tag{7}$$

$$u(x_1,\ c,\ T) = 0, \qquad x_1 \in [a,\ b]. \tag{8}$$

The relation between problems (3) - (5) and (6) - (8) is established in the following statement.

Theorem 3. (Fredholm Alternative *Suppose that the conditions* $h,\ h_t,\ h_{x_2} \in C^{\alpha,\ \frac{\alpha}{2}}(\overline{\Omega}_T)$, $h(x,\ 0) = 0,\ \ x \in \partial\Omega,\ \ h(a,\ x_2,\ 0) = h(b,\ x_2,\ 0) = 0,\ \ (a,\ x_2),\ (b,\ x_2) \in \partial\Omega,\ \ |h(x_1,\ c,\ T)| \geq h_T > 0$ *are satisfied for the function* $h(x,\ t)$. *Then there exists a number* $\nu_0 > 0$ *such that if the condition* $b - a < \nu_0$ *is fulfilled, then for the inverse problem (3) - (5) one of the following statements (alternative) holds:*

(i) The homogeneous inverse problem (6) - (8) has a finite number of linearly independent solutions (in other words, the homogeneous problem has a nontrivial solution)

(ii) For any functions $\varphi \in C(\overline{\Omega})$, $\mu \in C(\Gamma_T)$ *satisfying the matching conditions* $\mu(x,\ 0) = \varphi(x)$, $x \in \partial\Omega$ *and any functions* $\chi \in C^{2,\alpha}[a,\ b]$ *the inhomogeneous problem (3) - (5) has an unique solution.*

The proof of Theorem 3 is based on reducing the problem of the solvability of the inverse problem (3) - (5) to the question of the solvability of a linear equation of the second kind with a compact operator in the Banach space $C^{\alpha}[a,\ b]$ and using of the Riesz-Schauder theorem for this equation.

As a corollary of Theorem 2 and Theorem 3, we obtain a theorem that gives a sufficient condition for the existence of a unique solution of the inverse problem (3) - (5).

Theorem 4. *Suppose that the conditions* $h,\ h_t,\ h_{x_2} \in C^{\alpha,\ \frac{\alpha}{2}}(\overline{\Omega}_T)$, $h(x,\ 0) = 0,\ \ x \in \partial\Omega$, $h(a,\ x_2,\ 0) = h(b,\ x_2,\ 0) = 0,\ \ (a,\ x_2),\ (b,\ x_2) \in \partial\Omega,\ \ |h(x_1,\ c,\ T)| \geq h_T > 0,\ \ x_1 \in [a,\ b]$ *are satisfied for the function* $h(x,\ t)$. *Then there exists a number* $\nu_0 > 0$ *such that for any segment* $\overline{\omega} \in \Omega$ *such that* $b - a < \nu_0$ *and any functions* $\varphi \in C(\overline{\Omega})$, $\ \mu \in C(\Gamma_T)$ *satisfying the matching conditions* $\mu(x,\ 0) = \varphi(x)$, $\ x \in \partial\Omega$ *and any functions* $g \in C^{\alpha,\ \frac{\alpha}{2}}(\overline{\Omega}_T)$, $\ \chi \in C^{2,\alpha}[a,\ b]$ *the inverse problem (3) - (5) has a unique solution.*

Acknowledgments

This work was supported by the Program of Competitiveness Increase of the National Research Nuclear University MEPhI (Moscow Engineering Physics Institute); project No 02.a03.21.0005.27.08.2013.

References

[1] Ladyjenskaya O A, Solonnikov V A and Ural'ceva N N 1967 Linear and quasi linear equations of parabolic kind (Moscow: Nauka) 736
[2] Denisov A M 2016 Computational Mathematics and Mathematical Physics Uniqueness and non-uniqueness of the solution heat equation source determination inverse problem **56** 1754-59
[3] Denisov A M 1994 Introduction into the inverse problems theory (Moscow: Moscow State University Press) 207
[4] Lavrent'ev M M 1973 Conditionally well-posed problems for differential equations (Novosibirsk: Novosibirsk State University Press) 71

[5] Prilepko A I, Orlovsky D G and Vasin I A 2000 Methods for solving inverse problems in mathematical physics (New York: Basel: Marcel Dekker Inc.) 723
[6] Isakov V 2006 Inverse problems for partial differential equations (New York: Springer) 344
[7] Kostin A B 2013 Sbornik: Mathematics Source inverse problem for a parabolic equation by non-local observation condition **204** **N10** 3-46

Using sobol sequences for planning computer experiments

I N Statnikov, G I Firsov

Blagonravov Mechanical Engineering Research Institute of RAS, Moscow, 101990, Russian

E-mail: firsovgi@mail.ru

Abstract. Discusses the use for research of problems of multicriteria synthesis of dynamic systems method of Planning LP-search (PLP-search), which not only allows on the basis of the simulation model experiments to revise the parameter space within specified ranges of their change, but also through special randomized nature of the planning of these experiments is to apply a quantitative statistical evaluation of influence of change of varied parameters and their pairwise combinations to analyze properties of the dynamic system.Start your abstract here...

1. Introduction

Starting from the second half of the 20th century, when solving problems of designing modern dynamic systems, the research emphasis is shifted to mathematical modeling. At the same time, the data collection process becomes a complex problem both because of the possibility of obtaining them in large quantities, and the difficulties that arise for this reason in interpreting the results of the computational experiment. In this case, according to R. Fisher, the first problem can be solved with the help of planning experiments, and the second - with the help of the logic of scientific inference. The method of the planned LP search (PLP-search) developed by the Blagonravov Mechanical Researh Institute of the Russian Academy of Sciences [1] makes it possible to solve both of the above-mentioned problems at the same time.

2. Procedure of PLP-search

Developed by I.M. Sobol uniformly distributed sequences [2] of binary-rational pseudo-random numbers, originally intended for the calculations of multidimensional integrals, began to be used in search procedures, realizing the ideology of "blind" search, as is typical for the methods of the Monte Carlo family. The structure of these sequences allowed the combinatorial way [1] to construct in the J-dimensional space of variable parameters of the lattice (the matrix of planned experiments), in calculating the functions under study it was possible to obtain answers to the following questions:

- which of the varied parameters with a given (required) probability have a significant effect on the values of the function (the quality criterion of the system, in other words, to statistically evaluate the derivative of the quality criterion for each variable parameter;

- on the chosen metric between the current value of the quality criterion and its extreme value, known either as a priori or determined by the results of computational experiment, to determine the areas of concentration of the best (achievable) quality criterion values;

- construct a Pareto set in the multidimensional space quality criterion or, if a compromise scheme is specified between the quality criterion, select a subdomain containing the largest concentration of compromise solutions in the J-dimensional parameter space;
- approximate the selected subdomains of quality criterion values by some types of regression dependencies.

The essence of PLP search is that the method combines ideas of discrete review of the space of analyzed parameters and the theory of planning of mathematical experiments [3]. The PLP-search method, probabilistic in nature, adequately corresponds to modern principles of simulation and heuristic modeling. Method allows you to build a matrix of planned experiments and the simulation model simulations to implement quasi-uniform in probability view space varied parameters within specified ranges of their changes, and also as a result of the special nature of these experiments planning randomized apply quantitative statistical) evaluation of the impact of variable parameters and their paired combinations to analyze properties of investigated (projected) object. If this turns out to be possible when installed (assigned) probability estimates received as result analysis of the properties of an object to its analysis of the same properties, but according to the "medium" from each variable parameter, i.e., to analyze the "one-dimensional" dependencies that psychologically and practically becomes very useful to the researcher (designer), so as to significantly overcome the curse of dimensionality "on Bellman. At the same time, the presence of the dependencies mean values of quality criteria, as well as the ability to determine the influence of parameters on quality criteria with the required confidence probability, largely solve the problem of integrating a huge numerical information obtained in the experiment, in clear and quantifiable logical analysis (up to "common sense") features.

At the same time, by constructing approximation models of criteria depending on the variable parameters, it is possible to estimate the sensitivity of the criteria on average for these parameters. The effectiveness of the plans for experiments in the PLP search is determined not only by the possibility of using them in dispersion analysis. These plans prove to be effective both in constructing regression dependencies, and generally in regression analysis, both in the computational aspect and in terms of a number of optimality criteria for these plans.

In particular, for the case of linear, quadratic and cubic regression obtained the value of the determinant of the Fisher information matrix and the limits of change of dispersion of the predicted values. Formula analysis showed that the increasing number of experiments in the series, a number of series of experiments and number of varied parameters values of the Fisher information matrix determinant grow, thereby making the plans similar properties to orthogonal; all correlation coefficients regression models for each of the regressions have good convergence to zero. For example, in the case of ten series of experiments, eight experiments in series and three varied parameters correspond to linear, quadratic and cubic regression constitute 22500, 27000 and 18750. Any of a series built plan experiments will be D-optimal.

3. Examples of PLP-search

Below are some of the examples of PLP-search in carrying out the tasks of designing various dynamic systems [4]. Here MM - mathematical model, NODE is a nonlinear ordinary differential equation, LODE is a linear differential equation, PDE - partial differential equation.

1) Rotary indexing table with hydro-mechanical drive. MM: 3 NODE of the second order. $J = 9$. $K = 3$. Result: found scope for compromise solutions, the volume of which amounted to ~ 0.2% of the originally specified.

2) High accuracy pressure pneumatic regulator. MM: 4 NODE of the second order. $J = 4$. $K = 1$. Result: found better solutions area with up to 0.5% of the originally specified.

3) Pneumatic shaking machine [5-7]. MM: 4 NODE of the second order. $J = 8$. $K = 1$. Result: 4 influential parameter; the selection amounted to 5% of the originally specified.

4) Multi-circuit planetary gear train. MM: 23 LODE second order nonhomogeneous $J = 25$. $K = 6$. Result: 8 parameters, simultaneously affected all criteria.

5) Sewing machine. MM: 5 LODE inhomogeneous second order. J = 6. K = 5. Result: in areas built regression dependences of natural frequencies of parameters mm.

6) Resonant converter for ship shafting. MM: 2 NODE of the second order. J = 6. K = 1. The result: two influential parameter; the criterion value improved 5.2 times compared with the same in the source area.

7) Transmission main drive working stand rolling mill. MM: 5 NODE of the second order. J = 5. K = 5. Result: found an area of compromise, representing ~ 3.5% of the originally specified.

8) Heat exchanger. MM: 1 PDE. J from 8 to 18. K = 4. The result: defined for each J significant parameters and built area of compromise.

4. Conclusion

Thus, the PLP-search method not only makes it possible, on the basis of simulation simulations, to perform a quasi-uniform scanning of the parameter space in given ranges of their changes, but also as a result of a special randomized nature of the planning of these experiments, to use quantitative statistical estimates of the effect of variation of variable parameters and their pairwise combinations on analyzed properties of the dynamical system under consideration.

Proving the effectiveness of PLP search in comparison with the "blind" way of searching for extremes, it is not suggested to discard the latter. However, it is obvious that PLP search not only can help speedy search for extremes if it is required, but also provides information on the influence of variable parameters and, what is also important, contributes to issues related to the study of the space of variable parameters (in particular, the results of the PLP search can help to choose an effective compromise scheme). So, it can be argued that if the decision is made to investigate the formulated problem first in a discrete way (which is useful even in the case when it is possible to obtain analytical dependencies, but very complex ones), PLP search seems to be a very effective method of computer technologies in the sense of the previously mentioned second component.

References

[1] Statnikov I N, Andreenkov E V 2006 *PLP-search-heuristic method of solving tasks of mathematical programming.* (Moscow, IIP MGUDT) (In Russian).

[2] Sobol I M 1998 *Math. Comput. Simulation* **47** 103.

[3] Montgomery D C 2001 *Design and Analysis of Experiments* (Hoboken, NJ, John Wiley and Sons).

[4] Statnikov I N, Firsov G I 2016 *Vestnik of Scientific and Technical Development* **3(103)** 38. (In Russian).

[5] S.Yu. Misyurin, V.I. Ivlev, V.M. Bozrov, and A.P. Nelyubin Parameterization of an Air Motor Based on Multiobjective Optimization and Decision Support // ISSN 105-6188, Journal of Machinery Manufacture and Reliability, 2013, Vol. 42, No. 5, pp. 353–358.

[6] G.V.Kreinin and S.Yu.Misyurin. Phased Synthesis of a Mechatronic System ISSN 1028_3358, Doklady Physics, 2014, Vol. 59, No. 11, pp. 539–543.

[7] S. Yu. Misyurin and G. V. Kreinin. On Choosing the Drive Type for the Power Unit of a Mechatronics System. Journal of Machinery Manufacture and Reliability, 2015, *Vol. 44, No. 4, pp. 305–311.*

On the Localization Principle for Laplace Operator with Piecewise Constant Coefficient in Domains without Discontinuity Points

M V Suchkov and V P Trifonenkov

National Research Nuclear University MEPhI (Moscow Engineering Physics Institute), 31 Kashirskoe shosse, 115409 Moscow, Russia

E-mail: mv_suchkov@mail.ru

Abstract. Some properties of spectral expansions that correspond to eigenfunction system of Laplace operator with discontinuous coefficient are studied. The authors prove that for the case of arbitrary domains of dimension $N = 2$ or $N = 3$, localization principle in domains without discontinuity points is preserved despite the discontinuity of the coefficient. If $N \geq 5$, then localization principle is violated in points located "far" from the surface of discontinuity of the coefficient, even in case of infinite smoothness of expanded function.

1. Introduction

This article deals with Laplace operator with discontinuous coefficient. Laplace operator with piecewise constant coefficient in N-dimensional domain g with bound Γ is considered.

Let suppose that C is closed smooth surface of discontinuity of Laplace operator coefficient is inside domain g. Thus domain g is divided by surface C into pair of subdomains: g_1 closed by C and g_2 .

Then eigenfunction problem in ($g + \Gamma$) is considered:

$$
\begin{aligned}
& k_1 \Delta u + \lambda u = 0;\ x \in g_1 \\
& k_2 \Delta u + \lambda u = 0;\ x \in g_2 \\
& u\big|_{C-0} = u\big|_{C+0}; \\
& k_1 \frac{\partial u}{\partial n}\bigg|_{C-0} = k_2 \frac{\partial u}{\partial n}\bigg|_{C+0}; \\
& u\big|_{\Gamma} = 0
\end{aligned}
\qquad (1)
$$

where k_1 and k_2 are positive constants; $C, \Gamma \in C^{2,\alpha}$ ($\alpha > 0$).

So, function $u(x)$ as non-zero solution of problem (1) where

$$u(x) \in C^1(g_1 + C) \wedge C^2(g_1);$$
$$u(x) \in C^1(g_2 + C + \Gamma) \wedge C^2(g_2);$$

$u(x) \in C(g+\Gamma)$

is known as a classical eigenfunction of the problem.

As well known [1], there is a complete system of quasiclassical eigenfunctions of this self-adjoint problem, the system is orthonormal in $L_2(g)$ and, at the same time, it is a system of generalized eigenfunctions of this problem (i.e. eigenfunctions satisfy a certain integral identity).

Then properties of spectral expansions that correspond to Laplace operator with discontinuous coefficient are studied. As proved in the article, for the case of arbitrary domains of dimension $N=2$ or $N = 3$, localization principle [2] in domains without discontinuity points is preserved despite the discontinuity of the coefficient. If $N \geq 5$, then localization principle is violated in points located "far" from the surface of discontinuity of the coefficient, even in case of infinite smoothness of expanded function and it's finiteness relatively to domain g.

The proof of the result is relied upon V.A. Il'in technique of investigation of general fundamental systems of Laplace operator and Lions theorem about a spaces interpolation.

2. Problem formulation and solution

Let Ω is strictly contained in g_1 or g_2. In the following let use the term fundamental system of functions (FSF) of operator $k_1\Delta u$ $(k_2\Delta u)$ in domain Ω for complete orthonormal in $L_2(g)$ system of functions that satisfy equation $k_1\Delta u+\lambda u=0$ ($k_2\Delta u+\lambda u=0$) in Ω.

Consider system of classic eigenfunctions of the problem (1) as FSF and prove the following statement, basing on V.A. Il'in theorem [2].

Theorem. Let $f(x) \in \overset{\circ}{W}{}_2^1(g)$, $N=3$ or $f(x) \in \widetilde{W}_2^{1/2}(g)$, $N=2$; ($\widetilde{W}_2^{1/2}(g)$ is class of functions from $W_2^{1/2}(R_2)$ with support in $(g+\Gamma)$). Then localization principle for FSF of problem (1) is true for domain Ω (i.e. if $f(x)$ equals zero in domain Ω then it's Fourier series in terms of the FSF considered converges uniformly to zero in each subdomain strictly contained in Ω).

Proof. The validity of the theorem statement follows from V.A. Il'in theorem [2] and, in the case $N=3$, from estimation $\sum_{n=1}^{\infty} f_n^2\lambda_n \leq C\cdot\|f\|^2_{\overset{\circ}{W}{}_2^1}$ (where f_n is coefficient of Fourier series of function $f(x)$ in terms of the FSF of operator $k_1\Delta u$ $(k_2\Delta u)$ in Ω, λ_n is eigenvalues of the problem (1), $C>0$ is constant). Since FSF of problem (1) is the same as the system of generalized eigenfunctions of the problem (1) then the estimation for system of eigenfunctions considered is known [5]. In the case $N=2$, the estimation $\sum_{n=1}^{\infty} f_n^2\lambda_n^{1/2} < \infty$ is based on Lions theorem about a spaces interpolation (one pair of spaces in this theorem is spaces of functions with convergent series $\sum_{n=1}^{\infty} f_n^2\lambda_n$ and $\sum_{n=1}^{\infty} f_n^2$ accordingly and another pair of spaces is $\overset{\circ}{W}{}_2^1(g)$ and $L_2(g)$).

Thus, proved that for the case of arbitrary domains of dimension $N=2$ or $N=3$, localization principle in domains without discontinuity points is preserved despite the discontinuity of the coefficient.

Then let prove that if $N \geq 5$, then localization principle is violated in points located "far" from the surface of discontinuity of the coefficient, even in case of infinite smoothness of expanded function. Consider domains g_1 and g as concentric balls of radii r_0 and R relatively and center in the origin of coordinates. Then C is sphere of radius r_0; Γ is sphere of radius R.

After separation of variables in the spherical coordinate system we deduce that radial eigenfunction $e(r)$ is solution of the problem:

$$e''(r)+\frac{N-1}{r}\cdot e'(r)+\frac{\lambda}{k_1}e(r)=0;\ 0<r<r_0$$

$$e''(r)+\frac{N-1}{r}\cdot e'(r)+\frac{\lambda}{k_2}e(r)=0;\ r_0<r<R \tag{2}$$

$$e\big|_{r_0-0}=e\big|_{r_0+0};\ k_1\frac{de}{dr}\bigg|_{r_0-0}=k_2\frac{de}{dr}\bigg|_{r_0+0};\ |e(r)|<\infty;\ 0\le r\le R;\ e(R)=0.$$

As proved in [5], eigenfunctions of problem (2) form a complete orthonormal system of functions in the space $L_2(0,R;r^{N-1})$, r^{N-1} is weight function.

Discuss the example of function that violates localization principle. Consider $f(r)$ is function of special kind depends on radius r,

$$f(r)=\begin{cases}0; & 0\le r\le R_1<r_0;\quad R_1>0\\ a_0+a_1r^2+\ldots+a_{4s+3}r^{8s+6}; & R_1\le r\le R_2;\quad r_0<R_2<R;\\ 0; & R_2\le r\le R;\end{cases}$$

s – is sufficiently large natural number. Constant values a_i, $i=0,1,\ldots,4s+3$; such that: $f(r)\in C^{2s}[0,R]$; $f_r'(r_0)\neq 0$. Existence of a_i is proved in [2]. Let Γ_i – is sphere of radius R_i; $i=1,2$; C is sphere of radius r_0; Ω_i is spherical layer bounded by C and Γ_i; $i=1,2$.

Denote:

$$T_n(r)=\frac{C_{1n}J_{\frac{N-2}{2}}(\alpha_n r)}{r^{\frac{N-2}{2}}};\ 0\le r\le r_0;$$

$$Q_n(r)=\frac{C_{2n}J_{\frac{N-2}{2}}(\beta_n r)}{r^{\frac{N-2}{2}}}+\frac{C_{3n}Y_{\frac{N-2}{2}}(\beta_n r)}{r^{\frac{N-2}{2}}};\ r_0<r\le R;\ n=1,2,\ldots$$

Then eigenfunctions of problem (2) $e_n(r)$ are given by:

$$e_n(r)=\begin{cases}T_n(r); & 0\le r\le r_0;\\ Q_n(r); & r_0<r\le R.\end{cases}$$

Calculate coefficients of Fourier series of function $f(r)$ in terms of eigenfunctions $e_n(r)$. Use the Green formula in domains Ω_1 and Ω_2 s times (taking into account that

$k_1\Delta T_n(r)+\lambda_n T_n(r)=0;\ 0<r<r_0$,

$k_2\Delta Q_n(r)+\lambda_n Q_n(r)=0;\ r_0<r<R$,

and therefore, functions $T_n(r)$ and $Q_n(r)$ are expressed in terms of Laplace operator). Then:

$$-f_n=\left[\frac{k_1^s}{\lambda_n^s}\int_{\Omega_1}(\Delta^s f)T_n\,d\Omega_1+\frac{k_2^s}{\lambda_n^s}\int_{\Omega_2}(\Delta^s f)Q_n\,d\Omega_1\right](-1)^{s+1}+$$

$$+\sum_{i=1}^{s}\left\{(-1)^{i-1}\left[\frac{(k_2^i-k_1^i)}{\lambda_n^i}\int_C\frac{\partial}{\partial n}(\Delta^{i-1}f)T_n\,dC-\right.\right.$$

$$-\frac{1}{\lambda_n^i}\int\limits_C\left(\Delta^{i-1}f\right)\left(k_2^i\frac{\partial Q_n}{\partial n}-k_1^i\frac{\partial T_n}{\partial n}\right)dC+\frac{k_2^i}{\lambda_n^i}\int\limits_{\Gamma_2}\left(\left(\Delta^{i-1}f\right)\frac{\partial Q_n}{\partial n}-Q_n\frac{\partial\left(\Delta^{i-1}f\right)}{\partial n}\right)d\Gamma_2-$$

$$-\frac{k_2^i}{\lambda_n^i}\int\limits_{\Gamma_1}\left(\Delta^{i-1}f\cdot\frac{\partial T_n}{\partial n}-T_n\frac{\partial}{\partial n}\left(\Delta^{i-1}f\right)\right)d\Gamma_1\Bigg]\Bigg\} \quad (3)$$

Take into account that $\frac{\partial}{\partial n}=\frac{d}{dr}$, all subintegral functions depend on radius only, conditions:

$$f\big|_{\Gamma_1}=\frac{df}{dr}\Big|_{\Gamma_1}=\ldots=\frac{d^{2s}f}{dr^{2s}}\Big|_{\Gamma_1}=0;\; f\big|_{\Gamma_2}=\frac{df}{dr}\Big|_{\Gamma_2}=\ldots=\frac{d^{2s}f}{dr^{2s}}\Big|_{\Gamma_2}=0,$$

matching condition for functions $T_n(r)$ and $Q_n(r)$ on the surface C and estimations [5]:

$$S_1n^2\le\lambda_n\le S_2n^2,\quad T_1\lambda_n^{1/4}\le\left|C_{1n}\right|\le T_2\lambda_n^{1/4},\quad \left|C_{2n}\right|,\left|C_{3n}\right|\le T_3\lambda_n^{1/4}. \quad (4)$$

Then summand with $i=1$ is dominant in sum (3). This means that:

$$f_n=\frac{(k_1-k_2)}{\lambda_n}C_{1n}\frac{J_{\frac{N-2}{2}}(\alpha_n r_0)}{r_0^{\frac{N-2}{2}}}{}'f_r'(r_0)\,\omega_{N_{r_0}}+\underline{\underline{O}}\left(\frac{1}{\lambda_n^{\frac{3}{2}}}\right).$$

Then Fourier component in the point $r=0$ is expressed as follows (use asymptotics of cylindrical functions):

$$f_ne_n(0)=f_nT_n(0)=\frac{M\cdot C_{1n}^2(k_2-k_1)}{\lambda_n^{5/4}}\cdot\lambda_n^{\frac{N-2}{4}}\cdot\sin\left(\alpha_n r_0-\left(\frac{N+1}{4}\right)\pi\right)+$$

$$+\lambda_n^{\frac{N-2}{4}}\cdot\underline{\underline{O}}\left(\frac{1}{\lambda_n^{5/4}}\right)=M(k_2-k_1)C_{1n}^2\lambda_n^{\frac{N-7}{4}}\cdot\sin\left(\alpha_n r_0-\left(\frac{N+1}{4}\right)\pi\right)+\underline{\underline{O}}\left(\lambda_n^{\frac{N-7}{4}}\right), \quad (5)$$

where $M=\text{const}$.

From estimation (4) follows $T_1\lambda_n^{1/4}\le\left|C_{1n}\right|\le T_2\lambda_n^{1/4}$, and then, $T_1^2\lambda_n^{\frac{N-5}{4}}\le C_{1n}^2\lambda_n^{\frac{N-7}{4}}\le T_2^2\lambda_n^{\frac{N-5}{4}}$ for sufficiently large values of n.

Then for $N\ge5$ from (5) follows, that for $f_ne_n(0)$ to tend to zero as $n\to\infty$, it is necessary that $\sin\left(\alpha_n r_0-\left(\frac{N+1}{4}\right)\pi\right)\to0$. But from the estimation [5]:

$$\alpha_n r_0=\left(\frac{N+1}{8}\right)\pi+\frac{\pi n}{2}\left(1+\overset{=}{o}(1)\right)$$

follows that

$$\alpha_n r_0-\left(\frac{N+1}{4}\right)\pi=-\left(\frac{N+1}{8}\right)\pi+\frac{\pi n}{2}\left(1+\overset{=}{o}(1)\right)$$

and that is why $\sin\left(\alpha_n r_0-\left(\frac{N+1}{4}\right)\pi\right)$ does not tend to zero as $n\to\infty$, and this concludes the example of function that violates localization principle.

3. Conclusion

We studied some properties of spectral expansions that correspond to eigenfunction system of Laplace operator with discontinuous coefficient and proved that for the case of domain dimension $N\ge5$, localization principle in domains without points of discontinuity is violated in points located "far" from the surface of discontinuity of the coefficient, even in case of infinite smoothness of expanded function.

References

[1] Il'in V A 1961 O sisteme klassicheskih sobstvennyh funkcij linejnogo samosoprjazhennogo ellipticheskogo operatora s razryvnymi koefficientami *Doklady Akademii Nauk SSSR* **137(2)** 272–275

[2] Il'in V A 1968 Problemy lokalizacii i skhodimosti dlya ryadov Fur'e po fundamental'nym sistemam funkcij operatora Laplasa *Uspehi Matematicheskih Nauk* **23(2)** 61–120

[3] Soldatova M A 2002 Some order-of-magnitude estimates for nonorthonormalized fundamental (in Il'in's sense) system of functions of the Laplace operator *Differential Equations* **38(4)** 541–546

[4] Suchkov M and Trifonenkov V 2015 Estimations of eigenfunctions of ordinary fourth-order differential operator and their use when checking the basis property condition by V A Il'in in Lp spaces *Bulletin of Moscow Region State University Series: Physics and Mathematics* **4** 36–43

[5] Suchkov M and Trifonenkov V 2017 On the localization principle for Laplace operator with discontinuous coefficient in domains without discontinuity points *Bulletin of Moscow Region State University Series: Physics and Mathematics* **1** 8–17

Cluster analysis of *S. Cerevisiae* nucleosome binding sites

Y Suvorova[1] and E Korotkov [1,2]
[1] Institute of Bioengineering, Research Center of Biotechnology of the Russian Academy of Sciences, Moscow, Russia
[2] National Research Nuclear University MEPhI (Moscow Engineering Physics Institute). Moscow, Russia
E-mail: suvorovay@gmail.com

Abstract. It is well known that major part of a eukaryotic genome is wrapped around histone proteins forming nucleosomes. It was also demonstrated that the DNA sequence itself is playing an important role in the nucleosome positioning process. In this work, a cluster analysis of 67 517 nucleosome binding sites from the *S. Cerevisiae* genome was carried out. The classification method is based on the self-adjusting dinucleotides position weight matrix. As a result, 135 significant clusters were discovered that contain 43225 sequences (which constitutes 64% of the initial set). The meaning of the found classes is discussed, as well as the possibility of the further usage.

1. Introduction

It is known that the length of a DNA molecule in its unfolded form is many times greater than the size of a cell nucleus, where it is located. There exists a complex system of DNA packaging into chromosomes for DNA placement in the nucleus and access to different part of the genome. Nucleosomes formation is one of the first stages of such packaging. The nucleosome consists of the histone proteins complex (two molecules of the H2A, H2B, H3 and H4 proteins), around which nearly 147 base pairs (bp) of DNA are wrapped, constituting 1.67 turns of a double helix. The distance between the adjacent nucleosomes varies from several bases to hundreds of bases. Sequences that are not occupied by nucleosomes are called linkers. In total, up to 80% of a genome could to be covered by nucleosomes. It was demonstrated that the location of nucleosomes in addition to packaging possesses a regulatory function. The location of nucleosomes regulates availability of DNA sequences for transcription factors and other regulated proteins. The part of DNA sequence, which is connected with the histone proteins complex, is called the nucleosome binding site.

Since the DNA molecule is wrapped around the histone, the local DNA rigidity plays an important role. This rigidity is different for different combinations of nucleotides. Therefore, the DNA sequences, where nucleosomes are positioned, are characterized by certain preferences. There are preferred pairs of nucleotides (dinucleotides) in the sites of direct interaction between the DNA and histone proteins. Under the optimum conditions, when forming a nucleosome, the bent DNA should be located uniformly. The strongest contacts of histone proteins with DNA in a nucleosome are separated by 10-11 nucleotides [1]. The hypothesis that the nucleosome positioning is determined by the nucleotide sequence itself was first put forward by E. Trifonov back in 1980 [2]; he also demonstrated that the optimal nucleosome sequences are characterized by the dinucleotides oscillations with the period of ~10. Similar patterns were found practically all studied eukaryotic genomes, including the human [3]. Later, E. Trifonov and his colleagues, as well as the other authors, examined in detail the periodicity of 10 nucleotide pairs and its relationship with various nucleosome sequences [4–7]. However, there are studies that claim that other factors, such as transcription factors and motor proteins, which are capable to move the nucleosomes along the DNA strand to a desired position, are *in vivo* playing the decisive role [8]. Discussions about which of the factors has major influence are still being conducted. Also, some assessments demonstrate that the 10 period appears more likely to be a group property of several nucleosome sequences rather than of a separate binding site [9].

In the past 30 years, a fairly large number of mathematical methods were developed to determine the possible position of a nucleosome along a DNA sequence. Methods of the nucleosome positioning prediction could be conditionally divided into two groups: statistical and biophysical methods [10]. Methods of the first group are based on the statistical properties of a DNA sequence, and consider it as a sequence of symbols; primarily such methods are based on a periodicity of 10 symbols [11,12]. Methods of the second group are based on calculating the flexibility of the double helix section composed of different nucleotides and of the corresponding nucleosome formation energies [13]. Besides, there are methods that combine both these two approaches. A detailed analysis and comparison of the methods could be found in works [10,14].

However, beside the described periodicity of 10 for different dinucleotides, no general properties pertaining to the nucleosome binding sites were identified. The standard methods of bioinformatics were still not able to construct a significant multiple alignment of the given sequences. There are few studies devoted to the nucleosomes multiple alignment using an extremely limited set of data. The authors of the work [15] constructed a multiple alignment for 177 nucleosome binding sites. Therefore, selection of classes and corresponding common properties of the nucleosome binding sites could appear to be the first step in searching for new common properties of these sequences.

2. Methods

2.1. Data preparation

In this paper we are considering the nucleosome binding sites as a symbolic sequence constructed from four-symbol alphabet {“A”, “C”, “T”, “G”}. A total of 67 517 unique nucleosome binding sites in accordance with the experimental map (unredundant map) [16] were extracted from the *S. cerevisiae* genome. This genome was chosen as the most studied from the point of view of the nucleosomes location. The nucleosome binding sites were having different lengths ranging from 146 to 152 nucleotides. In order to bring all the sequences to the same length, we cut each sequence down to 140 nucleotides preliminary aligning them in the center.

As a result, the set N of DNA sequences of the length $L = 140$ was obtained ($|N| = 67\,517$). Each sequence was translated into the 16-symbol dinucleotide alphabet {'AA', 'AC', 'AT' ... 'GG'} by intersecting pairs. And the corresponding sequences of 139 dinucleotides length were received. Alphabet of the 16 symbols size was chosen for two reasons. First, the increase of the size of the alphabet extends the space of the indicators being examined for clustering. Second, from the description of the nucleosome properties it becomes evident that primarily the pairs of nucleotides are playing an important role in the formation of nucleosomes.

In order to determine the level of statistical significance of values calculated in the work, each of the initial DNA sequences of the N set was randomly shuffled (using the pseudo-random numbers generator). The resulting mixed sequences were also translated into a 16-symbol alphabet and constituted a set of random sequences designated as R.

2.2. Method of classification

The clustering procedure is based on the self-learning weight matrices sized 16×139. This is the so-called position weight matrix; the columns of such a matrix correspond to the position in the sequence, and the rows - to one of 16 dinucleotides. At the initialization stage, the matrix is filled with the dinucleotide frequencies of one sequence (S_0) chosen from the initial set N. Let us denote the obtained initial matrix as $\boldsymbol{M_0}$. Then the matrix is normalized ($\boldsymbol{M_0}$*) using the special procedure, which is described below.

Further, for each dinucleotide sequence S_n from the set N a total weight was calculated using the $\boldsymbol{M_0}$* normalized matrix

$$F_n = \sum_{i=1}^{139} M_0^*[s_i, i] \quad (1)$$

Where s_i is the ith element of the sequence S_n, and M_0^* $[n,i]$ is the element of the $\boldsymbol{M_0}$*. In the same way, the weight for each sequence R_n from the shuffled set was calculated corresponding value: $Fr_n = \sum_{i=1}^{139} m[r_i, i]$. The weights of real sequences are normalized $Z_n' = (F_n - \bar{F})/\sqrt{D(F)}$. Where $\bar{F}$ and $D(F)$are the mean value and dispersion of the sequences weight calculated on the set of shuffled sequences R. The sequences S_n with the obtained weight exceeding the specified threshold value: $Z_n' > Z_0^k$were selected in a new subset N_1. Based on the sequences from N_1 a new frequency matrix $\boldsymbol{M_1}$ is filled, while the previous one is being deleted. The new matrix undergoes normalization; and the process is repeated. The process includes several iterations K with a gradual increase in the Z_0^k threshold.

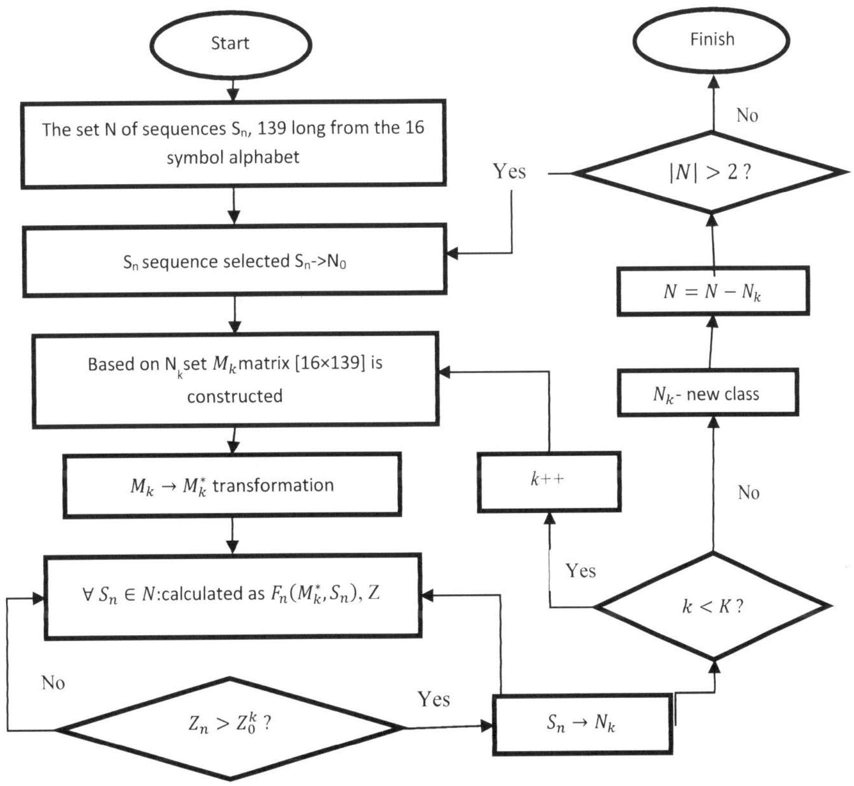

Figure 1*. The diagram showing the main algorithm for the nucleosome binding sites classification based on dinucleotide position weight matrix*

The set of sequences N_k, which weight that were calculated using the current matrix are above the threshold at the last iteration k=K, are registered as the new class N_k and are excluded from further consideration. The obtained profile matrix is also registered. Next, a new initial sequence is selected from the remaining $|N| = |N| - |N_k|$ set of sequences, and a new matrix is being initialized. The process is repeated from the very beginning. The procedure is repeated until the non-classified sequences remain in the initial N multiplicity. The diagram of the clustering procedure is presented in Figure 1. The procedure is carried out from different starting points (first sequence S_0); and the best result is selected corresponding to the maximum size of the resulting class.

2.3. Matrixes normalization procedure

The procedure for normalizing the position weight matrix of dinucleotides $\boldsymbol{M}$ in a general case is carried out, as follows. For each element of the matrix, the expected frequencies are calculated:$P_{km} = (S_k * S_m)/L^2$, where S_k and S_m are the corresponding marginal sums $S_m = \sum_{i=1}^{16} M[i,m]$; $S_k = \sum_{j=1}^{139} M[k,j]/$, and L is the sum of the matrix $\boldsymbol{M}$ elements. Afterwards, on the basis thereof mean and dispersion $V_{ij} = P_{ij} * (1 - P_{ij}) * L$ are calculated. As a result: $M^*[i,j] = (M[i,j] - M_{ij})/\sqrt{V_{ij}}$.

2.4. Matrixes normalization procedure taking into consideration the triplet periodicity

To normalize the dinucleotides position weight matrix $\boldsymbol{M_k}$ (16 × 139) considering the triplet periodicity of the sequences a special procedure was created. A triplet matrix $\boldsymbol{T}$ is filled according to the same nucleotide sequences from the N_k set, as $\boldsymbol{M_k}$ at the current iteration k. The matrix $\boldsymbol{T_k}$ is (4 × 3) frequency matrix. The matrix rows correspond to four types of nucleotides, and the columns - to three codon positions (disjoint triples). Then, the $\boldsymbol{T_k}$ matrix is transformed according to the following formula: $T[i,j] = T[i,j]/K_j$,where K_j is the sum of elements in the jth column of the $\boldsymbol{T_k}$ matrix.

After that, theoretical probabilities based on the triplets of the $\boldsymbol{T_k}$ matrix are calculated for each cell $M[i,j]$: $P_{ij} = P(n_1)/TPj * P(n_2)/TPj$, where P_1/TP is the probability of the first nucleotide in the pair in accordance with its position in the triplet in the initial nucleotide sequence, and P_2/TP is the probability of the second nucleotide. Thus: $P_{ij} = T[n_1,t_1] * T[n_2,t_2]$. Then the following expected values are calculated: $Mean_{ij} = P_{ij} * N$ and $D_{ij} = P_{ij} * (1 - P) * N$. Here N is the sum of elements of the M matrix corresponding column. Then the elements of the M matrix are transformed according to the following formula

$$M[i,j]^* = \frac{M[i,j] - Mean_{ij}}{\sqrt{D_{ij}}}. \quad (2)$$

2.5. Implementation

The software implementation of the described algorithms was created on the C ++ programming language for the Linux operating system and using the MPI library for parallel computations. The calculations were carried out on the Joint Supercomputer Center of the RAS and on the cluster of Institute of Bioengineering RAS. In order to calculate the Fourier transform, the fft library of the R programming language was employed.

3. Results and discussion

Classification analysis of the 67 517 nucleosome binding sites extracted from the *S. cerevisiae* genome was carried out with the developed iteration procedure based on the dinucleotides position weight matrices. Analysis of the set of random (shuffled) sequences of the same volume demonstrated that the group size of more than 80 sequences is statistically significant, and corresponds to the 5% significance level. As a result, 135 disjoint significant clusters were found at this level when processing real data. The clusters contained 43 225 of 67 517 sequences from the initial set N (about 64%). The average class size is about 400 sequences.

Many nucleosome binding sites in our study corresponds to protein coding regions. Previously, it was demonstrated that the triplet periodicity is inherent in the most protein coding sequences [17]; and the triplet periodicity of most known genes belongs to one of 2,5 thousand classes [18]. Thus, the triplet periodicity of different sequences could also affect the results of the classification carried out.

In order to check this hypothesis, an additional classification was carried out, which took into account the influence of the triplet periodicity at the matrix normalization stage (the procedure was described in Section 2.4.). For the new version of the algorithm, the minimum cluster size corresponding to the 5% statistical significance level was also determined on the set of the shuffled sequences *R*. As a result of this new classification 174 significant classes were discovered that contained 39 176 of 67 517 sequences (about 58%).

In order to compare the obtained result with the standard methods, the BLAST (The Basic Local Alignment Search Tool) program was employed to conduct analysis of the nucleosome binding sites in the initial *N* set [19]. The analysis was aimed at searching for pair similarities of nucleotide sequences within the set. As a result, 3200 cases of the sequences significant similarity inside the set (e-value threshold <1e-5) were found. This result means that at the level of the bioinformatics traditional tools there is practically no significant similarity between sequences of the initial set; and it is impossible to identify classes on the basis of such similarity.

As an example of the obtained class features the values of chi-square statistic for each of 139 dinucleotides positions of nucleosome is shown in Figure 2. The chi-square values illustrate the difference between the observed dinucleotide frequencies of the real class and the theoretical frequencies for the position. For clarity' sake all chi-square values were normalized ($Z(Chi)$). Plots for three sets are shown. The first one is the set of real nucleosome sequences of one of the resulting classes (*Real*). The volume of the set *Real* is 325 sequences. The second one is the set of randomly shuffled sequences of the class *Real* (*Shuffled*). And the last one is the set of nucleosomes of the same volume as the set *Real*, randomly chosen from the general set *N*, without the classification procedure (*Rand Class*). The figure shows that the specificity of the positions of the real class is superior to the rest of the two sets. The plots reveal that the center of nucleosome binding sites is more conservative, this property is noticeable even on randomly selected sequences and is greatly enhanced in the *Real* class.

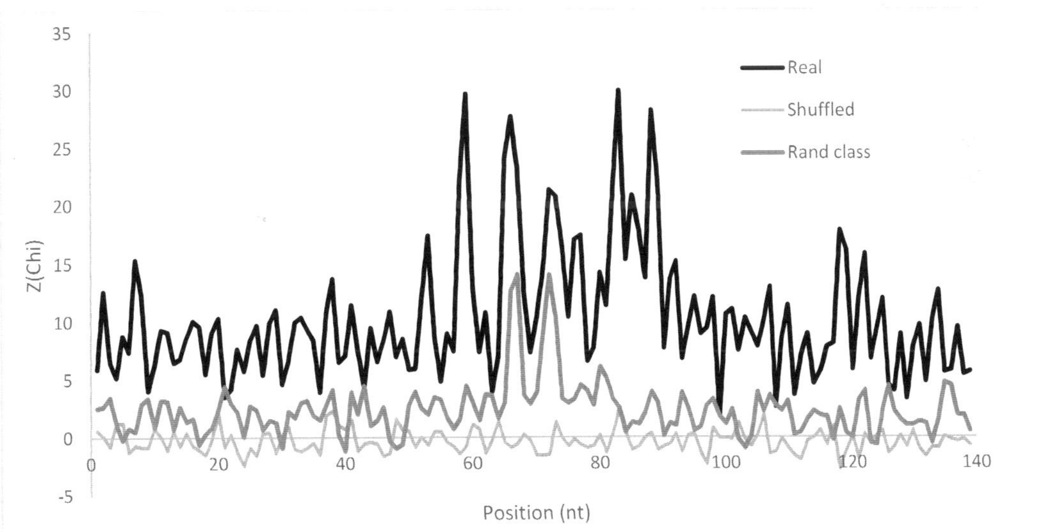

Figure 2. Normalized chi-square value along nucleosome profile the of a set of nucleosome sequence class. *Real* is set of real nucleosome sequences of one of the resulting classes *Shuffled* – is the set of randomly shuffled sequences of the class *Real*. *Rand Class* is the set of nucleosomes of the same volume as the set Real, chosen from the general set N in a random way.

Figure 3 demonstrates the Fourier transform of the chi-squared values of the three sets discussed above (see Figure 2). One can see, that only a set of sequences of the *Real* class has a significant periodicity. The periods 10, 20 that traditionally associated with nucleosome binding sequences. And the most significant period here is 47 symbols. This period has already been described in Trifonov's work on the study of nucleosomes and it is also associated with the period (10.4) [20].

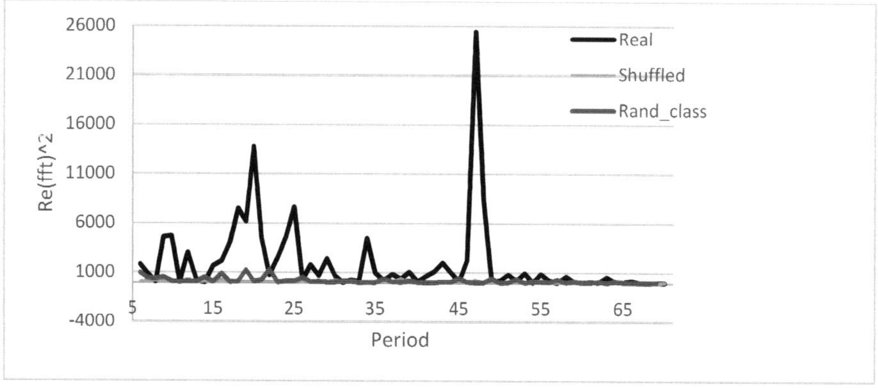

Figure 3. Fourier transform of the chi-square of the three sets. *Real* is set of real nucleosome sequences of one of the resulting classes *Shuffled* – is the set of randomly shuffled sequences of the class *Real*. *Rand Class* is the set of nucleosomes of the same volume as the set Real, chosen from the general set N in a random way.

The results obtained in the work demonstrate that there is a fuzzy similarity between the nucleosome binding sequences, which allows us to divide most of such sequences into the statistically significant classes. The obtained position weight matrixes received as the result of the study could be further employed in searching for nucleosome binding sequences using the profile analysis.

Classification procedure, which considers the triplet periodicity, showed that in the case the triplet periodicity has minor influence on the classification result. This could be explained by the fact that the difference in the classes of triplet periodicity is more often observed between genes from different genomes, than within a genome [21]. The result also suggests that there are other factors that play a more significant role in the classification of nucleosome binding sites.

The developed method is not suggesting registration of insertions and deletions inside the nucleosome binding sites, since it is supposed that the structure of a nucleosome itself does not allow such insertions [22]. However, it is known that the experimental methods that determining the coordinates of the nucleosome in a DNA sequence may not be entirely accurate. So, probably, taking into account the shift of the entire section plot to several bases relative to the center would help us to further improve the classification and the obtained position weight matrixes in the future.

Acknowledgements
This work was supported by a grant provided by the Russian Science Foundation (14-24-00175).

References

[1] Xu F and Olson W K 2010 Dna architecture, deformability, and nucleosome positioning *J. Biomol. Struct. Dyn.*
[2] Trifonov E N 1980 Sequence-dependent deformational anisotropy of chromatin DNA *Nucleic Acids Res.*
[3] Bettecken T and Trifonov E N 2009 Repertoires of the nucleosome-positioning dinucleotides *PLoS One*

[4] Radman-Livaja M and Rando O J 2010 Nucleosome positioning: How is it established, and why does it matter? *Dev. Biol.*
[5] Segal E, Fondufe-Mittendorf Y, Chen L, Thåström A, Field Y, Moore I K, Wang J-P Z and Widom J 2006 A genomic code for nucleosome positioning *Nature*
[6] Gabdank I, Barash D and Trifonov E N 2009 Nucleosome dna bendability matrix (c. elegans) *J. Biomol. Struct. Dyn.*
[7] Salih B, Tripathi V and Trifonov E N 2015 Visible periodicity of strong nucleosome DNA sequences *J. Biomol. Struct. Dyn.*
[8] Hughes A L and Rando O J 2014 Mechanisms Underlying Nucleosome Positioning In Vivo *Annu. Rev. Biophys.*
[9] Jin H, Rube H T and Song J S 2016 Categorical spectral analysis of periodicity in nucleosomal DNA *Nucleic Acids Res.*
[10] Teif V B 2016 Nucleosome positioning: Resources and tools online *Brief. Bioinform.*
[11] Xi L, Mittendorf Y F, Xia L, Flatow J, Widom J and Wang J P 2010 Predicting nucleosome positioning using a duration Hidden Markov Model *BMC Bioinformatics*
[12] Xing Y Q, Liu G Q, Zhao X J and Cai L 2013 An analysis and prediction of nucleosome positioning based on information content *Chromosom. Res*
[13] Morozov A V., Fortney K, Gaykalova D A, Studitsky V M, Widom J and Siggia E D 2009 Using DNA mechanics to predict in vitro nucleosome positions and formation energies *Nucleic Acids Res.*
[14] Liu H, Zhang R, Xiong W, Guan J, Zhuang Z and Zhou S 2013 A comparative evaluation on prediction methods of nucleosome positioning *Brief. Bioinform.*
[15] Wang J P Z and Widom J 2005 Improved alignment of nucleosome DNA sequences using a mixture model *Nucleic Acids Res*
[16] Brogaard K, Xi L, Wang J-P and Widom J 2012 A map of nucleosome positions in yeast at base-pair resolution *Nature*
[17] Yin C and Yau S S-T 2007 Prediction of protein coding regions by the 3-base periodicity analysis of a DNA sequence. *J. Theor. Biol.* **247** 687–94
[18] Frenkel F E and Korotkov E V 2008 Classification analysis of triplet periodicity in protein-coding regions of genes. *Gene* **421** 52–60
[19] Altschul S F, Gish W, Miller W, Myers E W and Lipman D J 1990 Basic local alignment search tool. *J. Mol. Biol.* **215** 403–10
[20] Salih F, Salih B and Trifonov E N 2008 Sequence structure of hidden 10.4-base repeat in the nucleosomes of c. elegans *J. Biomol. Struct. Dyn.*
[21] Suvorova Y M and Korotkov E V 2015 Study of triplet periodicity differences inside and between genomes *Stat. Appl. Genet. Mol. Biol.* **14**
[22] Ioshikhes I, Bolshoy A, Derenshteyn K, Borodovsky M and Trifonov E N 1996 Nucleosome DNA Sequence Pattern Revealed by Multiple Alignment of Experimentally Mapped Sequences *J. Mol. Biol.*

On a sufficient condition for the subharmonicity of functions, satisfying the discrete mean value inequality

D S Telyakovskii[1]

National Research Nuclear University MEPhI (Moscow Engineering Physics Institute), 31 Kashirskoe shosse, 115409 Moscow, Russia

E-mail: dtelyakov@mail.ru

Abstract. We obtain a sufficient condition for subharmonicity of functions of two variables that uses Laplace difference operator of Schwartz type. M.A. Kreines proved that an upper semi-continuous function is subharmonic, if for some $n \geq 3$ for every point ζ of its domain one can construct of a sufficiently small radius with the center at ζ such that the right n-gon can be inscribed in the circle, while the function value at the point ζ does not exceed the arithmetic mean of its values at the vertices of this n-gon. In this note the semi-continuity condition for the function is replaced by some conditions of function directional continuity an summability (with respect to the planar Lebesgue measure.)

We also provide the position of nodes for the Laplace difference operator such that the proposition similar to the Kreines's Theorem on a sufficient condition for the subharmonicity is valid for the functions of three variables.

In book of I.I. Privalov [1] is given the following result of M.A. Kreines that Kreines communicated to Privalov.

Theorem (Theorem (Kreines)). *Let function $u(z) = u(x, y)$ be upper semi-continuous in region $G \subset \mathbb{R}^2$. If for every point $\zeta_0 \in G$ for some natural $n \geq 3$ in every circle of sufficiently small radius with center at ζ_0 we can inscribe a right n-gon with vertices at points $\zeta_1, \dots, \zeta_n$ for which the following inequality holds*

$$u(\zeta_0) \leq \frac{1}{n} \sum_{j=1}^{n} u(\zeta_j), \tag{1}$$

then the function $u(z)$ is subharmonic in region G.

Inequality (1) is the discrete analog of the mean value inequality for the subharmonic functions. In the current paper this inequality will be written as a discrete relation condition of Schwartz type for the Laplace equation.

We note that in the formulation of the above theorem given in [1] the function $u(z)$ is assumed to be continuous, but the same proof would hold if the continuity is replaced by the upper semi-continuity of $u(z)$.

[1] This work is partially supported by the program of increasing of competitiveness of the National Nuclear Research University MEPhI, project No. 02.a03.21.0005 from 27.08.2013

In order to write the inequality (1) in the form of the difference relation condition we introduce the following notation: points $\zeta_0, \zeta_1, \dots, \zeta_n$ we call nodes of the difference relation, the point ζ_0 is the central node of the set of nodes $A_{\zeta_0} := \{\zeta_0, \zeta_1, \dots, \zeta_n\}$ and to simplify the notation we write $u_j := u(\zeta_j)$, $j = 0, 1, \dots, n$.

Define the difference relation $\Delta^{(*)}u(\zeta_0)$ that we consider in this paper

$$\Delta^{(*)}u(\zeta_0) = \Delta^{(*)}u(\zeta_0; A_{\zeta_0}) := \frac{1}{nr^2}\Big(\sum_{j=1}^{n} u_j - nu_0\Big) = \frac{1}{nr^2}\sum_{j=1}^{n}(u_j - u_0), \tag{2}$$

where r is the radius of the circle, the points $\zeta_1, \dots, \zeta_n$ are on that circle. Using the expression (2) the inequality (1) can be rewritten as $\Delta^{(*)}u(\zeta_0; A_{\zeta_0}) \geq 0$.

In this paper the expression (2) is applied to the set of nodes located not only in the plane, but in space too. In any case those nodes $\zeta_1, \dots, \zeta_n$ are located on the same distance r away from the central node ζ_0. In planar case nodes $\zeta_1, \dots, \zeta_n$ are always vertices of the right n-gon with the center at ζ_0, this fact will not be mentioned each time separately.

We use the following definitions and terminologies.

As a *star* S_ζ with center at the point ζ we will call the collection of several segments with the common end ζ. We will consider only those stars S_ζ, that inside of every half plane containing point ζ at its boundary, contain at least one ray, and do not contain rays in the angles of a quantity not exceeding $\pi/2$, formed by any two other rays. It is clear that under such conditions, a star can not contain less than three or more than seven rays and that the maximum angle between neighboring rays is greater $\pi/4$.

Let $h(t)$, $t \geq 0$, be a modulus of continuity type function and let function $u(z)$ be defined on some set B_ζ for which the point ζ is the limiting point. If for some value $L_\zeta > 0$ at every point $z \in B_\zeta$ the inequality $|u(z) - u(\zeta)| \leq L_\zeta h(|z - zeta|)$ holds, then we say that function $u(z)$ is *h-regular in ζ with respect to B_ζ*. In the theorem 1 as a condition of the continuity of the function $u(z)$ in points $\zeta \in G$ we impose the condition of its h-regularity with respect to some star S_ζ.

As usual, the positive and negative parts of the function $u(z)$ are denoted by $u^+(z) := \max\{u(z), 0\}$ and $u^-(z) := -\min\{u(z), 0\}$.

Theorem 1. *Let function $u(z)$ be defined in the domain $G \subset \mathbb{R}^2$, for each point $\zeta \in G$ the function $u(z)$ is h-regular with respect to some star S_ζ, for which the maximum angle between neighborhooding rays does not exceed $\alpha\pi$ and the function $|u(z)|^{\max\{2\alpha, 1\}}$ is locally summable in G. Then, if for each point $\zeta_0 \in G$ in any neighborhood U_{ζ_0} for some $n \geq 3$ there exists a set of nodes A_{ζ_0} for which the quantity*

$$\Big(\Delta^{(*)}u(\zeta_0; A_{\zeta_0})\Big)^{-} \to 0 \tag{3}$$

when the neighborhood U_{ζ_0} is contracted to the point ζ_0, then function $u(z)$ is subharmonic in region G.

Note that in this theorem the number of nodes in the set A_{ζ_0} may be different for different points ζ_0 and different neighborhoods U_{ζ_0}. If in the condition (3) we assume the tendency to zero not of the negative part of the difference relation $\Delta^{(*)}u(\zeta_0; A_{\zeta_0})$, but of its own, then the theorem 1 gives a sufficient condition for harmonicity.

The proof of the theorem 1 is similar to the proof of theorem from the author's work [2], so we do not give it. Note that the theorem, which was formulated in [2] as sufficient condition of harmonicity, in fact, gives a sufficient condition subharmonicity.

For upper semicontinuous functions of three variables, we can specify the location five nodes under which the fulfillment of the condition (3) implies subharmonicity. We will consider only sets of five nodes $A_{\boldsymbol{x}^0}$ in which the point $\boldsymbol{x}^0$ is the center of regular tetrahedron, and the remaining nodes $\boldsymbol{x}^j$ are located at the vertices of this tetrahedron. This location of nodes will not be discussed every time specially.

Theorem 2. *Let function $u(\boldsymbol{x}) = u(x_1, x_2, x_3)$ be upper semicontinuous in the domain $G \subset \mathbb{R}^3$. If for every point $\boldsymbol{x}^0 \in G$ in any its neighborhood $U_{\boldsymbol{x}^0}$ there exists a collection of nodes $A_{\boldsymbol{x}^0}$ for which the relation* (3)*, then the function $u(\boldsymbol{x})$ is subharmonic in the domain G.*

To prove this theorem we need the following lemma. We denote by $\operatorname{tr} d^2u$ the trace of the matrix of the quadratic form of the second differential d^2u.

Lemma 1. Assume that function $u(\boldsymbol{x})$ has the second Peano differential at point $\boldsymbol{x}^0 \in \mathbb{R}^3$

$$u(\boldsymbol{x}) = u(\boldsymbol{x}^0) + du(\boldsymbol{x}-\boldsymbol{x}^0) + \frac{1}{2}d^2u(\boldsymbol{x}-\boldsymbol{x}^0) + o(|\boldsymbol{x}-\boldsymbol{x}^0|^2) \quad \text{while } \boldsymbol{x} \to \boldsymbol{x}^0. \tag{4}$$

Then if collection of nodes $A_{\boldsymbol{x}^0}$ is contracted to the point $\boldsymbol{x}^0$

$$\Delta^{(*)}u(\boldsymbol{x}^0; A_{\boldsymbol{x}^0}) \to \frac{1}{3}\operatorname{tr} d^2u. \tag{5}$$

Proof. First we show that for an arbitrary first-order polynomial $p(\boldsymbol{x}) = p_0 + \sum p_k x_k$ the quantity $\Delta^{(*)}p(\boldsymbol{x}^0; A_{\boldsymbol{x}^0}) = 0$ at any point $\boldsymbol{x}^0$ and for any collection of nodes $A_{\boldsymbol{x}^0}$.

According to the definition of $\Delta^{(*)}$

$$\Delta^{(*)}p(\boldsymbol{x}^0; A_{\boldsymbol{x}^0}) = \frac{1}{4r^2}\sum_{j=1}^{4}(p(\boldsymbol{x}^j) - p(\boldsymbol{x}^0)) = \frac{1}{4r^2}\sum_{k=1}^{3} p_k \sum_{j=1}^{4}(x_k^j - x_k^0). \tag{6}$$

Each of the inner sums on the right-hand side of this formula is equal to the sum of the moments of system of unit masses located at the nodes $\boldsymbol{x}^j$ (that is, at vertices regular tetrahedron with center $\boldsymbol{x}^0$) with respect to the axis that passes through the point $\boldsymbol{x}^0$ and is parallel to the coordinate axis Ox_k. Since the central node $\boldsymbol{x}^0$ is the center of gravity of this mass system, then the expression in the formula (6) is zero.

Now let $p(\boldsymbol{x})$ be a second-order polynomial. According to what has been proved above we can assume that $p(\boldsymbol{x})$ contains only terms of the second order, and thus $p(\boldsymbol{x})$ is a quadratic form

$$p(\boldsymbol{x}) = p_{11}x_1^2 + p_{22}x_2^2 + p_{33}x_3^2 + 2p_{12}x_1x_2 + 2p_{13}x_1x_3 + 2p_{23}x_2x_3.$$

Changing the coordinate system is orthogonal coordinate transformation. Since the trace of the matrix of the quadratic form is invariant under the orthogonal transformation, the coordinate system can be chosen arbitrarily.

We choose a coordinate system with the origin at the point $\boldsymbol{x}^0$ in which the nodes $\boldsymbol{x}^1$, $\boldsymbol{x}^2$, $\boldsymbol{x}^3$ and $\boldsymbol{x}^4$ have respectively coordinates $(r, 0, 0)$, $\left(-\frac{r}{3}, 0, \frac{2\sqrt{2}\,r}{3}\right)$, $\left(-\frac{r}{3}, \frac{\sqrt{2}\,r}{\sqrt{3}}, -\frac{\sqrt{2}\,r}{3}\right)$ and $\left(-\frac{r}{3}, -\frac{\sqrt{2}\,r}{\sqrt{3}}, -\frac{\sqrt{2}\,r}{3}\right)$.

The values of the polynomial $p(\boldsymbol{x})$ at the nodes of the set $A_{\boldsymbol{x}^0}$ are equal, respectively:

$$p(\boldsymbol{x}^0) = 0, \qquad p(\boldsymbol{x}^1) = p_{11}r^2, \qquad p(\boldsymbol{x}^2) = \left(\frac{p_{11}}{9} + \frac{8p_{33}}{9} - \frac{4\sqrt{2}\,p_{13}}{9}\right)r^2,$$

$$p(\boldsymbol{x}^3) = \left(\frac{p_{11}}{9} + \frac{2p_{22}}{3} + \frac{2p_{33}}{9} - \frac{2\sqrt{2}\,p_{12}}{3\sqrt{3}} + \frac{2\sqrt{2}\,p_{13}}{9} - \frac{4p_{23}}{3\sqrt{3}}\right)r^2,$$

$$p(\boldsymbol{x}^4) = \left(\frac{p_{11}}{9} + \frac{2p_{22}}{3} + \frac{2p_{33}}{9} + \frac{2\sqrt{2}\,p_{12}}{3\sqrt{3}} + \frac{2\sqrt{2}\,p_{13}}{9} + \frac{4p_{23}}{3\sqrt{3}}\right)r^2.$$

Substituting these values into the expression (2) for $\Delta^{(*)}p$ we obtain that for each point $\boldsymbol{x}^0$ and each set of nodes $A_{\boldsymbol{x}^0}$

$$\Delta^{(*)}p(\boldsymbol{x}^0, A_{\boldsymbol{x}^0}) = \frac{1}{3}(p_{11} + p_{22} + p_{33}) = \frac{1}{3}\operatorname{tr} p(\boldsymbol{x}). \tag{7}$$

To prove the lemma it remains to verify that the remainder term in the formula (4) does not affect the value of the limit (5), but this is obvious. The Lemma is proved.

Let real-values functions $u(z)$ and $v(z)$ be defined in the domain G (are the $u(z)$ and $v(z)$ defined on the boundary of the domain isn't sufficient). Will say that at the boundary point $\zeta \in \partial G$ the function $v(z)$ majorizes function $u(z)$ if $\overline{\lim}\big(v(z) - u(z)\big) \geq 0$ when $z \to \zeta$, $z \in G$.

We will check the fulfillment of the following definition of subharmonicity.

Definition. A function $u(\boldsymbol{x})$ defined in a domain G is said to be *subharmonic*, if it is a) upper semicontinuous and not identically $-\infty$ in G and b) for each closed ball $D \subset G$ any harmonic function that majorizes the function $u(\boldsymbol{x})$ on the frontier ∂D also majorizes $u(\boldsymbol{x})$ inside D.

The proof of the theorem 2 follows the Privalov scheme [3], applied to prove the harmonicity of continuous functions, which satisfy the Laplace equation.

Proof of Theorem 2. Suppose that the theorem is not true and the function $u(\boldsymbol{x})$ is not subharmonic everywhere in G. Then there is a ball $D \subset G$ and harmonic inside of the ball D function $v(\boldsymbol{x})$, which majorizes the function $u(\boldsymbol{x})$ on ∂D, but does not majorizes $u(\boldsymbol{x})$ inside D. We set $w(\boldsymbol{x}) := u(\boldsymbol{x}) - v(\boldsymbol{x})$.

The function $w(\boldsymbol{x})$ is upper semicontinuous in D, so it reaches a maximum on D. Since at some interior points of D the function $w(\boldsymbol{x})$ is positive, this maximum is positive, and since on ∂D the function $w(\boldsymbol{x})$ does not exceed zero, the point of maximum $w(\boldsymbol{x})$ belongs to interior of D. Let $\max\limits_{D} w(\boldsymbol{x}) = w(\boldsymbol{x}^*)$.

Inside the ball D we denote as

$$q(\boldsymbol{x}) := w(\boldsymbol{x}) + \frac{w(\boldsymbol{x}^*)}{2\mathrm{diam}^2 D}\big((x_1{-}x_1^*)^2 + (x_2{-}x_2^*)^2 + (x_3{-}x_3^*)^2\big).$$

By definition, the function $q(\boldsymbol{x})$ is upper semicontinuous on the ball D, therefore, it reaches a maximum on D, and since $q(\boldsymbol{x}) \geq w(\boldsymbol{x})$, then $\max\limits_{D} q(\boldsymbol{x}) \geq \max\limits_{D} w(\boldsymbol{x}) = w(\boldsymbol{x}^*)$. Since the function $q(\boldsymbol{x})$ does not exceed $\dfrac{w(\boldsymbol{x}^*)}{2}$ on the frontier ∂D, then $q(\boldsymbol{x})$ takes the maximum value strictly inside the ball D. Let $\max\limits_{D} q(\boldsymbol{x}) = q(\boldsymbol{x}^0)$.

By the hypothesis of the Theorem, there exists a sequence of sets of nodes $\{A^n_{\boldsymbol{x}^0}\}$ lying inside the ball D that contracts to the point $\boldsymbol{x}^0$ for which $\big(\Delta^{(*)}u(\boldsymbol{x}^0, A^n_{\boldsymbol{x}^0})\big)^- \to 0$ as $n \to \infty$. Since the function $v(\boldsymbol{x})$ is harmonic in some neighborhood of the point $\boldsymbol{x}^0$, then, by virtue of the Lemma, for any sequence of sets of knots contracting to $\boldsymbol{x}^0$, including the sequence $\{A^n_{\boldsymbol{x}^0}\}$, is satisfied $\Delta^{(*)}v(\boldsymbol{x}^0, A^n_{\boldsymbol{x}^0}) \to 0$. By the definition $w(\boldsymbol{x}) = u(\boldsymbol{x}) - v(\boldsymbol{x})$, then $\Delta^{(*)}w(\boldsymbol{x}^0, A^n_{\boldsymbol{x}^0}) = \Delta^{(*)}u(\boldsymbol{x}^0, A^n_{\boldsymbol{x}^0}) - \Delta^{(*)}v(\boldsymbol{x}^0, A^n_{\boldsymbol{x}^0})$ and, therefore,

$$\lim_{n\to\infty}\Big(\Delta^{(*)}w(\boldsymbol{x}^0, A^n_{\boldsymbol{x}^0})\Big)^- = \lim_{n\to\infty}\Big(\Delta^{(*)}u(\boldsymbol{x}^0, A^n_{\boldsymbol{x}^0})\Big)^- = 0. \tag{8}$$

It follows from the equality (7) that for a polynomial of the second order $\dfrac{w(\boldsymbol{x}^*)}{2\mathrm{diam}^2 D}\big((x_1{-}x_1^*)^2 + (x_2{-}x_2^*)^2 + (x_3{-}x_3^*)^2\big)$ for each of the sets of nodes $A^n_{\boldsymbol{x}^0}$ the value of $\Delta^{(*)}$ is equal to the same positive number $\dfrac{w(\boldsymbol{x}^*)}{6\mathrm{diam}^2 D}$. From this and from (8) it follows that, starting with a sufficiently large number, the quantity $\Delta^{(*)} q(\boldsymbol{x}^0, A^n_{\boldsymbol{x}^0})$ is positive:

$$\Delta^{(*)} q(\boldsymbol{x}^0, A^n_{\boldsymbol{x}^0}) = \Delta^{(*)} w(\boldsymbol{x}^0, A^n_{\boldsymbol{x}^0}) + \frac{w(\boldsymbol{x}^*)}{6\mathrm{diam}^2 D} > 0.$$

Therefore, with such numbers, at least one term in the expression (2) for $\Delta^{(*)} q(\boldsymbol{x}^0, A^n_{\boldsymbol{x}^0})$ is positive. This contradicts the assumption that $\boldsymbol{x}^0$ is the maximum point of the function $q(\boldsymbol{x})$ in the ball D.

Therefore our assumption, that the function $u(\boldsymbol{x})$ is not subharmonic everywhere inside of the domain G is not true. The Theorem is proved.

References

[1] I I Privalov Subgarmonicheskie funkcii (Subharmonic functions). Moscow, Leningrad: ONTI Publ., 1937, 199 p. (in Russian).

[2] D S Telyakovskii A sufficient condition for the harmonicity of a function of two variables satisfying the Laplace difference equation. Proc. of IMM UB RAS, vol. 22, no 4, 2016, 269–283 (in Russian).

[3] I Priwaloff Sur les fonctions harmoniques. Mat. Sb., 1925, vol. 23, no. 3, pp. 464–471.

Computational model for fuel component supply into a combustion chamber of LRE

A V Teterev[1,2], P A Mandrik[1], L V Rudak[1] and N I Misyuchenko[1]

[1] Belarusian State University, Nezavisimosti Av., 4, 220030 Minsk, Belarus
[2] National Research Nuclear University "MEPhI", Kashirskoe Av., 31, 115409 Moscow, Russia

teterev@bsu.by

Abstract. A 2D-3D computational model for calculating a flow inside jet injectors that feed fuel components to a combustion chamber of a liquid rocket engine is described. The model is based on the gasdynamic calculation of compressible medium. Model software provides calculation of both one- and two-component injectors. Flow simulation in two-component injectors is realized using the scheme of separate supply of "gas-gas" or "gas-liquid" fuel components. An algorithm for converting a continuous liquid medium into a "cloud" of drops is described. Application areas of the developed model and the results of 2D simulation of injectors to obtain correction factors in the calculation formulas for fuel supply are discussed.

1. Introduction

Fuel is supplied to a mixing chamber by means of injectors. The main requirement for injectors is to provide, as much as possible, a finer and more uniform spray of fuel at a sufficiently small pressure drop on injectors [1, 2]. Usually, two main types of injectors are used in rocket engines: jet and centrifugal. The application of injectors of both types is possible. To cool combustion chamber walls, slit injectors are also used as a kind of jet injectors having a slot-like shape of the outlet. A jet injector is an outlet hole in the engine chamber head or a separate element shaped as a tube that communicates a fuel or oxidizer cavity with the inner volume of the combustion chamber. The main advantage of jet injectors is a large capacity of the head with jet injectors. In addition, the jet injector is smaller than the centrifugal one. This makes it possible to place more injectors on the unit surface of the head. As a result, with the same pressure drop, jet injectors can provide greater fuel component consumption per unit surface of the head, i.e., they have greater capacity. According to the aggregate state of supplied fuel components, there are liquid, gas and gas-liquid injectors; according to the number of injected fuel components they are divided into one-component and two-component injectors.

A uniform distribution of fuel components over the combustion chamber volume in the case of application of two-component injectors is provided by their design. In engines with one-component injectors, to ensure a good mixture formation, fuel and oxidizer injectors should uniformly alternate. In general, one-component injectors on the mixing head of the chamber are placed in concentric circles. In this case, either the belts of oxidizer and fuel injectors or the injectors of oxidant and fuel alternate in one concentric circle. Two-component injectors can be placed according to any scheme, but more often they are placed in concentric circles.

When designing a new engine, injectors must be calculated in order to ensure necessary fuel consumption. Numerical simulation of flow inside injectors and engine chambers is usually performed either in the project approximation [3-5], when analytical dependences and formulas are used, or in the gasdynamic approximation [6-8], when finite-difference methods are adopted to solve systems of partial differential equations.

2. The physical model of the fuel supply through jet injector

The developed model of injector calculation is intended for closed-type engines, when one or both fuel components pass through a gas generator. The scheme of organization of the working process in the combustion chamber of such engines is "gas-gas" or "gas-liquid". I.e., at least, one fuel component is in the gas phase. The mathematical model is a system of gasdynamic equations for compressible medium with given initial conditions and corresponding boundary conditions. These are the non-flow conditions on all rigid walls of an injector and the inflow conditions at the fuel tank inlet and the outflow conditions at the combustion chamber outlet. Input parameters for calculation are a selected type of fuel or oxidizer or both components, gasdynamic parameters and a mixture composition in the combustion chamber, fuel component consumption and injector geometries. The shape of the channel edge of a gas jet injector at its inlet can be sharp, radically rounded or with a chamfer. In addition, the injector can be cylindrical or cone-shaped.

When simulating the injectors supplying a fuel component into the liquid phase, the most complex flow occurs in the zone where its continuity undergoes discontinuity. In this zone, a lot of drops are born that seem to emerge from the continuous medium region and to continue moving in the form of discrete formations with definite physical characteristics. Based on the parameters of the continuous medium in the injector, the model describing this process should form a granulometric composition of drops, their thermodynamic parameters and should make a smooth transition of the flow description from the continuous medium approximation to the discrete one. A continuous medium of the fuel liquid component is transferred into a set of drops at a time when the medium density falls below a certain critical value. As the model for the interaction of drops and the surrounding gas flow, we used the Sand Bag model that was successfully used in describing the dynamics of condensed particles with a gas stream of combustion products in the Laval nozzle [9].

When constructing mathematical and computational models describing the flow of liquid or gaseous components of fuel into the combustion chamber of a liquid rocket engine, we will proceed from the following basic assumptions:

- Motion of liquid and gaseous media is described by non-stationary gasdynamic equations written for a multicomponent compressible medium.
- The transition of the liquid jet into a drop-like state is carried out depending on its parameters at the injector section or inside the combustion chamber.
- The interaction of the emerging droplet liquid component with the gaseous medium in the combustion chamber is described by a non-stationary system of ordinary differential equations.
- The destruction of a liquid jet inside the combustion chamber is modeled using the concentration method, which makes it possible to describe contact boundaries in multiply connected regions.
- Maintaining the initially set pressure drop on the injector and the pressures in the fuel tank and combustion chamber is accomplished by setting the appropriate boundary conditions

3. Mathematical model of motion of continuous media in injectors

The mathematical model constructed on these assumptions is a nonstationary system of partial differential equations describing the gas-dynamic flow of a multicomponent multiphase compressible medium. The conservative form of the non-stationary Euler equations in the cylindrical coordinate system (r, φ, z) can be written as follows:

$$\frac{\partial[r\mathbf{U}]}{\partial t}+\frac{\partial[r\mathbf{F}(\mathbf{U})]}{\partial r}+\frac{1}{r}\frac{\partial[r\mathbf{G}(\mathbf{U})]}{\partial \varphi}+\frac{\partial[r\mathbf{H}(\mathbf{U})]}{\partial z}=\mathbf{S}(\mathbf{U}) \tag{1}$$

Here the conservative state vector **U** is defined as

$$\mathbf{U}=(\rho,\rho v_r,\rho v_\varphi,\rho v_z,E) \tag{2}$$

The flux vectors **F(U)**, **G(U)** and **H(U)** and the source term **S(U)** are

$$\mathbf{F}(\mathbf{U})=\begin{pmatrix}\rho v_r\\ \rho v_r^2+p\\ \rho v_r v_\varphi\\ \rho v_r v_z\\ [E+p]v_r\end{pmatrix},\quad \mathbf{G}(\mathbf{U})=\begin{pmatrix}\rho v_\varphi\\ \rho v_r v_\varphi\\ \rho v_\varphi^2+p\\ \rho v_\varphi v_z\\ [E+p]v_\varphi\end{pmatrix},\quad \mathbf{H}(\mathbf{U})=\begin{pmatrix}\rho v_z\\ \rho v_r v_z\\ \rho v_\varphi v_z\\ \rho v_z^2+p\\ [E+p]v_z\end{pmatrix} \tag{3}$$

$$\mathbf{S}(\mathbf{U})=(0,p+\rho v_\varphi^2,-\rho v_\varphi v_r,0,0) \tag{4}$$

To close system (1)-(4), we use the caloric and thermal equations of state, as well as the expression for the total density of a medium in terms of the density of its components:

$$p=p(\rho,\varepsilon)\quad T=T(\rho,\varepsilon),\quad \rho=\sum_i \rho_i \tag{5}$$

The mathematical model of the interaction of the droplet fraction of the fuel components with the gaseous medium with respect to flows in the combustion chamber and Laval nozzle is described in [9], constructed on the basis of the model set forth in [10].

In addition, for multiphase regions condition

$$p_g=p_l \tag{6}$$

where p_g and p_l there are gas and liquid pressures, respectively. An iterative process ensuring that condition (6) is satisfied allows us to correct the volume concentrations of the liquid and gas components of the medium in the calculated cell.

In the case of the two-dimensional version of the model, the same system of equations in the cylindrical coordinate system (r, z) is used.

4. Statement of boundary conditions

The task statement is completed by setting the initial conditions and the corresponding boundary conditions. These are the non-flow conditions on all rigid walls of an injector and the inflow conditions at the fuel tank inlet and the outflow conditions at the combustion chamber outlet. Input parameters for calculation are a selected type of fuel or oxidizer or both components, gasdynamic parameters and a mixture composition in the combustion chamber, fuel component consumption and injector geometries. The shape of the channel edge of the gas jet injector at the inlet can be sharp, radically rounded or with a chamfer. In addition, the injector can be cylindrical or conical in shape.

The boundary conditions in subsonic flow tasks play an important role, since they are a constant source of flow disturbances in the computational domain. In order to maintain a constant pressure drop between the fuel tank and the combustion chamber, different boundary conditions were tested at the tank inlet and at the chamber outlet. In calculations, the following boundary conditions can be set:

inflow at a given mass flow rate

$$\dot{m}=\text{const},\ \rho_{\text{bnd}}=\rho_N+\Delta\rho,\ v_{\text{bnd}}=v_N+\Delta v,\ E_{\text{bnd}}=E_N+\Delta E \tag{7}$$

outflow at a given pressure

$$p_{\text{bnd}}=(p_N+p_{\text{ing}})\times 0.5,\ v_{\text{bnd}}=v_N-3v_{N-1} \tag{8}$$

outflow with a given hole

$$r_{\text{hole}} = (2 \div 3) r_{\text{inj}} \tag{9}$$

outflow with a semi-transparent hole

$$r_{\text{hole}} = (2 \div 3) r_{\text{inj}}, \; \dot{m}_{\text{out}} = \dot{m}_{\text{inj}} \tag{10}$$

At conditions (7) and (8), the index bnd means the boundary between the calculated cell N and the cell N+1. At all formulas the subscript inj means the values in the injector. The application of conditions (7) usually changes the fuel tank pressure and condition (9) as a rule decreases the combustion chamber pressure. The injector pressure drop can be most accurately maintained under boundary conditions of form (8) and (10).

5. Simulation results

To calculate the flow velocity of fuel liquid components from jet injectors, we use the theoretical flow velocity of incompressible liquid from the hole, which is determined by the formula [1]:

$$w = \left(\frac{2\Delta p_{\text{inj}}}{\rho_{\text{liq}}} \right)^{1/2} \tag{11}$$

where $\Delta p_{\text{inj}} = (p_{\text{inj}} - p_{\text{ch}})$ is the pressure drop at the injector, ρ_{liq} is the liquid density. In this case, the liquid flow rate through the injector with the hole cross-sectional area F_c is given by equation

$$\dot{m}_{\text{inj}} = \mu_{\text{inj}} F_c \sqrt{2\Delta p_{\text{inj}} \rho_{\text{liq}}} \tag{12}$$

where μ_{inj} is the flow rate coefficient taking into account the jet constriction and the flow velocity decrease in comparison with the theoretical one.

The quality of the factory produced injectors is checked by hydraulic tests, conducting the so-called "spillage" with water [2]. Figure 1 shows the dynamics of water consumption through a jet injector with a diameter of 5 mm at a pressure of 250 atm in the combustion chamber and at a pressure drop of 10 atm. It also illustrates the graph fragment, according to which the frequency of consumption oscillations is determined to be approximately 40 kHz, which corresponds to the frequency of pressure pulsations on the reservoir symmetry axis that come from the boundary, at which the fuel is supplied. The dash-dotted line corresponds to the water consumption calculated from equation (12) at μ=1.

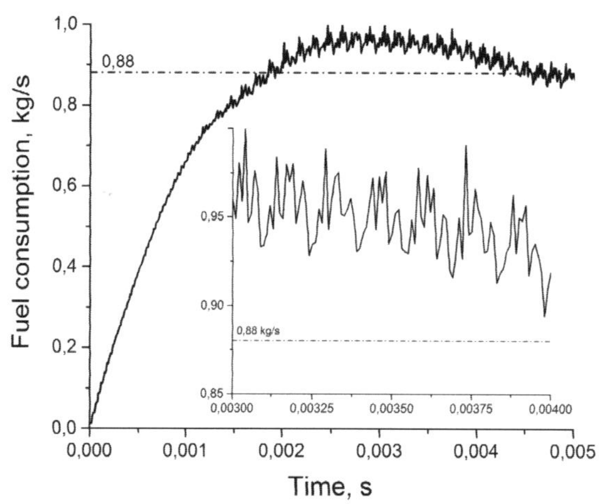

Figure 1. Dynamics of water consumption through the injector.

To identify the factors that affect the process of drop formation, an illustrative calculation was performed by the model with a contact boundary. A jet of liquid (water) flowing from the jet injector into the region of the cold compressed gas (air) was simulated. The initial gas parameters were:

temperature – 298 K, density – 292 kg/m^3, pressure – 250 atm. The pressure drop across the injector was $\Delta p = 35$ atm and its diameter was 0.5 mm. The flow dynamics can be seen from the fragments of density fields depicted in Figure 2 for three time moments. The time moment $t = 1.3 \cdot 10^{-4}$ s in Figure 2(a) corresponds to the beginning of the process of destruction of the already formed "large" drop, which is an expanded front part of the jet that propagates into the gas reservoir. By this time moment, the jet penetrates to a distance of ~ 3 mm from the injector section. Destruction begins at the back surface of the drop directly near the cylindrical jet surface. By the time $t = 1.73 \cdot 10^{-4}$ s in Figure 2(b), the jet penetrates to a distance of ~ 6 mm into the reservoir, and a "crack" spreads approximately to the half the length of the drop. It can be seen that the stern part of the separating part of the drop continues in turn to disintegrate, thus disintegrating into smaller formations spraying in the ambient air. By the time moment $t = 2.41 \cdot 10^{-4}$ s in Figure 2(c), when the jet penetrates to a depth of 7 mm, the side part of the drop is almost completely separated from the central part of the drop representing the initial jet.

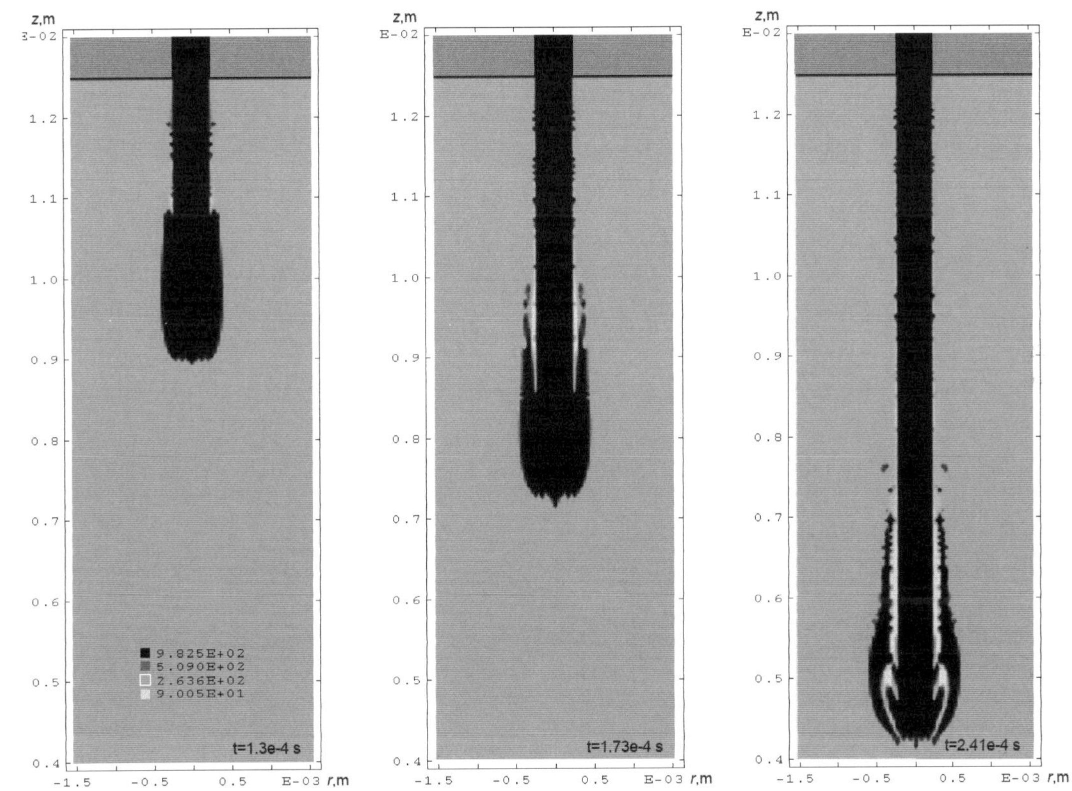

(a) (b) (c)

Figure 2. Fragments of the density fields at different time moments.

The described drop destruction process can be explained on gasdynamic grounds and, apparently, it is primarily associated with a vortex flow arising in the immediate vicinity of the side peripheral region of an elongated drop. The vortex dynamics on the side surface of the jet can be seen in the velocity vector fields. Developed vortex flow, arising in the lower part of the jet, intensively destroys a water flow. Similar destruction was observed in simulating the destruction in dense layers of the atmosphere of water formations that occur when small comets fall [11]

To calculate the rate of gas-generating gases through a jet injector, the expression [1] is used:

IOP Conf. Series: Journal of Physics: Conf. Series **937** (2017) 012054 doi:10.1088/1742-6596/937/1/012054

$$w=\left(\frac{2\gamma}{\gamma-1}RT_{\text{in}}\left(1-\left(\frac{p_{\text{out}}}{p_{\text{in}}}\right)^{\frac{\gamma-1}{\gamma}}\right)\right)^{1/2} \tag{13}$$

where R, $T_{\text{вх}}$ is the gas constant and the gas temperature in front of the injector respectively and the flow rate is given by formula

$$\dot{m}_{\text{inj}}=\mu_{\text{inj}}wF_{c}\rho_{\text{out}} \tag{14}$$

In Figure 3, the lines represent the velocities calculated by (13) as a function of pressure drop for two gas temperatures and two pressures in the combustion chamber.

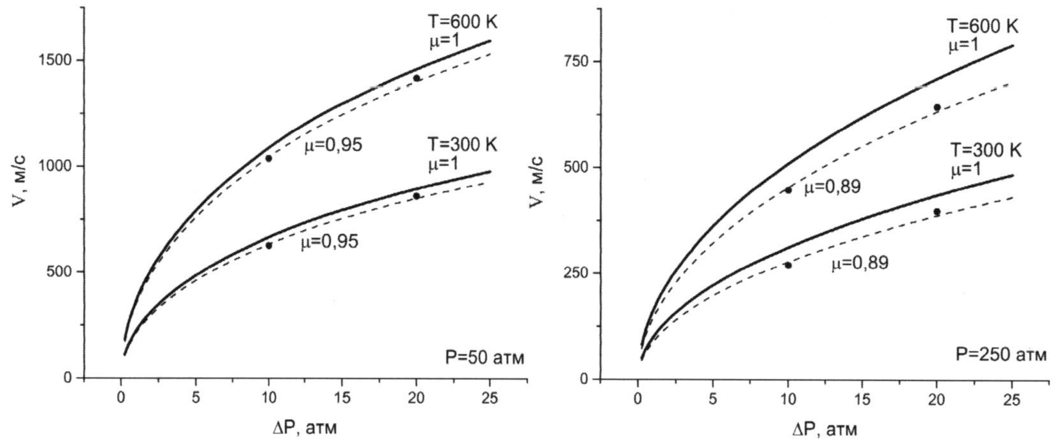

Figure 3. Hydrogen velocity at the injector outlet.

Here, the points stand for the numerical simulation results of hydrogen supply through an injector with a diameter of 5 mm and a length of 20 mm. The dotted lines with a corresponding flow rate coefficient are drawn through the points.

Figure 4 (left) shows the time dependences of the oxygen velocity at the injector outlet (left) and of the mass flow rate (right). The straight lines correspond to the mass flow rate calculated according to equation (13) with the corresponding coefficients.

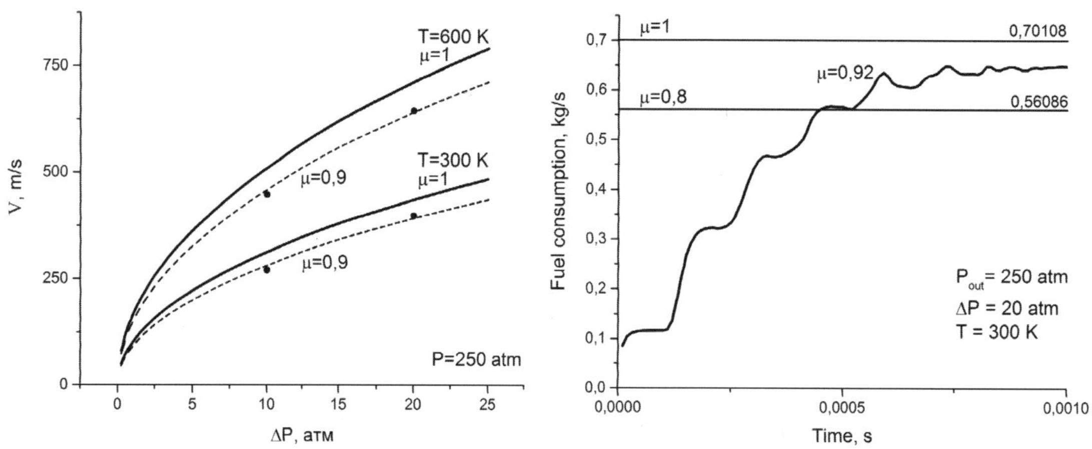

Figure 4. Oxygen velocity and flow rate at the injector outlet.

The oxygen velocity growth with increase in the pressure difference continues until it reaches the sound speed. At this moment, the "reservoir" flow is "locked", and a further increase in the difference does not affect the fuel consumption through the injector [12]. Such flow is shown in Figure 5, where the temperature fields are shown on the left and the density fields – on the right. They are displayed on the monitor screen during the calculation. In this version of calculation, the oxygen pressure in the tank was 250 atm, and the pressure in the combustion chamber was about 10 atm; the sound speed and the limiting flow rate occurred even at a combustion chamber pressure of 100 atm. But even at a pressure of 150 atm, the flow rate was almost maximum, although the outflow velocity was far from the sound speed. It should be noted that at a pressure of 50 atm and more, a weakly expanding jet outflows from the injector, and only with a further pressure decrease its shape is similar to the shape underexpanded jet from the Laval nozzle. A sharp expansion of the jet at the combustion chamber inlet causes strong cooling, and as a result of acceleration, the temperature reaches 500 K at an initial temperature of 300 K. The depicted flows can appear at full-flow start.

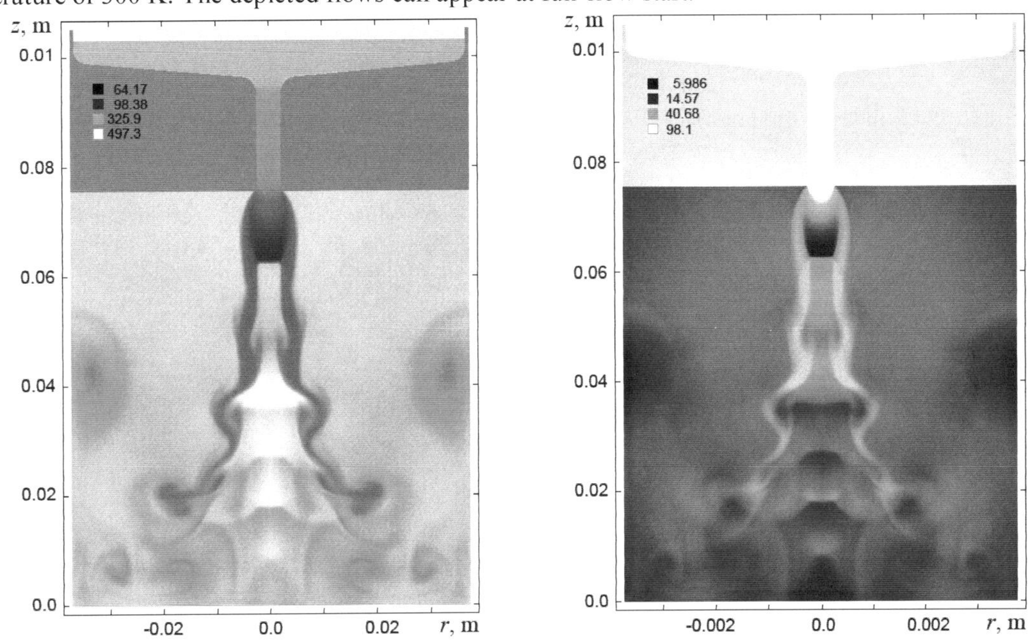

Figure 5. Temperature and density fields at a large pressure drop on the injector.

6. Conclusion

The developed model of flow calculation in injectors is designed to solve different simulation problems. First, it can be used in designing injectors for parametric calculations. This will allow one to more accurately determine sizes and parameters of designed injectors in comparison with analytical approximation calculations. In addition, this simulation makes it possible to investigate the flow features of fuel components within the unsteady flow regime and how it is influenced by gas parameter pulsations in the combustion chamber. Second, based on multiparametric calculations, it is possible to tabulate correction factors for the analytical expressions, in terms of which the fuel supply into the combustion chamber is calculated. This in turn will improve the simulation accuracy of a rocket engine. Third, with the appropriate software modification, this model can be directly used in simulating the flow in the combustion chamber. The essence of such a modification is to calculate injectors and combustion chamber on various three-dimensional grids. In this case, the calculation results of injectors will be the boundary conditions for fuel component inflow into the combustion chamber; the calculation results of the combustion chamber will in turn be the boundary conditions for calculation of flow in injectors. Such a calculation scheme will allow revealing the influence of fuel supply pulsations in the combustion chamber on the low-frequency instability of combustion.

Currently, the developed software is designed for flow simulations in jet injectors, the results of which will allow one to determine correction factors in the analytical expressions for fuel supply. As parameters, we use fuel consumption, thermodynamic parameters of fuel and oxidizer, as well as gas parameters in the combustion chamber.

References

[1] Dobrowolski M 2005 *Liquid rocket engine* (Moscow) p 488
[2] Barrere M A, Jaumotte B, Fracijs de Veubeke and Vandenkerckhove J 1960 *Rocket Propulsion* (Amsterdam, London, New York and Princeton)
[3] Egorychev V S 2011 *Calculation and design of a mixture in a liquid rocket engine* (Samara: Izd. SGAU) p 108
[4] Egorychev V S 2009 Method of design calculation of centrifugal emulsive duplex fuel nozzle *Bulletin of the Samara state aerospace university* **19** 248-52
[5] Egorychev V S 2011 Method of design calculation jet-centrifugal gas-liquid duplex fuel nozzle with internal mixing *Bulletin of the Samara state aerospace university* **27** 199-204
[6] Salich V L 2011 The numerical modelling of processes in combustion chambers of low thrust rocket engines *Bulletin of the Samara state aerospace university* **27** 120-5
[7] Matyunin O O, Bachev N L, Bulbovich R V 2015 Numerical simulation of the flow in two-component fuel injector *Bulletin of PNIPY Aerospace Engineering* **43** 19-34
[8] Strokach E A, Borovik I N 2016 Numerical simulation of the process of kerosene spraying with a centrifugal injector *Bulletin of MSTU. Ser. "Mechanical Engineering"* **3** 37-54
[9] Teterev A V, Mandrik P A, Rudak L V and Misuchenko N I (2017) Effect of the condensed particles on a flow of combustion products in a Laval nozzle *IOP Conf. Series: Journal of Physics: Conf. Series* **788** 012034
[10] Teterev A V (1999) Cratering model of asteroid and comet impact on a planetary surface *Int. J. Impact Eng.* **23** 921-932
[11] Teterev A V, Mandrik P A, Rudak L V 2009 The passage of large meteoroids through planetary atmospheres: theory and numerical simulation *Proc. XVII Int. Scientific and Technical Conference "Applied problems of mathematics and mechanics"* (Sevastopol)
[12] Loitcyanskii L G 1987 Mechanics of liquid and gas (Moskow, Nauka) p 840

Analytical properties of solutions to a class of third-order nonlinear dynamical dissipative systems with no chaotic behaviour

V Tsegel'nik

Belarusian State University of Informatics and Radioelectronics, P. Brovki Str., 6, Minsk, 220013, Belarus

E-mail: tsegvv@bsuir.by

Abstract. The singular analysis was performed for solutions of one class of third-order nonlinear dynamical systems with no chaotic behaviour. It is established that all systems of the given class (except one) are not the systems of Painlevé-type.

1. Introduction

One of the important events in classical nonlinear physics of mid-twentieth century was the realization that nonlinear deterministic equations (systems) could have chaotic solutions that exhibited both a sensitive dependence on initial conditions and long-term unpredictability. An interesting and remaining to be solved is the problem of identifying minimum necessary conditions for chaos. In [1], computer simulation was used to obtain 19 third-order dynamic systems with complex chaotic behavior, which are algebraically simpler than the known Lorentz and Rössler ones. The distinctive difference of the above Sprott systems is that their right-hand parts contain 6 components with one quadratic nonlinearity or 5 components with two quadratic nonlinearities. In [2], the class (including four families) of third-order nonlinear dynamical dissipative systems with no chaotic behaviour was found

$$\dot{x} = y^2 - x, \quad \dot{y} = z, \quad \dot{z} = x. \tag{1.1}$$

$$\dot{x} = y^2 + z, \quad \dot{y} = x, \quad \dot{z} = -z. \tag{1.2}$$

$$\dot{x} = yz - x, \quad \dot{y} = x, \quad \dot{z} = y. \tag{1.3}$$

$$\dot{x} = y^2, \quad \dot{y} = x + z, \quad \dot{z} = -z. \tag{1.4}$$

$$\dot{x} = y^2, \quad \dot{y} = z - y, \quad \dot{z} = x. \tag{1.5}$$

$$\dot{x} = y^2, \quad \dot{y} = z, \quad \dot{z} = x - z. \tag{1.6}$$

$$\dot{x} = yz, \quad \dot{y} = x, \quad \dot{z} = x - z. \tag{1.7}$$

$$\dot{x} = yz, \quad \dot{y} = x, \quad \dot{z} = y - z. \tag{1.8}$$

$$\dot{x} = y^2 + yz, \quad \dot{y} = x, \quad \dot{z} = -z. \tag{2.1}$$

$$\dot{x} = y^2 + z^2, \quad \dot{y} = x, \quad \dot{z} = -z. \tag{2.2}$$

$$\dot{x} = y^2 - x, \quad \dot{y} = xz, \quad \dot{z} = \varepsilon y. \tag{2.3}$$

$$\dot{x} = y^2 - x, \quad \dot{y} = xz, \quad \dot{z} = kz. \tag{2.4}$$

$$\dot{x} = y^2 - x, \quad \dot{y} = z^2, \quad \dot{z} = x. \tag{2.5}$$

$$\dot{x} = y^2 + y, \quad \dot{y} = xz, \quad \dot{z} = -z. \tag{2.6}$$

$$\dot{x} = y^2 + z, \quad \dot{y} = x^2, \quad \dot{z} = -z. \tag{2.7}$$

$$\dot{x} = y^2 + z, \quad \dot{y} = xz, \quad \dot{z} = -z. \tag{2.8}$$

$$\dot{x} = yz - x, \quad \dot{y} = x^2, \quad \dot{z} = \varepsilon x. \tag{2.9}$$

$$\dot{x} = yz - x, \quad \dot{y} = x^2, \quad \dot{z} = y. \tag{2.10}$$

$$\dot{x} = yz - x, \quad \dot{y} = x^2, \quad \dot{z} = kz. \tag{2.11}$$

$$\dot{x} = yz - x, \quad \dot{y} = \varepsilon xz, \quad \dot{z} = y. \tag{2.12}$$

$$\dot{x} = yz - x, \quad \dot{y} = z^2, \quad \dot{z} = \varepsilon x. \tag{2.13}$$

$$\dot{x} = \varepsilon y - x, \quad \dot{y} = xz, \quad \dot{z} = x^2. \tag{2.14}$$

$$\dot{x} = \varepsilon y - x, \quad \dot{y} = xz, \quad \dot{z} = y^2. \tag{2.15}$$

$$\dot{x} = y - x, \quad \dot{y} = z^2, \quad \dot{z} = x^2. \tag{2.16}$$

$$\dot{x} = y - x, \quad \dot{y} = z^2, \quad \dot{z} = xy. \tag{2.17}$$

$$\dot{x} = x^2 + yz, \quad \dot{y} = -2xy, \quad \dot{z} = -z. \tag{3.1}$$

$$\dot{x} = y^2 + yz, \quad \dot{y} = x^2, \quad \dot{z} = -z. \tag{3.2}$$

$$\dot{x} = y^2 + yz, \quad \dot{y} = \varepsilon xz, \quad \dot{z} = -z. \tag{3.3}$$

$$\dot{x} = y^2 + \varepsilon z^2, \quad \dot{y} = x^2, \quad \dot{z} = -z. \tag{3.4}$$

$$\dot{x} = y^2 + \varepsilon z^2, \quad \dot{y} = xz, \quad \dot{z} = -z. \tag{3.5}$$

$$\dot{x} = xy - x, \quad \dot{y} = xz, \quad \dot{z} = -yz. \tag{3.6}$$

$$\dot{x} = y^2 - x, \quad \dot{y} = xz, \quad \dot{z} = x^2. \tag{3.7}$$

$$\dot{x} = y^2 - x, \quad \dot{y} = xz, \quad \dot{z} = y^2. \tag{3.8}$$

$$\dot{x} = y^2 - x, \quad \dot{y} = z^2, \quad \dot{z} = x^2. \tag{3.9}$$

$$\dot{x} = y^2 - x, \quad \dot{y} = z^2, \quad \dot{z} = xy. \tag{3.10}$$

$$\dot{x} = yz - x, \quad \dot{y} = x^2, \quad \dot{z} = \varepsilon xy. \tag{3.11}$$

$$\dot{x} = yz - x, \quad \dot{y} = x^2, \quad \dot{z} = y^2. \tag{3.12}$$

$$\dot{x} = yz - x, \quad \dot{y} = \varepsilon xz, \quad \dot{z} = y^2. \tag{3.13}$$

$$\dot{x} = y^2 + yz, \quad \dot{y} = -y, \quad \dot{z} = x^2. \tag{3.14}$$

$$\dot{x} = \varepsilon + y^2, \quad \dot{y} = xz, \quad \dot{z} = -z. \tag{4.1}$$

$$\dot{x} = 1 + yz, \quad \dot{y} = x^2, \quad \dot{z} = -z. \tag{4.2}$$

$$\dot{x} = 1 - x, \quad \dot{y} = xz, \quad \dot{z} = y^2. \tag{4.3}$$

$$\dot{x} = y^2 - x, \quad \dot{y} = xz, \quad \dot{z} = 1. \tag{4.4}$$

$$\dot{x} = yz - x, \quad \dot{y} = x^2, \quad \dot{z} = 1 \tag{4.5}$$

with unknown functions x, y, z of the independent variable t, where k is a parameter ($k < 1$) and $\varepsilon^2 = 1$.

Each of the above systems is, as noted, dissipative and contains four components in the right-hand part. Right-hand parts of the systems (1.1)–(1.8), (2.1)–(2.17), (3.1)–(3.14) contain one, two, three quadratic nonlinearities correspondingly. Right-hand parts of the systems (4.1) – (4.5) contain one constant and two quadratic nonlinearities.

Assuming the independent variable t to be complex, let us determine whether the general solution of the above systems has no moving critical singular points, that is, whether the so called Painlevé property is fulfilled for them. Systems (equations) with the Painlevé property are called systems (equations) of the Painlevé-type or P-type.

1. The system (1.1) is equivalent to equation

$$\dddot{y} = y^2 - \ddot{y}. \tag{1}$$

The system (1.2) is equivalent to equation

$$\ddot{y} = y^2 + Ce^{-t}, \tag{2}$$

where C is an arbitrary constant.

The system (1.3) has a first integral $x + y - \frac{z^2}{2} = H$ (H is an arbitrary constant) and so is equivalent to equation

$$\ddot{z} + \dot{z} - \frac{z^2}{2} = H. \tag{3}$$

The system (1.4) also is equivalent to equation (2). Each of the systems (1.5), (1.6) is equivalent to equation (1). The system (1.8) is equivalent to equation

$$\dddot{z} + \ddot{z} - z\dot{z} - z^2 = 0. \tag{4}$$

Note, that the equations (2), (3) are not included in the list of equations in [3], general solutions of which have no movable critical singular points. The equation (4) is not included in [4] in the list of equations $\dddot{u} = P(t, u, \dot{u}, \ddot{u})$, where P is a polynomial in $u, \dot{u}, \ddot{u}$ with analytic in t coefficients, the general solutions are free from movable critical singular points. Because of this the system (1.8) is not the system of Painlevé-type.

The system (1.7) is equivalent to equation

$$y\dddot{y} + y\ddot{y} - \dot{y}\ddot{y} - \dot{y}\ddot{y} - y^2\dot{y} = 0. \tag{5}$$

To clarify the question of whether the system (1.7) is the system of Painlevé-type we use the Painlevé test. The formal Painlevé test refers to any algorithm, which verifies the fulfillment of the conditions required for the existence in differential equation (system) the Painlevé property. The system (1.7) study was conducted according to the same scheme as in [5].

Theorem 1 Systems (1.2), (1.4) are not the systems of P-type. However, componet z of the data systems does not have movable critical singular points at all.

Theorem 2 Systems (1.1), (1.3), (1.5), (1.6), (1.8) are not the systems of P-type.

Theorem 3 The system (1.7) is not the system of P-type. The validity of this assertion follows from the fact that for the system (1.7) is not fulfilled the condition of the 3-rd step in Painlevé test, namely: the system for determining constant solutions of the system (1.7) in formal Laurent expansions is not joint.

Corollary 1. Equation (5) is not a Painlevé-type equation.

2. Systems (2.1), (2.2), (2.4), (2.6)–(2.8), (2.11) due to the fact that the unknown function z is explicit (and does not have movable critical singular points) are reduced to nonlinear nonautonomous equations of the second order, the solutions of which according to [3] do not possess the Painlevé property. Thus, the following is true

Theorem 4 None of the systems (2.1), (2.2), (2.4), (2.6)—(2.8), (2.11) is of the Painlevé-type. At the same time, a component z of these systems has no movable critical singular points at all.

Theorem 5 None of the systems (2.5), (2.10), (2.16), (2.17) is of the Painlevé-type.

The proof of this theorem follows from the fact that each of these systems fails the first step of Painlevé test.

Systems (2.3), (2.9), (2.12), (2.15) are equivalent to equations

$$z\dddot{z} = \dot{z}\ddot{z} - z\ddot{z} + \varepsilon\dot{z}^2 z^2, \tag{6}$$

$$z\dddot{z} = \dot{z}\ddot{z} - z\ddot{z} + \varepsilon\dot{z}^2 z^2 + \dot{z}^2, \tag{7}$$

$$z\dddot{z} = \dot{z}\ddot{z} - z\ddot{z} + \varepsilon\dot{z}z^3, \tag{8}$$

$$x\dddot{x} = \dot{x}\ddot{x} - x\ddot{x} + \varepsilon x^2(\dot{x} + x)^2 + \dot{x}^2. \tag{9}$$

The system (2.13) and the system (2.14) are equiavalent to the equation

$$w\dddot{w} = \dot{w}\ddot{w} - w\ddot{w} + \dot{w}^2 + \varepsilon w^4 \tag{10}$$

from z and x correspondingly.

Theorem 6 None of the systems (2.3), (2.9), (2.12)–(2.15) is of the Painlevé-type. Applying to systems (2.3), (2.9), (2.13)–(2.15) the Painlevé-test shows, that for them the condition of the first step of the given test is not fulfilled. As for the system (2.12), for it the condition of the third step of the test is not fulfilled: the system for determining constant solutions of the system (2.12) in formal Laurent expansions is not joint.

Corollary 2. None of the equations (6) – (10) is of the Painlevé-type.

3. Analysis of the systems (3.1), (3.3), (3.5) shows, that each of them admits a reduction of order and they are reduced to nonlinear nonautonomous equations of the second order, which are not included in the list in [3] of equations of P-type. Consequently, the following is true

Theorem 7 None of the systems (3.1), (3.3), (3.5) is of the Painlevé-type. At the same time, a component z of these systems has no movable critical singular points at all.

Each of the systems (3.2), (3.4), (3.14) admits a reduction of order. In this connection the following is true

Theorem 8 None of the systems (3.2), (3.4), (3.14) is of the Painlevé-type. At the same time, one of the components (z for (3.2), (3.4)) and y for (3.14) has no movable critical singular points at all.

Theorem 9 The system (3.6) is of the Painlevé-type. Its general solution has the form

$$x = \frac{\dot{y}}{z},\ z = c_3 e^{c_1 t + c_2 e^{-t}},\ y = c_1 + c_2 e^{-t},$$

where c_1, c_2, c_3 are arbitrary constants.

Indeed, the system (3.6) $\dot{x} = xy - x, \dot{y} = xz, \dot{z} = -yz$ has the first integral $xz + y = c$, which allows to reduce it to the linear equation $\dot{y} + y = c$ and find the explicit form of y.

For systems (3.7), (3.9), (3.10), (3.12) the condition of the second step of Painlevé test is not fulfilled: the resonance equations have two complex conjugate roots with positive real part.

Systems (3.8), (3.11), (3.13) do not pass the third step of the Painlevé test:

1. The system for determining constant solutions of the system (3.8) in formal Laurent expansions is not joint.

2. Laurent expansions of the solutions of the system (3.11) contain 2 (instead 3) arbitrary constants. In this connection the resonance equation has a positive root of multiplicity 2. In this case the system (3.11) has the first integral

$$\frac{x^2}{2} + \frac{\varepsilon z^2}{2} + y = K,$$

where K is an arbitrary constant.

3. Laurent expansions of the solutions of the system (3.13) contain 2 (instead 3) arbitrary constants. The resonance equation has the same as in the case of the system (3.11) positive root of multiplicity 2.

Theorem 10 None of the systems (3.8), (3.11), (3.13) is of the Painlevé-type.

Systems (4.1) – (4.5) are reduced to two-dimensional non-autonomous systems, each of which does not pass the Painlevé test. Thus, the following is true

Theorem 11 None of the systems (4.1) – (4.5) is of the Painlevé-type. At the same time, one of the components (z for (4.1), (4.2), (4.4), (4.5) and x for (4.3)) has no movable critical singular points at all.

References

[1] Sprott J C 1994 Some simple chaotic flows *Phys. Rev. E.* **50**. pp 647-650.
[2] Heidel J and Zhang Fu 1999 Nonchaotic behaviour in three-dimensional quadratic systems *Nonlinearity* **10**. pp 1289-1303.
[3] Ince I L 1956 *Ordinary differential equations* (New York: Dover)
[4] Cosgrove C M 2000 Higher-order Painlevé equation in the polynomial class. I *Stud. Appl. Math.* **104**. no. 1. pp 1-65.
[5] Tsegelnik V V 2015 Analytical properties of solutions to autonomous systems of nonlinear third and fourth-order differential equations with chaotic behaviour *Vestnik NIYaU MIFI* **4**. no. 2. pp 101-106 (in Russian).

Reduced dimension model for heat transfer of ground heat exchanger in permafrost

M Vasilyeva[1], S Stepanov[1,2] and I Sirditov[1]

[1] North-Eastern Federal University, Yakutsk, Russia
[2] Melnikov Permafrost Institute SB RAS, Yakutsk, Russia

E-mail: sirditov@gmail.com

Abstract. In this work, we present reduced dimensional model for heat transfer processes of ground heat exchanger in permafrost. A mathematical model is described by a coupled system of equations for heat transfer in the ground subdomain and in heat exchanger (pipes). Because radius of the pipes is very small compared to the size of surrounding ground, we write heat transfer problem in pipes as reduced dimensional equation, where we consider pipes as one-dimensional lines. We present a computational algorithm and numerical results for model problem.

1. Introduction

For stabilization of foundations the methods of ground freezing are often used. Freezing of ground can be made by using of special installations-refrigerators or by seasonal cooling installations. They can significantly influence on temperature of ground and, consequently, their thermomechanical characteristics.

Seasonal cooling systems are installed for the purpose of forming a mass of frozen soil under building foundations and engineering constructions to prevent their thawing in summer-autumn period. In addition, there is another category of widely–spread piping systems — heat pumps. They are used for pumping heat out of ground to heat buildings (geothermal heat pump) with ability to switch from heating mode in winter period to condensing mode in summer period. Technical device in which heat exchange between two media having different temperatures is called a heat exchanger. Their mathematical models are also described by ground–pipe systems.

Ground–pipe systems use a temperature difference between the ground and the buried pipes, for heating and cooling buildings, for formation of a frozen rock solid that increase thermal mechanical characteristics and are applied to increase a stability of foundations under permafrost conditions. A coolant is used as a heat carrier in pipes. The coolant is a gas with a low boiling (evaporation) point. It remains in a gaseous state under normal conditions and becomes a liquid under high pressure.

A mathematical model for such processes consists of heat and mass transfer equation and considers a phase transition of porous moisture to ice. For such models, there are two approaches for phase transition. The first is when the boundary of the phase transition is fixed at the grid level and a calculation of heat transfer can occur in the region occupied by thawed zone, and the second approach is based on use of the two–phase Stefan model. The second approach is more appropriate. In a two–phase model a boundary of the phase transition is represented by the transition zone [1, 2, 3, 4].

In the computational implementation of models of heat transfer in permafrost, it is necessary to construct geometric domains and computational grids in such a way that the geometric domain of pipes is resolved by mesh. Let us note that computational grids for planning a construction and a pipe system location must be unstructured. The computational grids for such regions will have areas of local condensation in pipe areas and will lead to very large systems of equations at discretization. The alternative approach proposed in this paper is based on reducing of model dimensionality on the pipe and reducing it to a one-dimensional hydraulic model, and further under assumption of the continuity of temperature field, we will approximate it using the methods similar to those used in modeling of fractions in oil and gas layers, or to be precise, a method of discrete model of fraction [5, 6, 7, 8, 9]. The obtained discrete problem will contain a system of linear equations for each time layer for a temperature in each node of a computational grid. The system of equations obtained in this way for finding a temperature distribution can be complicated by adding a convective heat transfer (heat and mass transfer). To calculate the filtration velocity in the ground the authors use the Darcy's law and an equation of mass conservation for the case of incompressible fluid [10, 11]. To count a phase transition, i.e. an appearance of a frozen ground zone in the areas of pipes freezing and a surface of soil, where the ambient temperature is set by the Robin boundary condition, an ambient air temperature is periodic and well described by sinusoid. For numerical solution of the problem of pore water filtration in the ground, a method of fictitious regions in frozen ground is used which was studied by the authors in [12].

In this paper we constructed a new mathematical model and a computational algorithm for solving problems in ground–pipe systems. The work consists of three parts. In the first part the authors present a mathematical model for heat transfer in the ground taking into account the phase transition. In the second part we derive a reduced-dimensional mathematical model for describing a temperature distribution in the pipe. In the third part, an associated model and a computational algorithm for heat and mass transfer problem for ground–pipe system is given, taking into account a presence of filtration in ground. At the end the authors provide calculations of a model problem using proposed model and a computational algorithm.

2. Heat transfer in the ground

We consider a mathematical model describing the heat transfer processes in the frozen and thawed soils. This model takes into account the presence of phase transitions of pore water in the ground at a given temperature phase transition T^* in domain $\Omega = \Omega^- \cup \Omega^+$:

$$\Omega^+(t) = \{x | x \in \Omega, \quad T(x,t) > T^*\}, \quad \Omega^-(t) = \{x | x \in \Omega, \quad T(x,t) < T^*\}.$$

Here Ω^+ is a domain occupied by the liquid phase, where the temperature is higher than the phase transition temperature and Ω^- is a domain occupied by the solid phase, phase transition occurs at the interface S.

For the simulation of heat transfer processes with phase transitions, we use the Stefan model [1, 2]

$$\left(\alpha(\phi) + \rho^+ L \phi'\right) \frac{\partial T}{\partial t} - \operatorname{div}\left(\lambda(\phi) \operatorname{grad} T\right) = f, \tag{1}$$

where L is the specific heat of the phase transition, m is porosity, ρ^+, c^+, λ^+ and ρ^-, c^-, λ^- are density, specific heat capacity and the thermal conductivity of melted and frozen zones, respectively. We have the following coefficients

$$\alpha(\phi) = \rho^- c^- + \phi(\rho^+ c^+ - \rho^- c^-), \quad \lambda(\phi) = \lambda^- + \phi(\lambda^+ - \lambda^-),$$

$$c^-\rho^- = (1-m)c_{sc}\rho_{sc} + mc_i\rho_i, \quad \lambda^- = (1-m)\lambda_{sc} + m\lambda_i,$$
$$c^+\rho^+ = (1-m)c_{sc}\rho_{sc} + mc_w\rho_w, \quad \lambda^+ = (1-m)\lambda_{sc} + m\lambda_w.$$

and

$$\phi = \begin{cases} 0, & \text{when} \quad T < T^*, \\ 1, & \text{when} \quad T > T^*. \end{cases}$$

The indexes sc, w, i denote the solid skeleton, water, and ice, respectively.

In practice, the phase transitions occur in a small temperature range $[T^* - \Delta, T^* + \Delta]$. As the function ϕ we take ϕ_Δ:

$$\phi_\Delta = \begin{cases} 0, & T \le T^* - \Delta, \\ \dfrac{T - T^* + \Delta}{2\Delta}, & T^* - \Delta < T < T^* + \Delta, \\ 1, & T \ge T^* + \Delta. \end{cases}$$

Therefore, we get the following equation for nonlinear parabolic temperature in the domain Ω:

$$\left(\alpha(\phi_\Delta) + \rho_l L \phi'_\Delta\right) \frac{\partial T}{\partial t} - \operatorname{div}(\lambda(\phi_\Delta) \operatorname{grad} T) = f. \tag{2}$$

The equation (2) is supplemented with the initial condition:

$$T(\boldsymbol{x}, 0) = T_0(\boldsymbol{x}), \quad \boldsymbol{x} \in \Omega, \tag{3}$$

and boundary conditions

$$-k\frac{\partial T}{\partial n} = \alpha(T - T_{Air}), \quad x \in \Gamma_1, \quad -k\frac{\partial T}{\partial n} = 0, \quad x \in \Gamma_2, \quad T = T_1, \quad x \in \Gamma_3, \tag{4}$$

where α is heat convection coefficient, T_{air} is outdoor temperature.

3. Heat and mass transfer in the heat exchanger

To calculate the thermal regime in the pipe system (heat exchanger), we use a reduced dimension model. In this model, we assume that the coolant, that have a low freezing temperature, circulates in the cylindrical tube. We suppose, that the coolant does not freeze under the temperatures considered in the model. Then, for a heat transfer in a pipes, we have following equation in the cylindrical coordinates

$$c_p\rho_p \left(\frac{\partial T}{\partial t} + v_r \frac{\partial T}{\partial r} + v_\tau \frac{\partial T}{\partial \tau}\right) - \frac{1}{r}\frac{\partial}{\partial r}\left(r\lambda_p \frac{\partial T}{\partial r}\right) - \frac{\partial}{\partial \tau}\left(\lambda_p \frac{\partial T}{\partial \tau}\right) = 0, \tag{5}$$

where c_p, ρ_p and λ_p are specific heat capacity, density and the thermal conductivity of coolant, T is temperature of the coolant in pipe, v_r is velocity of the coolant in the radial direction and v_τ is velocity of the coolant in the direction of flow.

Since the flow in pipes is characterized by significant velocities of the coolant in longitudinal direction, it is possible to neglect the radial flow velocity, $v_r = 0$. Then we multiply by the radius r and integrate the resulting equation along the axis $r \in [0, R]$

$$\int_0^R c_p\rho_p r \frac{\partial T}{\partial t} dr + \int_0^R c_p\rho_p r v_\tau \frac{\partial T}{\partial \tau} dr - \int_0^R r \frac{\partial}{\partial \tau}\left(\lambda_p \frac{\partial T}{\partial \tau}\right) dr + \lambda_p r \frac{\partial T}{\partial r}\bigg|_0^R = 0. \tag{6}$$

Next, we determine the average temperature of the coolant in a pipe

$$T_p = \frac{2}{R^2}\int_0^R r\,T\,dr,$$

and suppose following interface condition between pipe and surrounding ground

$$\lambda_p r \frac{\partial T_p}{\partial r}\bigg|_0^R = -R\alpha_p(T - T_p).$$

Therefore, we obtain following reduced dimension (one–dimensional) equation in a pipe

$$\pi R^2 c_p\rho_p\left(\frac{\partial T_p}{\partial t} + v_\tau\frac{\partial T_p}{\partial \tau}\right) - \pi R^2\frac{\partial}{\partial \tau}\left(\lambda_p\frac{\partial T_p}{\partial \tau}\right) = 2\pi R\alpha_p(T - T_p). \tag{7}$$

As a boundary condition, we can use a fixed temperature on one of the boundaries$T_p = T_{in}$, where T_{in} is temperature of the coolant at the pipe inlet.

4. Coupled heat transfer problem and finite element approximation

A mathematical model is described by a coupled system of equations for heat transfer in the ground subdomain and in heat exchanger (pipes)

$$\left(\alpha(\phi_\Delta) + \rho_l L\phi'_\Delta\right)\frac{\partial T}{\partial t} - \operatorname{div}(\lambda(\phi_\Delta)\operatorname{grad} T) = f, \quad x \in \Omega \tag{8}$$

$$\pi R^2 c_p\rho_p\left(\frac{\partial T_p}{\partial t} + v_\tau\frac{\partial T_p}{\partial \tau}\right) - \pi R^2\frac{\partial}{\partial \tau}\left(\lambda_p\frac{\partial T_p}{\partial \tau}\right) = q_i, \quad x \in \gamma_i, \tag{9}$$

where γ_i is the i–th pipe.

For including thermal effects of the ground, we define following heat source at the pipe locations.

$$f = \sum_{i=1}^{n} q_i\delta(x - x_i), \tag{10}$$

where q_i is power unit length of ith heat sink, n is quantity of heat sinks and δ is the Dirac delta function determining the position of point sources in a ground. Here, we use the Dirac delta function, since the diameter of the pipes is much smaller than the ground size.

Assuming an ideal thermal contact between the outer surface of the pipes and the surrounding ground, the power of the heat sinks is written as follows

$$q_i = 2\,\pi\,R^i\,\alpha_p^i(T - T_p), \tag{11}$$

where α_p is heat convection coefficient between the pipe surface and surrounding gound, R is pipe radius, T and T_p^i are the ground and coolant temperatures.

For the numerical solution of the problem, we use a finite–element method [13]. Let $W = H_1(\Omega)$, then we have following variational formulation: find $T \in W$ such that

$$\int_\Omega \left(\alpha(\phi_\Delta) + \rho_l L\phi'_\Delta\right)\frac{\partial T}{\partial t}\,w\,dx + \int_\Omega \lambda(\phi_\Delta)\operatorname{grad} T\cdot\operatorname{grad} w\,dx$$
$$+\pi R^2\sum_{i=1}^{n}\int_{\gamma_i} c_p\rho_p\left(\frac{\partial T_p}{\partial t} + v_\tau\cdot\operatorname{grad}_\tau T_p\right)w\,ds + \pi R^2\sum_{i=1}^{n}\int_{\gamma_i}\lambda_p\operatorname{grad}_\tau T_p\cdot\operatorname{grad}_\tau w\,ds = 0, \quad w \in W, \tag{12}$$

IOP Conf. Series: Journal of Physics: Conf. Series **937** (2017) 012056 doi:10.1088/1742-6596/937/1/012056

where γ_i is the i–th pipe.

Here, for approximation of the coupled system of equations (8) – (9), we suppose that the temperature on the ground and pipes at the similar nodes are same. This approximaiton method is inspired by the the well–known discrete fracture model (DFM) that used for fluid flow problems in the fractured porous medium. In this method, we construct unstructured grid, and resolve one–dimentional pipe lines by construction conforming mesh [6, 7, 8, 9].

5. Numerical results and conclusion

In this section, we present results of the numerical simulation for the three–dimensional problem. As a model parameters, we use following properties of the ground

- for melt zone $\rho^+c^+ = 2.772 \cdot 10^6 \, [J/(m^3 \cdot K)]$, $\lambda^+ = 1.94 \, [W/(m \cdot K)]$,
- for frozen zone $\rho^-c^- = 2.052 \cdot 10^6 \, [J/(m^3 \cdot K)]$, $\lambda^- = 2.14 \, [W/(m \cdot K)]$,
- for phase transition $\rho^+L = 125.532 \cdot 10^6 \, [J/(m^3 \cdot K)]$ and $\delta = 1.0$.

For the pipe system, the following parameters is used: $v_\tau = 1.0 \, [m/s]$, $\rho_p c_p = 1.763 \cdot 10^6 \, [J/(m^3 \cdot K)]$, $R = 0.05 \, [m]$ and the temperature of the coolant at the inlet is set equal to the air temperature.

We perform numerical simulation in the computational domain with size 10 meters in the direction X, 10 meters in the direction Y and 10 meters in the direction Z. On the top boundary, we set boundary conditions that simulate the presence of the warm construction of dimensions 5×4 meters. This condition is given in the form of a Dirichlet boundary condition with a constant temperature $20°C$. At a depth of 1 meter under the structure installed a system of pipes with circulating coolant. We present results for the 1 year simulation of the heat processes with time step $\tau = 1$ day. For numerical simulations using finite–element method, we construct unstructured tetrahedral grid with 650k cells and 105k vertices (Fig. 1). The ground has an initial temperature $T_0 = -3.0^0$. On the top boundary, we set Robyn type boundary conditions that taking into account the amplitude of the air temperature fluctuation, which is of a regional characteristic and varies from $T_w = -50°C$ in winter to $T_s = 30°C$ in summer.

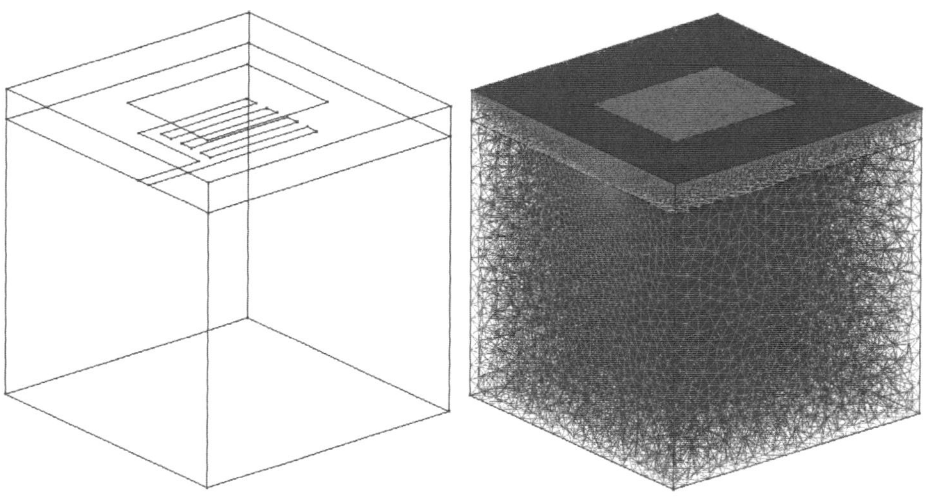

Figure 1. Computational geometry and unstructured grid

The numerical results of the heat transfer problem solution using proposed method are presented in Fig. 2 and 3. Fig. 2 illustrates the slice at the depth 1 meters. The slices

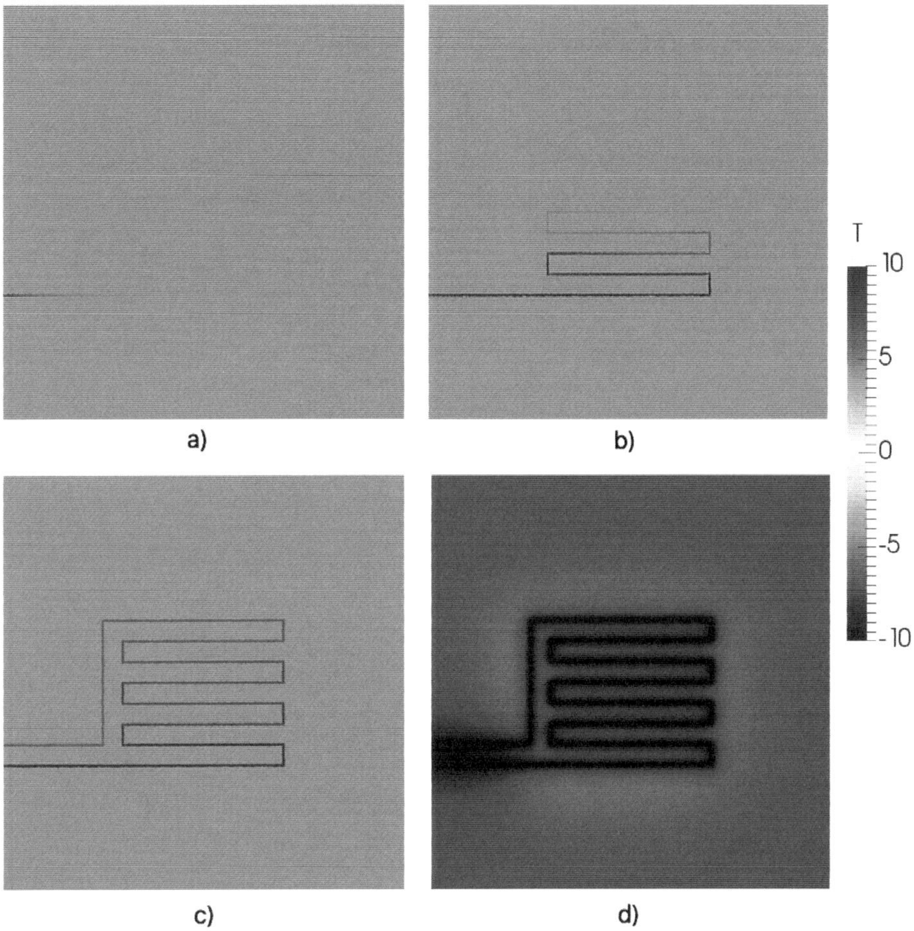

Figure 2. Temperature distribution is top view of slice at depth 1 meters a) $t = 1$ sec, b) $t =$ 15 sec, c) $t = 1$ min, d) $t = 10$ days

are presented for several initial time layers and illustrates the dynamics of the movement of the coolant in the pipe and the beginning of freezing of the surrounding ground. In Fig. 3, we shown a dynamics of freezing of soil at different simulation times. From the results, we see that the temperature under a construction remains negative for all simulation time and help to prevent soils melt.

The presented results illustrate the efficiency of the proposed mathematical model and numerical algorithm for heat transfer processes in the ground with pipes systems. This model help to significantly reduce the dimension of the computational problem that arises after space approximations, since the model for pipes is written in the form of a one-dimensional equation.

6. Acknowledgments

Sergei Stepanov and Ivan Sirditov works was supported by the mega-grant of the Russian Federation Government (N 14.Y26.31.0013). MV's part was supported by the grand of Russian Scientific Fund (N 17–71–20055).

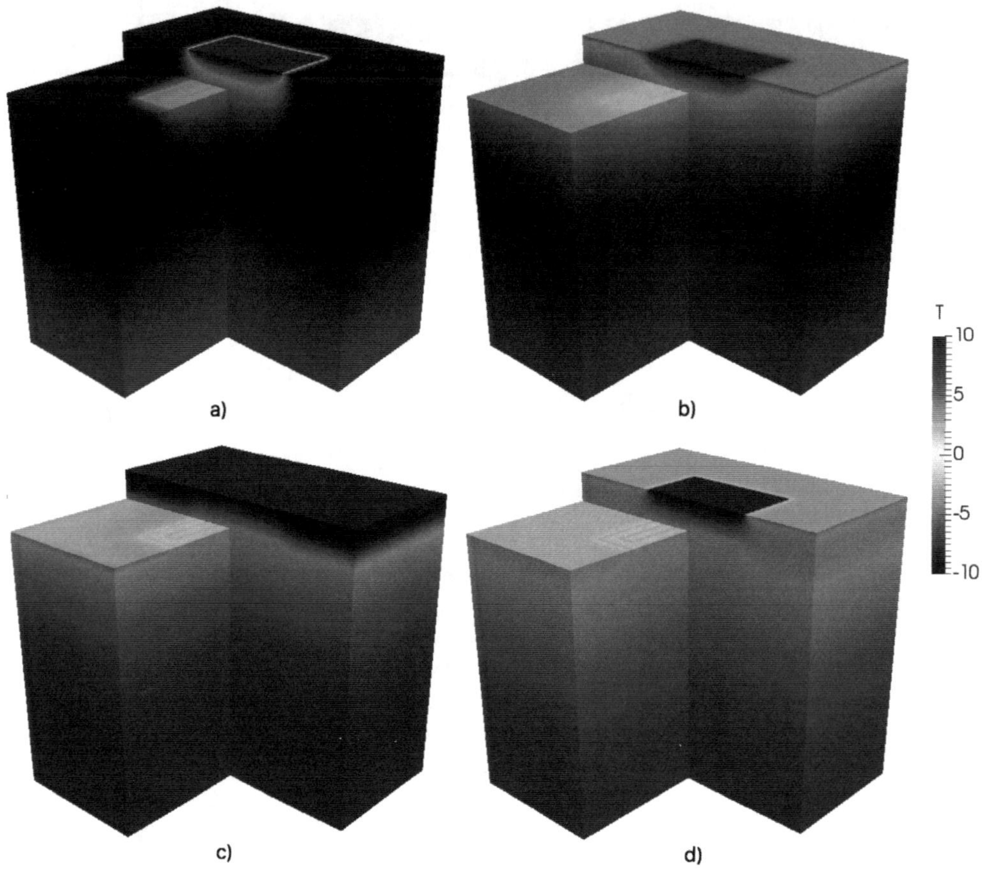

Figure 3. Temperature distribution a) winter, b) spring, c) summer, d) autumn

References

[1] Pavlova N V, Vabishchevich P N, Vasilyeva M V and Gornov V F 2014 *Computing technologies* **19** 19–31
[2] Gornov V, Stepanov S, Vasilyeva M, Vasilyev V and Todorov M D 2014 *AIP Conference Proceedings* vol 1629 (AIP) pp 424–431
[3] Stepanov S, Vabishchevich P, Varlamov S, Vasilieva M and Vasiliev V 2016 *Mathematical Models and Computer Simulations* **9** 110–124
[4] Stepanov S, Vasilyeva M, Vasilyev V and Todorov M D 2016 *AIP Conference Proceedings* vol 1773 (AIP Publishing) p 110011
[5] Kim J G and Deo M D 2000 *AIChE Journal* **46** 1120–1130
[6] Karimi-Fard M, Durlofsky L J, Aziz K *et al.* 2003 *SPE Reservoir Simulation Symposium* (Society of Petroleum Engineers)
[7] Karimi-Fard M, Firoozabadi A *et al.* 2001 *SPE annual technical conference and exhibition* (Society of Petroleum Engineers)
[8] Akkutlu I Y, Efendiev Y, Vasilyeva M and Wang Y 2017 *Journal of Natural Gas Science and Engineering*
[9] Chung E T, Efendiev Y, Leung T and Vasilyeva M 2017 *GEM-International Journal on Geomathematics* **8** 9–41
[10] Chung E T, Efendiev Y, Li G and Vasilyeva M 2015 *to appear in Applicable Analysis* **255** 1–15
[11] Chung E T, Efendiev Y and Lee C S 2015 *Multiscale Modeling & Simulation* **13** 338–366
[12] Vabishchevich P, Vasilyeva M and Pavlova N 2015 *Mathematical Models and Computer Simulations* **7** 154–164
[13] Brenner S and Scott L 2007 *The Mathematical Theory of Finite Element Methods* (New York: Springer-Verlag)

The approach to investigation of the the regions of self-oscillations

T R Velieva[1], **D S Kulyabov**[1,2], **A V Korolkova**[1], **I S Zaryadov**[1,3]

[1]Department of Applied Probability and Informatics,
Peoples' Friendship University of Russia (RUDN University),
6 Miklukho-Maklaya str., Moscow, 117198, Russia
[2]Laboratory of Information Technologies, Joint Institute for Nuclear Research,
6 Joliot-Curie, Dubna, Moscow region, 141980, Russia
[3]Institute of Informatics Problems, FRC CSC RAS, IPI FRC CSC RAS,
44-2 Vavilova str., Moscow, 119333, Russia

E-mail: velieva_tr@rudn.university, kulyabov_ds@rudn.university, korolkova_av@rudn.university, zaryadov_is@rudn.university

Abstract. Self-oscillating modes in control systems of computer networks quite negatively affect the characteristics of these networks. The problem of finding the areas of self-oscillations is actual and important as the study of parameters of self-oscillations. Due to the significant nonlinearity of control characteristics, the study of the self-oscillating modes presents certain difficulties. This paper describes the technique of research of self-oscillating modes.

1. Introduction

While modeling technical systems with control it is often required to study their characteristics. Also it is necessary to investigate the influence of system parameters on characteristics. In systems with control there is a parasitic phenomenon as a self-oscillating mode. We have previously carried out studies on the numerical determination of the region of occurrence of self-oscillations [1]. These studies provided information about the presence or absence of self-oscillations. However, the parameters of these self-oscillations the influence of various factors on the cause of self-oscillations was not investigated by as.

In this paper, we propose the research plan that allows us to obtain the parameters of self-oscillations and to investigate various factors of self-oscillations appearance. The proposed method uses the harmonic linearization. In addition, the studies suggest verification of the results obtained using the simulation software.

The section 2 describes the structure of the research. The model under investigation is introduced in the section 3. In the section 4, a brief description of the method of harmonic linearization is given. In the section 5, the harmonic linearization for Random Early Detection (RED) [2, 3, 4, 5] algorithm is described, as also as the verification of the obtained results.

2. Research structure

The general structure of the research is as follows.

2.1. Formulation of the problem
In network systems, in particular in a system with TCP (Transmission Control Protocol) sources and a traffic management module, the phenomenon of global synchronization has been experimentally revealed. In this case, each TCP source decreases and increases the transmission data rate simultaneously with other TCP sources. The effect of global synchronization leads to the inefficient use of bandwidth and the increased data latency. The global synchronization is implemented as a self-oscillating mode in a system consisting of a TCP source and a traffic management module.

One of the tasks of the RED algorithm was to eliminate the global synchronization mode. However, this problem was not solved fundamentally. With some parameters of the RED algorithm, there is a self-oscillating mode in the system. We assume that the form of the drop function has a decisive influence on the occurrence of an self-oscillating mode. We need to obtain phase and parametric portraits of the parameters of system (based on the threshold values of the RED algorithm) to determine the region of self-oscillations occurrence.

2.2. Mathematical model
The model of the system was previously proposed in the works [6, 7] as the system of nonlinear differential equations. The study of the model is complicated by the presence of a nonsmooth function (the drop function) in the equation. This function can be a discontinuous, for example, in the RED algorithm, or it can have the form of a hysteresis, for example, in the Adaptive RED (ARED) algorithm. In addition, there is a delay in the model. All this makes the study the system extremely difficult. We would like to highlight separately the influence of various conditions (such as the nature of the drop function and the delay) on the occurrence of a self-oscillating mode.

2.3. Harmonic linearization
We proposed [8] to linearize the initial model and investigate the behavior of the linearized model. However, the ordinary linearization destroys the information about the oscillating mode and makes the linearized model non-equivalent to the original one. For this type of model, it is possible to apply the so-called harmonic linearization, which allows to carry out a partial linearization of the model and to investigate the occurrence of a self-oscillating mode. Several criteria can be used to find out the presence and nature of the self-oscillating mode. One of them is more convenient to use in order to find out the presence of self-oscillations (Nyquist criterion [9]), others to determine the nature of self-oscillation stability (Hurwitz criterion [10]).

2.4. Verification of results
We suggest to verify the received solutions by using the NS-2 software simulator. This simulator is the reference implementation of network protocols. It is usually assumed that the simulation on NS-2 is equivalent to the behavior of a real system.

3. The RED congestion adaptive control mechanism
The RED algorithm uses a weighted queue length as a factor determining the probability of packet drop. As the average queue length grows, the probability of packets drop also increases (see Eq. (1)). The algorithm uses two threshold values of the average queue length to control the drop function:

$$p(\hat{Q}) = \begin{cases} 0, & 0 < \hat{Q} \leqslant Q_{\min}, \\ \dfrac{\hat{Q} - Q_{\min}}{Q_{\max} - Q_{\min}} p_{\max}, & Q_{\min} < \hat{Q} \leqslant Q_{\max}, \\ 1, & \hat{Q} > Q_{\max}. \end{cases} \tag{1}$$

Here $p(\hat{Q})$ is the packet drop function (drop probability), $\hat{Q}$ is the exponentially-weighted moving average of the queue size average, $Q_{\min}$ and $Q_{\max}$ are the thresholds for the weighted average of the queue length, $p_{\max}$ is the the maximum level of packet drop.

The RED algorithm is quite effective due to simplicity of its implementation in the network hardware, but it has a number of drawbacks. In particular, for some parameters values there is a steady oscillatory mode in the system, which negatively affects quality of service (QoS) indicators [11, 12, 13]. Unfortunately there are no clear selection criteria for RED parameters values, at which the system does not enter in self-oscillating mode.

To describe the RED algorithm we will use the following continuous model (see [6, 7, 14, 15, 16, 17]) with some simplifying assumptions: the model is written in the moments; the model describes only the phase of congestion avoidance for TCP Reno protocol; in the model the drop is considered only after reception of 3 consistent ACK confirmations.

$$\begin{cases} \dot{W}(t) = \dfrac{1}{T(Q,t)} - \dfrac{W(t)W(t-T(Q,t))}{2T(t-T(Q,t))} p(t-T(Q,t)); \\ \dot{Q}(t) = \dfrac{W(t)}{T(Q,t)} N(t) - C; \\ \dot{\hat{Q}}(t) = -w_q C\hat{Q}(t) + w_q CQ(t). \end{cases} \tag{2}$$

Here the following notation is used: W is the average TCP window size; Q is the average queue size; $\hat{Q}$ is the exponentially weighted moving average (EWMA) of the queue size average; C is the queue service intensity; T is the full round-trip time; $T = T_p + \frac{Q}{C}$, where T_p is the round-trip time for free network (excluding delays in hardware); $\frac{Q}{C}$ is the time which batch spent in the queue; N is the number of TCP sessions; p is the packet drop function.

4. Harmonic linearization method

The method of harmonic linearization is an approximate method. It is used for study of self-oscillation conditions and determination of the parameters of self-oscillations, for the analysis and evaluation of their sustainability, as well as for the study of forced oscillations. Harmonically-linearized system depends on the amplitudes and frequencies of periodic processes. The harmonic linearization differs from the common method of linearization (leading to purely linear expressions) and allows to explore the basic properties of nonlinear systems.

The method of harmonic linearization is used for systems of a certain structure (see Fig. 1). The system consists of linear part H_l and the nonlinear part, which is set by a function $f(x)$. It is generally considered a static nonlinear element.

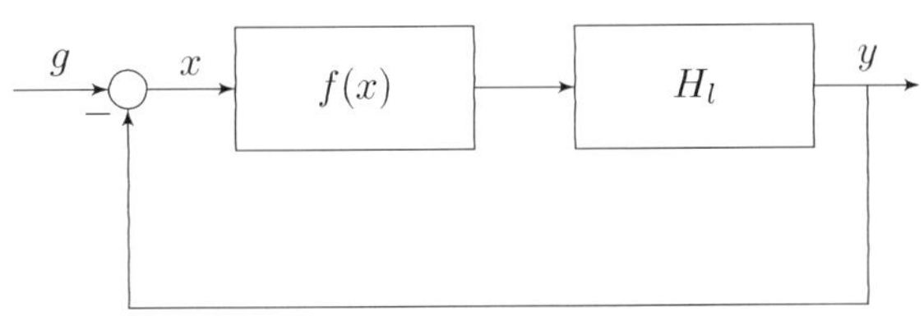

Figure 1. Block structure of the system for the harmonic linearization method

For the harmonic linearization method the free movement mode (input $g(t) = 0$) is assumed. The free harmonic oscillations are applied to the input of the nonlinear element:

$$x(t) = A\sin(\omega t). \tag{3}$$

On the output of the nonlinear element $f(x)$ we get a periodic signal. Let's expand it in a Fourier series.

The linear element is a low-pass filter, that is, when k is increasing the linear elements suppress higher harmonics. We will consider only the first harmonics:

$$f(x) = a_1 \sin(\omega t) + b_1 \cos(\omega t). \tag{4}$$

From (3) one can write:

$$\begin{aligned} x &:= A\sin(\omega t); \\ \cos(\omega t) &= \frac{1}{A\omega}\frac{\mathrm{d}x}{\mathrm{d}t} = \frac{1}{A\omega}\frac{\mathrm{d}}{\mathrm{d}t}x. \end{aligned} \tag{5}$$

Then we may rewrite (4) with respect (5):

$$f(x) = \left[\varkappa(A) + \frac{\varkappa'(A)}{\omega}\frac{\mathrm{d}}{\mathrm{d}t}\right]x \overset{\frac{\mathrm{d}}{\mathrm{d}t}\to \mathrm{i}\omega}{\longrightarrow} \left[\varkappa(A) + \mathrm{i}\varkappa'(A)\right]x = H_{nl}(A)x, \tag{6}$$

where $H_{nl}(A)$ is the approximate transfer function of the nonlinear unit, $\varkappa(A)$ and $\varkappa'(A)$ are the harmonic linearization coefficients:

$$\begin{aligned} \varkappa(A) &= \frac{a_1}{A} = \frac{1}{A\pi}\int_0^{2\pi} f(A\sin(\omega t))\sin(\omega t)\,\mathrm{d}(\omega t)\,; \\ \varkappa'(A) &= \frac{b_1}{A} = \frac{1}{A\pi}\int_0^{2\pi} f(A\sin(\omega t))\cos(\omega t)\,\mathrm{d}(\omega t)\,. \end{aligned} \tag{7}$$

After finding the coefficients of harmonic linearization for given nonlinear unit, it is possible to study the parameters of the oscillation mode. The existence of oscillation mode in a nonlinear system corresponds to the determination of oscillating boundary of stability for the linearized system. Then A and ω can be found by using stability criteria of linear systems (Mikhailov, Nyquist–Mikhailov [9], Routh–Hurwitz [10, 18]). Thus, the study of self-oscillation parameters can be done by one of the methods of determining the limits of stability of the linear systems.

4.1. The Nyquist–Mikhailov criterion

The Nyquist-Mikhailov criterion [9, 19] allows to judge about the stability of the open-loop automatic control system by using Nyquist plot (amplitude-phase characteristic) of the open-loop system.

Make the substitutions $\partial_t \to \mathrm{i}\omega$ and $s \to \partial_t \to \mathrm{i}\omega$ in the transfer function H_l and H_{nl}. Undamped sinusoidal oscillations with constant amplitude are determined by passing the amplitude-phase characteristics of the open-loop system through the point $(-1, \mathrm{i}0)$.

Thus we obtain:

$$H_l(\mathrm{i}\omega)H_{nl}(A, \mathrm{i}\omega) = -1. \tag{8}$$

Given by (6) from (8) the equality is obtained:

$$H_l(\mathrm{i}\omega) = -\frac{1}{\varkappa(A) + \mathrm{i}\varkappa'(A)}. \tag{9}$$

The left part of the equation (9) is the amplitude-phase characteristic of the linear unit, and the right part is the inverse of the amplitude-phase characteristic of the first harmonic non-linear level (with opposite sign). And the equation (9) is the equation of balance between the frequency and the amplitude.

4.2. Mikhailov criterion

Let us write the characteristic equation of the conservative system (12) (Fig. 1) with respect of $s \to \partial_t \to \mathrm{i}\omega$:

$$P_d(\mathrm{i}\omega) = 0. \tag{10}$$

In addition, if one explicitly allocates real and complex parts, then the equation (10) can be written in the following form:

$$\begin{aligned} \operatorname{Re}\{P_d(\mathrm{i}\omega)\} &= 0, \\ \operatorname{Im}\{P_d(\mathrm{i}\omega)\} &= 0. \end{aligned} \tag{11}$$

Thus, the parameters of self-oscillations can be determined from equation (11).

4.3. RouthHurwitz criterion

This criterion is an algebraic criterion of stability [18, 10].

The transfer function of the conservative system (Fig. 1) has the form:

$$H_c(s) = \frac{H_l(s)}{1 + H_l(s) H_{nl}(s)} := \frac{P_n(s)}{P_d(s)}. \tag{12}$$

The equation $P_d(s) = 0$ is the characteristic equation of the system. It can be presented in the polynomial form:

$$P_d(s) := a_0 s^n + a_1 s^{n-1} + \cdots + a_n.$$

Thus, the Hurwitz determinant will look like this [20]:

$$\Delta = \begin{vmatrix} a_1 & a_3 & a_5 & \dots & 0 \\ a_0 & a_2 & a_4 & \dots & 0 \\ 0 & a_1 & a_3 & \dots & 0 \\ 0 & a_0 & a_2 & \dots & 0 \\ \dots & \dots & \dots & \dots & \dots \\ \dots & \dots & \dots & \dots & a_n \end{vmatrix}.$$

By the Hurwitz criterion, in order for the dynamical system to be stable, it is necessary and sufficient that all n principal diagonal minors of the Hurwitz determinant be positive, provided that $a_0 > 0$. These minors are called determinants of Hurwitz. For $a_n = 0$, the system is at the boundary of aperiodic stability. If the penultimate Hurwitz determinant (the minor of the Routh–Hurwitz determinant) $\Delta_{n-1} = 0$, then the system is at the boundary of oscillatory stability. Actually, this expression is necessary for us to determine the conditions for the appearance and parameters of self-oscillations.

5. Linearization of the model

To rewrite the model (2) in the block-linear approach we need to linearize it.

We will carry out the linearization near the equilibrium point (the balance point is denoted by f index). At the equilibrium point time derivatives turn to zero, so the system of equations (2) will be as follows:

$$\begin{cases} 0 = \dfrac{1}{T_f} - \dfrac{W_f^2}{2T_f} p_f; \\ 0 = \dfrac{W_f}{T_f} N_f - C; \\ 0 = -w_q C \hat{Q}_f + w_q C Q_f. \end{cases} \tag{13}$$

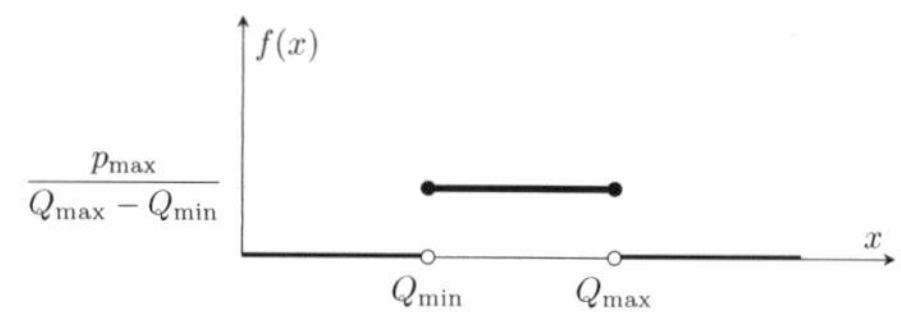

Figure 2. The function P_{RED}

From the system of equations (13) we get the bound equation for the equilibrium values of the variables:

$$\begin{cases} p_f = \dfrac{2}{W_f^2}; \\ W_f = \dfrac{CT_f}{N_f}; \\ \hat{Q}_f = Q_f. \end{cases}$$

The variation of the right part (13) for all variables in a neighborhood of the equilibrium point is:

$$\begin{cases} \delta W(s) = -\dfrac{1}{s + \frac{N}{CT_f^2}\left(1 + \mathrm{e}^{-sT_f}\right)} \dfrac{C^2 T_f}{2N^2} \mathrm{e}^{-sT_f}\, \delta p(s)\,; \\ \delta Q(s) = \dfrac{1}{s + \frac{1}{T_f}} \dfrac{N}{T_f}\, \delta W(s)\,; \\ \delta \hat{Q}(s) = \dfrac{1}{1 + \frac{s}{w_q C}}\, \delta Q(s)\,. \end{cases} \tag{14}$$

In addition, let us to linearize the drop function (1):

$$\delta p(\hat{Q}, t) = \begin{cases} 0, & 0 < \hat{Q} \leqslant Q_{\min}, \\ \dfrac{p_{\max}}{Q_{\max} - Q_{\min}}\, \delta \hat{Q}(t)\,, & Q_{\min} < \hat{Q} \leqslant Q_{\max}, \\ 0, & \hat{Q} > Q_{\max}. \end{cases} \tag{15}$$

The (15) may be denoted as

$$\delta p(\hat{Q}, t) = P_{\text{RED}}\, \delta \hat{Q}(t)\,;$$
$$P_{\text{RED}} := \begin{cases} 0, & 0 < \hat{Q} \leqslant Q_{\min}, \\ \dfrac{p_{\max}}{Q_{\max} - Q_{\min}}, & Q_{\min} < \hat{Q} \leqslant Q_{\max}, \\ 0, & \hat{Q} > Q_{\max}. \end{cases} \tag{16}$$

Considering the formula $\delta \hat{Q}(s)$ from the system of equations (14), we can write out (16) in the following form:

$$\delta p(s) = P_{\text{RED}} \frac{1}{1 + \frac{s}{w_q C}}\, \delta Q(s)\,. \tag{17}$$

The function P_{RED} has the form shown in Fig. 2

Based on (14) and (17) the block representation of the linearized RED model (Fig. 3) is constructed.

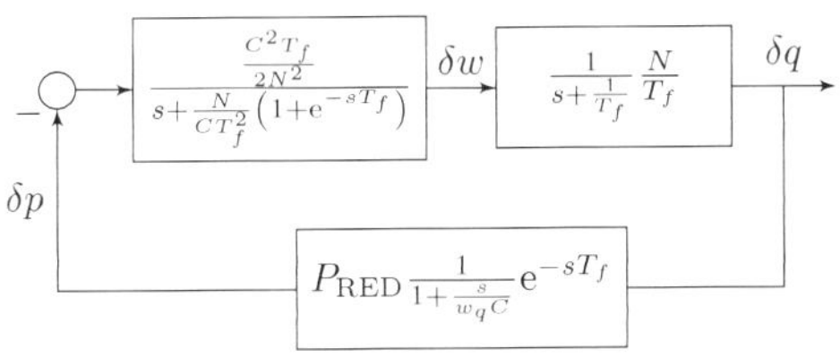

Figure 3. Block representation of the linearized RED model

5.1. Harmonic linearization of the linearized RED model

Let's reduce the block diagram of linearized model (Fig. 3) to the form required for harmonic linearization.

As a static nonlinear function we will use P_{RED} (16). The linear part we get from (14) and (17):

$$H_l = \frac{1}{s+\frac{N}{CT_f^2}\left(1+e^{-sT_f}\right)}\frac{C^2 T_f}{2N^2}e^{-sT_f} \times \frac{1}{s+\frac{1}{T_f}}\frac{N}{T_f} \times \frac{1}{1+\frac{s}{w_q C}} = \\ = \frac{1}{s+\frac{N}{CT_f^2}\left(1+e^{-sT_f}\right)}\frac{1}{s+\frac{1}{T_f}}\frac{1}{1+\frac{s}{w_q C}}\frac{C^2}{2N}e^{-sT_f}.$$

Let us compute the coefficients of harmonic linearization $\varkappa(A)$ and $\varkappa'(A)$ (7) for the static nonlinearity P_{RED}:

$$\varkappa(A) = \frac{4}{A\pi}\frac{p_{\max}}{Q_{\max}-Q_{\min}}\left(\sqrt{1-\frac{Q_{\min}^2}{A^2}}-\sqrt{1-\frac{Q_{\max}^2}{A^2}}\right);$$

$$\varkappa'(A) = \frac{4}{A\pi}\frac{p_{\max}}{Q_{\max}-Q_{\min}}\frac{Q_{\max}-Q_{\min}}{A} = \frac{4p_{\max}}{A^2\pi}.$$

For the example of the the calculation we have chosen the following parameters: $Q_{\min} = 100$ [packets], $Q_{\max} = 150$ [packets], $p_{\max} = 0.1$, $T_p = 0.0075$ s, $w_q = 0.002$, $C = 2000$ [packets]/s, $N = 60$ (the number of TCP sessions).

As a result we obtained the following values for the amplitude and the cyclic frequency: $A = 1.89$ [packets], $\omega = 16.55s^{-1}$.

The traffic behavior can be demonstrated by using the standard computer networks simulation software NS-2 [21, 22].

For selected parameters we will get the graph of the oscillations of the instantaneous queue length at a router under RED control (Fig. 4). With the help of fast Fourier transform we will study the results of NS-2 simulation (Fig. 5).

6. Conclusion

The authors proposed the technique for studying the self-oscillating modes of systems with control. The basic stages of the research process are described. The main element of this process is the harmonic linearization.

In the article, we demonstrated the technique for one set of parameters based on our methodology. Accordingly, only one pair of amplitude–frequency values is obtained. The obtained theoretical values are verified by using the NS-2 simulation software. In further research the authors plan to calculate the existence of self-oscillating modes for all threshold values for the RED algorithm. In addition, it is planned to conduct research for different RED-like algorithms.

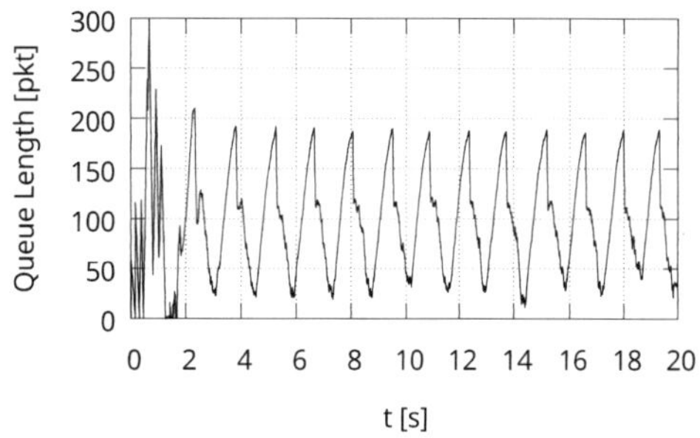

Figure 4. A router's queue oscillation under RED control

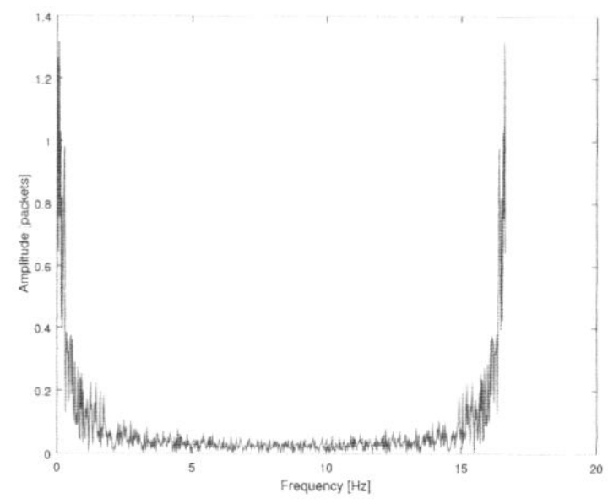

Figure 5. The spectrum of the oscillations

6.1. Acknowledgments
The work is partially supported by Russian Foundation for Basic Research (RFBR) grant No 16-07-00556. Also the publication was prepared with the support of the "RUDN University Program 5-100".

References

[1] Korolkova A V and Kulyabov D S 2010 *Bulletin of Peoples' Friendship University of Russia. Series "Mathematics. Information Sciences. Physics"* 54–64
[2] Floyd S and Jacobson V 1993 *IEEE/ACM Transactions on Networking* **1** 397–413
[3] Jacobson V 1988 *ACM SIGCOMM Computer Communication Review* **18** 314–329 (*Preprint* `arXiv:1011.1669v3`)
[4] Kushwaha V and Gupta R 2014 *Journal of Network and Computer Applications* **45** 62–78
[5] Adams R 2013 *IEEE Communications Surveys & Tutorials* **15** 1425–1476
[6] Misra V, Gong W B and Towsley D 1999 *Proceedings of PERFORMANCE* **99**
[7] Misra V, Gong W B and Towsley D 2000 *ACM SIGCOMM Computer Communication Review* **30** 151–160
[8] Kulyabov D S, Korolkova A V, Velieva T R, Eferina E G and Sevastianov L A 2018 *DepCoS-RELCOMEX 2017. Advances in Intelligent Systems and Computing* (*Advances in Intelligent Systems and Computing* vol 582) (Cham: Springer International Publishing) pp 215–224
[9] Nyquist H 1932 *Bell System Technical Journal* **11** 126–147
[10] Hurwitz A 1895 *Mathematische Annalen* **46** 273–284
[11] Jenkins A 2013 *Physics Reports* **525** 167–222 (*Preprint* `1109.6640`)
[12] Ren F, Lin C and Wei B 2005 *Computer Networks* **49** 580–592
[13] Lautenschlaeger W and Francini A 2015 *Proc. 16-th International Conference on High Performance Switching and Routing, IEEE HPSR 2015* (Budapest, Hungary) (*Preprint* `1602.05333`)
[14] Hollot C V V, Misra V, Towsley D and Wei-Bo Gong 2001 *Proceedings IEEE INFOCOM 2001. Conference on Computer Communications. Twentieth Annual Joint Conference of the IEEE Computer and Communications Society (Cat. No.01CH37213)* vol 3 (IEEE) pp 1726–1734
[15] Velieva T R, Korolkova A V and Kulyabov D S 2015 *6th International Congress on Ultra Modern Telecommunications and Control Systems and Workshops (ICUMT)* (IEEE Computer Society) pp 570–577 (*Preprint* `1504.02324`)
[16] Korolkova A V, Velieva T R, Abaev P A, Sevastianov L A and Kulyabov D S 2016 *Proceedings 30th European Conference on Modelling and Simulation* 685–691
[17] Brockett R 1999 *Proceedings of the 38th IEEE Conference on Decision and Control (Cat. No.99CH36304)* vol 3 (IEEE) pp 3077–3082
[18] Routh E J 1877 *A Treatise on the Stability of a Given State of Motion: Particularly Steady Motion* (Macmillan)
[19] Hsu J and Meyer A 1968 *Modern Control Principles and Applications* (McGraw-Hill)
[20] Gantmacher F R 1959 *The Theory of Matrices* (Chelsea Pub. Co.)
[21] Altman E and Jiménez T 2012 *Synthesis Lectures on Communication Networks* **5** 1–184
[22] Issariyakul T and Hossain E 2012 *Introduction to Network Simulator NS2*

IOP Conf. Series: Journal of Physics: Conf. Series **937** (2017) 012058 doi:10.1088/1742-6596/937/1/012058

Sufficient Conditions for the Basicity in L_p and Equiconvergence with the Trigonometric Series of Spectral Decompositions for the Second Order Ordinary Differential Operator

V E Volkov

National Research Nuclear University MEPhI (Moscow Engineering Physics Institute), 31, Kashirskoe shosse, 115409 Moscow, Russia

vv1720@mail.ru

Abstract. Theorems about sufficient conditions for the basicity for second order ordinary differential operator are proved.

Consider an arbitrary (finite or infinite) real interval G of an ordinary second-order differential operator

$$Lu = -u'' + q(x)u\,, \tag{1}$$

where $q(x)$ is a complex-valued function, $q(x) \in L_r^{loc}(G), r > 1$.

Suppose that the potential $q(x)$ admits the existence of a closed system and a minimal system $\{u_n(x)\}_{n=1}^{\infty}$ of eigenfunctions and associated functions of the operator (1) in $L_p(G)$.

The system $\{u_n(x)\}_{n=1}^{\infty}$, where $u_n(x) \in L_p(G), 1 \le p < \infty$ is called closed in $L_p(G)$ if an arbitrary function $f(x) \in L_p(G)$ can be approximated arbitrary accurately in the metric $L_p(G)$ by a finite linear combination of the elements of the system $\{u_n(x)\}_{n=1}^{\infty}$.

The system $\{u_n(x)\}_{n=1}^{\infty}$, where $u_n(x) \in L_p(G), 1 \le p < \infty$ is called complete in $L_p(G)$ where $p^{-1} + q^{-1} = 1$, if any function $f(x) \in L_p(G)$ orthogonal to all elements $u_n(x)$ is a zero element only. The closure of system $\{u_n(x)\}_{n=1}^{\infty}$ in $L_p(G)$ is equivalent to its completeness in $L_p(G)$.

Since the system $\{u_n(x)\}_{n=1}^{\infty}$ is closed in $L_p(G)$ and minimal, there exists a unique system $\{v_n(x)\}_{n=1}^{\infty}$, $v_n(x) \in L_q(G), p^{-1} + q^{-1} = 1$, biorthogonally conjugate to $\{u_n(x)\}_{n=1}^{\infty}$.

Let us set $\mu_n = \sqrt{\lambda_n} = \rho_n + \nu_n$, where $\rho_n \ge 0$, and for $\rho_n = 0$, $\nu_n \ge 0$, and let us we require the condition

$$\sum_{t \le \rho_n \le t+1} 1 \le C = const, \qquad \forall t \ge 0. \tag{2}$$

In this case, there are no restrictions on the numbers ν_n.

We enumerate eigenvalues λ_n in the nondecreasing of ρ_n and consider the partial sum of its spectral decomposition for an arbitrary function $f(x) \in L_p(G)$:

$$\sigma_\mu(x,t) = \sum_{\rho_n \le \mu} (f, v_n) u_n(x), \qquad \mu \ge 1. \tag{3}$$

Definition 1. The system $\{u_n(x)\}_{n=1}^{\infty}$ forms a basis in L_p on any compact set if $\forall f(x) \in L_p(G)$ and
$\forall K \in G$ is always true $\lim_{\mu\to\infty} \|\sigma_\mu(x,f) - f(x)\|_{L_{p(K)}} = 0.$

Theorem 1. Let $1 < p < \infty$ and $\{u_n(x)\}_{n=1}^{\infty}$ – be an arbitrary closed system in $L_p(G)$ and the minimal system of eigenfunctions and associated functions of the operator (1). Then if three conditions are satisfied:

1) $q(x) \in L_r^{loc}(G), r > 1$;
2) the numbers ρ_n satisfy the requirement (2);
3) for any compact $K \in G$ there exists a constant $C(K)$ such that

$$\|u_n\|_{L_{p(K)}} \|v_n\|_{L_{q(K)}} = C(K) \tag{4}$$

for $q = \frac{p}{p-1}$ and $n = 1, 2, \dots$ then the system $\{u_n(x)\}_{n=1}^{\infty}$ forms a basis on any compact set.

Suppose now that $\{u_n(x)\}_{n=1}^{\infty}$ is closed and minimal in $L_p(G)$, $1 \le p < \infty$.

Let an arbitrary compact K in the interval G is fixed and a positive number R such that it does not exceed the distance from the compact K to boundaries of the interval G. And along with decomposition (3) we consider a partial sum of trigonometric Fourier series of an arbitrary function $f(x) \in L_p(G)$:

$$S_\mu(x,t) = \frac{1}{\pi} \int_{|x-y|\le R} \frac{\sin(x-y)}{x-y} f(y)dy, \tag{5}$$

where $x \in K$, $y \in G$, $\mu \le 1$.

Definition 2. If for $\mu \to \infty$ the difference $\sigma_\mu(x,f) - S_\mu(x,f)$ tends to zero with respect to $x \in K \subset G$, then we say that the expansion of the function $f(x)$ into the biorthogonal series in the $\{u_n(x)\}_{n=1}^{\infty}$ and the expansion in the trigonometric Fourier series are equiconvergence uniformly on any compact K of the interval G.

Theorem 2. Let $1 < p < \infty$ and $\{u_n(x)\}_{n=1}^{\infty}$ – be an arbitrary closed in $L_p(G)$ minimal system of eigenfunctions and associated functions of the operator (1), and suppose that the three conditions of Theorem 1 are satisfied. Then the expansion of an arbitrary function $f(x) \in L_p(G)$ in the system $\{u_n(x)\}_{n=1}^{\infty}$ and in the trigonometric Fourier series are equiconvergence uniformly on any compact K of the interval G.

References

[1] Yoo I and Komornic V 1983 Acta Sci. Math. **46** 357-75
[2] Il'in V A 1980 Diff. Eq. **16** 980-1009
[3] Yoo I 1982 Acta Sci. Math. **44** 87-93
[4] Komornic V 1982 Acta Sci. Math. **44** 95-8
[5] Tikhomirov V V 1983 Dokl. AN SSSR 807-10
Tikhomirov V V 1983 Diff. Eq. **19** 1378-85
[6] Il'in V A and Yoo I 1979 Diff. Eq. **15** 1175-93
[7] Volkov V E 1984 Diff. Eq. **20** 1852-62

[8] Kritsov L V 2017 Diff. Eq. **53** 180-91
[9] Il'in V A and Kritsov L V 2003 J. of Math. Sci. **116** 3489-550
[10] Sadovnichaya I V 2016 Proc. Steklov Inst. Math. **293** 288-316
[11] Sadovnichaya I V 2010 Sbornic Mathematics **201** 1307-22

IOP Conf. Series: Journal of Physics: Conf. Series 937 (2017) 012059 doi:10.1088/1742-6596/937/1/012059

Application of thermal-hydraulic model of RBMK reactor fuel channel to correction of power and coolant flow measurements

A M Zagrebaev, A V Trifonenkov and V P Trifonenkov

National Research Nuclear University MEPhI (Moscow Engineering Physics Institute), 31 Kashirskoe shosse, 115409 Moscow, Russia

E-mail: AVTrifonenkov@mephi.ru

Abstract. The effectiveness and the security of RBMK reactor operation depends on the accuracy of the control over reactor's parameters and their limitations. The processing of operational parameters archive helps to adjust different mathematical models and significantly widen their field of use. Pressure differential between common pressure header and steam separator is the sum of calculated pressure differential and friction loss on flow control valve. There is known mathematical software, which allows to adapt such model for each fuel channel using the archive. In this research it is suggested not to replace the regular mechanism with such approach, but to use the adapted mathematical model to calculate corrected values of power and flow, which were measured by regular means. Mathematical expressions and procedures for such approach are given.

1. Introduction

The effectiveness and the security of RBMK reactor operation depends on the accuracy of the control over reactor's parameters and their limitations. Two of the most important parameters are the coolant flow through fuel channel and the energy emission, or power of the channel. Currently used means of control and algorithms of information processing surely give a reliable solution to the problem [1]. Nevertheless, research of new methods of accuracy increase for control systems is always a problem of interest. At present, such methods may involve using an archive of nuclear reactor operational parameters for a long period of time. In particular, the processing of operational parameters archive helps to adjust different mathematical models and significantly widen their field of use. In this research it is suggested to use thermal-hydraulic mathematical model of the pressure differential in fuel channel to correct experiment-calculated values of power and coolant flow, derived by regular methods.

2. Design features of RMBK reactor

The core of RBMK-1000 contains 1661 fuel channels [1]. During operation, at temperature of 270 °C and at pressure of approximately 7 MPa is delivered by main circulating pumps into the common pressure header. Then through group distribution headers coolant flows to fuel channels. After passing the reactor core steam-water mixture enters a steam separator, where is separated into liquid and gaseous phases. Steam then proceeds to a turbine, while liquid water returns back to the reactor. All

IOP Conf. Series: Journal of Physics: Conf. Series **937** (2017) 012059 doi:10.1088/1742-6596/937/1/012059

channels are equipped with flow meters of coolant, which take part in flow measurement process, and flow control valves, which help to change the flow.

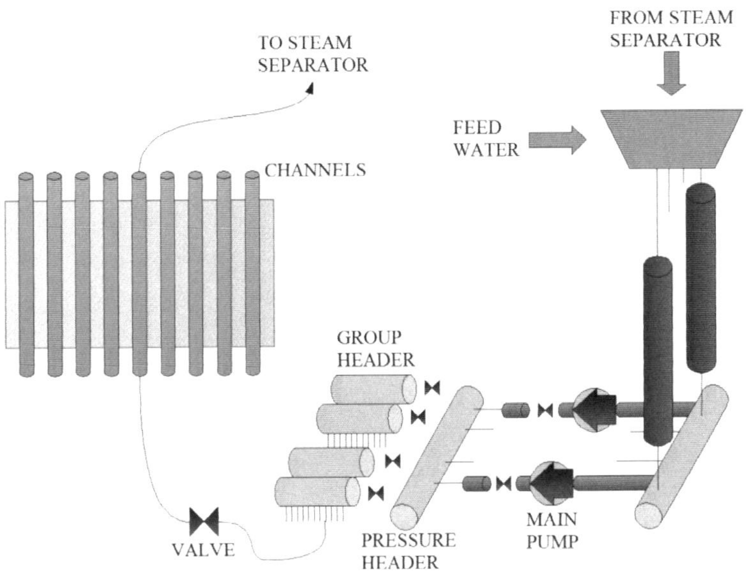

Figure 1. A part of repeated forced circulation circuit of RBMK reactor

Pressure differential between common pressure header and steam separator is the sum of calculated pressure differential and friction loss on flow control valve: $\Delta P_V = k_i \cdot G_i^2$. Thus:

$$\Delta P = \Delta P_{HDC}(G_i, W_i) + k_i G_i^2 \tag{1}$$

Here $\Delta P_{HDC}(G_i, W_i)$ is hydrodynamic characteristic (HDC) of fuel channel. It is derived as a result of hydrodynamic calculation [2], based on mathematical model, expressed by thermal-hydraulic equations for two-phased coolant flow: the continuity equation and equations for dynamics and energy. One important feature of any reactor of RBMK type is that fuel channels are separated into large groups, each forming an isolated system, within which channels have the same pressure header at the input and the same steam separator at the output. Thus, pressure differential between pressure header and steam separator for each channel of a group is the same and is measured by regular means of pressure control.

In [3, 4] mathematical procedure and software is given, which allows to find the coefficient k_i for each fuel channel using the archive. (In what follows the index is omitted, supposing that some fixed channel is always being considered).

Thus, in the equation for pressure differential (1) ΔP and k are given. Then, one of the parameters can be calculated using another, for example, with fuel channel power given, same channel's flow can be calculated, and vice versa. In such way the mathematical model became adapted with the use of operational parameters archive.

It is suggested not to replace the regular mechanism with such approach, but to use the adapted mathematical model to calculate corrected values of power and flow, which were measured by regular means.

3. The search of corrected values

Let W_{reg} and G_{reg} be values of power and flow calculated by regular means. These values are practically derived with given inaccuracy, which dispersions are D_W and D_G respectively.

Let the corrected values be calculated from the following conditions

$$\min_{W,G}\left[\frac{\left(W-W_{reg}\right)^2}{D_W}+\frac{\left(G-G_{reg}\right)^2}{D_G}\right] \tag{2}$$

$$\Delta P=\Delta P_{HDC}(G,W)+kG^2=0 \tag{3}$$

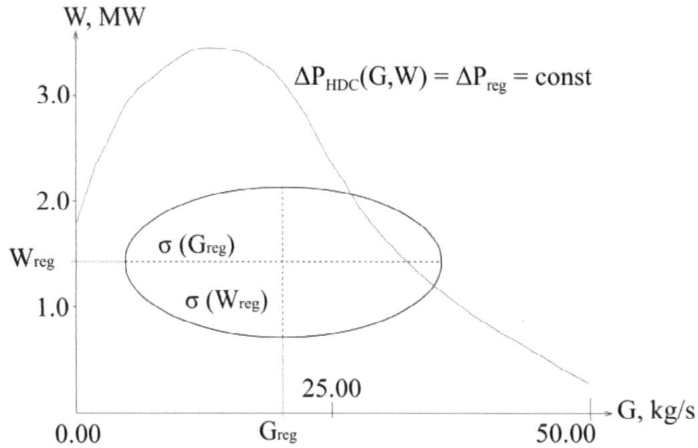

Figure 2. An example plot of curves (2) and (3), made for measurements of pressure, power and flow at some single instant

As for (3) the equality may be achieved by adjusting both W and G. But as both parameters are already approximately calculated, then they should be corrected in such way to ramain within declared inaccuracy.

Expression (2) describes an ellipsis and (3) describes some other curve, which lies within intersection of a surface and plane. Combined, (2) and (3) form the optimization problem with (3) representing limitation function. Optimal solution lies in tangency point $(\tilde{W},\tilde{G})$ of ellipsis and curve.

The equation for tangent to the curve (3) in point $(\tilde{W},\tilde{G})$ is

$$\frac{\partial\Delta P_{HDC}}{\partial W}\Big|_{\tilde{W},\tilde{G}}\left(\tilde{W}-W\right)+\left(\frac{\partial\Delta P_{HDC}}{\partial G}\Big|_{\tilde{W},\tilde{G}}+2kG\right)\left(\tilde{G}-G\right)=0 \tag{4}$$

The equation for tangent to the ellipsis (2) in the same point is

$$\frac{2\left(\tilde{W}-W_{ш}\right)}{D_W}\left(\tilde{W}-W\right)+\frac{2\left(\tilde{G}-G_{ш}\right)}{D_G}\left(\tilde{G}-G\right)=0 \tag{5}$$

The condition of concurrence of tangents gives the system of equations:

$$\begin{aligned}&\frac{\partial\Delta P_{HDC}}{\partial W}\Big|_{\tilde{W},\tilde{G}}=\frac{2\left(\tilde{W}-W_{reg}\right)}{D_W}\\&\frac{\partial\Delta P_{HDC}}{\partial G}\Big|_{\tilde{W},\tilde{G}}+2kG=\frac{2\left(\tilde{G}-G_{reg}\right)}{D_G}\end{aligned} \tag{6}$$

Corrected values are calculated within iteration process.

4. Simulation results

The solution of the optimization problem, formed by (2) and (3), is derived by several steps. In this article the algorithm suggested includes the calculation of adaptation coefficient, the calculation of

IOP Conf. Series: Journal of Physics: Conf. Series **937** (2017) 012059 doi:10.1088/1742-6596/937/1/012059

hydrodynamic characteristic and the search of optimum. The algorithm is assumed to be used together with the archive of operational parameters of RBMK reactor.

The actual algorithm used in the background of this article is applied to the set of measurements at instants $t_1,...,t_N$, represented by triples $(\Delta P_1, G_1, W_1),...,(\Delta P_N, G_N, W_N)$ of actual parameter values, where flow and power values require correction.

The adaptation coefficient, which is required for the calculation of HDC, adjusts proportionality between summands in (3). Given flow, power and pressure difference, such coefficient is uniquely defined at any instant. But the value of the coefficient can't be calculated precisely owing to inaccuracy of flow and power measurements. The procedure for calculating the coefficient as the parameter of channel flow control valve is described in [3, 4].

The calculation of HDC implies the process of search of the valid pairs of flow and power values, which conform to the given pressure difference according to mathematical model from [2]. The result of the search is the set of points, which represent the curve in fig.2, and then the search of optimal point is performed on it.

The optimal point is uniquely defined by (6), but for numerical calculations more practical approach is to use the equation for ellipsis as an expression for distance in compressed coordinates and choose the nearest point in such terms.

The example of the result of calculations is given in figure below:

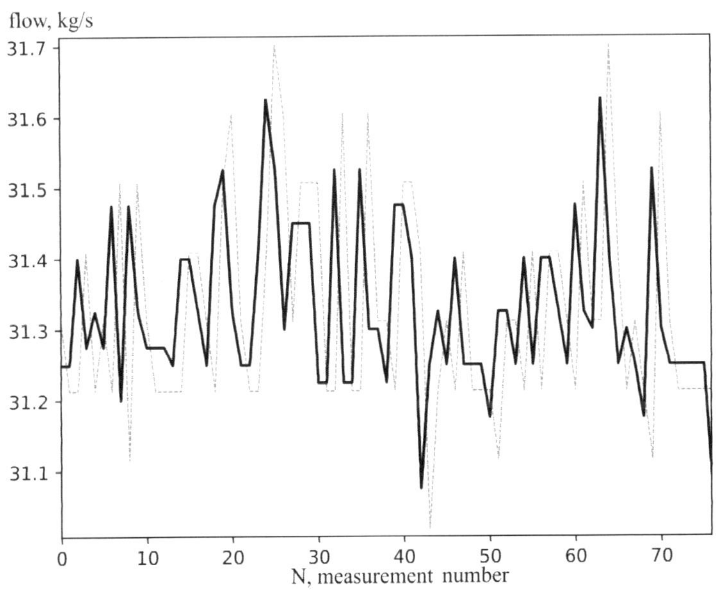

Figure 3. Comparison of coolant flow measurements at 75 sequential instants: original values, derived by regular means (grey dash line, shifted by one measurement for convenience) and calculated by the suggested algorithm (black solid line)

The difference between new and original values lies around 0.1% for flow values and 1.5% for power values for the given sample.

It is important to note, that the suggested algorithm sets additional interconnection between hydrodynamic parameters of the reactor fuel channel. The results of the procedure do not replace the results of calculation, performed by regular means. New values of the parameters accommodate given errors of regular means of calculation. The algorithm may serve as supplementary mathematical instrument to standard algorithms, and also be used together with the archive of the reactor's operational data to analyse and predict actual phase states of a reactor more precisely.

Conclusion

In this article method of correction of reactor hydrodynamic parameters is considered. The parameters taken into account are channel power, coolant flow and pressure difference. The method is based on

IOP Conf. Series: Journal of Physics: Conf. Series **937** (2017) 012059 doi:10.1088/1742-6596/937/1/012059

mathematical model, which conforms the parameters to each other. The correction algorithm is based on the procedure of the search of new parameters' values. The result of the correction algorithm for one instant is the set of corrected values, which conform to each other taking into account given theoretical correlations from mathematical model. The algorithm suggested may be used as supplementary mathematical instrument to standard algorithms. Also it may be used along with the reactor's operational data archive, which allows to analyze and predict reactor states more precisely.

References

[1] Abramov M A, Avdeev V I and Adamov E O 2006 *Kanalnyj energeticheskij yadernyj reaktor RBMK* (Moscow: GUP NIKIET)

[2] Rukovodjaschij tehnicheskij material 1979 RTM 95 1181–83: Metodika teplogidravlicheskogo raschiota sborok teplovydelyayushchikh elementov yzdernykh reaktorov tipa RBMK (Moscow: Minatoma)

[3] Zagrebayev A M, Kostanbayev S V and Ovsyannikova N V 2010 Avtomatizirovannaya sistema kontrolya stepeni otkrytiya zaporno-reguliruyushchikh klapanov v reaktore RBMK *Avtomatizatsiya v promyshlennosti* **2** 29–31

[4] Zagrebayev A M, Kostanbayev S V and Ovsyannikova N V 2009 Avtomatizirovannaya sistema kontrolya raskhoda teplonositelya v kanale reaktora RBMK po informatsii o perepade davleniya *Estestvenniye i tekhnicheskiye nauki* **44** 540–541

Institute of Physics
Dirac House, Temple Back
Bristol BS1 6BE UK

ISSN: 1742-6588
ISBN 978-1-5108-5517-5

9 781510 855175